Progress in Mathematics

Volume 251

Rigorous Quantum Field Theory

A Festschrift for Jacques Bros

Anne Boutet de Monvel
Detlev Buchholz
Daniel Iagolnitzer
Ugo Moschella
Editors

Birkhäuser Verlag
Basel · Boston · Berlin

Editors:

Anne Boutet de Monvel
Institut de Mathématiques de Jussieu
Université Paris 7
175 rue du Chevaleret
75013 Paris
France
e-mail: aboutet@math.jussieu.fr

Daniel Iagolnitzer
Service de Physique Théorique
CEA/Saclay
Orme des Merisiers
91191 Gif-sur-Yvette cedex
France
e-mail: daniel.iagolnitzer@cea.fr

Detlef Buchholz
Institut für Theoretische Physik
Universität Göttingen
Friedrich-Hund-Platz 1
37077 Göttingen
Germany
e-mail: buchholz@theorie.physik.uni-goettingen.de

Ugo Moschella
Dipartimento di Fisica e Matematica
Università dell'Insubria
Via Valleggio, 11
22100 Como
Italy
e-mail: ugo.moschella@uninsubria.it

2000 Mathematics Subject Classification 81Txx

Library of Congress Control Number: 2006934501

Bibliographic information published by Die Deutsche Bibliothek
Die Deutsche Bibliothek lists this publication in the Deutsche Nationalbibliografie; detailed bibliographic data is available in the Internet at <http://dnb.ddb.de>.

ISBN 978-3-7643-7433-4 Birkhäuser Verlag, Basel – Boston – Berlin

Part of Springer Science+Business Media
Printed on acid-free paper produced of chlorine-free pulp. TCF ∞
Printed in Germany
ISBN-10: 3-7643-7433-0
ISBN-13: 978-3-7643-7433-4
e-ISBN-10: 3-7643-7434-9
e-ISBN-13: 978-3-7643-7434-1

9 8 7 6 5 4 3 2 1

www.birkhauser.ch

Contents

1

Introduction

After the great scientific revolutions of Special Relativity and of Quantum Mechanics in the first half of the twentieth century, Relativistic Quantum Field Theory (QFT) was introduced to provide a synthesis of these two new paradigms (in Kuhn's words). As such, QFT is one of the major advances for theoretical physics in the second half of the last century. The main reason for emphasizing the importance of QFT lies in the impressive effort that it represents towards a unified understanding of the structure of matter at subatomic scales, as it emerges in the collision phenomena of particle physics in the range of energies presently explored in particle accelerators. But another reason is the extraordinary mathematical wealth of this theoretical framework, which makes it a fascinating domain of research for mathematical physics and may also stimulate the interest of mathematicians.

Foreseen by Dirac in 1933 and finaly established discovered by Feynman around 1949 for the treatment of quantum electrodynamics, the so-called "path-integral formalism" of quantum field theory is also now considered as a powerful tool for providing perturbative methods of computation in many problems of theoretical physics, namely in statistical mechanics and in string theory. However, the use of relativistic quantum fields as a basic concept of mathematical physics underlying all the phenomena of particle physics in a very large range of energies represents a much more ambitious program.

This program was indeed stimulated by the success of the quantum electrodynamics (QED) formalism for computing the electron-photon, electron-electron and electron-positron scattering amplitudes. Even today, it is by no means understood why the perturbative expansion of QED in powers of the coupling parameter, that is the electric charge of the electron, provides such a spectacular agreement with experimental data, although in practice it is reduced to the computation of the very first terms of the series (the only ones which are computable). This is even more surprising since we now know that the series cannot be convergent, so that, paraphrasing Wigner's words, one may wonder about the "unreasonable effectiveness" of perturbation theory in four-dimensional QED.

In the 1950s, however, the success of these computations was credited to the smallness of the coupling parameter of QED, and this situation stimulated research for other methods of investigation of QFT, which could apply to quantum field models with large coupling: it was in fact the very concept of a quantum field which appeared as the most powerful and promising one for explaining the phenomena of strong nuclear interaction of particle physics. During all that period the study of phenomena of weak nuclear interaction also benefitted, like QED, from the success of a perturbative QFT formalism reduced to very few terms.

The demand for a more general nonperturbative treatment of QFT motivated an important community of mathematical physicists to work out a model-independent *axiomatic approach to relativistic quantum field theory* [2,8,9,11,14]. Their first task was to provide a mathematically meaningful concept of a relativistic quantum field in terms of *"operator-valued distribution in the Hilbert space of quantum physical states"*, in such a way that the notion of ingoing and outgoing particle states could be introduced in terms of appropriate asymptotic forms of the field operators.

The precise mathematical formulation of fundamental physical principles of relativistic quantum theory in terms of field operators considered as basic entities of the system allowed the possibility to offer a general model-independent framework, in which one could propose a theoretical treatment of high-energy collision phenomena.

These phenomena involve not only the usual processes of "elastic scattering" preserving the number and nature of the particles, but also all possible processes of creation and annihilation of particles permitted by the kinematical laws of special relativity and the conservation laws of basic quantum numbers called "charges". In this mathematical framework, all the collision processes are encoded in a certain general unitary *scattering operator* S in the Hilbert space of states, according to the concept originally introduced by Heisenberg, and all the corresponding *scattering kernels* $S_{m,n}$ of $(m \to n)$-particle collision processes are related to general N-point structure functions of the quantum fields (with $N = m + n$) via the so-called LSZ *"reduction formulae"*. This formalism was fully justified at least for processes involving only massive particles; the inclusion of massless particles, such as the photons of QED, soon revealed the existence of hard mathematical problems of various types.

This general approach of QFT was followed by a conceptually important variant, in which the quantum fields are supposed to generate "algebras of local observables" attached to arbitrary regions of Minkowski spacetime. This is the *algebraic approach to quantum field theory* (Haag, Kastler 1964), whose most important developments between 1960 and 1990 have been presented in a book by Haag (1992) under the name of "local quantum physics" [6].

From both mathematical and physical viewpoints, the visionary works of Von Neumann, which were contemporary to the birth of quantum mechanics, are at the origin of all these developments; however, the algebraic structures involved in relativistic QFT refer to the most advanced developments of the theory of Von Neumann algebras. Just as an example, it is worthwhile to recall that, in contrast with the case of nonrelativistic quantum mechanics, one is necessarily confronted by factors of type III in the standard classification of Von Neumann algebras. This occurs already in the simplest models of relativistic QFT, namely the free-fields. Moreover, important conceptual progress was made by the algebraic QFT viewpoint with respect to the primitive quantum-mechanical notion of the Hilbert-space of states of a given physical system. It consists in introducing algebras of local observables of the system as *abstract C^*-algebras* (and even more general $*$-algebras) and then representing all possible physical states of the system as linear functionals on such algebras. In this more general viewpoint, various types of physically relevant Hilbertian representations of the algebras may be introduced mathematically via GNS-type (Gelfand-Naimark-Segal) [6] constructions. As a typical illustration of this viewpoint, one can mention the Borchers-Uhlmann algebraic presentation of the original Wightman axiomatic QFT [3].

In these various general approaches to relativistic QFT aiming to provide a theoretical understanding of particle physics, a central role is played by a certain field-theoretical formulation of *Einstein causality* together with the relativistic principle of positivity of energy (also called *spectral condition*) and an appropriate expression of the covariance of the system under the Poincaré group.

The interplay of these postulates has generated rigorous proofs of important physical properties, thanks to the discovery of very rich *analyticity properties* of the N-point structure functions of the fields (elaborated in successive works by Ruelle, Steinmann, Araki, Bros, Epstein, Glaser, Stora). As a matter of fact, all these functions enjoy *two associated analytic structures*, which coexist in the complexified spacetime variables z and in the Fourier-conjugate complexified energy-momentum variables k; this double analytic structure is equipped with a peculiar type of algebraic relations between boundary values from tube domains and relevant (multiple) discontinuities, including various interconnections by Fourier-Laplace transformations in several variables. In both z-space and k-space, nontrivial problems of holomorphy envelopes of the so-called "edge-of-the-wedge" type which emerge from this double analytic structure have only been very partially solved. As typical illustrations of the latter, one can say that on the one hand the exploitation of the z-space analytic structure has provided proofs in general QFT of such important properties of particle physics as the PCT-symmetry and the spin-statistics connection . On the other hand, the exploitation of the k-space analytic structure has brought a general justification in QFT of important analytic continuation properties of the scattering kernels on the so-called *complex mass shell manifold* of various collision processes. As a typical example, let us mention the crossing property between the "particle-particle" and "particle-antiparticle" scattering processes, which can be geometrically expressed as the existence of a domain of analyticity on the complex mass shell manifold relating two different "physical regions" on the *real mass shell submanifold* (Bros, Epstein and Glaser). Such successful results, which were mostly obtained in the 1960s, correspond to the "golden age" of the so-called *dispersion relations* in particle physics. It is rather remarkable that such model-independent Cauchy-type integral relations were found to be in satisfactory agreement with the experimental data of two-particle collision processes and that in this analytic framework, important "high-energy bounds" could also be derived from general principles in basic works by Froissart and Martin [10].

Of course, as a natural complement to this conceptual progress, there was (and there is) the need to produce *nontrivial models* of relativistic quantum fields: for that matter, a basic criterion of nontriviality is the requirement that the associated scattering operator S be different from the identity operator. This accounts for the occurrence of another school of mathematical physicists, called *"Constructive quantum field theory"* and founded essentially by Glimm and Jaffe in the late 1960s [5], whose purpose was to produce rigorous existence proofs together with properties of the corresponding solutions for models of relativistic QFT associated with Lagrangians of suitable type. Ingenious methods of *"cluster expansions"* were invented which allowed one to treat successfully various classes of nontrivial models in two and three-dimensional Minkowski spacetime, thus displaying that (at least for these dimensions) the axiomatic framework was not a pure abstraction, but it contained a lot of mathematically well-defined Lagrangian theories. It was even shown that, for such models, the traditional perturbation expansion, defined as a formal power series in terms of a small coupling parameter, had Borel-summability properties which allowed one to reach the exact solution of the model [5, 7].

However, it became patent that all the models of interacting relativistic quantum fields in the physical four-dimensional spacetime are afflicted by very hard mathematical problems which have to do with unavoidable short-distance as well as long-distance divergent behaviours of the structure functions of the fields. These difficulties are closely related to the *renormalization* problems in the standard perturbative approach of QFT. In the 1980s, in spite of the fact that no completely rigorous proof of it could be produced, there appeared a wide consensus among theoretical physicists that the QFT model with quartic interaction, that is the "simplest" Lagrangian model of scalar fields to be considered, probably did not exist in four-dimensional spacetime, but only reduced to a pure formal series in the sense of the perturbative approach.

The same consensual opinion also applies to QED, which therefore makes it still more thrilling to have in mind the physical success of its perturbation series. The additional intriguing aspect of QED is the fact that it incorporates the fundamental physical notion of *electric charge*. As a matter of fact, the horizons of QFT were considerably renewed in the 1970s by the generalization of the notion of *charge* in particle physics and the associated concept of gauge symmetry groups for the quantum field description of charged states. From this viewpoint, in comparison with all the models studied in constructive QFT, the Lagrangian of QED appeared as the simplest one to incorporate a *local gauge symmetry group*, namely the gauge group based on the abelian group $U(1)$. A very important event was then the discovery of more complicated Lagrangians of relativistic fields, symmetric under *nonabelian gauge groups*, in particular the model called *quantum chromodynamics (QCD)*. The successful exploitation of the latter for the phenomenology of particle physics opened a way that is still used nowadays in describing elementary particle interactions. In fact, for theoreticians of particle physics, the early success of QED in perturbation theory was renewed in the 1970s and extended to all the phenomena of weak and strong nuclear interactions: this was mainly due to the success of the so-called "renormalization group" treatment of QCD for the understanding of strong nuclear interactions, and of the *standard model*, whose fundamental set of boson and fermion quantum fields are supposed to take into account all the basic interactions of matter but gravitation, thus providing a unified treatment of the electroweak and strong interactions together. In particular, it is the discovery of the basic notion of *asymptotic freedom* which allowed one to give a sense to expansions of the field-theoretical structure functions in terms of a large coupling parameter, and therefore to renew to some extent with QCD the old success of QED in the spirit of perturbation theory.

In that regard, however, the point of view of mathematical physicists remained somewhat different and cultivated a still larger ambition in the line of its tradition. As ingenious as they are, the computational techniques of QCD and of the standard model treat the quantum fields of those models as very formal objects. In these approaches, whose computational efficiency is the main criterion to be fulfilled, the problem of the mathematical existence of fields has not been really considered. A pragmatic attitude consists in being satisfied with a certain type of approximate computations, whose recipes pertain to a weakened conceptual framework of quantum field theory, which is sometimes called "effective quantum field theory" by the tenants of these approaches. However, the impressive adequacy of some of the results obtained with the experimental data is stimulating for the mathematical physicist whose hope is to obtain a fully coherent understanding of this fundamental branch of physics in terms of mathematically well-defined objects. There are basic mathematical problems of the QFT framework which remain unsolved and whose solutions might shed new light on deep conceptual aspects of high-energy particle physics.

For a large part of the community of theoreticians of particle physics, increasing evidence has been accumulated since the 1970s that nonabelian gauge quantum field theory might be the only class of theories relevant for elementary particle physics, and whose mathematical existence in four-dimensional Minkowskian spacetime might hopefully be established some day. As an abelian gauge theory, QED might have a less favourable mathematical status as an isolated theory, and (according to certain arguments) should be more credibly thought as "embedded" in a larger nonabelian theory (such as the "standard model"). During the last thirty years, these considerations have been taken into account by the community of QFT mathematical physicists.

On the one hand, the problem of treating charged states in the case of a *global gauge symmetry group* (namely an internal symmetry group independent of spacetime localization) gave rise to deep conceptual progress in the general approach of algebraic QFT. This was due to the analysis of charge sectors in the Hilbert space of states by Doplicher, Haag and Roberts in the 1970s, and to the axiomatic introduction of nonlocal charge-carrying quantum fields such as the "string-localized fields" of Buchholz-Fredenhagen (1982). This contributed to enlarging considerably the concept of a quantum field and its relationship with the concept of a particle and thereby to opening new ways of research for QFT in the spirit of mathematical physics.

On the other hand, the mathematical study of QFTs involving a *local* gauge group (like QED and QCD) remains a much harder task. Indeed the very existence of a local gauge symmetry renders the axioms incompatible and one has to choose whether to work in local and covariant gauges and abandon Hilbertian positivity, and therefore a direct quantum mechanical interpretation, or to work in positive gauges where all the standard wisdom of QFT fails (in particular the expression of Einstein causality). A general approach to these problems was introduced and developed in the papers by Strocchi and Wightman in the 1970s. Still harder problems concern the construction of charged states for nonabelian gauge QFTs, the "confinement problem" being a particular instance of it [12] (see also further remarks below). The importance of this stream of research may be symbolized by the inscription of the study of Yang-Mills Quantum Field Theory (namely the "gluonic part" of QCD) in the seven Millennium Prize Problems proposed by the Clay Mathematical Institute. We refer to the presentation of this subject by A. Jaffe and E. Witten for a description of the type of hard mathematical problems to be solved in that context.

In their spirit, all the articles presented in this book consider quantum fields as genuine mathematical objects, whose various properties and relevant physical interpretations have to be studied in a well-defined mathematical framework. They therefore pertain to that tradition of "Rigorous Quantum Field Theory", which traces back to the basic axiomatic settings, supplemented by relevant constructive approaches, that we have previously mentioned. In this spirit, the most recent investigations of QFT may be characterized by the conceptual needs of extending the general axiomatic framework at least in two directions: i) taking into account the inclusion of internal (abelian and nonabelian) gauge groups as basic structures suggested by particle physics; ii) going beyond the Minkowskian spacetime of special relativity by incorporating more general notions concerning space and time. The latter investigations include, on the one hand, the consideration of QFT on *curved spacetime* manifolds, stimulated by various currents of research in general relativity, astrophysics and cosmology; in this context, unexpected relationships with basic concepts of "thermal QFT", namely field theory in a medium with a large number of particles, have also been exhibited. They also include, on the other hand, the recent development of QFT on *noncommutative spacetimes*, stimulated by new insights on the very-short-range structure of space and time. From these various viewpoints, it seems that one has now entered a new phase, namely the genesis of a deep renewal of the fundamental structures of QFT, in which the most advanced progress in pure mathematics may have its role to play.

We shall end this introduction with a very brief overview of the different topics which are represented in the various texts of this book. These topics can be grossly regrouped under seven titles, whose order is by no means significant and which do not exclude overlap:

1. *Analytic Structures of QFT.*
   The analytic structures of QFT in complex energy-momentum space offer microlocal aspects and global aspects. While the microlocal aspects concern the analytic wave-fronts of N-point functions and of general multiparticle collision amplitudes, the global aspects involve such general concepts as holomorphy envelopes in $\mathbf{C}^n$, Fredholm-type equations with floating cycles in complex manifolds (i.e., the so-called Bethe-Salpeter-type equations), polar singularities in the space $\mathbf{C}^2$ of the complex energy and angular momentum variables, interpreted as Regge particles in QFT.
2. *Renormalization group methods.*
   A version of rigorous renormalization theory based on the flow equations of the Wilson renormalization group is presented and illustrated in particular by considering the case of the scalar field with quartic coupling. More general results of the rigorous approach of renormalization group methods are also described. Moreover a new look at the renormalization group from the point of view of Algebraic Quantum Field Theory is proposed which results in a consistent definition of local algebras of observables and of interacting fields in renormalized perturbative QFT, involving an appropriate use of the classical action of the models.
3. *New investigations and results related to Gauge QFT.*
   Some basic unsolved problems of Quantum Electrodynamics which concern the formulations of that theory in different gauges are investigated from the viewpoint of the Wightman functions of the fields.

   Yang-Mills field equations are the starting point of a far-reaching mathematical study on "Yang-Mills algebra", "quadratic self-duality algebras" and "super" versions of them as applications of the theory of homogeneous algebras.

4. *New methods and results in Constructive QFT.*
   a) The invention of new procedures for constructing models of local relativistic quantum fields starting with a given scattering operator gives successful results for the case of factorizing S-matrices in two-dimensional spacetime. An intermediate step in the construction procedure makes use of nonlocal fields, called "Polarization-Free Generators" whose affiliated algebras are localized in wedge-shaped regions of spacetime rather than in bounded regions. The hope of obtaining similar results when starting from more general collision operators in four-dimensional spacetime by also incorporating the concept of "lightfront holography" is also discussed. In the same spirit, a general procedure for constructing stringly-localized fields from Wigner representations of the Poincaré group in d-dimensional spacetime is also presented.

   b) Thermal Quantum Field Theory, whose axiomatic status has been implemented in recent years on the basis of the KMS condition (starting from the general analysis of Haag, Hugenholtz and Winnink), presents a characteristic analytic structure in complex spacetime. This structure, which involves a periodicity in imaginary times given by the inverse of the temperature, is an alternative to the one prescribed by the spectral condition for the usual theories with a ground state (or zero-energy "vacuum"), thus appearing as the case of zero temperature. The construction of thermal field models in two-dimensional spacetime, associated with polynomial Lagrangians (of the type considered by the shool of constructive QFT in the 1970s) has been performed very recently, and it is proven here that they satisfy a relativistic form of the KMS-condition introduced earlier at the axiomatic level by Bros and Buchholz.
5. *Stability properties in QFT and extensions of the Axiomatic Framework of QFT.*
   Apart from the standard postulates of QFT expressed respectively by the spectral condition for theories with a ground state, and by the KMS-condition for thermal QFT, have both been proved to result from a general stability criterion called "passivity" (according to a basic work of Pusz and Woronowicz). Another type of stability condition is presented here, which is formulated in terms of "quantum energy inequalities" to be satisfied by the energy-momentum tensor of the theory. The links between this type of condition with the passivity condition are investigated, as also those with the important concept of "nuclearity" introduced in QFT by Buchholz.

   Another direction of research concerns the concepts of Anosov flows and of Kolmogorov systems, the important point being that they can be translated from classical to quantum systems. With some modifications necessary to keep the same clustering behavior as the typical one for classical Anosov systems, Anosov structure then appears rather naturally in a typeIII_1 algebra. Here Anosov structure and Kolmogorov structure with respect to modular evolution are even equivalent. The Rindler wedge of quantum field theory offers a typical example.

6. *QFT on Models of Curved Spacetimes.*
   Quantum field theory in curved spacetime is one important issue of modern theoretical physics, in that, while waiting for new high energy physics experiments in terrestrial laboratories, most of the new data come from observational cosmology. In particular, the evidence for a nonzero cosmological constant indicates that the de Sitter geometry plays and will play in future an important role. The study of soluble models on the two-dimensional de Sitter spacetime should be based on the two-dimensional massless scalar field. An original study of that model is presented here, that puts in evidence the anomaly of the equations of motion and fully characterizes the charge structure of the model.

   Anti-de Sitter Quantum Field theory is nowadays very popular because of its appearence in the context of string theory with the so-called "AdS-CFT correspondence". Euclidean AdS QFT can also provide a covariant regularization of flat QFT. The difficulty in studying AdS-QFT lies mainly in the absence of global hyperbolicity, caused by the presence of a boundary at spacelike infinity. A general rigorous approach to AdS QFT that bypasses these problems is presented here together with a collection of structural results that are implied by the chosen axioms.
7. *QFT on Noncommutative Minkowski Spacetime.*
   One of the fundamental questions in field theory on noncommutative spacetimes is how to find suitable generalizations of the local interaction terms we know from ordinary field theory. By construction, the notion of a point loses its meaning on such a spacetime and, not surprisingly, the principle of locality has to be modified. Depending on how this is done, various interaction terms are obtained. One can replace the idea of coinciding points in a way compatible with the uncertainty relations. The resulting interaction term resembles a point-split regularized product of fields and leads to an ultraviolet finite perturbation theory.

   The Wick reduction of *non-locally* time-ordered products of Wick monomials can be performed in a quantum field theory with a certain nonlocal self–interaction as introduced by Doplicher, Fredenhagen and Roberts, and simple Dyson diagrams are discussed.

References

1. N.D. Birrell and P.C.W. Davies: *Quantum Fields in Curved Space.* Cambridge University Press, Cambridge, 1982.
2. N.N. Bogoliubov, A.A. Logunov, A.I. Oksak and I.T. Todorov: *General Principles of Quantum Field Theory.* Kluwer, Dordrecht, 1990.
3. H.-J. Borchers: On Structure of the Algebra of Field Operators. *Nuovo Cimento* 24:214 (1962).

4. K. Fredenhagen: Quantum Field Theory on non-trivial space-times. In: *Mathematical Physics towards the* 21st *century* (R.N. Sen and A. Gersten, eds.). Ben-Gurion Univ. Negev Press, Beer-Sheva, 1993. See also all the articles of the section *Quantum Field Theory* in the Encyclopaedia of Mathematical Physics, Elsevier, Oxford, 2006.
5. J. Glimm and A. Jaffe: *Quantum Physics. A Functional Integral Point of View.* Springer, Berlin, 1981.
6. R. Haag: *Local Quantum Physics.* Springer, Berlin, 1992.
7. D. Iagolnitzer: *Scattering in quantum field theories: The Axiomatic and constructive approaches.* Princeton University Press, Princeton, 1992.
8. R. Jost: *The general theory of quantized fields.* AMS, Providence, 1965.
9. H. Lehmann, K. Symanzik and W. Zimmermann: On the formulation of quantized field theories. *Nuovo Cimento* 1:205–225 (1955) and 6:319–333 (1957).
10. A. Martin: *Scattering Theory: Unitarity, Analyticity and Crossing.* Lecture Notes in Physics, Springer, Berlin, 1969.
11. R.F. Streater and A.S. Wightman: *PCT, Spin and Statistics, and all that.* Princeton University Press, Princeton, 1964.
12. F. Strocchi: Selected topics on the general properties of quantum field theory. *World Sci. Lect. Notes Phys.* 51:1 (1993).
13. R.M. Wald: *Quantum field theory in curved space-time and black hole thermodynamics.* Chicago University Press, Chicago, 1995.
14. A.S. Wightman and L. Garding: Fields as operator-valued distributions in relativistic quantum theory. *Arkiv For Fysik* 28:129 (1964).

2

Local Counterterms on the Noncommutative Minkowski Space

Dorothea Bahns

Max-Planck-Institut für Gravitationsphysik (Albert-Einstein-Institut)
Am Mühlenberg 1, D-14476 Golm, Germany
dorothea.bahns@aei.mpg.de

Summary. The quasiplanar Wick products on the noncommutative Minkowski space defined in [5] by subtracting only local counterterms are reviewed. In addition, it is shown that ordinary Wick products on the noncommutative Minkowski space indeed involve nonlocal counterterms. A discussion of Wightman functions on the noncommutative Minkowski space is included and a cluster decomposition property (i.e. causality at large distances) is proved.

One of the fundamental questions in field theory on noncommutative space-times is how to find suitable generalizations of the local interaction terms we know from ordinary field theory. While in the latter case the principle of locality along with Lorentz and gauge invariance uniquely fixes the possible form of the interaction, some freedom remains in its definition on noncommutative space-times.

The appearance of this additional freedom follows from the very motivation and construction of such space-times as defined in [9]. Gravity along with the principles of quantum theory suggests that it should be impossible to measure all coordinates of an event in space-time simultaneously with an arbitrarily high precision. Therefore, noncommuting coordinate operators were introduced in [9] which entail restrictions on the possible precision of a coordinate measurement by means of uncertainty relations. By construction, the notion of a point loses its meaning on such a space-time and, not surprisingly, the principle of locality has to be modified. Depending on how this is done, we end up with various interaction terms.

One route we have pursued is based on replacing the concept of coinciding points in a way compatible with the uncertainty relations (using what we have called the quantum diagonal map). The resulting interaction term resembles a point-split regularized product of fields and leads to an ultraviolet finite perturbation theory [4].

Another possibility (see [5]) is based on mimicking the ordinary subtraction formulas (i.e. renormalizing using counterterms). Here, the crucial input is the definition of the properties an allowed counterterm should have. We have shown that for theories with noncommuting time variable it is possible to define products of massive fields by subtracting only counterterms which are local in a particular sense. They differ from the ordinary Wick products and are referred to as quasiplanar Wick products.

The purpose of this paper is threefold. For one thing, our results on quasiplanar Wick products from [5] are rewritten in a way which I hope to be more easily accessible and more effective in actual calculations than the compact notation we employed there. Secondly, I included a concrete example to show how on the noncommutative Minkowski space, the ordinary Wick subtraction leads to nonlocal counterterms. And since the quasiplanar Wick products may be understood in terms of vacuum expectation values, I have also included a discussion of vacuum expectation values of free fields on the noncommutative Minkowski space. Their most surprising feature is that they still fulfill the cluster decomposition property, i.e. causality at large distances.

2.1 Free fields and perturbation theory

The noncommutative space-time considered in this paper, the so-called noncommutative Minkowski space, was defined in [9]. Here, the noncommuting coordinate operators are subject to the following relations ($\mu = 0, \ldots, 3$, Einstein's convention for sums is employed, and I is the identity):

$$[q^\mu, q^\nu] = iQ^{\mu\nu}\ , \qquad Q^{\mu\nu}Q_{\mu\nu} = 0\ , \quad \tfrac{1}{8}\,Q^{\mu\nu}\epsilon_{\mu\nu\rho\sigma}Q^{\rho\sigma} = \pm \mathrm{I}\ , \quad [q^\mu, Q^{\nu\rho}] = 0\ .$$

Though formulated in terms of a specific set of coordinates, the construction is invariant under the action of the full Poincaré group, since by definition the spectrum Σ of the mutually commuting operators $Q^{\mu\nu}$ is fixed in a Poincaré-invariant way. Therefore, the above can be said to define a noncommutative version of Minkowski space[1]. As was shown in [9], any state in the domain of the commutators fulfills the desired uncertainty relations.

In what follows, we will consider the commutation relations in the stronger Weyl form

$$e^{ikq}\, e^{ipq} = e^{-\frac{i}{2}kQp}\, e^{i(k+p)q}\ , \tag{2.1}$$

where $kq = k_\mu q^\mu$, $kQp = k_\mu Q^{\mu\nu} p_\nu$. The exponential $e^{-\frac{i}{2}kQp}$ is referred to as the twisting.

[1]The elements of Σ are antisymmetric 4×4-matrices σ with real entries which – in the same manner as the field strength tensor of electrodynamics – can be parametrized by an "electric" and a "magnetic" component, $\sigma_{0i} = e_i$, $\sigma_{ij} = \epsilon_{ijk}m_k$ for $\vec{e}, \vec{m} \in \mathbb{R}^3$ with $\vec{e}^2 - \vec{m}^2 = 0$, $\vec{e}\vec{m} = \pm 1$ (by the defining relations). As a topological space, Σ was shown to be homeomorphic to two copies of TS^2 (which is not compact). For the purposes of this paper the reader who is more familiar with models with fixed noncommutativity matrix may choose a specific element of Σ, as the use of the full spectrum is not crucial in the construction of the quasiplanar Wick products. In this case, however, Lorentz invariance is lost.

2.1.1 Free fields

In [9], "functions" on the noncommutative Minkowski space were defined in terms of the Weyl correspondence,

$$W(f) \equiv f(q) := \int dk\; e^{ikq} \check{f}(k) = (2\pi)^{-4} \int dk\; e^{-ikq} \hat{f}(k)\;,$$

where $\check{f}(k) = (2\pi)^{-4} \int dx\; f(x) e^{-ikx}$, $\hat{f}(k) = (2\pi)^4 \check{f}(-k)$. Using the Weyl relations (2.1) we can calculate products of these expressions and refer to the resulting algebra as $\mathcal{E}$ (for a more extensive review see [8] in this volume). By construction, the *full* Poincaré group acts as an automorphism on $\mathcal{E}$. By analogy with the above, the free field ϕ on the quantum space-time was defined in [9] by the heuristic formula

$$\phi(q) = \int dk\; \check{\varphi}(k) \otimes e^{ikq}\;,$$

where φ is the free field on Minkowski space and $\check{\varphi}$ is its inverse Fourier transform. $\phi(q)$ is to be thought of as a (formal) element of the tensor product $\mathfrak{F} \otimes \mathcal{E}$, where $\mathfrak{F}$ is the algebra of polynomials of the free field. In other words, a quantum field on the noncommutative Minkowski space is an affine functional on a suitable subset of the state space $S(\mathcal{E})$ of $\mathcal{E}$, taking values in $\mathfrak{F}$: For suitably chosen states ω, $k \mapsto \omega(e^{ikq}) =: \hat{\psi}_\omega(k)$ defines a suitable testfunction of k, such that

$$\phi(\omega) \equiv \int dk\; \check{\varphi}(k) \omega(e^{ikq}) = \int dx\; \varphi(x)\, \psi_\omega(x) = \varphi(\psi_\omega)\;,$$

and we recover the well-known expression of a field φ evaluated in a testfunction ψ_ω on Minkowski space. The state ω thus takes the role of a localizing function. Formally, we can now write down products of fields,

$$\phi(q)\phi(q) = \int dk_1 dk_2\; \check{\varphi}(k_1)\check{\varphi}(k_2) \otimes e^{ik_1 q} e^{ik_2 q}\;.$$

However, despite the appearance of a twisting in the product of Weyl operators, such products are ill-defined, since

$$(k_1, k_2) \mapsto \omega(e^{ik_1 q} e^{ik_2 q}) = \int d\mu(\sigma)\; e^{-\frac{i}{2}\sum k_i \sigma k_j}\, \psi(\sigma, \sum_j k_j)$$

with some function ψ and some measure[2] $d\mu$ on Σ, fails to be strongly decreasing. Contrary to that, the expression

$$\phi_f^n(\omega) \equiv \varphi^{\otimes n}(\psi_\omega^{\otimes n} \times f)\;, \tag{2.2}$$

[2]Note that this measure may even be given by the point measure whichs maps Q to a particular point $\theta \in \Sigma$.

where $\times$ denotes the convolution and $\hat{\psi}_\omega^{\otimes n} := \omega(e^{ik_1 q} \cdots e^{ik_n q})$, is indeed well-defined for $f \in \mathcal{S}(\mathbb{R}^{4n})$ and suitable ω. Explicitly, we may write

$$\begin{aligned}\phi_f^n(\omega) &= \int dx_N\, \varphi(x_1) \cdots \varphi(x_n) (\psi_\omega^{\otimes n} \times f)(x_N) \\ &= \int dk_N\, \check{\varphi}(k_1) \cdots \check{\varphi}(k_n)\, \hat{f}(k_N)\, \omega(e^{ik_1 q} \cdots e^{ik_n q}) \,,\end{aligned}$$

where $N = (1, \ldots, n)$ and x_N denotes the tuple $(x_i)_{i \in N}$. We call (2.2) the *regularized product of n fields.* Note that this definition is consistent, as $\phi_f^n\, \phi_g^m = \phi_{f \otimes g}^{n+m}$ for $f \in \mathcal{S}(\mathbb{R}^{4n})$, $g \in \mathcal{S}(\mathbb{R}^{4m})$, and that the adjoint is given by $\phi_f^{n*} = \phi_{f^*}^n$, where $f^*(x_1, \ldots, x_n) = \overline{f(x_n, \ldots, x_1)}$. Replacing f by a δ-distribution, we indeed recover the (ill-defined) product of fields $\phi(q)^n$ evaluated in ω. For formal manipulations it turns out to be convenient to use the symbolic expressions

$$\begin{aligned}\phi_f^n(q) &= \int dk_N\, \check{\varphi}(k_1) \cdots \check{\varphi}(k_n)\, \hat{f}(k_N) \otimes e^{ik_1 q} \cdots e^{ik_n q} \\ &= \int dx_N\, f(x_N)\, \phi(q + x_1) \cdots \phi(q + x_n) \,,\end{aligned}$$

where $q + x$ is shorthand for $q + x\mathrm{I}$ (recall that the translations act as automorphisms on $\mathcal{E}$). In what follows, the tensor product symbol will be dropped.

Given a regular representation of $\mathcal{E}$ on some Hilbert space $\mathcal{H}$, the (formal) elements ϕ_f^n of $\mathfrak{F} \otimes \mathcal{E}$ can be represented by operators on a dense domain in $\mathcal{H}_\varphi \otimes \mathcal{H}$, where $\mathcal{H}_\varphi$ is the Fock space of the free field. In this sense we can take partial expectation values with respect to Fock vectors Ψ_1, Ψ_2 as follows:

$$\begin{aligned}\langle \Psi_1 | \phi_f^n(q) | \Psi_2 \rangle &= \int dx_N f(x_N) \langle \Psi_1 |\, \phi(q + x_1) \cdots \phi(q + x_n) | \Psi_2 \rangle \\ &= \int dx_N f(x_N) \int dk_N \langle \Psi_1 |\, \check{\varphi}(k_1) \cdots \check{\varphi}(k_n) | \Psi_2 \rangle \prod_{j=1}^{n} e^{ik_j (q + x_j)} .\end{aligned} \tag{2.3}$$

Such expressions are referred to as "expectation values" although we still have to evaluate them in a state on $\mathcal{E}$ before they take values in $\mathbb{C}$.

In particular, we find that the vacuum expectation value $\langle \cdot \rangle$ of two fields is equal to the ordinary 2-point function:

$$\begin{aligned}\langle \phi(q + x_1) \phi(q + x_2) \rangle &= \int dk_1 dk_2\, \langle \check{\varphi}(k_1) \check{\varphi}(k_2) \rangle\, e^{ik_1 (q + x_1)}\, e^{ik_2 (q + x_2)} \\ &= \int d\mu(k)\, e^{-ik(x_1 - x_2)} = \Delta_+(x_1 - x_2) \,,\end{aligned}$$

where $d\mu(k)$ denotes the usual $L_+^\uparrow$-invariant measure on the positive mass shell, $d\mu(k) = (2\pi)^{-3} \frac{d^3\mathbf{k}}{2\omega_\mathbf{k}}\Big|_{k^0 = \omega_\mathbf{k}}$, $\omega_\mathbf{k} = \sqrt{m^2 + \mathbf{k}^2}$. Note however, that unless the vacuum expectation value is taken, the commutator of two fields is not equal to the ordinary commutator function,

$$[\phi(q+x_1),\phi(q+x_2)] \neq \Delta(x_1-x_2)\ .$$

2.1.2 Vacuum expectation values

To get used to field theory on the noncommutative Minkowski space, we will now consider the properties of vacuum expectation values of free fields. First we recall the explicit form of vacuum expectation values on the ordinary Minkowski space. Evaluated in a testfunction g, a vacuum expectation value of an even number of fields (for an odd number the vacuum expectation values vanish) is given by

$$\begin{aligned}\int dx_N\, g(x_N)\,\langle\varphi(x_1)\cdots\varphi(x_n)\rangle &= \sum_{\substack{C\in\mathcal{C}(N)\\ |A|=\frac{n}{2}}} \int dx_N g(x_N) \int d\mu(k_A) e^{-ik_A(x_A-x_{\alpha(A)})}\\ &= \sum_{\substack{C\in\mathcal{C}(N)\\ |A|=\frac{n}{2}}} \int d\mu(k_A)\ (2\pi)^{4n}\ \check{g}(k_N)\big|_{k_{\alpha(A)}=-k_A}\ .\end{aligned}$$

Here, $N=(1,\ldots,n)$, and for any ordered index set $A=(a_1,\ldots,a_{|A|})$, k_A denotes the tuple $(k_{a_1},\ldots,k_{a_{|A|}})$, $d\mu(k_A)=\prod_{a\in A} d\mu(k_a)$, and for two ordered index sets A and B with $|A|=|B|$, $k_A x_B=\sum_{i=1}^{|A|} k_{a_i}x_{b_i}$. We use $\mathcal{C}(N)$ denotes the set of all contractions of N, each of which is given by a pair (A,α), where A is an ordered subset of N (generally including $A=\emptyset$), and α is an injective map $\alpha: A\to N\setminus A$ with $\alpha(a)>a$ (with respect to the order of N) for all $a\in A$. The set $\alpha(A)=(\alpha(a_1),\ldots,\alpha(a_{|A|}))$ inherits its order from A. In what follows, the letter U is reserved for the ordered set $N\setminus(A\cup\alpha(A))$ (of course, U is empty in the contractions considered above where $|A|=\frac{n}{2}$). For any given contraction, the corresponding momenta k_A and k_U are called the internal and external momenta, respectively.

The situation is very similar on the noncommutative Minkowski space, where by application of (2.3) we find $\langle\phi(q+x_1)\cdots\phi(q+x_n)\rangle\equiv 0$ for n odd and for n even,

$$\begin{aligned}\langle\phi_f^n(q)\rangle &= \sum_{\substack{C\in\mathcal{C}(N)\\ |A|=\frac{n}{2}}} \int dx_N f(x_N) \int d\mu(k_A)\ e^{-ik_A(x_A-x_{\alpha(A)})} \prod_{i<j\in N} e^{-\frac{i}{2}k_iQk_j}\big|_{k_{\alpha(A)}=-k_A}\\ &= \sum_{\substack{C\in\mathcal{C}(N)\\ |A|=\frac{n}{2}}} \int d\mu(k_A)\ \Big((2\pi)^{4n}\ \check{f}(k_N) \prod_{i<j\in N} e^{-\frac{i}{2}k_iQk_j}\Big)\big|_{k_{\alpha(A)}=-k_A}\ .\end{aligned}$$

We see that the vacuum expectation values on the noncommutative Minkowski space depend only on the commutators $Q^{\mu\nu}$, not on the noncommuting coordinates themselves. Using the notation

$$W_Q(x_N) = \langle \phi(q+x_1)\cdots\phi(q+x_n)\rangle \quad \text{and} \quad W(x_N) = \langle \varphi(x_1)\cdots\varphi(x_n)\rangle\,,$$

we conclude

$$\int dx_N\; f(x_N)\, W_Q(x_N) = \int dx_N\; f_Q(x_N)\, W(x_N)\,,$$

where f_Q is the Fourier transform of $\check{f}(k_N)\prod_{i<j\in N} e^{-\frac{i}{2}k_i Q k_j}$. In fact, f_Q is a Schwartz function if f is Schwartz (the twisting being bounded and smooth). Note however, that f_Q takes values in $\mathcal{Z}$, the centre of the multiplier algebra of $\mathcal{E}$, which is identifiable with the bounded functions on Σ. In what follows, functions will be understood to take values in $\mathcal{Z}$ without further notice whenever necessary.

It is sometimes convenient to symbolize the elements of $\mathcal{C}(N)$ by graphs as follows. Draw $|N|$ points in a horizontal line and for each pair $(a,\alpha(a))$ of a contraction $C \in \mathcal{C}(N)$ connect the corresponding points by a connecting curve in the upper half plane (the so-called internal lines). For example, for $N = (1,\ldots,6)$, $A = (1,3)$, $\alpha(A) = (4,5)$, we draw the graph  . Using these graphs, we can easily read off the twisting of a contraction. First we rewrite the twisting as follows (see [1,5]),

$$\prod_{i<j\in N} e^{-\frac{i}{2}k_i Q k_j}\Big|_{k_{\alpha(A)}=-k_A} = e^{-i\langle k_A, I k_A\rangle - i\langle k_A, E k_U\rangle}\,,$$

with a $(4|A| \times 4|A|)$-matrix I (called the *intersection matrix*) and a $(4|A| \times 4|U|)$-matrix E (called the *enclosure matrix*). Now, each 4×4 block $I_{aa'}$ is given by the matrix $Q^{\mu\nu}$ if $a < a' \in A$ and the connecting curve of the corresponding graph which starts from a intersects the one starting from a'; otherwise $I_{aa'} = 0$. Likewise, each 4×4 block E_{au}, $a \in A$, $u \in U$, is given by the matrix $Q^{\mu\nu}$ if in the corresponding graph the point u (a so-called uncontracted point) is enclosed by the connecting curve starting from a.

It is also possible to describe properties of contractions directly in terms of these graphs, instead of using complicated notation. For instance, a connected component of a contraction is given by all internal lines which intersect each other, along with *all* elements[3] of N. A contraction with only one connected component (which then automatically coincides with the full contraction itself) is called a *connected contraction.* The contraction discussed above is connected, whereas the contraction 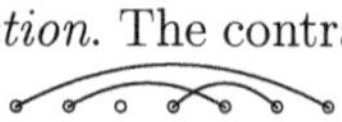 possesses the two connected components  and  .

Let us collect some properties of the n-point functions $W_Q(x_N)$ which follow directly from the definition and the classic results for fields on the ordinary Minkowski space.

1. The n-point functions behave as ordinary Wightman functions under complex conjugation,

[3] This is the convention we chose in [5], which differs from the one used in [1].

$$\overline{W_Q(x_1,\ldots,x_n)} = W_Q(x_n,\ldots,x_1)\ .$$

2. They are invariant under Poincaré transformations, since the fields transform covariantly,

$$U(\Lambda,a)\,\phi(q+x)\,U(\Lambda,a)^{-1} = \phi(\Lambda(q+x)+a)\ ,$$

the vacuum vector is invariant as usual, and the spectrum of the $Q^{\mu\nu}$ remains unaffected by the transformation (by definition, for all $\sigma \in \Sigma$, $\Lambda\sigma\Lambda^t$ is again an element of Σ). Note however, that after evaluation in a state ω (i.e. after integration against some measure $d\mu$ on Σ) the expression is no longer invariant but merely covariant. The reason is that the Lorentz group is not amenable such that there cannot be a finite measure on Σ whose support is invariant. Instead, the transformed measure will define a different state.
3. The n-point functions are translation-invariant with respect to the commutative arguments x_i, and hence their Fourier transforms have support only in $\sum_{i=1}^n p_i = 0$. Moreover, since we treat only free fields, the spectral property holds by definition.
4. After evaluation in a state ω, the n-point functions are positive definite in the usual sense ($\phi^* = \phi$),

$$\sum_{j,k} \int dx_1 \ldots dx_j\, dy_1 \ldots dy_k\, \overline{f_j(x_1,\ldots,x_j)}\,\cdot$$
$$\cdot\,\omega\big(\langle\phi(q+x_j)\cdots\phi(q+x_1)\,\phi(q+y_1)\cdots\phi(q+y_k)\rangle\big)\ f_k(y_1,\ldots,y_k) \geq 0$$

for any sequence of testfunctions $\{f_j\}$, $f_j \in \mathcal{S}(\mathbb{R}^{4j})$ with $f_j = 0$ except for a finite number of j. This follows as usual from the fact that the left-hand side of this inequality is given by the product of the state

$$f_0\,|\Omega\rangle + \int dx_1\ \phi(q+x_1)\,f_1(x_1)\,|\Omega\rangle$$
$$+ \int dx_1 dx_2\ \phi(q+x_1)\,\phi(q+x_2)\,f_2(x_1,x_2)\,|\Omega\rangle + \cdots$$

with its conjugate, and subsequent evaluation in ω.
5. Up to this point, no serious deviation from ordinary field theory has arisen. However, apart from the lowest possible order, the n-point functions (for n even) are not microcausal ("local" in the sense of Wightman theory). While for the 2-point function

$$W_Q(x_1,x_2) = \Delta_+(x_1-x_2)\ ,$$

which is the only n-point function (with n even) independent of Q, we have $W_Q(x_1,x_2) = W_Q(x_2,x_1)$ for $(x_1-x_2)^2 < 0$, this ceases to be the case when more fields are involved, i.e. for $(x_i-x_j)^2 < 0$ we find

$$W_Q(x_1,\ldots,x_i,x_j,\ldots,x_n) \neq W_Q(x_1,\ldots,x_j,x_i,\ldots,x_n)\ , \qquad n > 2\ .$$

Quite surprisingly in view of this last point, we do find some remnant of causality at large distances: the cluster decomposition property does indeed remain valid on the noncommutative Minkowski space.

Lemma 1. "Cluster decomposition property" *Let $b \in \mathbb{R}^4$ be a space-like vector, and $\lambda > 0$. Then*

$$W_Q(x_1, \ldots, x_j, x_{j+1} + \lambda b, \ldots, x_n + \lambda b) \\ \rightarrow W_Q(x_1, \ldots, x_j)\, W_Q(x_{j+1}, \ldots, x_n) \tag{2.4}$$

for $\lambda \rightarrow \infty$ in the sense of tempered distributions.

Proof. Let $J = (1, \ldots, j)$. After evaluation in a testfunction $f \in \mathcal{S}(\mathbb{R}^{4n})$, the left-hand side of (2.4) is given by

$$\sum_{\substack{C \in \mathcal{C}(N) \\ |A| = \frac{n}{2}}} \int dx_N\, f(x_N) \int d\mu(k_A) \prod_{a \in A} e^{-ik_a(x_a - x_{\alpha(a)} - \lambda_a b)} \prod_{i<j \in N} e^{-\frac{i}{2} k_i Q k_j} \Big|_{k_{\alpha(A)} = -k_A},$$

where $\lambda_a = 0$, if both a and $\alpha(a)$ are either in J or in $N \setminus J$, and $\lambda_a = \lambda$ otherwise. From the discussion on page 16 it follows that the twisting in a contraction with $\lambda_a = 0$ for all $a \in A$ (i.e. for all $a \in A$, both a and $\alpha(a)$ are either in J or in $N \setminus J$) consists of two distinct parts, one containing only momenta k_a with $a \in J \cap A$, the other containing only k_a with $a \in (N \setminus J) \cap A$. Therefore, these indeed constitute the right hand side of (2.4). To see that the remaining contractions, i.e. those where $\lambda_a = \lambda$ for at least one $a \in A$, vanish in the limit $\lambda \rightarrow \infty$, we first note that after the integration over x_N is performed, such a contribution yields

$$\int d\mu(k_A) \prod_{a \in A} e^{-ik_a \lambda_a b} \Big((2\pi)^{4n}\, \check{f}(k_N) \prod_{i<j \in N} e^{-\frac{i}{2} k_i Q k_j} \Big) \Big|_{k_{\alpha(A)} = -k_A},$$

where $\lambda_a \neq 0$ for at least one $a \in A$. As in the calculation on page 16, we may now define $\check{f}_Q(k_N) = e^{-\frac{i}{2} \sum k_i Q k_j}\, \check{f}(k_N)$. The twisting being bounded and smooth, this is again a Schwartz function (taking values in $\mathcal{Z}$), and the claim indeed follows since the 2-point function decreases exponentially for space-like argument. □

Note that in the proof locality was not required, since we only treated free fields.

2.1.3 Perturbation theory

As we have pointed out elsewhere [3], there are two approaches to define unitary perturbation theory on the noncommutative Minkowski space (with noncommuting time). As was shown in [1], these approaches are inequivalent unless the time variable commutes with the space variables, in which case they coincide with the so-called modified Feynman rules. The first possibility is the Yang-Feldman approach. Here, the field equation is solved recursively using ordinary retarded propagators. Explicit rules may be found in [1]. The second approach was proposed already in [9] and is based on the introduction of an interaction Hamiltonian $H_I(t) = \mu\Big(\int_{q^0=t} d^3q \, \mathcal{H}_I(\phi(q))\Big)$ where μ denotes a positive map with $\mu(e^{ikQp}) = \int d\mu(\sigma)\, e^{ik\sigma p}$ with some measure $d\mu$ on Σ and where the "integral" stands symbolically for a positive weight on $\mathcal{E}$. Expectation values are calculated from Dyson's series (for one possible set of rules see for instance [1]). In the Hamiltonian approach, we have investigated different possibilities to define a suitable generalization $\mathcal{H}_I$ of the ordinary interaction Hamiltonian density, one based on the quantum diagonal map [4], the others based on using $:\phi(q)^n:$ as the bare interaction term (see also [2]). For a review and a particularly lucid set of rules based on general considerations on Wick's reduction formulas see [8] in this volume.

2.2 Local counterterms

It is quite natural to ask whether the ordinary subtraction procedure which defines Wick products on the ordinary Minkowski space should be applied to the noncommutative situation to define products of fields at coinciding points. As we have argued in [5] (see also [1]), this is not the case, since nonlocal counterterms arise. In the first of the two following subsections a summary of our results from [5] is given. The second subsection treats the nonlocality of a typical counterterm in more detail.

2.2.1 Subtraction formulas

We first note that the ordinary Wick product on ordinary Minkowski space is given by the following subtraction formula, which is of course equivalent to the prescription to "put all annihilation operators to the right",

$$\begin{aligned} :\varphi_g^n: &= \sum_{C\in\mathcal{C}(N)} (-1)^{|A|} \int dx_N \, g(x_N) \int d\mu(k_A) \prod_{a\in A} e^{-ik_a(x_a - x_{\alpha(a)})} \prod_{u\in U} \varphi(x_u) \\ &= \sum_{C\in\mathcal{C}(N)} (-1)^{|A|} \varphi^{|U|}_{\gamma_0^C(g)} = \varphi_g^n + \sum_{\substack{C\in\mathcal{C}(N)\\ A\neq\emptyset}} (-1)^{|A|} \varphi^{|U|}_{\gamma_0^C(g)}\,, \end{aligned}$$

where for each C, $\gamma_0^C : \mathcal{S}(\mathbb{R}^{4n}) \to \mathcal{S}(\mathbb{R}^{4|U|})$ is the linear continuous map defined by $\gamma_0^C(g)(x_U) = \int dx_A dx_{\alpha(A)} \prod_{a\in A} \Delta_+(x_a - x_{\alpha(a)})\, g(x_N)$. To shorten the notation, we have defined $\varphi^{|U|}_{\gamma_0^C(g)} = \gamma_0^C(g)$ for $U=\emptyset$. It is well known that for $g(x_N) \to G(x_1)\prod_{j>2}\delta(x_1 - x_j)$, the Wick product remains well defined.

We now look for suitably subtracted products of fields on the noncommutative Minkowski space:

$$\vdots\phi^n_f\vdots = \sum_{\substack{C\in\mathcal{C}(N)\\ C \text{ with certain}\\ \text{properties}}} (-1)^{\kappa_C}\, \phi^{|U|}_{\gamma^C(f)}\ , \tag{2.5}$$

where $(-1)^{\kappa_C} = \pm 1$ denotes some possibly necessary sign, and where for each C which appears in the sum, $\gamma^C : \mathcal{S}(\mathbb{R}^{4n}) \to \mathcal{S}(\mathbb{R}^{4|U|})$ is a continuous linear map such that

1. when $f \to \delta$, the limit of $\vdots\phi^n_f\vdots$ exists as an affine $\mathfrak{F}$-valued functional on some dense subset of $S(\mathcal{E})$;
2. (locality condition): the maps γ^C can be chosen to be local in the sense that
$$\mathrm{supp}\gamma^C(f) \subset P_U\, \mathrm{supp} f\ ,$$
where P_U is the projection $\mathbb{R}^{4n} \mapsto \mathbb{R}^{4|U|}$ given by $P_U(x_N) = x_U$.

Note that in condition 1, the function ψ_ω corresponding to $\omega \in S(\mathcal{E})$, takes the role of the function G from the previous page. We first try to apply the ordinary Wick ordering and check whether condition 2 holds. Compared to the situation on the ordinary Minkowski space we find that internal and external momenta k_A and k_U are now linked via the twisting:

$$\begin{aligned}
:\phi^n_f(q): &= \sum_{C\in\mathcal{C}(N)} (-1)^{|A|} \int dx_N\, f(x_N) \int dk_N \prod_{a\in A} \langle \check{\varphi}(k_a)\check{\varphi}(k_{\alpha(a)})\rangle \\
&\quad \cdot \int dk_U \prod_{u\in U} \check{\varphi}(k_u) \prod_{i<j\in N} e^{-\frac{i}{2}k_i Q k_j}\ e^{i\sum_N k_j(q+x_j)} \\
&= \sum_{C\in\mathcal{C}(N)} (-1)^{|A|} \int dx_N\, f(x_N) \int d\mu(k_A) \prod_{a\in A} e^{-ik_a(x_a - x_{\alpha(a)})} \\
&\quad \cdot \int dk_U \prod_{u\in U} \check{\varphi}(k_u) \Big(\prod_{i<j\in N} e^{-\frac{i}{2}k_i Q k_j}\Big)\Big|_{k_{\alpha(A)}=-k_A} e^{i\sum_U k_u(q+x_u)} \\
&= \sum_{C\in\mathcal{C}(N)} (-1)^{|A|} \int dx_U\, \gamma^C(f)(x_U) \prod_{u\in U} \phi(q+x_u)
\end{aligned}$$

where $\gamma^C(f)$, the so-called *contraction map*, is the Fourier transform of

$$(\gamma^C(f))\check{\ }(k_U) = (2\pi)^{8|A|} \int d\mu(k_A)\, e^{-i\langle k_A, Ik_A\rangle - i\langle k_A, Ek_U\rangle}\, \check{f}(k_N)|_{k_{\alpha(A)}=-k_A}\ .$$

It is pointed out that the contraction map maps a Schwartz function f to a Schwartz function $\gamma_C(f)$ taking values in $\mathcal{Z}$. Note that, using the contraction map, we can rewrite the vacuum expectation values as

$$\int dx_N\, W_Q(x_N)\, f(x_N) = \sum_{\substack{C \in \mathcal{C}(N) \\ |A| = \frac{n}{2}}} \gamma^C(f)\,. \tag{2.6}$$

Likewise, the connected part of the n-point function, denoted $W_Q(x_N)^c$, is given by the above, with the sum now running only over connected contractions.

2.2.2 The simplest example

It is now shown that on the noncommutative Minkowski space, the subtractions which lead to the ordinary Wick product do not fulfill condition 2. on page 20. Consider the third Wick power $:\phi_f^3:$ which in terms of graphs is given by the following sum of contractions 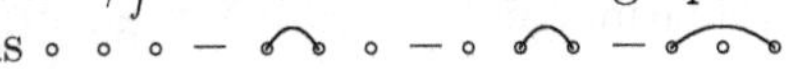 .
The last contraction yields

$$\gamma^C(f)(x_2) = \int dx_1 dx_3 \int d\mu(k)\, e^{-ik(x_1 - x_3)}\, f(x_1, x_2 + Qk, x_3)$$

where we have performed the coordinate transformation $x_2 \to x_2 + \sigma k$ (for all $\sigma \in \Sigma$).

Lemma 2. *Let $f = h \otimes h \otimes h$ be a smooth function on $\mathbb{R}^{12}$ with compact support. Then $\gamma^C(f)$ as above will in general not have compact support. Furthermore, we can find smooth functions h of compact support such that* $\operatorname{supp} \gamma^C(f) \cap \operatorname{supp} h = \emptyset$.

Proof. We first perform the integrations over x_1 and x_3 which yields the product $\hat{h}(k)\hat{h}(-k)$. Since h has compact support, $\hat{h}$ is analytic and therefore cannot vanish in an open subset without being identically 0. We can thus find h such that $H_m^+ \subset \operatorname{supp} \hat{h}$. Since the multiplication with an element of Σ will not confine the possible momenta to a compactum, we conclude that in this case, $\gamma^C(f)$ will not have compact support.

Now, let the support of h be contained in the box $[-a, a] \subset \mathbb{R}^4$, where $a \in \mathbb{R}^4$, $0 < a^\mu < \frac{1}{3} \inf_{\sigma \in \Sigma} \inf_{k \in H_m^+} |\sigma^{\mu\nu} k_\nu|$ for $\mu = 0, \ldots, 3$. Note that $a \neq 0$ since k is on the positive mass shell and σ is nondegenerate. It clearly follows that $\operatorname{supp} \gamma^C(f) \cap \operatorname{supp} h = \emptyset$. $\square$

Lemma 2 clearly implies that the map γ^C as above violates the locality condition from page 20. Let us now consider $\gamma^C(f)$ in the limit of coinciding points. Take a sequence $\{h_r\}$ of compactly supported functions centered around $0 \in \mathbb{R}^4$ with $\operatorname{supp} h_r \subset h_{r+1}$ and $h_r \to \delta$. Then $\gamma^C(h_r)(x)$ does not converge to $\delta(x)$ (in the appropriate sense). In fact, it follows from Lemma 2 that the limit of $\gamma^C(h_r)(x)$ is zero when evaluated on testfunctions which vanish outside a sufficiently small neighbourhood of 0. As was shown in [1], no addition from the kernel of the free field, i.e. a function of the form $(\Box + m^2)f$, can serve to make γ^C local.

If we hold on to our wish to preserve locality in the sense of not enlarging the support of testfunctions, we therefore must not subtract the counterterm  . Luckily, we do not have to subtract this term since it correponds to a *finite* renormalization. To see this, we evaluate γ^C in $f(x, y, z) = \delta(x)\delta(y)\delta(z)$ and calculate

$$\begin{aligned}\phi_{\gamma^C(f)} &= \int dx\, \gamma^C(f)(x)\phi(q + x) = \int dx \int d\mu(k)\, \delta(x + Qk)\, \phi(q + x) \\ &= \int dp \int d\mu(k)\, \check{\varphi}(p)\, e^{ip(q-Qk)} = \int dp\, \Delta_+(Qp)\check{\varphi}(p)\, e^{ipq}\ .\end{aligned}$$

Now, p is on the positive or negative mass shell, hence σp is space-like for all $\sigma \in \Sigma$ and therefore $\Delta_+(Qp)$ is smooth and quickly decreasing. Therefore, evaluating this expression in a suitable state ω yields an element of $\mathfrak{F}$.

Last but not least it is pointed out that the compactly supported testfunctions in the above lemma merely serve to illustrate the nonlocality of γ^C, but that such functions do not in fact provide a suitable space of testfunctions [6]. However, the locality condition is also violated when h (notation as above) is not compactly supported but instead given by, say,

$$h(x) \;=\; e^{-x_0^2-x_1^2-x_3^2}\, e^{-x_2^4/(x_2-a)^2}\, \chi_a(x_2)\ , \quad a > 0,$$

with the characteristic function χ_a of the interval $(-\infty, a]$. In this case, choosing the particular element $\sigma \in \Sigma$ given by the usual canonical symplectic matrix, $\sigma = \begin{pmatrix} 0 & -1 \\ 1 & 0 \end{pmatrix}$, we find that for all $n \geq 2$, $V_n \overset{\text{def}}{=} \{x \in \mathbb{R}^4 | \, x_2 \in [a+\frac{m}{n}, a+m-\frac{m}{n}]\}$, where m is the mass of the field, is contained in $\operatorname{supp} \gamma^C(g)$, but $V_n \cap \operatorname{supp} h = \emptyset$.

2.3 Quasiplanar Wick products

In the previous section we have seen seen that some of the ordinary counterterms used in the definition of Wick products cease to be local on the noncommutative Minkowski space. The consequence we have drawn was to define modified Wick products by subtracting only local counterterms. In this section, the definitions and combinatorial aspects of these products as discussed in [5] are rewritten in a form more similar to the conventions used in [1]. It is to be hoped that they will be more immediately accessible to the reader.

We start by observing that, for a contraction $C \in \mathcal{C}(N)$ with trivial enclosure matrix, the contraction map is given explicitly as

$$\gamma^C(f)(x_U) = \int d\mu(k_A)\, e^{-i\langle k_A, Ik_A\rangle}\, (\mathfrak{F}^{-1}_{N\setminus U} f)(\xi_N(k_A, x_U)) \tag{2.7}$$

with $\mathcal{F}^{-1}_{N\setminus U} f$ denoting the inverse Fourier transform of f with respect to the variables $x_{N\setminus U}$, and with $\xi_a = k_a$, $\xi_{\alpha(a)} = -k_a$ for $a \in A$, and $\xi_u = x_u$ for $u \in U$. It follows immediately that for such C the locality condition holds.

Contractions with trivial enclosure matrix are called *quasiplanar* contractions[4]. The set of contractions whose *connected components* are quasiplanar is denoted $\mathcal{C}_{qp}(N)$. Note that a contraction may have trivial enclosure matrix (i.e. be quasiplanar) but still not be in $\mathcal{C}_{qp}(N)$. As an example consider the quasiplanar contraction  which has one connected component that is not quasiplanar.

Using these definitions, we have given a recursive definition of products of fields on the noncommutative Minkowski space where only quasiplanar, i.e. local, counterterms are subtracted (see [5]). Let $h \in \mathcal{S}(\mathbb{R}^4)$, $g \in \mathcal{S}(\mathbb{R}^{4n})$, then

$$\vdots\phi^{1+|N|}_{h\otimes g}\vdots = \phi_h \;\vdots\phi^{|N|}_g\vdots \; - \sum_{\substack{C\in\mathcal{C}_{qp}((1)\sqcup N)\\ C\text{connected},1\in A}} \vdots\phi^{|U|}_{\gamma^C(h\otimes g)}\vdots \tag{2.8}$$

Here, $A\sqcup B$ denotes the ordered union of two ordered sets A and B, where B is appended to A. As we have shown in [5], this recursive definition is equivalent to

$$\vdots\phi^{|N|}_f\vdots = \sum_{C\in\mathcal{C}_{qp}(N)} (-1)^\kappa \phi^{|U|}_{\gamma^C(f)}\,, \tag{2.9}$$

where κ is the number of connected components of C. As desired, this is indeed of the form (2.5).

Lemma 3. *Let $M = (1, \ldots, n+1)$. Then (2.8) can be rewritten as follows:*

$$\vdots\prod_{j\in M}\phi(q+x_j)\vdots = \phi(q+x_1)\;\vdots\prod_{i=2}^{n+1}\phi(q+x_i)\vdots$$
$$- \sum_{a=1}^{[\frac{n+1}{2}]} W_Q(x_1,\ldots,x_{2a})^c \;\vdots\prod_{i=2a+1}^{n+1}\phi(q+x_i)\vdots\,,$$

where $W_Q(x_1,\ldots,x_{2a})^c$ denotes the connected contributions to the vacuum expectation values of $2a$ fields.

Proof. The sum over all connected contractions $C \in \mathcal{C}_{qp}(M)$ with $1 \in A$ from the right-hand side of (2.8) can be rewritten as follows ($f \in \mathcal{S}(\mathbb{R}^{4(n+1)})$):

$$\sum_{\substack{C\in\mathcal{C}_{qp}(M)\\ C\text{connected},1\in A}} \vdots\phi^{|U|}_{\gamma^C(f)}\vdots = \sum_{a=1}^{[\frac{n+1}{2}]} \sum_{\substack{C'\in\mathcal{C}(M)\\ C'\text{connected}\\ U=(2a+1,\ldots,n+1)}} \vdots\phi^{|U|}_{\gamma^{C'}(f)}\vdots\,.$$

[4]Since the corresponding contraction maps may still depend on the twisting (via the intersection matrix I), the term *planar contraction* is not suitable.

Now, by definition, C' is quasiplanar and $\gamma^{C'}(f)$ is given explicitly by (2.7) with C connected and with $\xi_N(k_A, x_U) = (k_{N\setminus U}, x_U)$ where $k_{\alpha(A)} = -k_A$. The claim now follows from (2.6). □

Now consider the quasiplanar Wick theorem from [5],

$$\vdots\phi_f^n\vdots\,\vdots\phi_g^m\vdots = \sum_{C\in\mathcal{C}(N,M)} \vdots\phi^{|U|}_{\gamma^C(f\otimes g)}\vdots\ ,$$

where $\mathcal{C}(N,M)$ denotes the set of all quasiplanar contractions $C \in \mathcal{C}(N \sqcup M)$ which have the property that every connected component C' of C connects N and M, in the sense that $\alpha(A_{C'} \cap N) \cap M \neq \emptyset$. The explicit form of the quasiplanar Wick theorem reads

$$\begin{aligned} \vdots\prod_{i=1}^n \phi(q+x_i)\vdots\,\vdots\prod_{j=1}^m \phi(q+y_j)\vdots \ &= \vdots\prod_{i=1}^n \phi(q+x_i)\prod_{j=1}^m \phi(q+y_j)\vdots \\ +\sum_{l=1}^n\sum_{k=1}^m W_Q(x_{n-l+1},\ldots,x_n,y_1,\ldots,y_k)^{(l,k)} &\vdots\prod_{i=1}^{n-l} \phi(q+x_i)\prod_{j=k+1}^m \phi(q+y_j)\vdots\ . \end{aligned}$$

Here, $W_Q(x_1,\ldots,x_{l+k})^{(l,k)}$ denotes the sum of those contributions to the $(l+k)$-point function which contain in every connected component at least one contraction $(a,\alpha(a))$ with $a \in (1,\ldots,l)$ and $\alpha(a) \in (l+1,\ldots,l+k)$.

As we have shown in [5] (see also [1]), quasiplanar Wick products can be reexpressed in terms of ordinary Wick products by the following formula:

$$\vdots\phi_f^{|N|}\vdots = \sum_{C\in\mathcal{C}_{ap}(N)} :\phi^{|U|}_{\gamma^C(f)}: \ = \ :\phi_f^{|N|}: + \sum_{\substack{C\in\mathcal{C}_{ap}(N)\\ |A|\neq 0}} :\phi^{|U|}_{\gamma^C(f)}:\ , \tag{2.10}$$

where $\mathcal{C}_{ap}(N)$ is the set of all *aplanar* contractions of N. A contraction is called aplanar if for *every* connected component the corresponding part of the enclosure matrix is nontrivial. Equivalently, $C \in \mathcal{C}_{ap}(N)$ if and only if for all $a \in A$ either $\exists\, u \in U$ such that $E_{au} \neq 0$ or $\exists\, u \in U$ and $\exists\, a_1,\ldots,a_k \in A$, $k \geq 1$, such that

$$\tilde{I}_{aa_1}\tilde{I}_{a_1a_2}\cdots\tilde{I}_{a_{k-1}a_k}E_{a_ku} \neq 0\ ,$$

where $\tilde{I}_{ab} = I_{ab}$ if $a < b$ and $\tilde{I}_{ab} = I_{ba}$ if $a > b$. In other words, if we define a set $\mathcal{C}_A(C)$ of components of C to consist of all internal lines which intersect each other, along with all uncontracted points enclosed by any of these internal lines, then C is aplanar if and only if every component in $\mathcal{C}_A$ contains at least one uncontracted point. An example for an aplanar contraction is 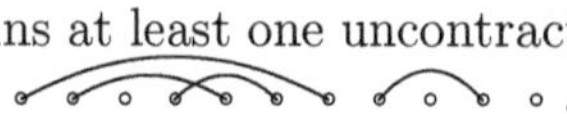 . Note that the empty contraction is quasiplanar and aplanar, and that contractions may be neither in $\mathcal{C}_{qp}$ nor in $\mathcal{C}_{ap}$ (as an example consider ).

Using (2.10) we have given an argument in [5] why the quasiplanar Wick products are well-defined even in the limit $f \to \delta$. The idea is the same as in the simplest case of three fields discussed in this paper, namely to show that the terms which we do not subtract compared to the ordinary Wick product remain finite in this limit.

Roughly speaking, the quasiplanarity of the counterterms we find from Lemma 3 ensures that, in this limit, the counterterms are always of the form $c_Q\,\phi(q)^n$ with (infinite) renormalization "constants" c_Q depending on Q. The details of the proof (domains of definition etc.) will be published elsewhere [6].

2.4 Outlook

The quasiplanar Wick products can (and in fact should) be applied both in the Hamiltonian approach and in the context of the Yang-Feldman equation. In the first case we use $\vdots\phi(q)^n\vdots$ as the interaction term. Note that since $\vdots\phi(q)^2\vdots = :\phi(q)^2:$, the free theory remains unchanged by the same argument as used in [9] for the ordinary Wick product. It is also straightforward how to calculate the interacting field within the Yang-Feldman approach using quasiplanar Wick products (see [1]). However, to calculate expectation values of fields (either directly within the Hamiltonian approach or using the interacting field from the Yang-Feldman approach), it seems to be most efficient to rewrite the quasiplanar Wick products in terms of ordinary Wick products using (2.10). Efficient rules remain to be developed.

The full proof that theories of massive self-interacting fields are renormalizable by subtraction of quasiplanar counterterms only is not yet available, but lower order calculations indicate that this is indeed the case. Moreover, the construction still has to be investigated for massless fields.

Last but not least I would like to point out that on space-times with commuting time variable, nonlocal counterterms are indeed needed [5].

References

1. D. Bahns: *Perturbative methods on the noncommutative Minkowski space.* Doctoral Thesis, Universität Hamburg, DESY-THESIS-2004-004, 2003. [www-library.desy.de/preparch/desy/thesis/desy-thesis-04-004.ps.gz].
2. D. Bahns: Ultraviolet finiteness of the averaged Hamiltonian on the noncommutative Minkowski space (2004). [arXiv:hep-th/0405224].
3. D. Bahns, S. Doplicher, K. Fredenhagen and G. Piacitelli: On the unitarity problem in space/time noncommutative theories. *Phys. Lett. B* 533:178 (2002). [arXiv:hep-th/0201222].
4. D. Bahns, S. Doplicher, K. Fredenhagen and G. Piacitelli: Ultraviolet finite quantum field theory on quantum space-time. *Commun. Math. Phys.* 237:221 (2003). [arXiv:hep-th/0301100].
5. D. Bahns, S. Doplicher, K. Fredenhagen and G. Piacitelli: Field theory on noncommutative spacetimes: Quasiplanar Wick products. *Phys. Rev. D* 71:025022 (2005). [arXiv:hep-th/0408204].

6. D. Bahns, S. Doplicher, K. Fredenhagen and G. Piacitelli: To appear (2005).
7. S. Doplicher, K. Fredenhagen and J.E. Roberts: The Quantum structure of space-time at the Planck scale and quantum fields. *Commun. Math. Phys.* 172:187 (1995). [arXiv:hep-th/0303037].
8. G. Piacitelli: DFR Perturbative Quantum Field Theory on Quantum Space Time, and Wick Reduction. This volume, Chapter 16 (2006).

3

Massless Scalar Field in a Two-dimensional de Sitter Universe*

Marco Bertola[1], Francesco Corbetta[2], Ugo Moschella[23]

[1] Concordia University, Department of Mathematics and Statistics, Loyola Campus, HB-237, 7141 Sherbrooke W., H4B 1R6, Montreal (QC), Canada
bertola@CRM.UMontreal.ca

[2] Dipartimento di Fisica e Matematica, Università dell'Insubria, 22100 Como, Italia
chicco-c@libero.it, ugo.moschella@uninsubria.it

[3] INFN, Sez. di Milano, Via Celoria 2, Milano, Italia

Summary. We study the massless minimally coupled scalar field on a two-dimensional de Sitter space-time in the setting of axiomatic quantum field theory. We construct the invariant Wightman distribution obtained as the renormalized zero-mass limit of the massive one. Insisting on gauge invariance of the model we construct a vacuum state and a Hilbert space of physical states which are invariant under the action of the whole de Sitter group. We also present the integral expression of the conserved charge which generates the gauge invariance and propose a definition of dual field.

3.1 de Sitter geometry and the massive scalar field

The de Sitter space-time may be represented as a d-dimensional one-sheeted hyperboloid

$$X_d(R) = X_d = \left\{x \in \mathbb{R}^{d+1} \ : \ x^{(0)^2} - x^{(1)^2} - \cdots - x^{(d)^2} = -R^2\right\} \qquad (3.1)$$

embedded in a Minkowski ambient space $\mathbb{R}^{d+1}$ with scalar product $x \cdot y = x^{(0)}y^{(0)} - x^{(1)}y^{(1)} - \cdots - x^{(d)}y^{(d)}$ (and $x^2 = x \cdot x$). The manifold X_d is then equipped with a causal ordering relation induced by that of $\mathbb{R}^{d+1}$. The invariance group of the de Sitter space-time $G_d = SO_0(1,d)$. The G_d-invariant volume form on X_d will be denoted by $\mathrm{d}\sigma(x)$. Let us introduce the complex hyperboloid

$$X_d^{(c)}(R) = X_d^{(c)} = \left\{z = x + iy \in \mathbb{C}^{d+1} \ : \ z^{(0)^2} - z^{(1)^2} - \cdots - z^{(d)^2} = -R^2\right\}, \qquad (3.2)$$

*Dedicated to Jacques Bros for his birthday

equivalently characterized as the set

$$X_d^{(c)} = \{(x,y) \in \mathbb{R}^{d+1} \times \mathbb{R}^{d+1} \ : \ x^2 - y^2 = -R^2, \ x \cdot y = 0\}.$$

We define the following open subset of $X_d^{(c)}$:

$$\mathcal{T}^+ = T^+ \cap X_d^{(c)}, \qquad \mathcal{T}^- = T^- \cap X_d^{(c)}, \tag{3.3}$$

where $T^\pm = \mathbb{R}^{d+1} + iV^\pm$ are the so-called *forward* and *backward tubes* in $\mathbb{C}^{d+1}$ [4] (here $V_\pm$ are the standard future/past lightcones in the ambient Minkowskian manifold). These are the (minimal) analyticity domains of the Fourier–Laplace transforms of tempered distributions $\widetilde{f}(p)$ with support contained in $\overline{V^+}$ or in $\overline{V^-}$.

The rôle of Wightman axioms [10] for a QFT in Minkowski space are replaced by other axioms of similar nature ([3], [4] and [9]) where the spectral condition is replaced by the notion of minimal analyticity. Specifically the axioms to be satisfied are (for a quasi-free theory):

1. Positivity: $\int_{X_d \times X_d} \mathcal{W}(x,x') f(x) \overline{f}(x') \mathrm{d}\sigma(x) \mathrm{d}\sigma(x') \geq 0, \ \forall \ f \in \mathcal{D}(X_d)$.
2. Locality: $\mathcal{W}(x,x') = \mathcal{W}(x',x)$ for every space-like separated pair (x,x').
3. Covariance: $\mathcal{W}(gx,gx') = \mathcal{W}(x,x'), \ \forall g \in G_d$.
4. Normal analyticity: the two-point function $\mathcal{W}(x,x') = \langle \Omega, \phi(x), \phi(x')\Omega \rangle$ is the boundary value of a function which is analytic in the domain $\mathcal{T}_{12}$ of $X_d^{(C)} \times X_d^{(c)}$.

For a minimally coupled Klein-Gordon field the field equation may be written as

$$\Box \phi + \mu^2 \phi = 0, \tag{3.4}$$

where $\mu^2 = M^2 + \zeta \rho$ is the relevant mass parameter, and ζ is the coupling constant to the scalar curvature ρ. A plane-wave analysis similar in spirit to the Minkowskian case and subsequent evaluation of integrals [4] lead to the following two-point function for such a scalar field:

$$W_\nu(z,z') = \frac{2c_{d,\nu} e^{\pi\nu} \pi^{d/2}}{R^{d-1} \Gamma(d/2)} P^{(d+1)}_{-\frac{d-1}{2}+i\nu}(\lambda), \ \lambda := \frac{z \cdot z'}{R^2} \quad R^2\mu^2 = \frac{(d-1)^2}{4} + \nu^2 \tag{3.5}$$

with

$$P^{(d+1)}_{-\frac{d-1}{2}+i\nu}(\lambda) = \frac{\Gamma(d/2)}{\sqrt{\pi}\,\Gamma\left(\frac{d-1}{2}\right)} \int_0^\pi \left(\lambda + \sqrt{\lambda^2 - 1}\cos\theta\right)^{-\frac{d-1}{2}+i\nu} \sin^{d-2}\theta \ \mathrm{d}\theta,$$

$$c_{d,\nu} = \frac{\Gamma\left(\frac{d-1}{2} + i\nu\right) \Gamma\left(\frac{d-1}{2} - i\nu\right) e^{-\pi\nu}}{2^{d+1}\pi^d}. \tag{3.6}$$

This two-point function gives rise to a QFT satisfying all the previously mentioned axioms and can be shown to be unique.

3.2 Massless limit

We are interested in the two-dimensional ($d = 1$) massless limit $\mu^2 \to 0$ for the two-point function in Eq. 3.5, namely the Wightman function of a massless minimally coupled scalar field.

As the Minkowskian case shows [11] [7] [8], two-dimensional massless QFTs present us with infrared singularities. In the present context — being de Sitter compact in the spatial directions — we may expect a natural IR cutoff. This indeed will be the case at the price of violating the equation of motion *and* the positivity assumption. We will show that restoring the positivity by means of Krein topology automatically restores the *effective* equation of motion. This means that the equation of motion will be automatically satisfied on the physical states (*á la* Gupta-Bleuer).

The Krein construction ensures that the Hilbert space obtained by closing the space of local states is a "maximal" space [11].

Let us proceed and perform the massless limit on the Wightman two-point function. We can write the massive two-point function as

$$W_\nu(\lambda) = \frac{\Gamma(1-\alpha)\,\Gamma(\alpha)}{4\pi} F\left(1-\alpha,\alpha;1;\frac{1-\lambda}{2}\right) , \tag{3.7}$$

where $\alpha = \frac{1}{2} - i\nu$, $F(a,b;c;x)$ is the hypergeometric function and $\lambda = \frac{z\cdot z'}{R^2}$. For $\alpha \to 0$ (or $\alpha \to 1$), we have that $\mu^2 = 1/R^2\,\alpha(1-\alpha) \to 0$.

The gamma function $\Gamma(\alpha)$ has a simple pole at the origin so that we have to renormalize the massive two-point function in order to obtain a finite result. To this end we expand $W_\nu(\lambda)$ in power series of α:

$$W_\nu(\lambda) = \frac{\Gamma(1-\alpha)\,\Gamma(\alpha)}{4\pi} + \frac{\Gamma(1-\alpha)\,\Gamma(\alpha)\alpha}{4\pi}\sum_{n=1}^{\infty}\frac{1}{n}\left(\frac{1-\lambda}{2}\right)^n + o(1). \tag{3.8}$$

By subtraction of the divergence due to the first term we obtain

$$W_0 \equiv W_0(\lambda) \equiv \lim_{\alpha\to 0}\left[W_\alpha(\lambda) - \frac{\Gamma(1-\alpha)\,\Gamma(\alpha)}{4\pi}\right] = -\frac{1}{4\pi}\ln\left[-\frac{(z-z')^2}{R^2}\right]. \tag{3.9}$$

Note that this is the same function as in the flat case, but with the argument being the (hyperbolic for time-like separation) sine of the pseudo-distance in the de Sitter manifold.

Such a two-point function is analytic in the extended tube and invariant under the action of the complex de Sitter group $G_2^{(c)}$ [4]. The boundary value $\mathcal{W}_0$ of W_0 is therefore invariant for a de Sitter transformation g, it is local and analytic in the domain $\mathcal{T}_{12}$.

This renormalization procedure has however introduced an anomaly in the equation of motion; indeed taking the limit

$$\begin{aligned} \Box \mathcal{W}_0(x,x') &= \lim_{\alpha\to 0} \Box \left(\mathcal{W}_\alpha(x,x') - C_{2,\alpha}\right) = -\lim_{\alpha\to 0} \mu^2 \mathcal{W}_\alpha(x,x') \\ &= -\lim_{\alpha\to 0} \mu^2 C_{2,\alpha} = -\frac{1}{4\pi \mathrm{R}^2} \end{aligned} \tag{3.10}$$

we get the equation of motion

$$\Box \mathcal{W}_0(x,x') = -\frac{1}{4\pi \mathrm{R}^2}. \tag{3.11}$$

The renormalization procedure has introduced an *anomaly* in the equation of motion. Moreover the resulting kernel is **not positive** essentially due to the logarithmic singularity at short distances, another drawback of this renormalization procedure. This forces us to use the extended Wightman axioms which require the introduction of a Krein topology, restoring at the same time positivity and equation of motion on the physical states.

Remark 1. One may ask why the solution of the equation $\Box F(\lambda) = 0$, namely

$$F(\lambda) = \ln\left(\frac{1-\lambda}{1+\lambda}\right), \tag{3.12}$$

is not an acceptable two-point function (λ being the invariant $\frac{z\cdot z'}{R^2}$). The reason is that this function is invariant under the action of G_2 but it is not analytic in the full extended domain: it is analytic in $\mathbb{C}^2$ minus the cuts $(-\infty,-1)$ and $(1,\infty)$. In particular the cut $(1,\infty)$ violates locality for antipodal points on the de Sitter manifold.

3.3 Invariant Krein space

A two-point function which is invariant, analytic and causal, but not positive, allows us to construct, via a GNS construction, a linear space invariant under de Sitter transformation with an indefinite inner product. Were the two-point (semi-)positive it would define a pre-Hilbert structure, and the space of physical states would be identified with the closure under such a norm. If the Wightman functions fail to be positive, such identification is not natural and, in general, not unique. We have to look for a *minimal* Hilbert topology, *i.e.* such that the closure of the span of local states is maximal (*Krein topology*). We give here a direct construction tailored to our specific case.

The invariant two-point function $\mathcal{W}_0$ is positive on the subspace $\mathcal{D}_0$, the space of null integral test functions

$$\mathcal{D}_0 := \left\{ f \in \mathcal{D}(X_2) : \int_{X_2} \mathrm{d}\sigma(x) f(x) = 0 \right\}, \tag{3.13}$$

because on this subspace the renormalization procedure by which we obtained the two-point function is irrelevant. Let us choose and fix a real function $h \in \mathcal{D}(X_2)$ such that

$$\int_{X_2} d\sigma(x)h(x) = 1 , \qquad \langle h, h\rangle = 0 . \tag{3.14}$$

Then, for any $f \in \mathcal{D}(X_2)$ we can decompose

$$f(x) = f_0(x) + \left(\int_{X_2} d\sigma(y)f(y)\right) h(x),$$

and therefore the test function space $\mathcal{D}(X_2)$ is decomposed in the direct sum

$$\mathcal{D}(X_2) = \mathcal{D}_0 + H, \tag{3.15}$$

where H is the one-dimensional space generated by h. Following this decomposition the indefinite sesquilinear form given by $\mathcal{W}_0$ is written as

$$\langle f, g\rangle = \langle f_0, g_0\rangle + \left(\int_{X_2} d\sigma(x)\overline{f}(x)\right) \langle h, g\rangle + \left(\int_{X_2} d\sigma(x)g(x)\right) \langle f, h\rangle. \tag{3.16}$$

The Krein inner product on $\mathcal{D}(X_2)$ is then defined by

$$(f, g) = \langle f_0, g_0\rangle + \langle f, h\rangle\langle h, g\rangle + \int_{X_2} d\sigma(x)\overline{f}(x) \int_{X_2} d\sigma(y)g(y). \tag{3.17}$$

The inner product (3.17) is positive semidefinite and hence defines a structure of pre-Hilbert space. The null space of this pairing is the subspace

$$\mathcal{I}_h = \{f \in \mathcal{D}_0 \; : \; \langle f_0, f_0\rangle = 0, \; \langle h, f\rangle = 0\} \; . \tag{3.18}$$

By quotienting $\mathcal{D}(X_2)$ and then completing in the ensuing Hilbert topology we obtain a Hilbert space which we denote with $\mathcal{K}^{(1)}$.

Lemma 1. *The linear functional on $\mathcal{K}^{(1)}$ defined by*

$$F_h(f) = \langle h, f\rangle, \tag{3.19}$$

(where h is the previously fixed test function with unit integral) has norm equal to 1 *and therefore it defines a normalized element v_0 of $\mathcal{K}^{(1)}$, such that $(v_0, f) = F_h(f)$. Furthermore, for all $f \in \mathcal{D}(X_2)$,*

$$\langle v_0, f\rangle = \int_{X_2} d\sigma(x)f(x) \; . \tag{3.20}$$

Proof. Let $\mathcal{I}_h$ be the ideal of the test functions of zero-norm (3.18). Then $\mathcal{K}^{(1)} = \overline{\mathcal{D}(X_2)/\mathcal{I}_h}^{\|\cdot\|}$. Defining

$$v_0 \equiv -4\pi \mathrm{R}^2 \Box h \tag{3.21}$$

we have

$$(v_0, f) = \langle v_0, f_0 \rangle + \langle v_0, h \rangle \langle h, f \rangle + \int_{X_2} \mathrm{d}\sigma(x) v_0(x) \int_{X_2} \mathrm{d}\sigma(y) f(y) = \langle h, f \rangle \ , \tag{3.22}$$

since $\langle v_0, h \rangle = 1$ and $v_0 \in \mathcal{D}_0$. Thus v_0 is the element identified by the functional F_h as in the statement of the lemma.

We now prove uniqueness of the vector v_0. Consider a different real function h' in $\mathcal{D}(X_2)$ satisfying eqs. (3.14). Then the difference $h - h'$ belongs to $\mathcal{D}_0$ and hence $\Box(h - h')$ belongs to the ideal $\mathcal{I}_h$, so that the corresponding vectors v_0' and v_0 are the same in the quotient space.

Finally the norm of v_0 is given by

$$(v_0, v_0) = 16\pi^2 \mathrm{R}^4 \left|\langle \Box h, h \rangle\right|^2 = 1 \tag{3.23}$$

and thus $\|F_h\| = 1$. Moreover, we have that

$$\langle v_0, v_0 \rangle = 0 \tag{3.24}$$

and, for every function f in $\mathcal{D}(X_2)$,

$$\langle v_0, f \rangle = -4\pi \mathrm{R}^2 \langle \Box h, f \rangle = \int_{X_2} \mathrm{d}\sigma(x) f(x). \tag{3.25}$$

□

Remark 2. The state v_0 plays the same role as the infinitely delocalized state of [7]. However there such a state is a limit of test functions and belongs only to the Krein completion of the space: here on the contrary the state is a perfectly well-behaved test function obtained by application of the Laplace–Beltrami operator $\Box$ to the chosen h. This may seem not to be an invariant state, however — as it is proven in Prop. 2 — the action of an isometry of the space-time does not change its *equivalence class* modulo the ideal $\mathcal{I}_h$.

Proposition 1. *The Hilbert space $\mathcal{K}^{(1)}$ is a Krein space and can be written as a direct sum:*

$$\mathcal{K}^{(1)} = \overline{\mathcal{D}_0^\perp/\mathcal{I}_h}^{\langle\cdot,\cdot\rangle} \oplus V_0 \oplus H \ , \tag{3.26}$$

where $\mathcal{D}_0^\perp$ is the subspace of $\mathcal{D}_0$ orthogonal to (the eq. class of) v_0 spanning $V_0 = (\{\lambda v_0 \ : \ \lambda \in \mathbb{C}\} + \mathcal{I}_h)/\mathcal{I}_h$ and $H = (\{\lambda h \ : \ \lambda \in \mathbb{C}\} + \mathcal{I}_h)/\mathcal{I}_h$.

The metric operator $\eta^{(1)}$ defined by $\langle\cdot,\cdot\rangle = (\cdot, \eta^{(1)}\cdot)$ is given by

$$\eta^{(1)}_{|\overline{\mathcal{D}_0^\perp}^{\langle\cdot,\cdot\rangle}} = \mathbf{1}_{|\overline{\mathcal{D}_0^\perp}^{\langle\cdot,\cdot\rangle}} \ , \quad \eta^{(1)} h = v_0 \ , \quad \eta^{(1)} v_0 = h \ . \tag{3.27}$$

Note: *from now on we will use the same symbols h, v_0 to denote the functions and the equivalence classes in the quotient space.*

Proof. We first note that $\mathcal{I}_h$ is entirely contained in $\mathcal{D}_0^\perp$ since for any $f \in \mathcal{D}_0$,

$$(f, v_0) \propto \langle f, h \rangle \tag{3.28}$$

and hence the orthogonality to v_0 is one of the defining properties of the vector space $\mathcal{I}_h$ (see eq. 3.18).

Let us write the subspace of zero-integral test functions as $\mathcal{D}_0 = \mathcal{D}_0^\perp \oplus \mathbb{C}\{v_0\}$, where the orthogonality is with respect to the (pre-)Hilbert product $(\cdot, \cdot)$. If restricted to the space $\mathcal{D}_0^\perp$, the Krein product is equal to the indefinite sesquilinear form $\langle \cdot, \cdot \rangle$. Indeed, for every pair of test functions $f, g \in \mathcal{D}_0^\perp$ we have

$$(f, g) = \langle f, g \rangle + \langle f, h \rangle \langle h, g \rangle = \langle f, g \rangle + (f, v_0)(v_0, g) = \langle f, g \rangle \ . \tag{3.29}$$

The mono-dimensional space H is orthogonal to V_0 because $(v_0, h) = \langle h, h \rangle = 0$. Moreover we have $\mathcal{D}_0^\perp \perp H$:

$$(h, f) = \int_{X_2} d\sigma(x) h(x) \int_{X_2} d\sigma(y) f(y) = 0 \ . \tag{3.30}$$

This is true also for the Hilbert space $\overline{\mathcal{D}_0^\perp}^{\langle \cdot, \cdot \rangle}$ and thus we have proved the decomposition (3.26). The definition of the metric operator $\eta^{(1)}$ is now obvious. The Hilbert space $\mathcal{K}^{(1)}$ is then a Krein space and, being $\eta^{(1)^{-1}}$ continuous, the Hilbert majorant topology defined by the Krein norm is minimal and the Krein space is maximal. □

The de Sitter group acts naturally on $\mathcal{D}(X_2)$ by $\alpha_g f(x) = f(g^{-1}x)$ and the restriction to $\overline{\mathcal{D}_0^\perp}^{\langle \cdot, \cdot \rangle}$ is unitary since $\mathcal{W}_0$ is invariant. Moreover we have

Proposition 2. *The representation $U(g)$ of the de Sitter group has a unique continuous extension from $\mathcal{D}(X_2)$ to $\mathcal{K}^{(1)}$ and*

$$U(g) v_0 = v_0 \ . \tag{3.31}$$

Proof. The only non-obvious part is the invariance of v_0. For any element g of $SO_0(1,2)$ the function $\Box h(x) - \Box h(gx)$ belongs to $\mathcal{I}_h$ (note that $h(x) - h(gx)$ is a function in $\mathcal{D}_0$) and thus it is zero in $\mathcal{K}^{(1)}$, thus proving that $U(g) v_0 = v_0$. □

From the decomposition (3.26) follows that v_0 is the unique invariant vector in $\mathcal{K}^{(1)}$. From the construction of the one-particle Hilbert-Krein space $\mathcal{K}^{(1)}$ one can build as usual a Fock-Krein space $\mathcal{K}$ and extend there the metric operator η. Then we can prove

Proposition 3. *The space $\mathcal{K}$ contains an infinite-dimensional subspace $\mathcal{V}_0$ of vectors invariant under the de Sitter transformation. However, the vacuum is still* essentially *unique (i.e. any strictly positive subspace of invariant vectors is one-dimensional).*

Proof. The proof of this theorem follows [7]. □

3.4 Equation of motion and gauge invariance

In this section we investigate the equation of motion and the existence of a charge operator generating gauge transformations. As we have remarked above, the most important difference between the massless scalar field in Minkowski space-time and the covariant massless scalar field in de Sitter space-time is that the invariant state v_0 now belongs to the space of test functions $\mathcal{D}(X)$. Therefore the operator $\phi(v_0)$ belongs to the field algebra and its introduction does not require any extension of the latter. This invariant operator commutes with any other operator of $\mathcal{A}$:

$$[\phi(v_0), \phi(f)] = 0, \tag{3.32}$$

as follows from the explicit expression of $v_0 = -4\pi \mathrm{R}^2 \Box h$ and the equations of motion $\Box_x W(x,x') = \Box_{x'} W(x,x') = -\frac{1}{4\pi \mathrm{R}^2}$.

We begin by proving that the equation of motion for the two-point function, $\Box \mathcal{W} = -1/(4\pi \mathrm{R}^2)$ translates into the following equation of motion for the quantum field ϕ.

Proposition 4. *The quantum field $\phi(x)$ satisfies*

$$\Box \phi(x) = -\frac{1}{4\pi \mathrm{R}^2} \phi(v_0) \ , \tag{3.33}$$

where v_0 is the previously defined state represented by $v_0 = [-4\pi \mathrm{R}^2 \Box h]$.

Proof. For any test function f we have $f = f_0 + h \int f$, the decomposition in its $\mathcal{D}_0$ and H components. We can obviously write $\phi(f) = \phi(f_0) + \int f \phi(h)$. Recalling that $\Box \phi_{|\mathcal{D}_0} \equiv 0$, we have

$$\begin{aligned} < \Box\phi(f)\phi(g) >= \int f < \phi(\Box h)\phi(g) >= -\frac{1}{4\pi \mathrm{R}^2} \int f \int g \\ =: -\frac{1}{4\pi \mathrm{R}^2} \int f < \phi(v_0)\phi(g) > \ . \end{aligned} \tag{3.34}$$

This, together with the fact that ϕ is a free field, proves the statement. □

In the homogeneous case the equation $\Box \phi = 0$ is clearly invariant under the transformation $\phi \to \phi + \lambda \mathbf{1}$. This invariance is still present as the following lemma shows

Lemma 2. *The equation of motion for the quantum field ϕ is invariant under the* gauge transformation $\gamma^\lambda(\phi) := \phi + \lambda$, $\lambda \in \mathbb{C}$.

Proof. The identity operator $\mathbf{1}$, as an operator-valued distribution, associates to a test function its total integral. In the equation of motion we then have

$$\Box \gamma^\lambda(\phi) = \Box \phi = -\frac{1}{4\pi \mathrm{R}^2} \phi(v_0) = -\frac{1}{4\pi \mathrm{R}^2} \gamma^\lambda(\phi)(v_0) \ ,$$

where the last equality follows from the fact that the state $v_0 = [4\pi \mathrm{R}^2 \Box h]$ has zero total integral. □

We can then define a *gauge transformation* as the automorphism of the field algebra generated by

$$\gamma^\lambda : \phi \to \phi + \lambda, \quad \lambda \in \mathbb{C}. \tag{3.35}$$

Let us now introduce the operator

$$Q = i\left[\phi^+(v_0) - \phi^-(v_0)\right], \tag{3.36}$$

where the operators $\phi^\pm(f)$ are defined as creators–annihilators of states v_0 in the Fock-Krein space $\mathcal{K}$. This is a continuous operator from the n-particle space $\mathcal{K}^{(n)}$ to $\mathcal{K}^{(n\pm1)}$. Moreover it is easily verified that Q satisfies the following commutation relations:

$$[Q, \phi(f)] = -i \int \mathrm{d}\sigma(x) f(x). \tag{3.37}$$

One can compute directly that — exactly as in the flat case —

$$[\phi_\pm(v_0), \phi(f)] = \mp\frac{1}{2}\int \mathrm{d}\sigma(x) f(x)\ ; \qquad [\phi_+(v_0), \phi_-(v_0)] = 0\ . \tag{3.38}$$

The natural question now arises as to whether this *charge* Q is the integral of a local expression. The classical charge of any solution of the wave equation in $d+1$ dimensions $\Box\varphi = 0$ is defined by integrating along any space-like d-surface the "time-like" derivative along the direction orthogonal to the surface

$$\int_\Sigma \mathrm{d}v_\Sigma \partial_{\hat{n}}\varphi. \tag{3.39}$$

Such an expression is "conserved" (i.e. independent of the spacelike surface Σ) because of the equation of motion. The integrand in eq. (3.39) is the Hodge dual of the 1-form $\mathrm{d}\varphi$, i.e. (in Lorentzian signature) the d-form defined by

$$(\mathrm{d}\varphi)^* := \sqrt{-g}\,\epsilon_{\mu_1,\ldots,\mu_d,\nu} g^{\nu\rho}\partial_\rho\varphi \mathrm{d}x^{\mu_1}\cdots \mathrm{d}x^{\mu_d}\ . \tag{3.40}$$

Such a d-form can be integrated on any d-surface in an intrinsic way. Now, such a form is *closed* iff the function φ satisfies the wave equation $\Box\varphi = 0$ because

$$\mathrm{d}(\mathrm{d}\varphi)^* \propto \Box\varphi\, \mathrm{d}vol\ .$$

The integral of a closed d-form on a d-surface is independent of continuous deformations of the surface (i.e. the choice of the "time-slice" in our setting).

This preamble shows clearly that we are to expect some problems from the fact that our quantum field ϕ does not satisfy the homogeneous wave equation.

Let us have a closer look at what happens in the case at hand: the explicit expression of the Hodge dual is now the $d = 1$-form (see also later the discussion about the dual field)

$$(\mathrm{d}\phi)^* = -\partial_\tau \phi \mathrm{d}\theta - \partial_\theta \phi \mathrm{d}\tau$$

for our (quantum) field ϕ (in conformal coordinates for clarity). The equation of motion (3.33) implies that this operator-valued form is not closed

$$\mathrm{d}(\mathrm{d}\phi)^* = \frac{1}{4\pi \mathrm{R}^2}\phi(v_0)\, \mathrm{d}vol = \frac{1}{4\pi \mathrm{R}^2}\phi(v_0)\frac{\mathrm{d}\tau \wedge \mathrm{d}\theta}{\cos^2(\tau/\mathrm{R})} .$$

Therefore the charge defined as in the flat case (with all the additional technical details which we now omit) [7,8] by

$$Q \propto \int \mathrm{d}x^1\, \partial_{x^0}\phi$$

could not be possibly conserved (in our case the expression would look more like $\int \mathrm{d}\theta\, \partial_\tau \phi$).

Proposition 5. *The charge* $Q = \frac{i}{2}\left(\phi^+(v_0) - \phi^-(v_0)\right)$ *is the integral of a local current, namely*

$$Q = -\frac{1}{8\pi}\int_\Sigma \left((\mathrm{d}\phi)^* + \frac{1}{4\pi \mathrm{R}} \tan(\tau/\mathrm{R})\phi(v_0)\mathrm{d}\theta \right) , \tag{3.41}$$

where Σ *is any space-like* $d = 1$*-surface (i.e. a curve, e.g.* $\tau = const$*). The integral is independent of the space-like (closed, simple) curve* Σ *because the 1-form in the integrand is closed.*

Remark 3. The integrand of eq. (3.41) is the differential of a **dual field** $\widetilde{\phi}$ up to an exact form: indeed we *define* the dual field by the equation

$$\mathrm{d}\widetilde{\phi} = (\mathrm{d}\phi)^* - \frac{\tan(\tau/\mathrm{R})}{4\pi \mathrm{R}}\left(\phi(v_0)\mathrm{d}\theta + \widetilde{\phi}(v_0)\mathrm{d}\tau\right) , \tag{3.42}$$

which differs from the integrand in eq. 3.41 by the exact operator-valued form

$$\frac{\tan(\tau/\mathrm{R})}{4\pi \mathrm{R}}\widetilde{\phi}(v_0)\mathrm{d}\tau = \frac{1}{4\pi}\widetilde{\phi}(v_0)\mathrm{d}\ln(\cos(\tau/\mathrm{R})) , \tag{3.43}$$

which does not contribute to the integral. In this notation the above proposition would read

$$Q = -\frac{1}{8\pi}\int_\Sigma \mathrm{d}\widetilde{\phi} . \tag{3.44}$$

The 1-form $\mathrm{d}\widetilde{\phi}$ is (classically) closed but not exact, i.e. the dual field $\widetilde{\phi}$ naturally lives on the universal covering of X_2. In other words, the dual field $\widetilde{\phi}$ carries a topological charge w.r.t. ϕ (see later).

Proof (Sketch). The proof consists in showing that the integrand is indeed a closed form (which is straightforward) and that the commutator of the integral with the fields $\phi(f)$ reproduces the correct commutator of Q as given in (3.37), which amounts to a direct manipulation of the integrals. □

From the relation

$$\frac{\mathrm{d}}{\mathrm{d}\lambda}\gamma^\lambda(\phi(f)) = \int_{X_2} \mathrm{d}\sigma(x) f(x) = i\,[Q, \phi(f)] \tag{3.45}$$

follows that the automorphism γ^λ is generated by the operator Q.

Theorem 1. *The automorphism γ^λ is implementable in the Krein space $\mathcal{K}$ by the η-unitary operator*

$$\gamma^\lambda = e^{i\lambda Q}. \tag{3.46}$$

The not difficult proof will be detailed elsewhere.

The Wightman function $\mathcal{W}_0(x, x')$ is not invariant under the gauge transformation: indeed, for $\phi \to \phi + \lambda$, we get $\mathcal{W}_0 \to \mathcal{W}_0 + \lambda^2$. For this reason some authors ([5, 6]) do not use Wightman formalism to construct a QFT for the massless minimally coupled field in de Sitter space-time. The construction is nonetheless still feasible because the vacuum is not gauge invariant but it is mapped to a non-physical state (*i.e.* a zero-norm state). If restricted to the physical space of states $\mathcal{H}_{phys}$ the two-point function is positive, analytic and invariant under de Sitter and gauge transformation.

3.5 Conclusions

We have constructed a QFT of a massless minimally coupled scalar field in a bidimensional de Sitter space-time. Although we have studied this particular case, the results are valid also for the four-dimensional de Sitter universe. The renormalized two-point function for a massless minimally coupled scalar field in X_4 is not defined to be positive, and one can construct a full de Sitter invariant vacuum with the same techniques we have described in section 3.3.

This seems to disagree with Allen's theorem [1] which states that in a four-dimensional de Sitter space-time there cannot exist a de Sitter invariant vacuum for a massless scalar field. As we have already pointed out, the problem is analogous to the case of the vacuum of the massless scalar field in a bidimensional Minkowski space-time: it is impossible to find a two-point function that is positive by definition, analytic, Lorentz invariant and local [11]; if one insists on Lorentz invariance one necessarily loses positivity.

If one uses the normal modes to construct the two-point function, its positivity is a natural consequence of the construction itself and, in agreement with Allen's theorem, de Sitter invariance is lost. With the normal modes construction, not only dS invariance but also gauge invariance is lost, being the Hilbert space of the states non-invariant under gauge transformation.

Another result is that, the vacuum being de Sitter invariant, the linear dependence on the time coordinate of the vacuum fluctuation on the non-invariant vacuum is lost [1, 2]. This result could have some consequence for the inflationary model, where this time dependence is used to explain the roll-over of the inflation fields responsible for the inflation. It is important to point out that these results are valid only for a de Sitter universe that exists forever and we do not yet know the implication for a de Sitter universe with a finite life.

References

1. B. Allen: *Phys. Rev. D* 32:3136 (1985).
2. B. Allen and A. Folacci: *Phys. Rev. D* 35:3771 (1987).
3. J. Bros, J.-P. Gazeau and U. Moschella: *Phys. Rev. D* 73:1746 (1994).
4. J. Bros and U. Moschella: *Rev. Math. Phys.* 8:327 (1996).
5. S. De Bièvre and J. Renaud: *Phys. Rev. D* 57:6230 (1998).
6. J.-P. Gazeau, J. Renaud and M.V. Takook: arXiv:gr-qc/9904023.
7. G. Morchio, D. Pierotti and F. Strocchi: *J. Math. Phys.* 31:1467 (1990).
8. G. Morchio, D. Pierotti and F. Strocchi: *J. Math. Phys.* 33:777 (1991).
9. U. Moschella: *Ann. Inst. Henri Poincaré* 63:411 (1995).
10. R.F. Streater and A.S. Wightman: *PCT, Spin and Statistics, and all that.* W.A. Benjamin, New York, 1964..
11. F. Strocchi: Selected topics on the general properties of quantum field theory. In: *Lecture notes in physics* 51, World Scientific, Singapore, 1993.

4

Locally Covariant Quantum Field Theories

Romeo Brunetti

II. Institut für Theoretische Physik, Luruper Chaussee 149, D-22761 Hamburg, Germany
romeo.brunetti@desy.de

Ausculta fili verba magistri
Benedetto (480-547), incipit from "Regola".

Dedicated to Jacques Bros

4.1 Locality and general covariance

In his famous "Missed Opportunities" paper [8], Freeman Dyson attempted to establish a *generally* covariant approach to quantum field theory. His aim was to unify the different perspectives on QFT of Tomonaga, Schwinger and Feynman, but the question remained unsettled, mainly because he made some strong assumtpions that were unjustified. But the important point is that he considered the framework of algebraic QFT of Araki, Haag, Kastler [1, 9] as the right one for implementation. A main aim of this article is to convince the reader that, with the appropriate generalizations, Dyson's aim can indeed be fullfilled.

The core of the generalization originates from a collaboration with Klaus Fredenhagen [5], dating back to the mid 1990s. The main tool that we shall be using is the language of category theory.

The idea that QFT can be described by the use of category theory is certainly not new. In the eighties, something similar was advocated by Sir Atiyah [2], for the sake of topological QFT, and along the same lines, by G. Segal [13] for conformal QFT. Moreover, Dimock has perhaps been the pioneer of the use of categories for the algebraic description of free QFT [7].

I wish to substantiate the idea with a claim that, unlike other attempts, originates from physics [6]. The claim is that it would be better to work locally, to describe (relativistic) quantum theories, where here locality is intended in the geometric sense, not as causality. The result is that the best things one can do for QFT in a general covariant setting is to quantize simultaneously in all space-times of a certain family. The recipe can be condensed into the following

Categorical Imperative: Do nothing that you cannot do on any manifold.

I will be dealing with the general structure at the algebraic level and present some examples. For stronger results dealing with the Wightman approach on specific space-times such as de Sitter, you should better look at the papers written by Jacques Bros [3].

4.2 Quantum Field Theory as a Functor

Rigorous implementation of the generally covariant locality principle uses the language of category theory.

The following two categories are used :

Loc The class of objects $\mathrm{obj}(\mathsf{Loc})$ is formed by all (smooth) d-dimensional ($d \geq 2$ is held fixed), globally hyperbolic Lorentzian space-times M which are oriented and time-oriented. Given any two such objects M_1 and M_2, the morphisms $\psi \in \hom_{\mathsf{Loc}}(M_1, M_2)$ are taken to be the isometric embeddings $\psi : M_1 \to M_2$ of M_1 into M_2 but with the following constraints;
  (i) if $\gamma : [a,b] \to M_2$ is any causal curve and $\gamma(a), \gamma(b) \in \psi(M_1)$, then the whole curve must be in the image $\psi(M_1)$, i.e., $\gamma(t) \in \psi(M_1)$ for all $t \in]a,b[$;
  (ii) any morphism preserves orientation and time-orientation of the embedded space-time.
  Composition is composition of maps; the unit element in $\hom_{\mathsf{Loc}}(M,M)$ is given by the identical embedding $\mathrm{id}_M : M \mapsto M$ for any $M \in \mathrm{obj}(\mathsf{Loc})$.

Obs The class of objects $\mathrm{obj}(\mathsf{Obs})$ is formed by all C*-algebras possessing unit elements, and the morphisms are faithful (injective) unit-preserving $*$-homomorphisms. The composition is again defined as the composition of maps; the unit element in $\hom_{\mathsf{Obs}}(\mathcal{A},\mathcal{A})$ is for any $\mathcal{A} \in \mathrm{obj}(\mathsf{Obs})$ given by the identical map $\mathrm{id}_{\mathcal{A}} : A \mapsto A$, $A \in \mathcal{A}$.

One may change categories according to particular needs, as for instance in perturbation theory where instead of C*-algebras general topological *-algebras are better suited. Or one may use von Neumann algebras, in case particular states are selected. On the other hand, one might consider for Loc bundles over space-times, or one might (in conformally invariant theories) admit conformal embeddings as morphisms. In case one is interested in space-times which are not globally hyperbolic, i.e. AdS, one could look at the globally hyperbolic subregions (where attention has to be payed to the causal convexity condition (i) above), and fix the target. Otherwise one might look at a larger family, as for instance the family composed of all asymptotically AdS space-times of the same dimension and look for embeddings that remain between globally hyperbolic subregions.

Now we define the concept of locally covariant quantum field theory.

Definition 1 (i) *A* **locally covariant quantum field theory** *is a covariant functor* $\mathscr{A}$ *from* Loc *to* Obs *and (writing* α_ψ *for* $\mathscr{A}(\psi)$*) with the covariance properties*

$$\alpha_{\psi'} \circ \alpha_\psi = \alpha_{\psi' \circ \psi}\,, \quad \alpha_{\mathrm{id}_M} = \mathrm{id}_{\mathscr{A}(M)}\,,$$

for all morphisms $\psi \in \hom_{\mathsf{Loc}}(M_1, M_2)$*, all morphisms* $\psi' \in \hom_{\mathsf{Loc}}(M_2, M_3)$ *and all* $M \in \mathrm{obj}(\mathsf{Loc})$*.*

(ii) *A locally covariant quantum field theory described by a covariant functor* $\mathscr{A}$ *is called* **causal** *if the following holds: Whenever there are morphisms* $\psi_j \in \hom_{\mathsf{Loc}}(M_j, M)$*,* $j = 1, 2$*, so that the sets* $\psi_1(M_1)$ *and* $\psi_2(M_2)$ *are causally separated in* M*, then one has*

$$[\alpha_{\psi_1}(\mathscr{A}(M_1)), \alpha_{\psi_2}(\mathscr{A}(M_2))] = \{0\}\,,$$

where the element-wise commutation makes sense in $\mathscr{A}(M)$*.*

(iii) *We say that a locally covariant quantum field theory given by the functor* $\mathscr{A}$ *obeys the* **time-slice axiom** *if*

$$\alpha_\psi(\mathscr{A}(M)) = \mathscr{A}(M')$$

holds for all $\psi \in \hom_{\mathsf{Loc}}(M, M')$ *such that* $\psi(M)$ *contains a Cauchy-surface for* M'*.*

Thus, a quantum field theory is an assignment of C*-algebras to (all) globally hyperbolic space-times so that the algebras are identifiable when the space-times are isometric, in the indicated way. This is a precise description of the **generally covariant locality principle**.

4.2.1 The traditional approach in Minkowski space-time

The traditional framework of algebraic quantum field theory, in the Araki-Haag-Kastler sense, on a fixed globally hyperbolic space-time can be recovered from a locally covariant quantum field theory, i.e. from a covariant functor $\mathscr{A}$ with the properties listed above. The change is done at the level of the first category by choosing a full subcategory in which a target manifold is fixed and one considers all globally hyperbolic space-times of the target.

Indeed, let M be an object in obj(Loc). We denote by $\mathcal{K}(M)$ the set of all open subsets in M which are relatively compact and contain with each pair of points x and y also all g-causal curves in M connecting x and y (cf. condition (i) in the definition of Loc). $O \in \mathcal{K}(M)$, endowed with the metric of M restricted to O and with the induced orientation and time-orientation, is a member of obj(Loc), and the injection map $\iota_{M,O} : O \to M$, i.e. the identical map restricted to O is an element in $\hom_{\mathsf{Loc}}(O, M)$. With this notation it is easy to prove the following assertion:

Theorem 1. *Let $\mathscr{A}$ be a covariant functor with the above stated properties, and define a map $\mathcal{K}(M) \ni O \mapsto \mathcal{A}(O) \subset \mathscr{A}(M)$ by setting*

$$\mathcal{A}(O) \doteq \alpha_{\iota_{M,O}}(\mathscr{A}(O)) .$$

Then the following statements hold:

(a) *The map fulfills isotony, i.e.*

$$O_1 \subset O_2 \Rightarrow \mathcal{A}(O_1) \subset \mathcal{A}(O_2) \qquad \text{for all} \qquad O_1, O_2 \in \mathcal{K}(M) .$$

(b) *The group G of isometric diffeomorphisms $\kappa : M \to M$ (so that $\kappa_* g = g$) preserving orientation and time-orientation, is represented by C^*-algebra automorphisms $\alpha_\kappa : \mathscr{A}(M) \to \mathscr{A}(M)$ such that*

$$\alpha_\kappa(\mathcal{A}(O)) = \mathcal{A}(\kappa(O)) , \quad O \in \mathcal{K}(M) .$$

(c) *If the theory given by $\mathscr{A}$ is additionally causal, then it holds that*

$$[\mathcal{A}(O_1), \mathcal{A}(O_2)] = \{0\}$$

for all $O_1, O_2 \in \mathcal{K}(M)$ with O_1 causally separated from O_2.

These properties are just the basic assumptions of the Araki-Haag-Kastler framework. It is certainly not a big advance to replace three axioms by two. We are confident that the more general description may open the way to a sharper understanding.

4.3 Beyond simple functoriality: Equivalence, Dynamics, Fields, Scattering, and more.

The real substance in category theory is the notion of natural transformations. So far, we have scratched only the surface of the new framework, and we will see now that introduction of the natural transformations allows us to reach a higher descriptive level. Let me then remind you of the general definition of natural trasformations: Consider two categories, say A and B, and two functors between them $\mathscr{F}$ and $\mathscr{G}$; then a natural transformation $\mathscr{N}$ is a "functor over functors", i.e., a function $N : \mathscr{F} \longrightarrow \mathscr{G}$ that assigns to each object a in A an arrow N_a of B and such that any arrow $A : a \to a'$ in A yields a commutative diagram such as

$$\begin{array}{ccccc} a & \quad & \mathscr{F}a & \xrightarrow{N_a} & \mathscr{G}a \\ \downarrow{\scriptstyle A} & & {\scriptstyle \mathscr{F}A}\downarrow & & \downarrow{\scriptstyle \mathscr{G}A} \\ a' & & \mathscr{F}a' & \xrightarrow{N_{a'}} & \mathscr{G}a' . \end{array}$$

The fact that the diagram is commutative is the heart of the matter. One may equivalently describe the naural transformation as a family of functors, i.e. $\{N_a\}_{a \in \mathrm{obj}(\mathscr{A})}$.

Let us then see the usefulness of this notion.

1. It allows a neat description of *equivalence* between theories, something which is certainly beyond reach in the more traditional formulations. Indeed, if we have two theories $\mathscr{F}_1$, $\mathscr{F}_2$, in the sense described above of covariant functors, then we may say the theories are equivalent whenever there exists a natural tranformation $\{N_M\}_{M\in \mathsf{obj}(\mathsf{Loc})}$ such that each element of the family is a *-bijective map. Examples and counterexamples are easily given, see e.g. [6]. Another, yet to be proved, equivalence may be discussed using the concept of PCT operator as a natural transformation. Here, we say that a theory $\mathscr{A}$ satisfies the PCT symmetry whenever a natural transformation Θ exists (again composed of a family of $*$-bijective maps) and links $\mathscr{A}$ to the theory $\mathscr{A}^*$ for which the space-times M have opposite time-orientations, notationally $\overline{M}$, and where the C*-algebras are complex conjugates, i.e. $\Theta(\mathscr{A}(M)) = \mathscr{A}(\overline{M})^*$.
2. The most important use of natural transformations comes with the definition of *locally covariant fields.* Here, such a field is described by a family of (linear) continuous maps $A_M : \mathscr{D}(M) \to \mathscr{A}(M)$, where $\mathscr{D}$ is the functor of a test-function space, and now $\mathscr{A}$ has to be interpreted in the larger sense of topological algebras, i.e., such that they conform to the definition above and therefore satisfy a commutativity rule
$$\mathscr{A}\psi \circ A_M = A_N \circ \mathscr{D}\psi \ ,$$
where $\mathscr{D}\psi$ denotes the push-forward on test function spaces. The importance of such a definition is that it provides a way for any theory to compare observations based on different portions of a single space-time or even better on different space-times.
3. In the general setting described in these notes one natural question is the description of *dynamics.* Usually this is done at the algebraic level by invoking space-time isometries, and especially time translations, whose generator is considered to be the Hamiltonian. Here, we do not have such a possibility. Indeed, in a generic space-time we do not have any one-parameter group of time translation. Hence, what can be done? An answer to this question takes advantage of the timeslice property, invoked in the first part, and the notion of a locally covariant field. Indeed, this allows us to compare "time-evolutions" on different space-times. Namely, let $N_\pm$ be two isometric neighbourhoods of two Cauchy surfaces in M_1, resp. M_2, and let $\psi_{i,\pm}$ be the corresponding embeddings. Then, using that the corresponding algebraic embeddings $\mathscr{A}\psi_{i,\pm} \doteq \alpha_{i,\pm}$ are surjective, one can define an automorphism of the first algebra $\mathscr{A}(M_1)$ with
$$\beta = \alpha_{1,+}\alpha_{2,+}^{-1}\alpha_{2,-}\alpha_{1,-}^{-1} \ .$$
This automorphism describes the effect of a change of the metric between the two Cauchy surfaces and, because of that, called the "relative Cauchy evolution." To see that this is related to the field definition, one may argue as follows. Suppose we introduce, for any space-time M, the set $\mathscr{L}(M)$

of Lorentzian metrics that differ from the metric of M only in a compact region. Let then β_{M_h} be the relative Cauchy evolution between M and the space-time obtained by replacing the metric with $h \in \mathscr{L}(M)$. Let now, $\beta_M : \mathscr{L}(M) \to \mathrm{Aut}(\mathscr{A}(M))$ be defined as

$$\beta_M(h) = \beta_{M_h} \ .$$

The family $\{\beta_M\}_{M \in \mathsf{obj}(\mathsf{Loc})}$ is then a natural transformation, as the commutation relation can be easily checked.

4. A last example, also in the nonlinear case, is related to the scattering theory. Here we mean scattering in the sense of the local S-matrices approach of Stückelberg-Bogoliubov-Epstein-Glaser, although not necessarily in the perturbative sense. These are unitaries $S_M(\lambda)$ with $M \in \mathsf{Loc}$ and $\lambda \in \mathcal{D}(M)$ which satisfy the conditions

$$S_M(0) = 1 \ , \qquad S_M(\lambda + \mu + \nu) = S_M(\lambda + \mu) S_M(\mu)^{-1} S_M(\mu + \nu)$$

for $\lambda, \mu, \nu \in \mathcal{D}(M)$ such that the supports of λ and ν can be separated by a Cauchy surface of M with suppλ in the future of the surface. These S-matrices can be used to define a new quantum field theory. The new theory is locally covariant if the original theory was and if the local S-matrices satisfy the local covariance condition. A perturbative construction was completed in this way by Hollands and Wald [10], based on previous work done by me and Fredenhagen [5].

4.3.1 Prolegomena to Generally Covariant Scaling Algebras

At this point it is a rather trivial task to generalize to the generally covariant situation the nice description of the renormalization group idea in the algebraic setting pioneered by Buchholz and Verch. Here, as in the previous section, we use natural transformations. One needs a slightly more general version of the locality category to include also a set of physical parameters, a generalization that is rather trivial and that will not be discussed in detail. So our starting point will be the family of all pairs given by a globally hyperbolic space-time and a set of physical parameters like masses, charges, couplings etc. as (M, p) (we assume that the parameters are the same for all space-times).

The main idea is that we are furnished with a one-parameter family $\mathscr{A}_\lambda$ of covariant functors between the same categories as before (with the locality category slightly generalized according to the above) where $\lambda \in]0,1]$, for instance, and is such that it acts as follows:

$$\mathscr{A}_\lambda : (M, p) \longrightarrow \mathscr{A}(M_\lambda, p_\lambda)$$

where M_λ means that the manifold is still M but its metric gets scaled by a factor λ, i.e., $g_\lambda \doteq \lambda^2 g$, and the parameters p are scaled accordingly. Now, we may define scaling transformations as natural ones such that given any element in the supplemented locality category (M, p) we may define maps $R_\lambda^{(M,p)}$ such that

$$R_\lambda^{(M,p)} : \mathscr{A}(M,p) \longrightarrow \mathscr{A}(M_\lambda, p_\lambda)$$

is a C*-isomorphism, and choosing a different element in Loc the appropriate commutative diagrams for the scaling w.r.t. the isometric embeddings are obtained, i.e., in functional form

$$R_\lambda^{(M',p)} \circ \alpha_\psi = \alpha_\psi^\lambda \circ R_\lambda^{(M,p)}$$

where α_ψ^λ isometrically embeds (M_λ, p_λ) into (M'_λ, p_λ).

In order to compare with the Buchholz-Verch analysis we need some form of dynamics and to this extent we assume the time-slice axiom and make use of the relative Cauchy evolution. It is now very easy to check that we have scaling transformations acting as intertwiners of the relative Cauchy evolutions,

$$R_\lambda^{(M,p)} \circ \beta_g = \beta_{g_\lambda} \circ R_\lambda^{(M,p)} \ .$$

If we call $\tilde{g}$ the compactly supported perturbation to the metric g and $\delta g = g - \tilde{g}$ their difference, then we may use the following as a substitute for the "uniformity" condition of Buchholz-Verch, namely

$$\limsup_{\lambda \in]0,1]} \| \beta_{g_\lambda}(R_\lambda^{(M,p)}(A)) - R_\lambda^{(M,p)}(A) \| \xrightarrow[\delta g \searrow 0]{} 0 \tag{4.1}$$

as long as δg tends to zero in the topology of test functions and for any element $A \in \mathscr{A}(M,p)$.

The scaling algebras will be formed by all uniformly bounded elements $R_\lambda^{(M,p)}(A)$ for which the latter condition holds true. Hence, we consider the functions

$$\underline{\mathscr{A}}^{(M,p)} : \lambda \in]0,1] \rightarrow \mathscr{A}(M_\lambda, p_\lambda) \tag{4.2}$$

and make all these functions into unital C*-algebras denoted by $\underline{\mathscr{A}}(M,p)$ by the usual definitions for linear and multiplicative properties and definition of the norm. We have now again a covariant functor between the same categories as before, but which represents the generally covariant scaling algebras of Buchholz and Verch $\underline{\mathscr{A}}$ and that underlies the algebraic description of the renormalization group, i.e.,

$$\underline{\mathscr{A}} : (M,p) \longrightarrow \underline{\mathscr{A}}(M,p)$$

where each unital C*-algebra contains the above functions (4.2) obeying the uniformity relation (4.1) and to each isometric embedding there is associated a lifting to the new algebras of the previous C*-monomorphisms. Accordingly, state spaces can be defined in the usual manner and the treatment of the scaling limit proceeds along the same line of the Buchholz and Verch treatment. More details will be presented elsewhere.

4.4 Conclusions and Outlook

The generally covariant treatment of QFT discussed in this paper is based on the first principle that ensures equivalence of observable algebras based on isometric regions of different space-times. That's all one needs to proceed, at the conceptual level. Important developments are those connected to the works of Hollands and Wald, Verch, Hollands, Ruzzi, and one easily foresees applications of the framework to interesting situations, such as those related to AdS space-time, or in general theories on space-times with boundaries, to the exploitation of the renormalization group at the algebraic level and its possible use towards a clarification of the role of the conformal anomaly in the treatment of theories on asymptotically AdS space-times. Another, perhaps more important topic, is that related to background independent formulation of perturbative quantum gravity. We hope to report on these soon.

References

1. H. Araki: *Mathematical theory of quantum fields.* Oxford University Press, Oxford, 1999.
2. M. Atiyah: Topological quantum field theories. *Inst. Hautes Études Sci. Publ. Math.* 68:175-186 (1989).
3. J. Bros, H. Epstein and U. Moschella, Towards a General Theory of Quantized Fields on the Anti-de Sitter Space-Time. *Communications in Mathematical Physics* 231:481-528 (2002).
4. R. Brunetti and K. Fredenhagen, Microlocal analysis and interacting quantum field theories: Renormalization on physical backgrounds. *Communications in Mathematical Physics* 208:623-661 (2000).
5. R. Brunetti, K. Fredenhagen and M. Köhler, The microlocal spectrum condition and Wick polynomials on curved spacetime. *Communications in Mathematical Physics* 180:633 (1996).
6. R. Brunetti, K. Fredenhagen and R. Verch, The generally covariant locality principle – A new paradigm for local quantum field theory. *Communications in Mathematical Physics* 237:31-68 (2003).
7. J. Dimock, Algebras of local observables on a manifold. *Communications in Mathematical Physics* 77:219 (1980) 12; and, Dirac quantum fields on a manifold. *Transactions of the American Mathematical Society* 269:133 (1982).
8. F. Dyson, Missed Opportunities, *Bulletin of the American Mathematical Society* 78:635 (1972).
9. R. Haag, *Local Quantum Physics.* Springer-Verlag, Berlin, Heidelberg, New York, 2nd ed., 1996.
10. S. Hollands and R.M. Wald, Local Wick polynomials and time ordered products of quantum fields in curved spacetime. *Communications in Mathematical Physics* 223:289-326 (2001); and, Existence of local covariant time ordered products of quantum field in curved spacetime. *Communications in Mathematical Physics* 231:309-345 (2003).
11. S. Hollands, PCT theorem for the operator product expansion in curved spacetime. *Communications in Mathematical Physics* 244:209-244 (2004).

12. G. Ruzzi, Punctured Haag duality in locally covariant quantum field theories, to be published in *Communications in Mathematical Physics* (2005), math-ph/0412014.
13. G. Segal, The definition of conformal field theory. *Topology, geometry and quantum field theory,* London Math. Soc. Lecture Note Ser. 308, Cambridge Univ. Press, Cambridge, pp. 421-577, 2004.
14. R. Verch, A spin-statistics theorem for quantum fields on curved spacetime manifolds in a generally covariant framework. *Communications in Mathematical Physics* 223:261 (2001).

5

Asymptotic Abelianness and Braided Tensor C^*-Categories

Detlev Buchholz[1], Sergio Doplicher[2], Giovanni Morchio[3], John E. Roberts[4] and Franco Strocchi[5]

[1] Institut für Theoretische Physik der Universität Göttingen, D-37077 Göttingen, Germany
buchholz@theorie.physik.uni-goettingen.de
[2] Dipartimento di Matematica, Università di Roma "La Sapienza", I-00185 Roma, Italy
dopliche@mat.uniroma1.it
[3] Dipartimento di Fisica dell'Università and INFN, I-56126 Pisa, Italy
morchio@difi.unipi.it
[4] Dipartimento di Matematica, Università di Roma "Tor Vergata", I-00133 Roma, Italy
roberts@mat.uniroma2.it
[5] Scuola Normale Superiore and INFN, I-56126 Pisa, Italy
f.strocchi@sns.it

Summary. By introducing the concepts of asymptopia and bi-asymptopia, we show how braided tensor C^*-categories arise in a natural way. This generalizes constructions in algebraic quantum field theory by replacing local commutativity by suitable forms of asymptotic Abelianness.

5.1 Introduction

The occurrence of superselection sectors in quantum theories was first recognized by Wick, Wightman and Wigner [20] who gave two important examples of situations leading to sectors, the univalence rule and the electric charge. For a while it looked as if little would be touched by their fundamental discovery. The unrestricted superposition principle had to be abandoned, but it at least remained valid within certain subspaces, the coherent subspaces. However the point of view persisted that pure states of the theory were described by the projective space associated to a preassigned Hilbert space, or rather a subset of that space to account for the new phenomenon.

But a drastic change in the basic picture allowed the new phenomenon to spawn new ideas and results: as was stressed very early by Haag, in quantum

*Supported in part by the MURST, CNR-GNAFA, INdAM-GNAMPA and the European Union under contract HPRN-CT2002-00280.

field theory the local observables are fundamental and generate an algebra. The different sectors provide inequivalent irreducible realizations of that algebra [13]. Furthermore, starting just from the vacuum sector and analyzing the structure of the algebra, we can in principle determine all sectors.

For superselection sectors describing localizable charges, this is possible by a selection criterion [6] singling out those irreducible representations. The mathematical object that emerges and describes the structure of the sectors is a symmetric tensor C^*-category with conjugates and irreducible unit. It was later shown that any such category is isomorphic to a category of unitary representations of some compact group, unique up to isomorphism [8]. Furthermore there is a canonical field net with ordinary Bose–Fermi commutation relations at spacelike separations [9] where this group, the gauge group, acts as automorphisms with the original net as fixed-point net.

The above selection criterion does not select all the relevant representations in every case. Although weaker physically significant conditions covering a large class of theories with short range forces have been analyzed with success [5], there is no known, or even plausible proposed criterion for singling out the relevant representations in all cases. In fact, it suffices to take the case of the electric charge, one of the key examples of [20], to realize that there are still unresolved problems. Essentially one would like in this case to arrive at a simple picture where the sectors are labelled by the electric charge corresponding to a gauge group $U(1)$. However for each value of the electric charge there are myriads of representations differing by their infrared clouds, cf. [2,10]and references therein. To find such a simple picture one would either have to take equivalence classes or choose a suitable subset of representations. Previous work on the problem of sectors in quantum electrodynamics include [1, 3, 4, 11, 12].

This paper has been a result of our attempts to describe the sector structure of quantum electrodynamics and of similar models exhibiting long range effects. In view of the central role played by the symmetric tensor C^*-category, we here propose a method of constructing such categories which might prove to be applicable in these cases but the scheme seems of interest in its own right.

In the simplest case of strictly localizable charges [6], one passes from the selected representations and their intertwiners to endomorphisms and their intertwiners using Haag duality. The endomorphisms of a C^*-algebra and their intertwiners form a tensor C^*-category, the endomorphisms being the objects and the intertwiners the arrows. The symmetry properties of the category can be deduced from analyzing the commutation properties of intertwiners.

The proposal described here allows for intertwiners not lying in the algebra where the endomorphisms act. In addition, our endomorphisms are not required to be locally inner but only asymptotically inner. We will show in Section 5.2 how an appropriate form of asymptotic Abelianness allows one to construct a tensor C^*-category from this structure without the additional input of Haag duality. Similarly, the symmetries on our derived tensor C^*-

categories, discussed in Section 5.3, will reflect asymptotic rather than purely local commutation properties of intertwiners; yet they are *exact symmetries* although one might have anticipated that asymptotic Abelianness would just lead to an asymptotic notion of symmetry.

The general mechanism presented here might still be too limited to be directly applicable to quantum electrodynamics, but might elucidate some important aspects of that theory. As a matter of fact, it covers examples of theories exhibiting long range effects which go beyond the limits of the approaches previously studied.

One such example is provided by the model of charges of electromagnetic type discussed in [3], which actually stimulated the study of the more general structure discussed in the present paper. We briefly discuss at the end of Section 5.4 how that model fits into the scheme presented here.

Another interesting example is concerned with the superselection structure of Bose sectors of the gauge invariant part of the algebra of a free massive Fermi field on the two-dimensional space-time. Those sectors are usually described by a tensor category of localized automorphisms with a trivial Bose symmetry. The breakdown of Haag duality in this model, however, allows for different descriptions: for each real parameter λ, there is a tensor category describing the same sectors, where the intertwiners do not belong to the algebra, but fulfil the asymptotic Abelianness condition of Section 5.2. That category is equipped with a braiding which is not a symmetry, unless $\lambda = 0$, and arises from a bi-asymptopia as described in Section 5.3. For $\lambda = 0$, one gets back the usual symmetric tensor category of localized automorphisms [18].

5.2 Asymptotically Abelian Intertwiners

One of the very first steps in the theory of superselection sectors is to show how a tensor C^*-category may be obtained by passing from a C^*-category of representations to a C^*-category of endomorphisms. In this step duality plays a fundamental role. The aim here is to describe an alternative mechanism, in suitable mathematical generality and abstraction.

Let $\mathcal{A} \subset \mathcal{B}$ be an inclusion of unital C^*-algebras and Δ a semigroup of endomorphisms of $\mathcal{A}$. Given $\rho, \sigma \in \Delta$, we consider the corresponding space (ρ, σ) of intertwiners

$$(\rho, \sigma) = \{R \in \mathcal{B} : R\rho(A) = \sigma(A)R, A \in \mathcal{A}\}$$

and obtain in this way a C^*-category $\mathcal{T}$, where the composition of arrows (intertwiners) is denoted by $\circ$.

This framework is supposed to model the situation of a set of representations and their intertwiners. Restricting attention to representations described by endomorphisms does not seem restrictive on the mathematical side. In the case of separable simple C^*-algebras $\mathcal{A}$ acting irreducibly on some Hilbert space $\mathcal{H}$, any other irreducible representation can be obtained (up to equivalence) by the action of some automorphism. One can choose that automorphism to be asymptotically inner in the sense that it is the limit of inner automorphisms induced by a continuous one-parameter family of unitaries of the same algebra (cf. [16], and references in there for previous results). If the algebra is also nuclear and not type I, any cyclic representation is obtained in a similar way by asymptotically inner endomorphisms [15]. In the non-separable case, customary in quantum field theory, the physically motivated split property [7] implies that $\mathcal{A}$ is generated by a type I_∞ funnel. Then all locally normal irreducible representations can also be described (up to equivalence) in one such representation by automorphisms [19].

The intertwiners between the representations considered, however, do not belong to the given C^*-algebra in general, so $\mathrm{End}\,\mathcal{A}$ does not model the category of representations, in contrast to the case of localizable charges in quantum field theory fulfilling duality. If duality fails and $\mathcal{B}$ denotes the C^*-algebra generated by the dual net, our category $\mathcal{T}$ models again a category of representations with localizable charges. More generally, whenever the above mentioned theorems apply and $\mathcal{A}$ acts irreducibly on $\mathcal{H}$, we may always take $\mathcal{B} = \mathcal{B}(\mathcal{H})$. We take these facts as motivation for studying our more general structure.

The descriptions of representations given by the above mentioned general mathematical results, however, either ignore or take only partially into account the local structure of $\mathcal{A}$ in quantum field theory. Yet certain aspects of locality are needed in order to turn $\mathcal{T}$ into a *tensor* C^*-category. Note that $\mathcal{T}$ does not yet have a tensor structure, for the arrows lie in $\mathcal{B}$ and each $\rho \in \Delta$ is defined only on $\mathcal{A}$.[6] So what is required is an extension of Δ to a semigroup of endomorphisms of the C^*-algebra $\mathcal{A}_\Delta \subset \mathcal{B}$ generated by $\mathcal{A}$ and all the intertwiner spaces (ρ, σ) in $\mathcal{B}$.

Two basic ingredients will help us to accomplish this task. First, we assume that any $\rho \in \Delta$ is asymptotically inner in $\mathcal{B}$, i.e. there are unitaries $U_m \in \mathcal{B}$ such that

$$U_m^* A U_m \to \rho(A) \tag{5.1}$$

in norm as $m \to \infty$ for each $A \in \mathcal{A}$. Secondly, letting V_n denote the analogues of the U_m for σ instead of ρ, equation (5.1) implies that for all $R \in (\rho, \sigma)$, $A \in \mathcal{A}$

$$[V_n R U_m^*, A] \to 0$$

in norm as $m, n \to \infty$. As we want to extend our morphisms to $\mathcal{A}_\Delta$, we require that this asymptotic Abelianness holds for intertwiners too, i.e. given $R \in (\rho, \sigma)$ and $R' \in (\rho', \sigma')$,

$$[V_n R U_m^*, R'] \to 0 \tag{5.2}$$

[6] $\mathcal{T}$ acquires a tensor structure if the tensor product of endomorphisms is defined by composition and if the tensor product of intertwiners $R \in (\rho, \rho')$ and $S \in (\sigma, \sigma')$ given by $R \times S \doteq R\rho(S)$ is meaningful as an element of $(\rho\sigma, \rho'\sigma')$.

in norm as $m, n \to \infty$. Indeed, condition (5.2) will allow us to use (5.1) as a definition of the desired extension of each $\rho \in \Delta$ to $\mathcal{A}_\Delta$.

Rather than spell out our assumptions in detail, we will first introduce some standard formalism from the theory of C^*-algebras enabling us to replace a sequence of unitaries in $\mathcal{B}$ by a unitary in a larger C^*-algebra $\mathbf{\mathcal{B}}$. This simplifies what are already simple proofs by eliminating the indices.

Now the bounded sequences in $\mathcal{B}$ with pointwise operations and norm defined as the supremum of the norms of elements of the sequence form a C^*-algebra and the subset of sequences that tend to zero in norm is a two-sided ideal in that algebra. $\mathbf{\mathcal{B}}$ is then the quotient C^*-algebra and, passing from an element of $\mathcal{B}$ to the corresponding constant sequences, $\mathcal{B}$ can and will be canonically identified with a C^*-subalgebra of $\mathbf{\mathcal{B}}$, denoted again by $\mathcal{B}$.

We will use the following notation: generic elements of $\mathbf{\mathcal{B}}$ will be denoted by bold face letters $\boldsymbol{B}$, whereas elements of $\mathcal{A}, \mathcal{B}$ and of their canonical images in $\mathbf{\mathcal{B}}$ are denoted by A, B. As indicated, the elements $\boldsymbol{B} \in \mathbf{\mathcal{B}}$ are equivalence classes of bounded sequences modulo sequences that tend to zero. If $\{B_n\}_{n\in\mathbb{N}}$ represents $\boldsymbol{B}$, then $\|\boldsymbol{B}\| = \limsup_n \|B_n\|$. Obviously any subsequence of a bounded sequence is again bounded and we say that a subset $\mathbf{\mathcal{S}} \subset \mathbf{\mathcal{B}}$ is *stable* if it is closed under taking subsequences.

Lemma 1. *A subset of $\mathbf{\mathcal{B}}$ consisting of a single element $\boldsymbol{B}$ is stable if and only if $\boldsymbol{B} = B \in \mathcal{B}$.*

Proof. Let $\{B_n \in \mathcal{B}\}_{n\in\mathbb{N}}$ be a sequence representing $B \in \mathcal{B}$, then $B_n - B \to 0$. Given any subsequence $\{B_{n(i)}\}_{i\in\mathbb{N}}$, then $B_{n(i)} - B \to 0$ and $\{B_{n(i)}\}_{i\in\mathbb{N}}$ again represents B, so the subset consisting just of $\boldsymbol{B}$ is stable. Conversely, if $\{B_n\}_{n\in\mathbb{N}}$ does not represent an element of $\mathcal{B}$, then $\{B_n\}_{n\in\mathbb{N}}$ is not a Cauchy sequence. But then there is an $\varepsilon > 0$ and for each $i \in \mathbb{N}$ an $n(i), n'(i) \geq i$ with $\|B_{n(i)} - B_{n'(i)}\| \geq \varepsilon$. Thus the subset consisting just of $\boldsymbol{B}$ is not stable. □

Lemma 2. *If $\boldsymbol{U}$ is a unitary from $\mathbf{\mathcal{B}}$, then there is a representing sequence consisting of unitaries.*

Proof. If $\{B_{n(i)}\}_{i\in\mathbb{N}}$ is a representing sequence for $\boldsymbol{U}$, then $B_n^* B_n$ represents $\boldsymbol{U}^*\boldsymbol{U} = I$ so that $B_n^* B_n \to 1$. Similarly $B_n B_n^* \to 1$. Thus $|B_n|$ is invertible for all sufficiently large n and if we then set $U_n \doteq B_n |B_n|^{-1}$, $U_n^* U_n = 1$ whilst $U_n U_n^* = B_n |B_n|^{-2} B_n^*$ is a projection tending in norm to 1. Consequently, U_n is unitary for all sufficiently large n and, setting $U_n = 1$ for other values of n to make it unitary, the result follows. □

Now if $\mathcal{U}(\mathbf{\mathcal{B}})$ denotes the set of unitaries in $\mathbf{\mathcal{B}}$, and $\boldsymbol{U} \in \mathcal{U}(\mathbf{\mathcal{B}})$, then

$$\tau(A) \doteq \boldsymbol{U}^* A \boldsymbol{U}, \quad A \in \mathcal{A}$$

is a morphism of $\mathcal{A}$ into $\mathbf{\mathcal{B}}$ and we set

$$\mathcal{U}(\mathbf{\mathcal{B}})_\tau \doteq \{\boldsymbol{U} \in \mathcal{U}(\mathbf{\mathcal{B}}) : \mathrm{Ad}\boldsymbol{U}^* \upharpoonright \mathcal{A} = \tau\}.$$

Lemma 3. $\tau(\mathcal{A}) \subset \mathcal{B}$ *if and only if* $\mathcal{U}(\mathbf{\mathcal{B}})_\tau$ *is stable.*

Proof. If $\boldsymbol{U} \in \mathcal{U}(\boldsymbol{\mathcal{B}})_\tau$ and $\mathcal{U}(\boldsymbol{\mathcal{B}})_\tau$ is stable then $\boldsymbol{U}^* A \boldsymbol{U}$ is stable and hence, by Lemma 1, an element of $\mathcal{B}$ for each $A \in \mathcal{A}$. Conversely if $\tau(\mathcal{A}) \subset \mathcal{B}$ then $\mathcal{U}(\boldsymbol{\mathcal{B}})_\tau$ is stable from the definition. □

Note that if we consider the C^*-category $\mathcal{C}$ whose objects are the morphisms of $\mathcal{A}$ into $\boldsymbol{\mathcal{B}}$ inner in $\boldsymbol{\mathcal{B}}$ and arrows their intertwiners, then all objects are obviously unitarily equivalent since $\boldsymbol{U} \in \mathcal{U}(\boldsymbol{\mathcal{B}})_\tau$ is a unitary in $\mathcal{C}$ from τ to the identity automorphism ι. Now $\mathcal{A}' \cap \boldsymbol{\mathcal{B}}$ is the set of arrows from ι to ι in $\mathcal{C}$ so if $\boldsymbol{U} \in \boldsymbol{\mathcal{U}}_\tau$, $\boldsymbol{V} \in \boldsymbol{\mathcal{U}}_\upsilon$, then $\boldsymbol{R} \in (\tau, \upsilon)$ if and only if $\boldsymbol{V}\boldsymbol{R}\boldsymbol{U}^* \in \mathcal{A}' \cap \boldsymbol{\mathcal{B}}$.

Obviously if $\rho(\mathcal{A}) \subset \mathcal{A}$ so that we are dealing with endomorphisms of $\mathcal{A}$ then $\mathcal{U}(\boldsymbol{\mathcal{B}})_\rho \mathcal{U}(\boldsymbol{\mathcal{B}})_\sigma \subset \mathcal{U}(\boldsymbol{\mathcal{B}})_{\sigma\rho}$ for two such endomorphisms ρ and σ. Now by Lemma 2, saying that an endomorphism of $\mathcal{A}$ is unitarily implemented in $\boldsymbol{\mathcal{B}}$ is the same as saying that it is asymptotically inner in $\mathcal{B}$. So the set of such endomorphisms of $\mathcal{A}$ obviously forms a semigroup. We begin here by studying a subsemigroup Δ and the associated C^*-category $\mathcal{T}$ of intertwiners in $\mathcal{B}$. Just as we defined $\boldsymbol{\mathcal{B}}$ from $\mathcal{B}$ so we can define a C^*-category $\boldsymbol{\mathcal{T}}$ by taking bounded sequences of arrows from $\mathcal{T}$ with pointwise composition $\circ$ and the supremum norm and quotient by the ideal of sequences of arrows that tend to zero in norm. The proof of Lemma 2 shows that a unitary arrow of $\boldsymbol{\mathcal{T}}$ has a representing sequence consisting of unitaries from $\mathcal{T}$. $\mathcal{T}$ can again be considered as a C^*-subcategory of $\boldsymbol{\mathcal{T}}$ in a natural way.

We now want to extend our semigroup of endomorphisms to the C^*-algebra $\mathcal{A}_\Delta \subset \mathcal{B}$ generated by $\mathcal{A}$ and the intertwiners $\mathcal{T} \subset \mathcal{B}$ so that the extended endomorphisms remain asymptotically inner in $\mathcal{B}$. The above discussion indicates what hypotheses will be necessary. We assume that for each $\rho \in \Delta$, we are given a stable subset $\boldsymbol{\mathcal{U}}_\rho \subset \mathcal{U}(\boldsymbol{\mathcal{B}})_\rho$ with $\boldsymbol{\mathcal{U}}_\rho \cap \boldsymbol{\mathcal{T}} \neq \emptyset$, where $\boldsymbol{\mathcal{T}}$ here refers to the union of its Hom-sets. Thus in each $\boldsymbol{\mathcal{U}}_\rho$ there is at least one sequence of intertwiners between morphisms belonging to Δ and

$$\boldsymbol{\mathcal{U}}_\rho \boldsymbol{\mathcal{U}}_\sigma \cap \boldsymbol{\mathcal{U}}_{\sigma\rho} \neq \emptyset.$$

Furthermore, we suppose that asymptotic Abelianness holds in the sense of Equation (5.2), *i.e.* given $R \in (\rho, \sigma) \subset \mathcal{B}$ and $R' \in (\rho', \sigma') \subset \mathcal{B}$ and $\boldsymbol{U} \in \boldsymbol{\mathcal{U}}_\rho$, $\boldsymbol{V} \in \boldsymbol{\mathcal{U}}_\sigma$,

$$[\boldsymbol{V}\boldsymbol{R}\boldsymbol{U}^*, R'] = 0.$$

Such a coherent assignment $\mathcal{U} : \rho \mapsto \boldsymbol{\mathcal{U}}_\rho$ for $\rho \in \Delta$ will be called an *asymptopia* for Δ. Note that there are in general several such asymptopias.

Theorem 4. *Let $\mathcal{A}_\Delta$ denote the C^*-subalgebra of $\mathcal{B}$ generated by $\mathcal{A}$ and $\mathcal{T}$, and let $\mathcal{U}$ be some asymptopia for Δ. Then every $\rho \in \Delta$ has a unique extension $\rho_{\mathcal{U}}$ to an endomorphism of $\mathcal{A}_\Delta$ such that*

$$\rho_{\mathcal{U}}(A) = \boldsymbol{U}^* A \boldsymbol{U}, \quad A \in \mathcal{A}_\Delta \quad \boldsymbol{U} \in \boldsymbol{\mathcal{U}}_\rho.$$

Furthermore $(\rho\sigma)_{\mathcal{U}} = \rho_{\mathcal{U}}\sigma_{\mathcal{U}}$ and $(\rho, \sigma) = (\rho_{\mathcal{U}}, \sigma_{\mathcal{U}})$. Thus $\mathcal{T}$ inherits the structure of a tensor C^*-category from End$\mathcal{A}_\Delta$.

We first prove the following lemma.

Lemma 5. *If $\rho, \sigma, \tau \in \Delta$ and $S \in (\sigma, \tau)$, then $\rho_{\mathfrak{U}}(S) \doteq \boldsymbol{U}^* S \boldsymbol{U}$ is independent of the choice of $\boldsymbol{U} \in \mathfrak{U}_\rho$ and is an element of $\mathcal{A}_\Delta$.*

Proof. If $\boldsymbol{U}, \boldsymbol{U}' \in \mathfrak{U}_\rho$,

$$\begin{aligned} \boldsymbol{U}^* S \boldsymbol{U} - \boldsymbol{U}'^* S \boldsymbol{U}' &= \boldsymbol{U}^*(S \boldsymbol{U}\boldsymbol{U}'^* - \boldsymbol{U}\boldsymbol{U}'^* S)\boldsymbol{U}' \\ &= \boldsymbol{U}^*[S, \boldsymbol{U} 1_\rho \boldsymbol{U}'^*]\boldsymbol{U}' = 0 \end{aligned}$$

as required. In particular, the subset consisting of the single element $\boldsymbol{U}^* S \boldsymbol{U}$ is stable since $\mathfrak{U}_\rho$ is stable. Thus by Lemma 1, $\rho_{\mathfrak{U}}(S) \in \mathcal{B}$ and since we can choose $\boldsymbol{U} \in \mathcal{T}$, $\rho_{\mathfrak{U}}(S) \in \mathcal{A}_\Delta$. □

Proof (of Theorem 4). Lemma 5 shows that each $\rho \in \Delta$ has the required unique extension to an endomorphism $\rho_{\mathfrak{U}}$ of $\mathcal{A}_\Delta$. The condition of asymptotic Abelianness shows that we do not lose any intertwiners in this way, *i.e.* that $(\rho, \sigma) = (\rho_{\mathfrak{U}}, \sigma_{\mathfrak{U}})$. Since $\boldsymbol{V}\boldsymbol{U} \in \mathfrak{U}_{\rho\sigma}$ for some $\boldsymbol{U} \in \mathfrak{U}_\rho$ and $\boldsymbol{V} \in \mathfrak{U}_\sigma$, $(\rho\sigma)_{\mathfrak{U}} = \rho_{\mathfrak{U}}\sigma_{\mathfrak{U}}$. Thus $\mathcal{T}$ can be identified with a full tensor C^*-subcategory of $\mathrm{End}\,\mathcal{A}_\Delta$ completing the proof of the theorem. □

For a given semigroup Δ of endomorphisms, the C^*-category $\mathcal{T}$ is determined by the inclusion $\mathcal{A} \subset \mathcal{B}$. Its tensor structure is determined by the choice of asymptopia $\mathfrak{U}$. Different asymptopias can lead to different tensor structures, in other words to different extensions of Δ to $\mathcal{A}_\Delta$, unless they are included in a common asymptopia. $\rho \mapsto \mathfrak{U}_{\rho_{\mathfrak{U}}}$, the set of all unitaries in $\mathfrak{B}$ inducing $\rho_{\mathfrak{U}}$ on $\mathcal{A}_\Delta$, is obviously a maximal asymptopia containing the given asymptopia. Thus two asymptopias lead to the same tensor structure if and only if there is an asymptopia containing both. If we consider the set of asymptopias ordered under inclusion, the different tensor structures correspond to the different path-components of this set.

Theorem 6 *Given an inclusion of unital C^*-algebras $\mathcal{A} \subset \mathcal{B}$ and a semigroup Δ of endomorphisms of $\mathcal{A}$, the path-components of the set of asymptopias for Δ are in natural 1–1 correspondence with the set of maximal asymptopias and with the set of extensions of Δ to a semigroup of asymptotically inner endomorphisms of the C^*-algebra $\mathcal{A}_\Delta$ generated by $\mathcal{A}$ and the intertwiners for Δ in $\mathcal{B}$.*

It is of interest that the above framework for studying semigroups of endomorphisms can be extended so as to cover the case of representations and their intertwiners. One then deals with a set Δ of morphisms of $\mathcal{A}$ into $\mathcal{B}$ and the associated C^*-category of intertwiners. We suppose that for each $\rho \in \Delta$, we are given a set $\mathfrak{U}_\rho$ of unitary operators in $\mathfrak{B}$ with $\mathfrak{U}_\rho \cap \mathfrak{T} \neq \emptyset$ such that

$$\rho(A) = \boldsymbol{U}^* A \boldsymbol{U}$$

for each $A \in \mathcal{A}$, $\boldsymbol{U} \in \mathfrak{U}_\rho$ and $\rho \in \Delta$. In particular, the inclusion mapping ι of $\mathcal{A}$ into $\mathcal{B}$ is in Δ with $\mathfrak{U}_\iota$ consisting just of the identity of $\mathfrak{B}$. Furthermore, we require that for each $\rho, \sigma \in \Delta$, there is a unique $\tau \in \Delta$ such that given $\boldsymbol{U} \in \mathfrak{U}_\rho$ there is a $\boldsymbol{V} \in \mathfrak{U}_\sigma$ with $\boldsymbol{VU} \in \mathfrak{U}_\tau$. As before, $R \in (\rho, \sigma)$ if and only if $\boldsymbol{V}R\boldsymbol{U}^* \in \mathcal{B} \cap \mathcal{A}'$. The intertwiners are again supposed to be asymptotically Abelian. Under these circumstances we have the following generalization of Theorem 4.

Theorem 7. *Let $\mathcal{A}_\Delta$ denote the C^*-subalgebra of $\mathcal{B}$ generated by $\mathcal{A}$ and $\mathcal{T}$, then every $\rho \in \Delta$ has a unique extension $\rho_\mathfrak{U}$ to an endomorphism of $\mathcal{A}_\Delta$ such that*

$$\rho_\mathfrak{U}(A) = \boldsymbol{U}^* A \boldsymbol{U}, \quad A \in \mathcal{A}_\Delta \quad \boldsymbol{U} \in \mathfrak{U}_\rho.$$

Furthermore $(\rho, \sigma) = (\rho_\mathfrak{U}, \sigma_\mathfrak{U})$ and the set $\Delta_\mathfrak{U}$ of the extended morphisms constitutes a unital semigroup of endomorphisms of $\mathcal{A}_\Delta$. Thus $\mathcal{T}$ inherits the structure of a tensor C^*-category from End$\mathcal{A}_\Delta$.

Proof. Note that Lemma 5 retains its validity in this new context and that $\rho(\mathcal{A}) \subset \mathcal{A}_\Delta$ since $\mathfrak{U}_\rho \cap \mathfrak{T} \neq \emptyset$. Thus $\rho_\mathfrak{U}(\mathcal{A}) \doteq \boldsymbol{U}^* \mathcal{A}_\Delta \boldsymbol{U} \subset \mathcal{A}_\Delta$. Now let $\rho, \sigma \in \Delta$. Choose $\tau \in \Delta$, $\boldsymbol{U} \in \mathfrak{U}_\rho$ and $\boldsymbol{V} \in \mathfrak{V}_\sigma$ such that $\boldsymbol{VU} \in \mathfrak{U}_\tau$. Since $\mathrm{Ad}\,\boldsymbol{U}^*\boldsymbol{V}^* = \mathrm{Ad}\boldsymbol{U}^*\mathrm{Ad}\boldsymbol{V}^*$, we conclude that $\rho_\mathfrak{U}\sigma_\mathfrak{U} = \tau_\mathfrak{U}$. Thus $\Delta_\mathfrak{U}$ is a semigroup with unit. Given $T \in (\rho, \sigma)$, then the set of $B \in \mathcal{B}$ such that

$$T\boldsymbol{U}^* B \boldsymbol{U} = \boldsymbol{V}^* B \boldsymbol{V} T,$$

$$T\boldsymbol{U}^* B^* \boldsymbol{U} = \boldsymbol{V}^* B^* \boldsymbol{V} T,$$

is a C^*-subalgebra of $\mathcal{B}$ containing $\mathcal{A}$. It also contains every arrow of $\mathcal{T}$ since intertwiners are asymptotically Abelian. This shows that $(\rho, \sigma) = (\rho_\mathfrak{U}, \sigma_\mathfrak{U})$.□

We conclude with the remark that all results of this section can be generalized in the following sense. Instead of considering an inclusion of unital C^*-algebras, we can consider a fixed unital C^*-algebra $\mathcal{L}$, say, and a set Δ of morphisms between unital C^*-subalgebras, closed under composition. Given $\rho, \sigma \in \Delta$, between two such subalgebras $\mathcal{A}$ and $\mathcal{B}$, we set

$$(\rho, \sigma) \doteq \{R \in \mathcal{L} : R\rho(A) = \sigma(A)R, \ A \in \mathcal{A}\},$$

and obtain in this way a C^*-category $\mathcal{T}$. In the presence of an asymptopia $\mathfrak{U}$ for Δ, the morphism ρ has a unique extension to a morphism $\rho_\mathfrak{U}$ between $\mathcal{A}_\Delta$ and $\mathcal{B}_\Delta$, the C^*-subalgebras of $\mathcal{L}$ generated by $\mathcal{A}$ and $\mathcal{T}$ and $\mathcal{B}$ and $\mathcal{T}$, respectively. In this way, $\mathcal{T}$ inherits the structure of a 2-C^*-category from the 2-C^*-category of morphisms and intertwiners between unital C^*-subalgebras of $\mathcal{L}$, cf. [14]. The reader should have no difficulty in enunciating and proving the analogue of the results of this section.

The above generalization may prove relevant to the theory of superselection sectors, first because it may prove advantageous to restrict a representation to a subnet of the observable net, and secondly because it may not be possible to get an endomorphism of the subnet by passing to an equivalent representation.

5.3 The Emergence of Braiding

In view of the results in the preceding section, we can now forget about the inclusion $\mathcal{A} \subset \mathcal{B}$, work with a single unital C^*-algebra $\mathcal{A}$ (corresponding to the previous algebra $\mathcal{A}_\Delta$) and suppose that we are given a tensor C^*-category $\mathcal{T}$ with Δ as its set of objects realized as a full tensor subcategory of $\mathrm{End}\,\mathcal{A}$. How it was obtained, in particular, which asymptopia, if any, was used to construct it, is for the moment quite irrelevant.

We also note that the C^*-category $\boldsymbol{\mathcal{T}}$ now becomes a tensor C^*-category in a natural way. The tensor product of objects is just the pointwise product of the individual sequences of endomorphisms, whilst if arrows $\boldsymbol{R}$ and $\boldsymbol{S}$ are represented by sequences $R_n \in (\rho_n, \rho_n')$ and $S_n \in (\sigma_n, \sigma_n')$, their tensor product $\boldsymbol{R} \times \boldsymbol{S}$ is represented by $R_n \rho_n(S_n) \in (\rho_n \sigma_n, \rho_n' \sigma_n')$, $n \in \mathbb{N}$. We further note that a sequence of endomorphisms ρ_n of $\mathcal{A}$ defines in a canonical way a unital morphism $\boldsymbol{\rho}$ of $\mathcal{A}$ into $\boldsymbol{\mathcal{A}}$, two sequences ρ_n and σ_n defining the same morphism if and only if $\rho_n(A) - \sigma_n(A) \to 0$ for every $A \in \mathcal{A}$. Adjoining intertwiners, we get a C^*-category $\mathrm{Mor}\,(\mathcal{A}, \boldsymbol{\mathcal{A}})$. There is now an obvious canonical *-functor F from $\boldsymbol{\mathcal{T}}$ to $\mathrm{Mor}\,(\mathcal{A}, \boldsymbol{\mathcal{A}})$ mapping a sequence of endomorphisms of $\mathcal{A}$ onto the induced morphism from $\mathcal{A}$ to $\boldsymbol{\mathcal{A}}$ and acting as the identity on intertwiners. The effect of F is to identify sequences of endomorphisms with the same asymptotic behaviour.

The task in this section is to show how to get a braiding and to develop criteria for deciding whether the braiding is a symmetry. The idea here is to define the braiding using suitable norm convergent sequences of unitaries and we will again realize these sequences as unitary operators in $\boldsymbol{\mathcal{A}}$. However, the conditions we need are now somewhat different. In particular, as a braiding is a function of two objects, we assume that there are for each object $\rho \in \Delta$ two stable sets of unitaries $\boldsymbol{\mathcal{U}}_\rho, \boldsymbol{\mathcal{V}}_\rho$ which are contained in $\boldsymbol{\mathcal{T}}$, each unitary $\boldsymbol{W}$ implementing ρ, *i.e.* $\rho(A) = \boldsymbol{W}^* A \boldsymbol{W}$, $A \in \mathcal{A}$. This means that $F(\boldsymbol{W})$ is an intertwiner in $\boldsymbol{\mathcal{A}}$ from ρ to the identity automorphism ι. Furthermore, we require the following notion of asymptotic Abelianness, where we work in the category $\mathrm{Mor}\,(\mathcal{A}, \boldsymbol{\mathcal{A}})$: given two intertwiners $R \in (\rho, \rho')$ and $S \in (\sigma, \sigma')$ of $\mathcal{T}$ and $\boldsymbol{U} \in \boldsymbol{\mathcal{U}}_\rho$, $\boldsymbol{U}' \in \boldsymbol{\mathcal{U}}_{\rho'}$, $\boldsymbol{V} \in \boldsymbol{\mathcal{V}}_\sigma$ and $\boldsymbol{V}' \in \boldsymbol{\mathcal{V}}_{\sigma'}$, then

$$F(\boldsymbol{U}' R \boldsymbol{U}^* \times \boldsymbol{V}' S \boldsymbol{V}^*) - F(\boldsymbol{V}' S \boldsymbol{V}^* \times \boldsymbol{U}' R \boldsymbol{U}^*) = 0. \tag{5.3}$$

This form of asymptotic Abelianness of the tensor product of arrows will allow us to define the universal rule for interchanging tensor products, in very much the same way as condition (5.2) allowed us to define the tensor product of arrows.

For that purpose this data should, in addition, be compatible with products in the sense that given $\rho, \rho' \in \Delta$, we can find $\boldsymbol{U} \in \boldsymbol{\mathcal{U}}_\rho$ and $\boldsymbol{U}' \in \boldsymbol{\mathcal{U}}_{\rho'}$ such that $\boldsymbol{U} \times \boldsymbol{U}' \in \boldsymbol{\mathcal{U}}_{\rho\rho'}$ and analogously for the second set of unitaries $\{\boldsymbol{\mathcal{V}}_\rho\}$. Any assignment $(\mathcal{U}, \mathcal{V}) : \rho \mapsto \boldsymbol{\mathcal{U}}_\rho, \boldsymbol{\mathcal{V}}_\rho$ with these properties will be referred to as a *bi-asymptopia* for Δ.

Theorem 8. *Let $(\mathcal{U}, \mathcal{V})$ be a bi-asymptopia for Δ. Given $\rho, \sigma \in \Delta$, then*

$$\varepsilon(\rho, \sigma) \doteq F(\boldsymbol{V}^* \times \boldsymbol{U}^*) \circ F(\boldsymbol{U} \times \boldsymbol{V})$$

is independent of the choice of $\boldsymbol{U} \in \mathcal{U}_\rho$ and $\boldsymbol{V} \in \mathcal{V}_\sigma$ and is in $(\rho\sigma, \sigma\rho)$. Furthermore, if $R \in (\rho, \rho')$ and $S \in (\sigma, \sigma')$, then

$$\varepsilon(\rho', \sigma') \circ R \times S = S \times R \circ \varepsilon(\rho, \sigma)$$

and if $\tau \in \Delta$, then

$$\varepsilon(\rho\sigma, \tau) = \varepsilon(\rho, \tau) \times 1_\sigma \circ 1_\rho \times \varepsilon(\sigma, \tau),$$

$$\varepsilon(\rho, \sigma\tau) = 1_\sigma \times \varepsilon(\rho, \tau) \circ \varepsilon(\rho, \sigma) \times 1_\tau.$$

In other words, ε is a braiding for the full subcategory of End$\mathcal{A}$ generated by Δ.

Proof. We have

$$\begin{aligned} & F(\boldsymbol{V}^* \times \boldsymbol{U}^*) \circ F(\boldsymbol{U} \times \boldsymbol{V}) - F(\boldsymbol{V}'^* \times \boldsymbol{U}'^*) \circ F(\boldsymbol{U}' \times \boldsymbol{V}') \\ & = F(\boldsymbol{V}^* \times \boldsymbol{U}^*) \circ \big(F((\boldsymbol{U} \circ \boldsymbol{U}'^*) \times (\boldsymbol{V} \circ \boldsymbol{V}'^*)) \\ & \quad - F((\boldsymbol{V} \circ \boldsymbol{V}'^*) \times (\boldsymbol{U} \circ \boldsymbol{U}'^*))\big) \circ F(\boldsymbol{U}' \times \boldsymbol{V}') = 0 \end{aligned}$$

proving independence of the choice of $\boldsymbol{U} \in \mathcal{U}_\rho$ and $\boldsymbol{V} \in \mathcal{V}_\sigma$. But then $F(\boldsymbol{V}^* \times \boldsymbol{U}^*) \circ F(\boldsymbol{U} \times \boldsymbol{V})$ is stable so that $\varepsilon(\rho, \sigma) \in (\rho\sigma, \sigma\rho)$. Furthermore,

$$\begin{aligned} & F(\boldsymbol{V}'^* \times \boldsymbol{U}'^*) \circ F(\boldsymbol{U}' \times \boldsymbol{V}') \circ R \times S \\ & \quad - S \times R \circ F(\boldsymbol{V}^* \times \boldsymbol{U}^*) \circ F(\boldsymbol{U} \times \boldsymbol{V}) \\ & = F(\boldsymbol{V}'^* \times \boldsymbol{U}'^*) \circ \big(F(\boldsymbol{U}' R \boldsymbol{U}^* \times \boldsymbol{V}' S \boldsymbol{V}^*) \\ & \quad - F(\boldsymbol{V}' S \boldsymbol{V}^* \times \boldsymbol{U}' R \boldsymbol{U}^*)\big) \circ F(\boldsymbol{U} \times \boldsymbol{V}) = 0, \end{aligned}$$

and we deduce that

$$\varepsilon(\rho', \sigma') \circ R \times S = S \times R \circ \varepsilon(\rho, \sigma).$$

Now, pick $\boldsymbol{U} \in \mathcal{U}_\rho$ and $\boldsymbol{V} \in \mathcal{U}_\sigma$ such that $\boldsymbol{U} \times \boldsymbol{V} \in \mathcal{U}_{\rho\sigma}$. Then, if $\boldsymbol{W} \in \mathcal{V}_\tau$,

$$F(\boldsymbol{W}^* \times \boldsymbol{U}^* \times \boldsymbol{V}^*) \circ F(\boldsymbol{U} \times \boldsymbol{V} \times \boldsymbol{W})$$

$$= F(\boldsymbol{W}^* \times \boldsymbol{U}^* \times 1_\sigma) \circ F(\boldsymbol{U} \times \boldsymbol{W} \times 1_\sigma) \circ F(1_\rho \times \boldsymbol{W}^* \times \boldsymbol{V}^*) \circ F(1_\rho \times \boldsymbol{V} \times \boldsymbol{W}),$$

and we conclude that

$$\varepsilon(\rho\sigma, \tau) = \varepsilon(\rho, \tau) \times 1_\sigma \circ 1_\rho \times \varepsilon(\sigma, \tau).$$

Now,

$$\varepsilon^{-1}(\rho, \sigma) \doteq \varepsilon(\sigma, \rho)^{-1} = F(\boldsymbol{U}^* \times \boldsymbol{V}^*) \circ F(\boldsymbol{V} \times \boldsymbol{U})$$

and as we have here just interchanged the roles of $\mathcal{U}$ and $\mathcal{V}$, we deduce that

$$\varepsilon(\rho, \sigma\tau) = 1_\sigma \times \varepsilon(\rho, \tau) \circ \varepsilon(\rho, \sigma) \times 1_\tau$$

as stated. □

There still remains the question of whether ε is in fact a *symmetry*, i.e. whether $\varepsilon = \varepsilon^{-1}$. But in the above proof, we have just seen that we pass from ε to ε^{-1} by interchanging the roles of $\mathcal{U}$ and $\mathcal{V}$. Hence one way of tackling the problem is to ask when two bi-asymptopias give rise to the same braiding and here we may follow our discussion in the case of asymptopias.

Two bi-asymptopias will obviously give rise to the same braiding if one is included in the other, or if their intersection is still a bi-asymptopia; thus if we order the set of bi-asymptopias under inclusion, we can divide that set into path-components, and the braiding associated to a bi-asymptopia will depend only on its path-component.

A difference from the case of asymptopias is worth noting here: whilst $\rho \mapsto \mathfrak{U}_{\rho\mathcal{U}}$ defines a maximal asymptopia, where in particular equation (5.2) is automatically satisfied, the notion of asymptotic Abelianness entering in the definition of bi-asymptopias requires a further mutual (asymptotic) commutativity, equation (5.3). Hence there is no a priori unique maximal bi-asymptopia containing a given one, just as in general there is no unique maximal Abelian subalgebra containing a given Abelian subalgebra. Thus in this case we cannot say a priori that *different* braidings correspond to different path components of asymptopias. But by the previous discussion we have

Theorem 9. *The bi-asymptopia $\{\mathcal{U}, \mathcal{V}\}$ gives rise to a symmetry if it lies in the same path-component as $\{\mathcal{V}, \mathcal{U}\}$.*

We close with a few comments. First, when does a braiding arise from a bi-asymptopia? Note that if ε is a braiding for $\mathcal{T}$, then this braiding extends in an obvious way to a braiding for $\boldsymbol{\mathfrak{T}}$. Writing $\boldsymbol{\rho}$ to denote an object of $\boldsymbol{\mathfrak{T}}$, *i.e.* a sequence of elements of Δ, $\boldsymbol{U} \in (\rho, \boldsymbol{\rho})$ and $\boldsymbol{V} \in (\sigma, \boldsymbol{\sigma})$, then

$$\boldsymbol{V} \times \boldsymbol{U} \circ \varepsilon(\rho, \sigma) = \varepsilon(\boldsymbol{\rho}, \boldsymbol{\sigma}) \circ \boldsymbol{U} \times \boldsymbol{V}.$$

So if we give ourselves a braiding ε for $\mathcal{T}$ then

$$\varepsilon(\rho, \sigma) = F(\boldsymbol{V}^* \times \boldsymbol{U}^* \circ \boldsymbol{U} \times \boldsymbol{V})$$

if and only if $F(\varepsilon(\boldsymbol{\rho}, \boldsymbol{\sigma})) = 1_\iota$. Next,

$$\varepsilon(\boldsymbol{\rho}', \boldsymbol{\sigma}') \circ (\boldsymbol{U}' R \boldsymbol{U}^*) \times (\boldsymbol{V}' S \boldsymbol{V}^*) = (\boldsymbol{V}' S \boldsymbol{V}^*) \times (\boldsymbol{U}' R \boldsymbol{U}^*) \circ \varepsilon(\boldsymbol{\rho}, \boldsymbol{\sigma}).$$

Finally, $F(\boldsymbol{\rho}) = F(\boldsymbol{\rho}') = F(\boldsymbol{\sigma}) = \iota$ and $F(\varepsilon(\boldsymbol{\rho}, \boldsymbol{\sigma})) = F(\varepsilon(\boldsymbol{\rho}', \boldsymbol{\sigma})) = 1_\iota$ implies $F(\boldsymbol{\rho}\boldsymbol{\rho}') = \iota$ and $F(\varepsilon(\boldsymbol{\rho}\boldsymbol{\rho}', \boldsymbol{\sigma})) = 1_\iota$. So we conclude

Theorem 10. *Let ε be a braiding for $\mathcal{T}$ and suppose given two mappings $\check{\mathcal{U}}, \check{\mathcal{V}}$ from the morphisms ρ into stable subsets of $\mathcal{U}(\mathfrak{B})_\rho$ such that each $\check{\mathfrak{U}}_\rho$ and each $\check{\mathfrak{V}}_\rho$ are non-empty. If, given any pair $\boldsymbol{U} \in (\rho, \boldsymbol{\rho})$ from $\check{\mathfrak{U}}_\rho$ and $\boldsymbol{V} \in (\sigma, \boldsymbol{\sigma})$ from $\check{\mathfrak{V}}_\sigma$, $F(\boldsymbol{\rho}) = F(\boldsymbol{\sigma}) = \iota$ $F(\varepsilon(\boldsymbol{\rho}, \boldsymbol{\sigma})) = 1_\iota$, then $\check{\mathcal{U}}, \check{\mathcal{V}}$ can be extended to a bi-asymptopia $\{\mathcal{U}, \mathcal{V}\}$ giving ε. Furthermore, we can even take $\mathcal{U}$ and $\mathcal{V}$ to be closed under composition on the right by unitary intertwiners and stable under tensor products.*

As a final comment, we note that our notion of asymptotic Abelianness of the $\times$-product implies the corresponding notion for the operator product. In fact, $F(\boldsymbol{U}'R\boldsymbol{U}^*) = F(\boldsymbol{U}'R\boldsymbol{U}^* \times 1_{\boldsymbol{\sigma}'})$ and

$$F(\boldsymbol{V}'S\boldsymbol{V}^*) = F(\boldsymbol{V}'S\boldsymbol{V}^* \times 1_{\boldsymbol{\rho}}) = F(\boldsymbol{V}'S\boldsymbol{V}^* \times \boldsymbol{U}1_\rho\boldsymbol{U}^*).$$

Hence

$$F(\boldsymbol{U}'R\boldsymbol{U}^*) \circ F(\boldsymbol{V}'S\boldsymbol{V}^*) = F(\boldsymbol{U}'R\boldsymbol{U}^* \times 1_{\boldsymbol{\sigma}'}) \circ F(\boldsymbol{U}1_\rho\boldsymbol{U}^* \times \boldsymbol{V}'S\boldsymbol{V}^*)$$

when we have a bi-asymptopia. Thus

$$F(\boldsymbol{U}'R\boldsymbol{U}^*) \circ F(\boldsymbol{V}'S\boldsymbol{V}^*) = F(\boldsymbol{U}'R\boldsymbol{U}^* \times \boldsymbol{V}'S\boldsymbol{V}^*)$$

and

$$F(\boldsymbol{U}'R\boldsymbol{U}^*) \circ F(\boldsymbol{V}'S\boldsymbol{V}^*) = F(\boldsymbol{V}'S\boldsymbol{V}^*) \circ F(\boldsymbol{U}'R\boldsymbol{U}^*)$$

as stated.

5.4 Algebraic Quantum Field Theory

We briefly outline the relations of the preceding general analysis to algebraic quantum field theory, giving in particular some examples of asymptopias.

To begin with, we consider the semigroup of localized morphisms ρ of an observable net $\mathfrak{A}$ on Minkowski space which are transportable as representations on the vacuum Hilbert space of the theory. The C^*-algebra $\mathcal{A}$ of Section 2 can then be thought of as the C^*-inductive limit of the net $\mathfrak{A}(\mathcal{O})$ of local algebras associated with double cones $\mathcal{O}$, $\mathcal{A} \doteq \overline{\bigcup_{\mathcal{O}} \mathfrak{A}(\mathcal{O})}$. The intertwiners in the sense of representations belong to the dual net $\mathfrak{A}^d$,

$$\mathfrak{A}^d(\mathcal{O}) \doteq \bigcap\nolimits_{\mathcal{O}_1 \subset \mathcal{O}'} \mathfrak{A}(\mathcal{O}_1)',$$

where as usual $\mathcal{O}'$ denotes the spacelike complement of the double cone $\mathcal{O}$. The role of $\mathcal{B}$ is played by the C^*-algebra generated by the net $\mathfrak{A}^d$, $\mathcal{B} \doteq \overline{\bigcup_{\mathcal{O}} \mathfrak{A}^d(\mathcal{O})}$. We know that if $U_m \in (\rho, \rho_m)$ is unitary and ρ_m is localized in $\mathcal{O}_m$, then

$$\rho(A) = U_m^* A U_m, \quad A \in \mathfrak{A}(\mathcal{O}), \quad \mathcal{O}_m \subset \mathcal{O}'.$$

Hence our endomorphisms are asymptotically inner in $\mathcal{B}$. If we assume Haag duality we have $\mathcal{A} = \mathcal{B}$, so we do not need to extend our endomorphisms and $\rho \mapsto \mathcal{U}(\boldsymbol{\mathcal{B}})_\rho$ is the unique maximal asymptopia.

If we just assume essential duality, then in space-time dimension $d > 2$, we know that if $R_i \in (\rho_i, \sigma_i)$, $i = 1, 2$, then $R_1 R_2 = R_2 R_1$ if ρ_1 and σ_1 are localized spacelike to R_2. Thus we can define an asymptopia $\mathcal{U}$ by taking $\boldsymbol{\mathcal{U}}_\rho$ to consist of all unitaries of $\mathcal{U}(\boldsymbol{\mathcal{B}})$ with representing sequences $\boldsymbol{U}_m \in (\rho, \rho_m)$, the ρ_m being localized in double cones $\mathcal{O}_m$ tending spacelike to infinity. In this case, $\mathcal{A}_\Delta = \mathcal{B}$ and our Theorem 2 is just a variant of a known result cf. [17, §3.4.6].

The reader's attention is also drawn to the case of charges localizable in spacelike cones [5] where the intertwiners likewise do not lie in the observable algebra and where the treatment in [9] has aspects in common with the present paper, cf. Lemma 5.5 of [9].

In space-time dimension $d = 2$, the spacelike complement of a double cone has two path-components, a spacelike left and a spacelike right. As far as the commutation properties of intertwiners $R_i \in (\rho_i, \sigma_i)$ go, we merely know then that $R_1 R_2 = R_2 R_1$ if ρ_1 and σ_1 are both localized left spacelike to R_2 or both localized right spacelike to R_2. Thus we can define two asymptopias $\mathcal{U}^\ell$ and $\mathcal{U}^r$ as above by letting $\mathcal{O}_m$ tend spacelike to left infinity or right infinity, respectively. The restricted commutation properties of intertwiners show that these two asymptopias lead to different tensor structures, in general. The vacuum representation of the observable net then induces a solitonic representation of the corresponding field net.

In replacing $\mathfrak{A}$ by the algebra $\mathcal{A}$, we are, on the one hand, simplifying the mathematical setting by suppressing the net structure and avoiding all reference to spacetime. But we also have in mind applications where the endomorphisms are no longer strictly localized but only asymptotically inner.

For an example going beyond the standard setting of strictly local or cone-localized charges, we turn to the model expounded in [3] and based on the free massless scalar field. $\mathcal{A}$ is here the C^*-algebra generated by Weyl operators $W(f)$, $f \in \mathfrak{L}$,

$$\mathfrak{L} \doteq \omega^{-\frac{1}{2}}\mathcal{D}(\mathbb{R}^3) + i\omega^{\frac{1}{2}}\mathcal{D}(\mathbb{R}^3).$$

Here $\mathcal{D}(\mathbb{R}^3)$ denotes the space of smooth real-valued functions with compact support and ω the energy operator. $\mathfrak{L}$ is equipped with the scalar product

$$(f, f') \doteq \int d^3\boldsymbol{x}\,\overline{f(\boldsymbol{x})} f'(\boldsymbol{x}),$$

determining the symplectic form

$$\sigma(f, f') = -\Im(f, f')$$

and the usual vacuum state:

$$\omega(W(f)) = e^{-\frac{1}{4}(f,f)}.$$

The C^*-algebra $\mathcal{B}$ can be taken to be the algebra of all bounded operators on the vacuum Hilbert space. For Δ, we take the group Γ of automorphisms of $\mathcal{A}$ generated by the space

$$\mathfrak{L}_\Gamma \doteq \omega^{-\frac{1}{2}}\mathcal{D}(\mathbb{R}^3) + i\omega^{-\frac{3}{2}}\mathcal{D}(\mathbb{R}^3).$$

For convenience, we use the same symbol γ to denote the element of $\mathfrak{L}_\Gamma$, parametrized by smooth functions g and h,

$$\gamma = i\omega^{-\frac{3}{2}}g + \omega^{-\frac{1}{2}}h, \quad g, h \in \mathcal{D}(\mathbb{R}^3),$$

and the automorphism it generates so that

$$\gamma(W(f)) = e^{i\sigma(\gamma,f)}W(f),$$

where the symplectic form σ is defined on $\mathcal{L}_\Gamma$ so as to extend that on the subspace $\mathcal{L}$ by

$$\sigma(\gamma,\gamma') = \int d^3\boldsymbol{p}\,\omega^{-2}(\tilde{g}(-\boldsymbol{p})\tilde{h}'(\boldsymbol{p}) - \tilde{g}'(-\boldsymbol{p})\tilde{h}(\boldsymbol{p})).$$

The sectors are characterized by the charge

$$\int d^3\boldsymbol{x}\, g(\boldsymbol{x}).$$

They are translation invariant and if γ_a denotes the translate of γ by a, we have unitary intertwiners $U_a \in (\gamma, \gamma_a)$ unique up to a phase. We define $\mathfrak{U}_\gamma$ to be the set of equivalence classes of sequences of unitaries $U_a \in (\gamma,\gamma_a)$ for which a tends spacelike to infinity and $a_0/|\boldsymbol{a}| \to 0$. Then

$$U_a^* W(f) U_a = e^{i\sigma(\gamma-\gamma_a,f)}W(f) \to \gamma(W(f)),$$

as follows from the asymptotic behaviour of the symplectic form, Theorem 3 of [3]. Obviously, $\mathfrak{U}_\gamma\mathfrak{U}_\delta = \mathfrak{U}_{\delta\gamma}$. Hence to show that the assignment $\gamma \mapsto \mathfrak{U}_\gamma$ defines an asymptopia, it suffices to check that the intertwiners are asymptotically Abelian. Now if $(\gamma,\delta) \neq 0$, it consists of multiples of $W(\delta - \gamma)$. Hence the intertwiners are asymptotically Abelian if

$$\lim_{a,b}\,[W(\delta_b - \gamma_a), W(\delta' - \gamma')] = 0,$$

whenever γ and δ are equivalent and γ' and δ' are equivalent. The norm of this commutator of Weyl operators is

$$|e^{i\sigma(\delta_b-\gamma_a,\delta'-\gamma')} - 1|$$

and asymptotic Abelianness follows from Theorem 3 of [3].

In order to define a bi-asymptopia, we may take $\mathfrak{U}_\gamma$ to consist of representing sequences $U_a \in (\gamma, \gamma_a)$, where a tends to spacelike infinity inside some (open) spacelike cone $\mathcal{S}$, and define $\mathfrak{V}_\gamma$ similarly using the spacelike cone $-\mathcal{S}$. In view of our previous computations, we need only verify the condition of asymptotic Abelianness. Since every non-zero element of (γ,γ') is a multiple of $W(\gamma' - \gamma)$, it suffices to check that

$$W(\gamma'_{a'} - \gamma_a) \times W(\delta'_{b'} - \delta_b) - W(\delta'_{b'} - \delta_b) \times W(\gamma'_{a'} - \gamma_a)$$

tends to zero as a, a', $-b$ and $-b'$ go spacelike to infinity in $\mathcal{S}$. But the norm of this expression is

$$|e^{i\sigma(\gamma'_{a'},\delta'_{b'})}e^{i\sigma(\delta_b,\gamma_a)} - 1|$$

and tends to zero as required by Theorem 3 of [3].

It is easy to see by direct computation that the braiding determined by this bi-asymptopia is given by

$$\varepsilon(\gamma,\delta) = e^{-i\sigma(\gamma,\delta)},$$

cf. Sect. 5 of [3], and is therefore a symmetry, but it is instructive to derive this from Theorem 9. Obviously, if we replace the spacelike cone $\mathcal{S}$ in the definition of the bi-asymptopia by a smaller spacelike cone $\mathcal{S}_1$ we remain in the same path-component. The same is therefore true if $\mathcal{S} \cap \mathcal{S}_1 \neq \emptyset$. By a sequence of such moves, we may interchange $\mathcal{S}$ and $-\mathcal{S}$ so that by Theorem 9 our braiding is a symmetry.

Acknowledgements

SD would like to thank the II. Institut für Theoretische Physik, Universität Hamburg, and SD and JER would like to thank the Institut für Theoretische Physik, Universität Göttingen, for their kind hospitality in the final stage of this collaboration. Both gratefully acknowledge the financial support of the Alexander von Humboldt Foundation, that made this collaboration possible.

References

1. D. Buchholz: The Physical State Space of Quantum Electrodynamics. *Comm. Math. Phys.* 85:49-71 (1982).
2. D. Buchholz and S. Doplicher: Exotic Infrared Representations of Interacting Systems. *Ann. Inst. H. Poincaré* 40:175-184 (1984).
3. D. Buchholz, S. Doplicher, G. Morchio, J.E. Roberts and F. Strocchi: A Model for Charges of Electromagnetic Type. In: Proceedings of the Conference on *Operator Algebras and Quantum Field Theory*, Rome 1966 (ed.: S. Doplicher et al.), International Press, 1997.
4. D. Buchholz, S. Doplicher, G. Morchio, J.E. Roberts and F. Strocchi: The Quantum Delocalization of the Electric Charge. *Ann. of Phys.* 290:53-66 (2001).
5. D. Buchholz and K. Fredenhagen: Locality and the Structure of Particle States. *Comm. Math. Phys.* 84:1-54 (1982).
6. S. Doplicher, R. Haag and J.E. Roberts: Local Observables and Particle Statistics I. *Comm. Math. Phys.* 23:199-230 (1971) and II. *Comm. Math. Phys.* 35:49-85 (1974).
7. S. Doplicher and R. Longo: Standard and Split Inclusions of von Neumann Algebras. *Invent. Math.* 75:493-536 (1984).
8. S. Doplicher and J.E. Roberts: A New Duality Theory for Compact Groups. *Invent. Math.* 98:157-218 (1989).
9. S. Doplicher and J.E. Roberts: Why There is a Field Algebra With a Compact Gauge Group Describing the Superselection Structure in Particle Physics. *Comm. Math. Phys.* 131:51-107 (1990).

10. S. Doplicher and M. Spera: Representations Obeying the Spectrum Condition. *Comm. Math. Phys.* 84:505-513 (1982).
11. J. Fröhlich: The Charged Sectors of Quantum Electrodynamics in a Framework of Local Observables. *Comm. Math. Phys.* 66:223-265 (1979).
12. J. Fröhlich, G. Morchio and F. Strocchi: Charged Sectors and Scattering States in Quantum Electrodynamics. *Ann. Phys.* 119:241-284 (1979).
13. R. Haag and D. Kastler: An Algebraic Approach to Quantum Field Theory. *J. Math. Phys.* 5:848-861 (1964).
14. R. Longo and J.E. Roberts: A Theory of Dimension. *K-Theory* 11:103-159 (1997).
15. A. Kishimoto: The Representations and Endomorphisms of a Separable Nuclear C^*-algebra. *Internat. J. Math.* 14:313-326 (2003).
16. A. Kishimoto, N. Ozawa and S. Sakai: Homogeneity of the Pure State Space of a Separable C^*-Algebra. *Canad. Math. Bull.* 4:365-372 (2003).
17. J.E. Roberts: Lectures on Algebraic Quantum Field Theory. In: *The Algebraic Theory of Superselection Sectors. Introduction and Recent Results* (ed.: D. Kastler), pp. 1-112. World Scientific, Singapore, 1990.
18. D. Salvitti: Generalized Particle Statistics in Two-Dimensions: Examples from the Theory of Free Massive Dirac Field. e-print: hep-th/0507107, and Doctoral Thesis, Rome, 2004.
19. M. Takesaki: Algebraic Equivalence of Locally Normal Representations. *Pacific J. Math.* 34:807–816 (1970).
20. J.C. Wick, A.S. Wightman and E.P. Wigner: The Intrinsic Parity of Elementary Particles. *Phys. Rev.* 88:101-105 (1952).

6

Yang–Mills and Some Related Algebras

Alain Connes[1] and Michel Dubois-Violette[2]

[1] Collège de France, 3 rue d'Ulm, 75 005 Paris, France, Institut des Hautes Etudes Scientifiques, 91440 Bures-sur-Yvette, France, and Vanderbilt University, USA
connes@ihes.fr

[2] Laboratoire de Physique Théorique, UMR 8627, Université Paris XI, Bâtiment 210, F-91 405 Orsay Cedex, France
Michel.Dubois-Violette@th.u-psud.fr

Dedicated to Jacques Bros

Summary. After a short introduction on the theory of homogeneous algebras we describe the application of this theory to the analysis of the cubic Yang–Mills algebra, the quadratic self-duality algebras, their "super" versions as well as to some generalization.

6.1 Introduction

Consider the classical Yang–Mills equations in $(s+1)$-dimensional pseudo Euclidean space $\mathbb{R}^{s+1}$ with pseudo metric denoted by $g_{\mu\nu}$ in the canonical basis of $\mathbb{R}^{s+1}$ corresponding to coordinates x^λ. For the moment the signature plays no role so $g_{\mu\nu}$ is simply a real nondegenerate symmetric matrix with inverse denoted by $g^{\mu\nu}$. In terms of the covariant derivatives $\nabla_\mu = \partial_\mu + A_\mu$ $(\partial_\mu = \partial/\partial x^\mu)$ the Yang–Mills equations read

$$g^{\lambda\mu}[\nabla_\lambda, [\nabla_\mu, \nabla_\nu]] = 0 \tag{6.1}$$

for $\nu \in \{0, \dots, s\}$. By forgetting the detailed origin of these equations, it is natural to consider the abstract unital associative algebra $\mathcal{A}$ generated by $(s+1)$ elements ∇_λ with the $(s+1)$ cubic relations (6.1). This algebra will be referred to as the Yang–Mills algebra. It is worth noticing here that Equations (6.1) only involve the product through commutators so that, by its very definition the Yang–Mills algebra $\mathcal{A}$ is a universal enveloping algebra.

Our aim here is to present the analysis of the Yang–Mills algebra and of some related algebras based on the recent development of the theory of homogeneous algebras [2, 4]. This analysis is only partly published in [10].

In the next section we recall some basic concepts and results on homogeneous algebras which will be used in this paper.

Section 6.3 is devoted to the Yang–Mills algebra. In this section we recall the definitions and the results of [10]. The proofs are omitted since these are in [10] and since very similar proofs are given in Sections 4 and 6. Instead, we describe the structure of the bimodule resolution of the Yang–Mills algebra and the structure of the corresponding small bicomplexes which compute the Hochschild homology.

In Section 6.4 we define the super Yang–Mills algebra and we prove for this algebra results which are the counterpart of the results of [10] for the Yang–Mills algebra.

In Section 6.5 we define and study the super self-duality algebra. In particular, we prove for this algebra the analog of the results of [10] for the self-duality algebra and we point out a very surprising connection between the super self-duality algebra and the algebras occurring in our analysis of noncommutative 3-spheres [9, 11].

In Section 6.6 we describe some deformations of the Yang–Mills algebra and of the super Yang–Mills algebra.

6.2 Homogeneous algebras

Although we shall be concerned in the following with the cubic Yang–Mills algebra $\mathcal{A}$, the quadratic self-duality algebra $\mathcal{A}^{(+)}$ [10] and some related algebras, we recall in this section some constructions and some results for general N-homogeneous algebras [2, 4]. All vector spaces are over a fixed commutative field $\mathbb{K}$.

A *homogeneous algebra of degree N or N-homogeneous algebra* is an algebra of the form

$$\mathcal{A} = A(E, R) = T(E)/(R)$$

where E is a finite-dimensional vector space, R is a linear subspace of $E^{\otimes^N}$ and where (R) denotes the two-sided ideal of the tensor algebra $T(E)$ of E generated by R. The algebra $\mathcal{A}$ is naturally a connected graded algebra with graduation induced by the one of $T(E)$. To $\mathcal{A}$ is associated another N-homogeneous algebra, *its dual* $\mathcal{A}^! = A(E^*, R^\perp)$ with E^* denoting the dual vector space of E and $R^\perp \subset E^{\otimes^N *} = E^{*\otimes^N}$ being the annihilator of R, [4]. The N-complex $K(\mathcal{A})$ of left $\mathcal{A}$-modules is then defined to be

$$\cdots \stackrel{d}{\to} \mathcal{A} \otimes \mathcal{A}^{!*}_{n+1} \stackrel{d}{\to} \mathcal{A} \otimes \mathcal{A}^{!*}_n \stackrel{d}{\to} \cdots \stackrel{d}{\to} \mathcal{A} \to 0 \tag{6.2}$$

where $\mathcal{A}^{!*}_n$ is the dual vector space of the finite-dimensional vector space $\mathcal{A}^!_n$ of the elements of degree n of $\mathcal{A}^!$ and where $d : \mathcal{A} \otimes \mathcal{A}^{!*}_{n+1} \to \mathcal{A} \otimes \mathcal{A}^{!*}_n$ is induced by the map $a \otimes (e_1 \otimes \cdots \otimes e_{n+1}) \mapsto ae_1 \otimes (e_2 \otimes \cdots \otimes e_{n+1})$ of $\mathcal{A} \otimes E^{\otimes^{n+1}}$ into $\mathcal{A} \otimes E^{\otimes^n}$, remembering that $\mathcal{A}^{!*}_n \subset E^{\otimes^n}$, (see [4]). In (6.2) the factors $\mathcal{A}$

are considered as left $\mathcal{A}$-modules. By considering $\mathcal{A}$ as right $\mathcal{A}$-module and by exchanging the factors one obtains the N-complex $\tilde{K}(\mathcal{A})$ of right $\mathcal{A}$-modules

$$\cdots \xrightarrow{\tilde{d}} \mathcal{A}^{!*}_{n+1} \otimes \mathcal{A} \xrightarrow{\tilde{d}} \mathcal{A}^{!*}_n \otimes \mathcal{A} \xrightarrow{\tilde{d}} \cdots \xrightarrow{\tilde{d}} \mathcal{A} \to 0 \tag{6.3}$$

where now $\tilde{d}$ is induced by $(e_1 \otimes \cdots \otimes e_{n+1}) \otimes a \mapsto (e_1 \otimes \cdots \otimes e_n) \otimes e_{n+1}a$. Finally one defines two N-differentials $d_{\mathbf{L}}$ and $d_{\mathbf{R}}$ on the sequence of $(\mathcal{A}, \mathcal{A})$-bimodules, i.e. of left $\mathcal{A}\otimes\mathcal{A}^{opp}$-modules, $(\mathcal{A}\otimes\mathcal{A}^{!*}_n\otimes\mathcal{A})_{n\geq 0}$ by setting $d_{\mathbf{L}} = d\otimes I_{\mathcal{A}}$ and $d_{\mathbf{R}} = I_{\mathcal{A}} \otimes \tilde{d}$ where $I_{\mathcal{A}}$ is the identity mapping of $\mathcal{A}$ onto itself. For each of these N-differentials $d_{\mathbf{L}}$ and $d_{\mathbf{R}}$ the sequences

$$\cdots \overset{d_{\mathbf{L}},d_{\mathbf{R}}}{\longrightarrow} \mathcal{A} \otimes \mathcal{A}^{!*}_{n+1} \otimes \mathcal{A} \overset{d_{\mathbf{L}},d_{\mathbf{R}}}{\longrightarrow} \mathcal{A} \otimes \mathcal{A}^{!*}_n \otimes \mathcal{A} \overset{d_{\mathbf{L}},d_{\mathbf{R}}}{\longrightarrow} \cdots \tag{6.4}$$

are N-complexes of left $\mathcal{A} \otimes \mathcal{A}^{opp}$-modules and one has

$$d_{\mathbf{L}} d_{\mathbf{R}} = d_{\mathbf{R}} d_{\mathbf{L}} \tag{6.5}$$

which implies that

$$d^N_{\mathbf{L}} - d^N_{\mathbf{R}} = (d_{\mathbf{L}} - d_{\mathbf{R}}) \left(\sum_{p=0}^{N-1} d^p_{\mathbf{L}} d^{N-p-1}_{\mathbf{R}} \right) = \left(\sum_{p=0}^{N-1} d^p_{\mathbf{L}} d^{N-p-1}_{\mathbf{R}} \right) (d_{\mathbf{L}} - d_{\mathbf{R}}) = 0 \tag{6.6}$$

in view of $d^N_{\mathbf{L}} = d^N_{\mathbf{R}} = 0$.

As for any N-complex [13] one obtains from $K(\mathcal{A})$ ordinary complexes $C_{p,r}(K(\mathcal{A}))$, *the contractions of* $K(\mathcal{A})$, by putting together alternatively p and $N-p$ arrows d of $K(\mathcal{A})$. Explicitly $C_{p,r}(K(\mathcal{A}))$ is given by

$$\cdots \overset{d^{N-p}}{\longrightarrow} \mathcal{A} \otimes \mathcal{A}^{!*}_{Nk+r} \overset{d^p}{\longrightarrow} \mathcal{A} \otimes \mathcal{A}^{!*}_{Nk-p+r} \overset{d^{N-p}}{\longrightarrow} \mathcal{A} \otimes \mathcal{A}^{!*}_{N(k-1)+r} \overset{d^p}{\longrightarrow} \cdots \tag{6.7}$$

for $0 \leq r < p \leq N-1$, [4] . These are here chain complexes of free left $\mathcal{A}$-modules. As shown in [4] the complex $C_{N-1,0}(K(\mathcal{A}))$ coincides with the *Koszul complex* of [2]; this complex will be denoted by $\mathcal{K}(\mathcal{A}, \mathbb{K})$ in the sequel. That is one has

$$\mathcal{K}_{2m}(\mathcal{A}, \mathbb{K}) = \mathcal{A} \otimes \mathcal{A}^{!*}_{Nm}, \quad \mathcal{K}_{2m+1}(\mathcal{A}, \mathbb{K}) = \mathcal{A} \otimes \mathcal{A}^{!*}_{Nm+1} \tag{6.8}$$

for $m \geq 0$, and the differential is d^{N-1} on $\mathcal{K}_{2m}(\mathcal{A}, \mathbb{K})$ and d on $\mathcal{K}_{2m+1}(\mathcal{A}, \mathbb{K})$. If $\mathcal{K}(\mathcal{A}, \mathbb{K})$ is acyclic in positive degrees then $\mathcal{A}$ will be said to be a *Koszul algebra*. It was shown in [2] and this was confirmed by the analysis of [4] that this is the right generalization for N-homogeneous algebra of the usual notion of Koszulity for quadratic algebras [16, 17]. One always has $H_0(\mathcal{K}(\mathcal{A}, \mathbb{K})) \simeq \mathbb{K}$ and therefore if $\mathcal{A}$ is Koszul, then one has a free resolution $\mathcal{K}(\mathcal{A}, \mathbb{K}) \to \mathbb{K} \to 0$ of the trivial left $\mathcal{A}$-module $\mathbb{K}$, that is the exact sequence

$$\cdots \overset{d^{N-1}}{\longrightarrow} \mathcal{A} \otimes \mathcal{A}^{!*}_{N+1} \overset{d}{\to} \mathcal{A} \otimes R \overset{d^{N-1}}{\longrightarrow} \mathcal{A} \otimes E \overset{d}{\to} \mathcal{A} \overset{\varepsilon}{\to} \mathbb{K} \to 0 \tag{6.9}$$

of left $\mathcal{A}$-modules where ε is the projection on degree zero. This resolution is a minimal projective resolution of $\mathcal{A}$ in the graded category [3].

One defines now the chain complex of free $\mathcal{A}\otimes\mathcal{A}^{opp}$-modules $\mathcal{K}(\mathcal{A},\mathcal{A})$ by setting

$$\mathcal{K}_{2m}(\mathcal{A},\mathcal{A})=\mathcal{A}\otimes\mathcal{A}^{!*}_{Nm}\otimes\mathcal{A},\quad \mathcal{K}_{2m+1}(\mathcal{A},\mathcal{A})=\mathcal{A}\otimes\mathcal{A}^{!*}_{Nm+1}\otimes\mathcal{A} \tag{6.10}$$

for $m\in\mathbb{N}$ with differential δ' defined by

$$\delta'=d_{\mathbf{L}}-d_{\mathbf{R}}:\mathcal{K}_{2m+1}(\mathcal{A},\mathcal{A})\to\mathcal{K}_{2m}(\mathcal{A},\mathcal{A}) \tag{6.11}$$

$$\delta'=\sum_{p=0}^{N-1}d^p_{\mathbf{L}}d^{N-p-1}_{\mathbf{R}}:\mathcal{K}_{2(m+1)}(\mathcal{A},\mathcal{A})\to\mathcal{K}_{2m+1}(\mathcal{A},\mathcal{A}) \tag{6.12}$$

the property $\delta'^2=0$ following from (6.6). *This complex is acyclic in positive degrees if and only if $\mathcal{A}$ is Koszul*, that is if and only if $\mathcal{K}(\mathcal{A},\mathbb{K})$ is acyclic in positive degrees, [2] and [4] . One always has the obvious exact sequence

$$\mathcal{A}\otimes E\otimes\mathcal{A}\xrightarrow{\delta'}\mathcal{A}\otimes\mathcal{A}\xrightarrow{\mu}\mathcal{A}\to 0 \tag{6.13}$$

of left $\mathcal{A}\otimes\mathcal{A}^{opp}$-modules where μ denotes the product of $\mathcal{A}$. It follows that if $\mathcal{A}$ is a Koszul algebra, then $\mathcal{K}(\mathcal{A},\mathcal{A})\xrightarrow{\mu}\mathcal{A}\to 0$ is a free resolution of the $\mathcal{A}\otimes\mathcal{A}^{opp}$-module $\mathcal{A}$ which will be refered to as *the Koszul resolution of $\mathcal{A}$*. This is a minimal projective resolution of $\mathcal{A}\otimes\mathcal{A}^{opp}$ in the graded category [3].

Let $\mathcal{A}$ be a Koszul algebra and let $\mathcal{M}$ be an $(\mathcal{A},\mathcal{A})$-bimodule considered as a right $\mathcal{A}\otimes\mathcal{A}^{opp}$-module. Then, by interpreting the $\mathcal{M}$-valued Hochschild homology $H(\mathcal{A},\mathcal{M})$ as $H_n(\mathcal{A},\mathcal{M})=\mathrm{Tor}_n^{\mathcal{A}\otimes\mathcal{A}^{opp}}(\mathcal{M},\mathcal{A})$ [6], the complex $\mathcal{M}\otimes_{\mathcal{A}\otimes\mathcal{A}^{opp}}\mathcal{K}(\mathcal{A},\mathcal{A})$ computes the $\mathcal{M}$-valued Hochschild homology of $\mathcal{A}$, (i.e. its homology is the ordinary $\mathcal{M}$-valued Hochschild homology of $\mathcal{A}$). We shall refer to this complex as *the small Hochschild complex of $\mathcal{A}$* with coefficients in $\mathcal{M}$ and denote it by $\mathcal{S}(\mathcal{A},\mathcal{M})$. It reads

$$\cdots\xrightarrow{\delta}\mathcal{M}\otimes\mathcal{A}^{!*}_{N(m+1)}\xrightarrow{\delta}\mathcal{M}\otimes\mathcal{A}^{!*}_{Nm+1}\xrightarrow{\delta}\mathcal{M}\otimes\mathcal{A}^{!*}_{Nm}\xrightarrow{\delta}\cdots \tag{6.14}$$

where δ is obtained from δ' by applying the factors d_L to the right of $\mathcal{M}$ and the factors d_R to the left of $\mathcal{M}$.

Assume that $\mathcal{A}$ is a Koszul algebra of finite global dimension D. Then the Koszul resolution of $\mathbb{K}$ has length D, i.e. D is the largest integer such that $\mathcal{K}_D(\mathcal{A},\mathbb{K})\neq 0$. By construction, D is also the greatest integer such that $\mathcal{K}_D(\mathcal{A},\mathcal{A})\neq 0$ so the free $\mathcal{A}\otimes\mathcal{A}^{opp}$-module resolution of $\mathcal{A}$ has also length D. Thus *for a Koszul algebra, the global dimension is equal to the Hochschild dimension.* Applying then the functor $\mathrm{Hom}_{\mathcal{A}}(\bullet,\mathcal{A})$ to $\mathcal{K}(\mathcal{A},\mathbb{K})$ one obtains the cochain complex $\mathcal{L}(\mathcal{A},\mathbb{K})$ of free right $\mathcal{A}$-modules

$$0\to\mathcal{L}^0(\mathcal{A},\mathbb{K})\to\cdots\to\mathcal{L}^D(\mathcal{A},\mathbb{K})\to 0$$

where $\mathcal{L}^n(\mathcal{A},\mathbb{K}) = \mathrm{Hom}_{\mathcal{A}}(\mathcal{K}_n(\mathcal{A},\mathbb{K}),\mathcal{A})$. The Koszul algebra $\mathcal{A}$ is *Gorenstein* iff $H^n(\mathcal{L}(\mathcal{A},\mathbb{K})) = 0$ for $n < D$ and $H^D(\mathcal{L}(\mathcal{A},\mathbb{K})) = \mathbb{K}$ (= the trivial right $\mathcal{A}$-module). This is clearly a generalisation of the classical Poincaré duality and this implies a precise form of Poincaré duality between Hochschild homology and Hochschild cohomology [5, 20, 21]. In the case of the Yang–Mills algebra and its deformations which are Koszul Gorenstein cubic algebras of global dimension 3, this Poincaré duality gives isomorphisms

$$H_k(\mathcal{A},\mathcal{M}) = H^{3-k}(\mathcal{A},\mathcal{M}), \quad k \in \{0,1,2,3\} \tag{6.15}$$

between the Hochschild homology and the Hochschild cohomology with coefficients in a bimodule $\mathcal{M}$.

6.3 The Yang–Mills algebra

Let $(g_{\lambda\mu}) \in M_{s+1}(\mathbb{K})$ be an invertible symmetric $(s+1)\times(s+1)$-matrix with inverse $(g^{\lambda\mu})$, i.e. $g_{\lambda\mu}g^{\mu\nu} = \delta^\nu_\lambda$. The *Yang–Mills algebra* is the cubic algebra $\mathcal{A}$ generated by $s+1$ elements ∇_λ ($\lambda \in \{0,\dots,s\}$) with the $s+1$ relations

$$g^{\lambda\mu}[\nabla_\lambda,[\nabla_\mu,\nabla_\nu]] = 0, \quad \nu \in \{0,\dots,s\}$$

that is $\mathcal{A} = A(E,R)$ with $E = \oplus_\lambda \mathbb{K}\nabla_\lambda$ and $R \subset E^{\otimes^3}$ given by

$$\begin{aligned} R &= \sum_\nu \mathbb{K} g^{\lambda\mu}[\nabla_\lambda,[\nabla_\mu,\nabla_\nu]_\otimes]_\otimes \\ &= \sum_\rho \mathbb{K}(g^{\rho\lambda}g^{\mu\nu} + g^{\nu\rho}g^{\lambda\mu} - 2g^{\rho\mu}g^{\lambda\nu})\nabla_\lambda\otimes\nabla_\mu\otimes\nabla_\nu. \end{aligned} \tag{6.16}$$

In [10] the following theorem was proved.

Theorem 1. *The cubic Yang–Mills algebra $\mathcal{A}$ is Koszul of global dimension 3 and is Gorenstein.*

The proof of this theorem relies on the computation of the dual cubic algebra $\mathcal{A}^!$ which we now recall.

The dual $\mathcal{A}^! = A(E^*, R^\perp)$ of the Yang–Mills algebra is the cubic algebra generated by $s+1$ elements θ^λ ($\lambda \in \{0,\dots,s\}$) with relations

$$\theta^\lambda\theta^\mu\theta^\nu = \frac{1}{s}(g^{\lambda\mu}\theta^\nu + g^{\mu\nu}\theta^\lambda - 2g^{\lambda\nu}\theta^\mu)\mathbf{g}$$

where $\mathbf{g} = g_{\alpha\beta}\theta^\alpha\theta^\beta$. These relations imply that $\mathbf{g} \in \mathcal{A}^!_2$ is central in $\mathcal{A}^!$ and that one has $\mathcal{A}^!_0 = \mathbb{K}\mathbb{1} \simeq \mathbb{K}$, $\mathcal{A}^!_1 = \oplus_\lambda\mathbb{K}\theta^\lambda \simeq \mathbb{K}^{s+1}$, $\mathcal{A}^!_2 = \oplus_{\mu\nu}\mathbb{K}\theta^\mu\theta^\nu \simeq \mathbb{K}^{(s+1)^2}$, $\mathcal{A}^!_3 = \oplus_\lambda\mathbb{K}\theta^\lambda\mathbf{g} \simeq \mathbb{K}^{s+1}$, $\mathcal{A}^!_4 = \mathbb{K}\mathbf{g}^2 \simeq \mathbb{K}$ and $\mathcal{A}^!_n = 0$ for $n \geq 5$. From this, one obtains the description of [10] of the Koszul complex $\mathcal{K}(\mathcal{A},\mathbb{K})$ and the proof of the above theorem. It also follows that the bimodule resolution $\mathcal{K}(\mathcal{A},\mathcal{A}) \stackrel{\mu}{\to} \mathcal{A} \to 0$ of $\mathcal{A}$ reads

$$0 \to \mathcal{A}\otimes\mathcal{A} \stackrel{\delta'_3}{\to} \mathcal{A}\otimes\mathbb{K}^{s+1}\otimes\mathcal{A} \stackrel{\delta'_2}{\to} \mathcal{A}\otimes\mathbb{K}^{s+1}\otimes\mathcal{A} \stackrel{\delta'_1}{\to} \mathcal{A}\otimes\mathcal{A} \stackrel{\mu}{\to} \mathcal{A} \to 0 \quad (6.17)$$

where the components δ'_k of δ' in the different degrees can be computed by using the description of $\mathcal{K}(\mathcal{A},\mathbb{K}) = C_{2,0}$ given in Section 3 of [10] and are given by

$$\begin{cases} \delta'_1(a\otimes e_\lambda\otimes b) = a\nabla_\lambda\otimes b - a\otimes\nabla_\lambda b \\ \delta'_2(a\otimes e_\lambda\otimes b) = (g^{\alpha\beta}\delta^\gamma_\lambda + g^{\beta\gamma}g^\alpha_\lambda - 2g^{\gamma\alpha}\delta^\beta_\lambda)\times \\ \qquad \times(a\nabla_\alpha\nabla_\beta\otimes e_\gamma\otimes b + a\nabla_\alpha\otimes e_\gamma\otimes\nabla_\beta b + a\otimes e_\gamma\otimes\nabla_\alpha\nabla_\beta b) \\ \delta'_3(a\otimes b) = g^{\lambda\mu}(a\nabla_\mu\otimes e_\lambda\otimes b - a\otimes e_\lambda\otimes\nabla_\mu b) \end{cases} \quad (6.18)$$

where $a,b\in\mathcal{A}$, e_λ $(\lambda = 0,\dots,s)$ is the canonical basis of $\mathbb{K}^{s+1}$ and ∇_λ are the corresponding generators of $\mathcal{A}$.

Let $\mathcal{M}$ be a bimodule over $\mathcal{A}$. By using the above description of the Koszul resolution of $\mathcal{A}$ one easily obtains the one of the small Hochschild complex $\mathcal{S}(\mathcal{A},\mathcal{M})$ which reads

$$0 \to \mathcal{M} \stackrel{\delta_3}{\to} \mathcal{M}\otimes\mathbb{K}^{s+1} \stackrel{\delta_2}{\to} \mathcal{M}\otimes\mathbb{K}^{s+1} \stackrel{\delta_1}{\to} \mathcal{M} \to 0 \quad (6.19)$$

with differential δ given by

$$\begin{cases} \delta_1(m^\lambda\otimes e_\lambda) = m^\lambda\nabla_\lambda - \nabla_\lambda m^\lambda = [m^\lambda,\nabla_\lambda] \\ \delta_2(m^\lambda\otimes e_\lambda) \\ \quad = ([\nabla_\mu,[\nabla^\mu,m^\lambda]] + [\nabla_\mu,[m^\mu,\nabla^\lambda]] + [m^\mu,[\nabla_\mu,\nabla^\lambda]])\otimes e_\lambda \\ \delta_3(m) = g^{\lambda\mu}(m\nabla_\mu - \nabla_\mu m)\otimes e_\lambda = [m,\nabla^\lambda]\otimes e_\lambda \end{cases} \quad (6.20)$$

with obvious notation. By using (6.20) one easily verifies the duality (6.15). For instance $H_3(\mathcal{A},\mathcal{M})$ is $\mathrm{Ker}(\delta_3)$ which is given by the $m\in\mathcal{M}$ such that $\nabla_\lambda m = m\nabla_\lambda$ for $\lambda = 0,\dots,s$, that is such that $am = ma$, $\forall a\in\mathcal{A}$, since $\mathcal{A}$ is generated by the ∇_λ and it is well known that this coincides with $H^0(\mathcal{A},\mathcal{M})$. Similarily $m^\lambda\otimes e_\lambda$ is in $\mathrm{Ker}(\delta_2)$ if and only if $\nabla_\lambda\mapsto D(\nabla_\lambda) = g_{\lambda\mu}m^\mu$ extends as a derivation D of $\mathcal{A}$ into $\mathcal{M}$ ($D\in\mathrm{Der}(\mathcal{A},\mathcal{M})$) while $m^\lambda\otimes e_\lambda = \delta_3(m)$ means that this derivation is inner $D = ad(m)\in\mathrm{Int}(\mathcal{A},\mathcal{M})$ from which $H_2(\mathcal{A},\mathcal{M})$ identifies with $H^1(\mathcal{A},\mathcal{M})$, and so on.

Assume now that $\mathcal{M}$ is graded in the sense that one has $\mathcal{M} = \oplus_{n\in\mathbb{Z}}\mathcal{M}_n$ with $\mathcal{A}_k\mathcal{M}_\ell\mathcal{A}_m\subset\mathcal{M}_{k+\ell+m}$. Then the small Hochschild complex $\mathcal{S}(\mathcal{A},\mathcal{M})$ splits into subcomplexes $\mathcal{S}(\mathcal{A},\mathcal{M}) = \oplus_n\mathcal{S}^{(n)}(\mathcal{A},\mathcal{M})$ where $\mathcal{S}^{(n)}(\mathcal{A},\mathcal{M})$ is the subcomplex

$$0 \to \mathcal{M}_{n-4} \overset{\delta_3}{\to} \mathcal{M}_{n-3} \otimes \mathbb{K}^{s+1} \overset{\delta_2}{\to} \mathcal{M}_{n-1} \otimes \mathbb{K}^{s+1} \overset{\delta_1}{\to} \mathcal{M}_n \to 0 \tag{6.21}$$

of (6.19). Assume furthermore that the homogeneous components $\mathcal{M}_n$ are finite-dimensional vector spaces, i.e. $\dim(\mathcal{M}_n) \in \mathbb{N}$. Then one has the following Euler–Poincaré formula

$$\begin{aligned} &\dim(H_0^{(n)}) - \dim(H_1^{(n)}) + \dim(H_2^{(n)}) - \dim(H_3^{(n)}) = \\ &\dim(\mathcal{M}_n) - (s+1)\dim(\mathcal{M}_{n-1}) + (s+1)\dim(\mathcal{M}_{n-3}) - \dim(\mathcal{M}_{n-4}) \end{aligned} \tag{6.22}$$

for the homology $H^{(n)}$ of the chain complex $\mathcal{S}^{(n)}(\mathcal{A}, \mathcal{M})$.

In the case where $\mathcal{M} = \mathcal{A}$, it follows from the Koszulity of $\mathcal{A}$ that the right-hand side of (6.22) vanishes for $n \neq 0$. Denoting as usual by $HH(\mathcal{A})$ the $\mathcal{A}$-valued Hochschild homology of $\mathcal{A}$ which is here the homology of $\mathcal{S}(\mathcal{A}, \mathcal{A})$, we denote by $HH^{(n)}(\mathcal{A})$ the homology of the subcomplex $\mathcal{S}^{(n)}(\mathcal{A}, \mathcal{A})$. Since $\mathcal{A}_n = 0$ for $n < 0$, one has $HH_0^{(n)}(\mathcal{A}) = 0$ for $n < 0$, $HH_1^{(n)}(\mathcal{A}) = 0$ for $n \leq 0$, $HH_2^{(n)}(\mathcal{A}) = 0$ for $n \leq 2$ and $HH_3^{(n)}(\mathcal{A}) = 0$ for $n \leq 3$. Furthermore one has

$$HH_0^{(0)}(\mathcal{A}) = HH_3^{(4)}(\mathcal{A}) = \mathbb{K}, \tag{6.23}$$

$$HH_0^{(1)}(\mathcal{A}) = HH_1^{(1)}(\mathcal{A}) = HH_2^{(3)}(\mathcal{A}) = \mathbb{K}^{s+1}, \tag{6.24}$$

$$HH_0^{(2)}(\mathcal{A}) = HH_1^{(2)}(\mathcal{A}) = \mathbb{K}^{\frac{(s+1)(s+2)}{2}}, \tag{6.25}$$

and the Euler–Poincaré formula reads here

$$\dim(HH_0^{(n)}(\mathcal{A})) + \dim(HH_2^{(n)}(\mathcal{A})) = \dim(HH_1^{(n)}(\mathcal{A})) + \dim(HH_3^{(n)}(\mathcal{A})) \tag{6.26}$$

for $n \geq 1$ which implies

$$\dim(HH_0^{(3)}(\mathcal{A})) + (s+1) = \dim(HH_1^{(3)}(\mathcal{A})) \tag{6.27}$$

for $n = 3$ while for $n = 1$ and $n = 2$ it is already contained in (6.24) and (6.25).

The complete description of the Hochschild homology and of the cyclic homology of the Yang–Mills algebra will be given in [12].

6.4 The super Yang–Mills algebra

As pointed out in the introduction, the Yang–Mills algebra is the universal enveloping algebra of a Lie algebra which is graded by giving degree 1 to the generators ∇_λ (see in [10]). Replacing the Lie bracket by a super Lie bracket, that is replacing in the Yang–Mills equations (6.1) the commutator by the anticommutator whenever the 2 elements are of odd degrees, one obtains a super version $\tilde{\mathcal{A}}$ of the Yang–Mills algebra $\mathcal{A}$. In other words one defines the *super Yang–Mills algebra* to be the cubic algebra $\tilde{\mathcal{A}}$ generated $s+1$ elements S_λ ($\lambda \in \{0,\dots,s\}$) with the relations

$$g^{\lambda\mu}[S_\lambda, \{S_\mu, S_\nu\}] = 0, \quad \nu \in \{0,\dots,s\} \tag{6.28}$$

that is $\tilde{\mathcal{A}} = \mathcal{A}(\tilde{E},\tilde{R})$ with $\tilde{E} = \oplus_\lambda \mathbb{K} S_\lambda$ and $\tilde{R} \subset \tilde{E}^{\otimes^3}$ given by

$$\tilde{R} = \sum_\rho \mathbb{K}(g^{\rho\lambda}g^{\mu\nu} - g^{\nu\rho}g^{\lambda\mu}) S_\lambda \otimes S_\mu \otimes S_\nu. \tag{6.29}$$

Relations (6.28) can be equivalently written as

$$[g^{\lambda\mu}S_\lambda S_\mu, S_\nu] = 0, \quad \nu \in \{0,\dots,s\} \tag{6.30}$$

which mean that $g^{\lambda\mu}S_\lambda S_\mu \in \tilde{\mathcal{A}}_2$ is central in $\tilde{\mathcal{A}}$.

It is easy to verify that the dual algebra $\tilde{\mathcal{A}}^! = \mathcal{A}(\tilde{E}^*, \tilde{R}^\perp)$ is the cubic algebra generated by $s+1$ elements ξ^λ ($\lambda \in \{0,\dots,s\}$) with the relations

$$\xi^\lambda \xi^\mu \xi^\nu = -\frac{1}{s}(g^{\lambda\mu}\xi^\nu - g^{\mu\nu}\xi^\lambda)\mathbf{g}$$

where $\mathbf{g} = g_{\alpha\beta}\xi^\alpha\xi^\beta$. These relations imply that $\mathbf{g}\xi^\nu + \xi^\nu \mathbf{g} = 0$, i.e.

$$\{g_{\lambda\mu}\xi^\lambda\xi^\mu, \xi^\nu\} = 0, \quad \nu \in \{0,\dots,s\} \tag{6.31}$$

and that one has $\tilde{\mathcal{A}}^!_0 = \mathbb{K}\mathbb{1} \simeq \mathbb{K}$, $\tilde{\mathcal{A}}^!_1 = \oplus_\lambda \mathbb{K}\xi^\lambda \simeq \mathbb{K}^{s+1}$, $\tilde{\mathcal{A}}^!_2 = \oplus_{\mu\nu}\mathbb{K}\xi^\mu\xi^\nu \simeq \mathbb{K}^{(s+1)^2}$, $\tilde{\mathcal{A}}^!_3 = \oplus_\lambda \mathbb{K}\xi^\lambda \mathbf{g} \simeq \mathbb{K}^{s+1}$, $\tilde{\mathcal{A}}^!_4 = \mathbb{K}\mathbf{g}^2 \simeq \mathbb{K}$ and $\tilde{\mathcal{A}}^!_n = 0$ for $n \geq 5$.

The Koszul complex $\mathcal{K}(\tilde{\mathcal{A}}, \mathbb{K})$ of $\tilde{\mathcal{A}}$ then reads

$$0 \to \tilde{\mathcal{A}} \xrightarrow{S^t} \tilde{\mathcal{A}}^{s+1} \xrightarrow{N} \tilde{\mathcal{A}}^{s+1} \xrightarrow{S} \tilde{\mathcal{A}} \to 0$$

where S means right multiplication by the column with components S_λ, S^t means right multiplication by the row with components S_λ and N means right multiplication (matrix product) by the matrix with components

$$N^{\mu\nu} = (g^{\mu\nu}g^{\alpha\beta} - g^{\mu\alpha}g^{\nu\beta})S_\alpha S_\beta$$

with $\lambda, \mu, \nu \in \{0,\dots,s\}$. One has the following result.

Theorem 2. *The cubic super Yang–Mills algebra $\tilde{\mathcal{A}}$ is Koszul of global dimension 3 and is Gorenstein.*

Proof. By the very definition of $\tilde{\mathcal{A}}$ by generators and relations, the sequence

$$\tilde{\mathcal{A}}^{s+1} \xrightarrow{N} \tilde{\mathcal{A}}^{s+1} \xrightarrow{S} \tilde{\mathcal{A}} \xrightarrow{\varepsilon} \mathbb{K} \to 0$$

is exact. On the other hand it is easy to see that the mapping $\tilde{\mathcal{A}} \overset{S^t}{\to} \tilde{\mathcal{A}}^{s+1}$ is injective and that the sequence

$$0 \to \tilde{\mathcal{A}} \overset{S^t}{\to} \tilde{\mathcal{A}}^{s+1} \overset{N}{\to} \tilde{\mathcal{A}}^{s+1} \overset{S}{\to} \tilde{\mathcal{A}} \overset{\varepsilon}{\to} \mathbb{K} \to 0$$

is exact which implies that $\tilde{\mathcal{A}}$ is Koszul of global dimension 3. The Gorenstein property follows from the symmetry by transposition. □

The situation is completely similar to the Yang–Mills case, in particular $\tilde{\mathcal{A}}$ has Hochschild dimension 3 and, by applying a result of [15], $\tilde{\mathcal{A}}$ has the same Poincaré series as $\mathcal{A}$ i.e. one has the formula

$$\sum_{n\in\mathbb{N}} \dim(\tilde{\mathcal{A}}_n)t^n = \frac{1}{(1-t^2)(1-(s+1)t+t^2)} \tag{6.32}$$

which, as will be shown elsewhere, can be interpreted in terms of the quantum group of the bilinear form $(g_{\mu\nu})$ [14] by noting the invariance of Relations (6.30) by this quantum group. For $s=1$ the Yang–Mills algebra and the super Yang–Mills algebra are particular cubic Artin–Schelter algebras [1] whereas for $s \geq 2$ these algebras have exponential growth as follows from Formula (6.32).

6.5 The super self-duality algebra

There are natural quotients $\mathcal{B}$ of $\mathcal{A}$ and $\tilde{\mathcal{B}}$ of $\tilde{\mathcal{A}}$ which are connected with parastatistics and which have been investigated in [15]. The *parafermionic algebra* $\mathcal{B}$ is the cubic algebra generated by elements ∇_λ ($\lambda \in \{0,\dots,s\}$) with relations

$$[\nabla_\lambda, [\nabla_\mu, \nabla_\nu]] = 0$$

for any $\lambda, \mu, \nu \in \{0,\dots,s\}$, while the *parabosonic algebra* $\tilde{\mathcal{B}}$ is the cubic algebra generated by elements S_λ ($\lambda \in \{0,\dots,s\}$) with relations

$$[S_\lambda, \{S_\mu, S_\nu\}] = 0$$

for any $\lambda, \mu, \nu \in \{0,\dots,s\}$. In contrast to the Yang–Mills and the super Yang–Mills algebras $\mathcal{A}$ and $\tilde{\mathcal{A}}$ which have exponential growth whenever $s \geq 2$, these algebras $\mathcal{B}$ and $\tilde{\mathcal{B}}$ have polynomial growth with Poincaré series given by

$$\sum_n \dim(\mathcal{B}_n)t^n = \sum_n \dim(\tilde{\mathcal{B}}_n)t^n = \left(\frac{1}{1-t}\right)^{s+1} \left(\frac{1}{1-t^2}\right)^{\frac{s(s+1)}{2}}$$

but they are not Koszul for $s \geq 2$, [15].

In a sense, the algebra $\mathcal{B}$ can be considered to be somehow trivial from the point of view of the classical Yang–Mills equations in dimension $s+1 \geq 3$

although the algebras $\mathcal{B}$ and $\tilde{\mathcal{B}}$ are quite interesting for other purposes [15]. It turns out that in dimension $s+1=4$ with $g_{\mu\nu}=\delta_{\mu\nu}$ (Euclidean case), the Yang–Mills algebra $\mathcal{A}$ has non-trivial quotients $\mathcal{A}^{(+)}$ and $\mathcal{A}^{(-)}$ which are quadratic algebras refered to as the *self-duality algebra* and the *anti-self-duality algebra* respectively [10]. Let $\varepsilon = \pm$, the algebra $\mathcal{A}^{(\varepsilon)}$ is the quadratic algebra generated by the elements ∇_λ ($\lambda \in \{0,1,2,3,\}$) with relations

$$[\nabla_0, \nabla_k] = \varepsilon[\nabla_\ell, \nabla_m]$$

for any cyclic permutation (k,ℓ,m) of (1,2,3). One passes from $\mathcal{A}^{(-)}$ to $\mathcal{A}^{(+)}$ by changing the orientation of $\mathbb{K}^4$ so one can restrict attention to the self-duality algebra $\mathcal{A}^{(+)}$. This algebra has been studied in [10] where it was shown in particular that it is Koszul of global dimension 2. For further details on this algebra and on the Yang–Mills algebra, we refer to [10] and to the forthcoming paper [12]. Our aim now in this section is to define and study the super version of the self-duality algebra.

Let $\varepsilon = +$ or $-$ and define $\tilde{\mathcal{A}}^{(\varepsilon)}$ to be the quadratic algebra generated by the elements S_0, S_1, S_2, S_3 with relations

$$i\{S_0, S_k\} = \varepsilon[S_\ell, S_m] \tag{6.33}$$

for any cyclic permutation (k,ℓ,m) of (1,2,3). One has the following.

Lemma 1. *Relations* (6.33) *imply that one has*

$$[\sum_{\mu=0}^{3}(S_\mu)^2, S_\lambda] = 0$$

for any $\lambda \in \{0,1,2,3\}$. *In other words,* $\tilde{\mathcal{A}}^{(+)}$ *and* $\tilde{\mathcal{A}}^{(-)}$ *are quotients of the super Yang–Mills algebra* $\tilde{\mathcal{A}}$ *for* $s+1=4$ *and* $g_{\mu\nu}=\delta_{\mu\nu}$.

The proof which is a straightforward verification makes use of the Jacobi identity (see also in [18]). Thus $\tilde{\mathcal{A}}^{(+)}$ and $\tilde{\mathcal{A}}^{(-)}$ play the same role with respect to $\tilde{\mathcal{A}}$ as $\mathcal{A}^{(+)}$ and $\mathcal{A}^{(-)}$ with respect to $\mathcal{A}$. Accordingly they will be respectively called the *super self-duality algebra* and the *super anti-self-duality algebra*. Again $\tilde{\mathcal{A}}^{(+)}$ and $\tilde{\mathcal{A}}^{(-)}$ are exchanged by changing the orientation of $\mathbb{K}^4$ and we shall restrict attention to the super self-duality algebra in the following, i.e. to the quadratic algebra $\tilde{\mathcal{A}}^{(+)}$ generated by S_0, S_1, S_2, S_3 with relations

$$i\{S_0, S_k\} = [S_\ell, S_m] \tag{6.34}$$

for any cyclic permutation (k,ℓ,m) of (1,2,3). One has the following result.

Theorem 3. *The quadratic super self-duality algebra* $\tilde{\mathcal{A}}^{(+)}$ *is a Koszul algebra of global dimension* 2.

Proof. One verifies that the dual quadratic algebra $\tilde{\mathcal{A}}^{(+)!}$ is generated by elements $\xi^0, \xi^1, \xi^2, \xi^3$ with relations $(\xi^\lambda)^2 = 0$, for $\lambda = 0, 1, 2, 3$ and $\xi^\ell \xi^m = -\xi^m \xi^\ell = i\xi^0\xi^k = i\xi^k\xi^0$ for any cyclic permutation (k, ℓ, m) of (1,2,3). So one has $\tilde{\mathcal{A}}_0^{(+)!} = \mathbb{K}\mathbb{1} \simeq \mathbb{K}$, $\tilde{\mathcal{A}}_1^{(+)!} = \oplus_\lambda \mathbb{K}\xi^\lambda \simeq \mathbb{K}^4$, $\tilde{\mathcal{A}}_2^{(+)!} = \oplus_k \mathbb{K}\xi^0\xi^k \simeq \mathbb{K}^3$ and $\tilde{\mathcal{A}}_n^{(+)!} = 0$ for $n \geq 3$ since the above relations imply $\xi^\lambda\xi^\mu\xi^\nu = 0$ for any $\lambda, \mu, \nu \in \{0, 1, 2, 3\}$. The Koszul complex $K(\tilde{\mathcal{A}}^{(+)}) = \mathcal{K}(\tilde{\mathcal{A}}^{(+)}, \mathbb{K})$ (quadratic case) then reads

$$0 \to \tilde{\mathcal{A}}^{(+)^3} \xrightarrow{D} \tilde{\mathcal{A}}^{(+)^4} \xrightarrow{S} \tilde{\mathcal{A}}^{(+)} \to 0$$

where S means right matrix product with the column with components S_λ ($\lambda \in \{0, 1, 2, 3\}$) and D means right matrix product with

$$D = \begin{pmatrix} iS_1 & iS_0 & S_3 & -S_2 \\ iS_2 & -S_3 & iS_0 & S_1 \\ iS_3 & S_2 & -S_1 & iS_0 \end{pmatrix}. \tag{6.35}$$

It follows from the definition of $\tilde{\mathcal{A}}^{(+)}$ by generators and relations that the sequence

$$\tilde{\mathcal{A}}^{(+)^3} \xrightarrow{D} \tilde{\mathcal{A}}^{(+)^4} \xrightarrow{S} \tilde{\mathcal{A}}^{(+)} \xrightarrow{\varepsilon} \mathbb{K} \to 0$$

is exact. On the other hand one shows easily that the mapping $\tilde{\mathcal{A}}^{(+)^3} \xrightarrow{D} \tilde{\mathcal{A}}^{(+)^4}$ is injective so finally the sequence

$$0 \to \tilde{\mathcal{A}}^{(+)3} \xrightarrow{D} \tilde{\mathcal{A}}^{(+)^4} \xrightarrow{S} \tilde{\mathcal{A}}^{(+)} \xrightarrow{\varepsilon} \mathbb{K} \to 0$$

is exact which implies the result. □

This theorem implies that the super self-duality algebra $\tilde{\mathcal{A}}^{(+)}$ has Hochschild dimension 2 and that its Poincaré series is given by

$$P_{\tilde{\mathcal{A}}^{(+)}}(t) = \frac{1}{(1-t)(1-3t)}$$

in view of the structure of its dual $\tilde{\mathcal{A}}^{(+)!}$ described in the proof. Thus everything is similar to the case of the self-duality algebra $\mathcal{A}^{(+)}$.

Let us recall that the *Sklyanin algebra*, in the presentation given by Sklyanin [18], is the quadratic algebra $\mathcal{S}(\alpha_1, \alpha_2, \alpha_3)$ generated by four elements S_0, S_1, S_2, S_3 with relations

$$i\{S_0, S_k\} = [S_\ell, S_m]$$

$$[S_0, S_k] = i\frac{\alpha_\ell - \alpha_m}{\alpha_k}\{S_\ell, S_m\}$$

for any cyclic permutation (k, ℓ, m) of (1,2,3). One sees that the relations of the super self-duality algebra $\tilde{\mathcal{A}}^{(+)}$ are the relations of the Sklyanin algebra which are independent from the parameters α_k. Thus one has a sequence of surjective homomorphisms of connected graded algebra

$$\tilde{\mathcal{A}} \to \tilde{\mathcal{A}}^{(+)} \to \mathcal{S}(\alpha_1, \alpha_2, \alpha_3).$$

On the other hand for *generic values of the parameters* the Sklyanin algebra is Koszul Gorenstein of global dimension 4 [19] with the same Poincaré series as the polynomial algebra $\mathbb{K}[X_0, X_1, X_2, X_3]$ and corresponds to the natural ambiant noncommutative 4-dimensional Euclidean space containing the noncommutative 3-spheres described in [9, 11] (their "homogeneisation"). This gives a very surprising connection between the present study and our noncommutative 3-spheres for generic values of the parameters. It is worth noticing here that in the analysis of [11] several bridges between noncommutative differential geometry in the sense of [7,8] and noncommutative algebraic geometry have been established.

6.6 Deformations

The aim of this section is to study deformations of the Yang–Mills algebra and of the super Yang–Mills algebra. We use the notation of Sections 6.3 and 6.4.

Let the dimension $s+1 \geq 2$ and the pseudo metric $g_{\lambda\mu}$ be fixed and let $\zeta \in P_1(\mathbb{K})$ have homogeneous coordinates $\zeta_0, \zeta_1 \in \mathbb{K}$. Define $\mathcal{A}(\zeta)$ to be the cubic algebra generated by $s+1$ elements ∇_λ ($\lambda \in \{0, \dots, s\}$) with relations

$$(\zeta_1(g^{\rho\lambda}g^{\mu\nu} + g^{\nu\rho}g^{\lambda\mu}) - 2\zeta_0 g^{\rho\mu}g^{\lambda\nu})\nabla_\lambda\nabla_\mu\nabla_\nu = 0$$

for $\rho \in \{0, \dots, s\}$. The Yang–Mills algebra corresponds to the element ζ^{YM} of $P_1(\mathbb{K})$ with homogeneous coordinates $\zeta_0 = \zeta_1$. Let ζ^{sing} be the element of $P_1(\mathbb{K})$ with homogeneous coordinates $\zeta_0 = \frac{s+2}{2}\zeta_1$; one has the following result.

Theorem 4. *For $\zeta \neq \zeta^{sing}$ the cubic algebra $\mathcal{A}(\zeta)$ is Koszul of global dimension 3 and is Gorenstein.*

Proof. The dual algebra $\mathcal{A}(\zeta)^!$ is the cubic algebra generated by elements θ^λ with relations

$$\theta^\lambda\theta^\mu\theta^\nu = \frac{1}{(s+2)\zeta_1 - 2\zeta_0}(\zeta_1(g^{\lambda\mu}\theta^\nu + g^{\mu\nu}\theta^\lambda) - 2\zeta_0 g^{\lambda\nu}\theta^\mu)\mathbf{g} \qquad (6.36)$$

for $\lambda, \mu, \nu \in \{0, \dots, s\}$ with $\mathbf{g} = g_{\alpha\beta}\theta^\alpha\theta^\beta$. This again implies that $\mathbf{g}$ is in the center and that one has $\mathcal{A}^!_0 = \mathbb{K}\mathbb{1} \simeq \mathbb{K}$, $\mathcal{A}^!_1 = \oplus_\lambda \mathbb{K}\theta^\lambda \simeq \mathbb{K}^{s+1}$, $\mathcal{A}^!_2 = \oplus_{\lambda,\mu}\mathbb{K}\theta^\lambda\theta^\mu \simeq \mathbb{K}^{(s+1)^2}$, $\mathcal{A}^!_3 = \oplus_\lambda \mathbb{K}\theta^\lambda\mathbf{g} \simeq \mathbb{K}^{s+1}$, $\mathcal{A}^!_4 = \mathbb{K}\mathbf{g}^2 \simeq \mathbb{K}$ while $\mathcal{A}^!_n = 0$ for $n \geq 5$, where we have set $\mathcal{A}^!_n = \mathcal{A}(\zeta)^!_n$. Setting $\mathcal{A} = \mathcal{A}(\zeta)$, the Koszul complex $\mathcal{K}(\mathcal{A}(\zeta), \mathbb{K})$ of $\mathcal{A}(\zeta)$ reads

$$0 \to \mathcal{A} \overset{\nabla^t}{\to} \mathcal{A}^{s+1} \overset{M}{\to} \mathcal{A}^{s+1} \overset{\nabla}{\to} \mathcal{A} \to 0$$

with the same conventions as before and M with components

$$M^{\mu\nu} = \frac{1}{(s+2)\zeta_1 - 2\zeta_0}(\zeta_1(g^{\mu\nu}g^{\alpha\beta} + g^{\mu\alpha}g^{\nu\beta}) - 2\zeta_0 g^{\mu\beta}g^{\nu\alpha})\nabla_\alpha\nabla_\beta,$$

$\mu,\nu, \in \{0,\dots,s\}$. The theorem follows then by the same arguments as before, using in particular the symmetry by transposition for the Gorenstein property. □

It follows that $\mathcal{A}(\zeta)$ has Hochschild dimension 3 and the same Poincaré series as the Yang–Mills algebra for $\zeta \neq \zeta^{sing}$.

Remark. One can show that the cubic algebra generated by elements ∇_λ with relations

$$(\zeta_1 g^{\rho\lambda}g^{\mu\nu} + \zeta_2 g^{\nu\rho}g^{\lambda\mu} - 2\zeta_0 g^{\rho\mu}g^{\lambda\nu})\nabla_\lambda\nabla_\mu\nabla_\nu = 0$$

cannot be Koszul and Gorenstein if $\zeta_1 \neq \zeta_2$ and $\zeta_0 \neq 0$ or if $(\zeta_1)^2 \neq (\zeta_2)^2$.

Let now $(B_{\lambda\mu}) \in M_{s+1}(\mathbb{K})$ be an arbitrary invertible $(s+1)\times(s+1)$-matrix with inverse $(B^{\lambda\mu})$, i.e. $B_{\lambda\mu}B^{\mu\nu} = \delta^\nu_\lambda$, and let $\varepsilon = +$ or $-$. We define $\mathfrak{A}(B,\varepsilon)$ to be the cubic algebra generated by $s+1$ elements E_λ with relations

$$(B^{\rho\lambda}B^{\mu\nu} + \varepsilon B^{\lambda\mu}B^{\nu\rho})E_\lambda E_\mu E_\nu = 0 \tag{6.37}$$

for $\rho \in \{0,\dots,s\}$. Notice that B is not assumed to be symmetric. If $B_{\lambda\mu} = g_{\lambda\mu}$ and $\varepsilon = -$, then $\mathfrak{A}(g,-)$ is the super Yang–Mills algebra $\tilde{A}$ ($E_\lambda \mapsto S_\lambda$) while if $B_{\lambda\mu} = g_{\lambda\mu}$ and $\varepsilon = +$, then $\mathfrak{A}(g,+)$ is $\mathcal{A}(\zeta^0)$ ($E_\lambda \mapsto \nabla_\lambda$) where ζ^0 has homogeneous coordinates $\zeta_1 \neq 0$ and $\zeta_0 = 0$. Thus $\mathfrak{A}(B,+)$ and $\mathfrak{A}(B,-)$ belong to deformations of the Yang–Mills and of the super Yang–Mills algebra respectively.

Theorem 5. *Assume that* $1 + \varepsilon B^{\rho\lambda}B^{\mu\nu}B_{\mu\lambda}B_{\rho\nu} \neq 0$, *then* $\mathfrak{A}(B,\varepsilon)$ *is Koszul of global dimension* 3 *and is Gorenstein.*

Proof. The Koszul complex $\mathcal{K}(\mathfrak{A}(B,\varepsilon),\mathbb{K})$ can be put in the form

$$0 \to \mathfrak{A} \xrightarrow{E^t} \mathfrak{A}^{s+1} \xrightarrow{L} \mathfrak{A}^{s+1} \xrightarrow{E} \mathfrak{A} \to 0$$

where $\mathfrak{A} = \mathfrak{A}(B,\varepsilon)$ and with the previous conventions, the matrix L being given by

$$L^{\mu\nu} = (B^{\mu\alpha}B^{\beta\nu} + \varepsilon B^{\nu\mu}B^{\alpha\beta})E_\alpha E_\beta \tag{6.38}$$

for $\mu,\nu \in \{0,\dots,s\}$. The arrow $\mathfrak{A} \xrightarrow{E^t} \mathfrak{A}^{s+1}$ is always injective and the exactness of $\mathfrak{A} \xrightarrow{E^t} \mathfrak{A}^{s+1} \xrightarrow{L} \mathfrak{A}^{s+1}$ follows from the condition $1+\varepsilon B^{\rho\lambda}B^{\mu\nu}B_{\mu\lambda}B_{\rho\nu} \neq 0$. On the other hand, by definition of $\mathcal{A}$ by generators and relations, the sequence $\mathfrak{A}^{s+1} \xrightarrow{L} \mathfrak{A}^{s+1} \xrightarrow{E} \mathfrak{A} \xrightarrow{\varepsilon} \mathbb{K} \to 0$ is exact. This shows that $\mathfrak{A}$ is Koszul of global dimension 3. The Gorenstein property follows from (see also in [1])

$$B^{\rho\lambda}B^{\mu\nu} + \varepsilon B^{\nu\rho}B^{\lambda\mu} = \varepsilon(B^{\nu\rho}B^{\lambda\mu} + \varepsilon B^{\mu\nu}B^{\rho\lambda})$$

for $\rho,\lambda,\mu,\nu \in \{0,\dots,s\}$. □

Remark. It is worth noticing here in connection with the analysis of [5] that for all the deformations of the Yang–Mills algebra (resp. the super Yang–Mills algebra) considered here which are cubic Koszul Gorenstein algebras of global dimension 3, the dual cubic algebras are Frobenius algebras with structure automorphism equal to the identity (resp. $(-1)^{\text{degree}}\times$ identity).

References

1. M. Artin and W.F. Schelter: Graded algebras of global dimension 3. *Adv. Math.* 66:171–216 (1987).
2. R. Berger: Koszulity for nonquadratic algebras. *J. Algebra* 239:705–734 (2001).
3. R. Berger: La catégorie graduée, 2002.
4. R. Berger, M. Dubois-Violette, and M. Wambst: Homogeneous algebras. *J. Algebra* 261:172–185 (2003).
5. R. Berger and N. Marconnet: Koszul and Gorenstein properties for homogeneous algebras. math.QA/0310070, 2003.
6. H. Cartan and S. Eilenberg: *Homological algebra.* Princeton University Press, 1973.
7. A. Connes: Non-commutative differential geometry. *Publ. IHES* 62:257–360 (1986).
8. A. Connes: *Non-commutative geometry.* Academic Press, 1994.
9. A. Connes and M. Dubois-Violette: Noncommutative finite-dimensional manifolds. I.Spherical manifolds and related examples. *Commun. Math. Phys.* 230:539–579 (2002).
10. A. Connes and M. Dubois-Violette: Yang-Mills algebra. *Lett. Math. Phys.* 61:149–158 (2002).
11. A. Connes and M. Dubois-Violette: Moduli space and structure of noncommutative 3-spheres. *Lett. Math. Phys.* 66:91–121 (2003).
12. A. Connes and M. Dubois-Violette: Yang-Mills algebra, II. *In preparation*, 2005.
13. M. Dubois-Violette: $d^N = 0$: Generalized homology. *K-Theory* 14:371–404 (1998).
14. M. Dubois-Violette and G. Launer: The quantum group of a non-degenerated bilinear form. *Phys. Lett.* 245B:175–177 (1990).
15. M. Dubois-Violette and T. Popov: Homogeneous algebras, statistics and combinatorics. *Lett. Math. Phys.* 61:159–170 (2002).
16. J.L. Loday: Notes on Koszul duality for associative algebras, 1999.
17. Yu.I. Manin: Some remarks on Koszul algebras and quantum groups. *Ann. Inst. Fourier, Grenoble* 37:191–205 (1987).
18. E.K. Sklyanin: Some algebraic structures connected with the Yang-Baxter equation. *Func. Anal. Appl.* 16:263–270 (1982).
19. S.P. Smith and J.T. Stafford: Regularity of the four dimensional Sklyanin algebra. *Compos. Math.* 83:259–289 (1992).
20. M. Van den Bergh: A relation between Hochschild homology and cohomology for Gorenstein rings. *Proc. Amer. Math. Soc.* 126:1345–1348 (1998).
21. M. Van den Bergh: Erratum. *Proc. Amer. Math. Soc.* 130:2809–2810 (2002).

7

Remarks on the Anti-de Sitter Space-Time

Henri Epstein

Institut des Hautes Etudes Scientifiques, 91440 Bures-sur-Yvette, France
epstein@ihes.fr

Summary. This is a short review of work done with Jacques Bros and Ugo Moschella. A general framework for QFT on AdS or its universal cover is given, with definitions of covariance, locality, and energy-momentum spectrum, and expressed in terms of analyticity properties of Wightman functions. The Bisognano-Wichmann analyticity follows (leading to the AdS-Unruh effect), as well as CTP.

Dedicated to Jacques Bros on the occasion of his seventieth birthday

7.1 Introduction

The de Sitter (dS) and Anti-de-Sitter (AdS) space-times have been the subjects of a huge literature, of which the bibliography of this paper (and of [4]) give only an extremely small and unbalanced sample. Here some remarks on QFT in AdS are given. They do not touch on the deeper questions, such as AdS-CFT correspondence in the sense of [8].

7.1.1 Real and complex $(d+1)$-Minkowski space-time

Recall that the real (resp. complex) $(d+1)$-Minkowski space-time is $\mathbf{R}^{d+1}$ (resp. $\mathbf{C}^{d+1}$) equipped with the scalar product

$$x \cdot y = x^0 y^0 - x^1 y^1 - \cdots - x^d y^d. \tag{7.1}$$

When no ambiguity arises,

$$x^2 \overset{\text{def}}{=} x \cdot x \ . \tag{7.2}$$

The future and past cones $V_\pm$ are given by

$$V_+ = -V_- = \{x \in \mathbf{R}^{d+1} \ : \ x^{(0)} > 0, \quad x \cdot x > 0\}, \tag{7.3}$$

and the future and past tubes by

$$\mathrm{T}_+ = -\mathrm{T}_- = \{x + iy \in \mathbf{C}^{d+1} \ : \ y \in V_+\}. \tag{7.4}$$

7.1.2 Real and complex d-AdS space-time

We always assume $d \geq 2$ in this paper. We identify the real (resp. complex) d-dimensional AdS space-time X_d (resp. $X_d^{(c)}$) with a hyperboloid imbedded in the "ambient space" E_{d+1} (resp. $E_{d+1}^{(c)}$):
$E_{d+1} = \mathbf{R}^{d+1}$, $E_{d+1}^{(c)} = \mathbf{C}^{d+1}$, with points denoted

$$x = (x^0,\ x^1,\ \dots,\ x^d) = (x^0,\ \vec{x},\ x^d) \tag{7.5}$$

and scalar product :

$$\begin{aligned}(x,y) &= x \cdot y \\ &= x^0y^0 - x^1y^1 - \cdots - x^{d-1}y^{d-1} + x^dy^d \\ &\stackrel{\text{def}}{=} x^\mu \eta_{\mu\nu} y^\nu, \end{aligned} \tag{7.6}$$

$$x^2 \stackrel{\text{def}}{=} x \cdot x \ \text{ if no ambiguity arises.} \tag{7.7}$$

Scalar product for tensors :

$$(A,B) = A^{\mu_1 \dots \mu_n} B_{\mu_1 \dots \mu_n} \ . \tag{7.8}$$

The basis $(e_0, \dots, e_d)$ in E_{d+1} is given by $e^\mu_\nu = \delta_{\mu\nu} = \eta^\mu_\nu$. A vector $x \in E_{d+1}$ is called time-like if $(x,x) > 0$, light-like if $(x,x) = 0$, space-like if $(x,x) < 0$. Given two vectors a and b in E_{d+1} (resp.$E_{d+1}^{(c)}$), $a \wedge b$ denotes the antisymmetric 2-tensor in E_{d+1} (resp.$E_{d+1}^{(c)}$) given by $a \otimes b - b \otimes a$, i.e. $(a \wedge b)^{\mu\nu} = a^\mu b^\nu - b^\mu a^\nu$. We note that

$$\frac{1}{2}(a \wedge b,\ a \wedge b) = (a,a)(b,b) - (a,b)^2\ , \tag{7.9}$$

$$\frac{1}{2}(e_0 \wedge e_d,\ a \wedge b) = a_0 b_d - a_d b_0\ . \tag{7.10}$$

The real and complex d-AdS space-times are defined by

$$X_d = \{x \in E_{d+1}\ :\ (x,x) = 1\}, \qquad X_d^{(c)} = \{z \in E_{d+1}^{(c)}\ :\ (z,z) = 1\}\ , \tag{7.11}$$

and

$$\widetilde{X}_d = \text{universal cover of } X_d\ . \tag{7.12}$$

7.1.3 Real and complex AdS transformations

We denote G (resp. $G^{(c)}$) the group of real (resp. complex) "AdS transformations", i.e. the set of real (resp. complex) linear transformations of E_{d+1} (resp $E_{d+1}^{(c)}$) which preserve the scalar product (7.6), G_0 and $G_0^{(c)}$ the connected components of the identity in these groups, $\widetilde{G}_0$ and $\widetilde{G}_0^{(c)}$ the corresponding universal covering groups. An element of G_0 (resp. $G_0^{(c)}$) will be called a proper

AdS transformation (resp. a proper complex AdS transformation). Denoting $M(n, \mathbf{R})$ (resp. $M(n, \mathbf{C})$) the set of all real (resp. complex) $n \times n$ matrices,

$$G = \{\Lambda \in M(d+1, \mathbf{R}) \ : \ (x,x) = (\Lambda x, \ \Lambda x) \ \forall x \in E_{d+1}\}, \tag{7.13}$$

$$G_0 = \text{connected component of 1 in } G, \tag{7.14}$$

$$G^{(c)} = \{\Lambda \in M(d+1, \mathbf{C}) \ : \ (z,z) = (\Lambda z, \ \Lambda z) \ \forall z \in E_{d+1}^{(c)}\}, \tag{7.15}$$

$$G_0^{(c)} = \text{connected component of 1 in } G^{(c)}, \tag{7.16}$$

$$\widetilde{G}_0, \ \widetilde{G}_0^{(c)} \ : \ \text{universal covering groups.} \tag{7.17}$$

The Lie algebra $\mathcal{G}$ of G_0 can be identified with

$$\mathcal{G} = \{M \in M(d+1, \mathbf{R}) \ : \ (x, \ Mx) = 0 \ \forall x \in E_{d+1}\}. \tag{7.18}$$

The linear isomorphism ℓ of the 2-contravariant tensors in E_{d+1} with linear operators in E_{d+1} is defined as

$$\ell(A)^{\mu}{}_{\nu} = A^{\mu\sigma}\eta_{\sigma\nu}. \tag{7.19}$$

$\mathcal{G}$ is then given as $\ell(\bigwedge_2(E_{d+1}))$, i.e.

$$\mathcal{G} = \{\ell(A) \ : A \text{ is an antisymmetric 2-contravariant tensor in } E_{d+1}\}. \tag{7.20}$$

A basis in $\mathcal{G}$ is provided by $\{M_{\mu\nu} \ : \ 0 \leq \mu < \nu \leq d\}$, with

$$M_{\mu\nu} = \ell(e_\mu \wedge e_\nu), \qquad (M_{\mu\nu})^{\rho}{}_{\sigma} = e_\mu^\rho e_{\nu\sigma} - e_\nu^\rho e_{\mu\sigma} \ . \tag{7.21}$$

Let $\mathcal{C}_1$ be the subset of $\mathcal{G}$ consisting of all elements of the form $\ell(a \wedge b)$ with $(a,a) = (b,b) = 1$, $(a,b) = 0$, and $(e_0 \wedge e_d, \ a \wedge b) > 0$. Then

$$\mathcal{C}_1 = \{\Lambda \, M_{0d} \, \Lambda^{-1} \ : \ \Lambda \in G_0\} \ . \tag{7.22}$$

We denote $\mathcal{C}_+$ the cone generated by $\mathcal{C}_1$ and $\widehat{\mathcal{C}}_+$ the convex hull of $\mathcal{C}_+$ in $\mathcal{G}$. In G_0, $\exp(2\pi M_{0d}) = 1$ and, by (7.22), $\exp(2\pi M) = 1$ for all $M \in \mathcal{C}_1$. The loop $t \mapsto \exp(2\pi t M_{0d})$, $t \in [0, \ 1]$ defines $\exp(2\pi M_{0d}) \in \widetilde{G}_0$, and since all the loops $t \mapsto \Lambda \exp(2\pi t M_{0d})\Lambda^{-1}$, $\Lambda \in G_0$, are homotopic, we have in $\widetilde{G}_0$,

$$\iota \stackrel{\text{def}}{=} \exp(2\pi M_{0d}) = \exp(2\pi M) \quad \forall M \in \mathcal{C}_1 \tag{7.23}$$

$$= \Lambda \exp(2\pi M_{0d})\Lambda^{-1} \quad \forall \Lambda \in \widetilde{G}_0 \ . \tag{7.24}$$

The last equation shows that ι belongs to the center of $\widetilde{G}_0$. $\iota^n \neq 1$ for all $n \in \mathbf{Z} \setminus \{0\}$, and $G_0 = \widetilde{G}_0 / H$, where H is the cyclic group generated by ι.

7.1.4 Angular coordinates

$$\chi(t,\ \vec{x}) = (\sqrt{1+\vec{x}^2}\ \sin t,\ \vec{x},\ \sqrt{1+\vec{x}^2}\ \cos t) \tag{7.25}$$

defines a diffeomorphism of $S^1 \times \mathbf{R}^{d-1}$ onto X_d, and it can be lifted into a diffeomorphism $\widetilde{\chi}$ of $\mathbf{R} \times \mathbf{R}^{d-1}$ onto $\widetilde{X}_d$. Thus $\widetilde{X}_d$ admits a global chart. Equation (7.25) can be rewritten as

$$\chi(t,\ \vec{x}) = (t, \vec{x}) \mapsto \exp(tM_{0d})(0,\ \vec{x},\ \sqrt{1+\vec{x}^2}), \tag{7.26}$$

where $t \in \mathbf{R}/2\pi\mathbf{Z} \simeq S^1$. The same formula also expresses $\widetilde{\chi}$ with t now running over $\mathbf{R}$ and the exponential interpreted as the exponential map of $\mathcal{G}$ into $\widetilde{G}_0$.

7.2 Some Lorentzian geometry in AdS

The pseudo-Riemannian metric $\eta_{\mu\nu}dx^\mu dx^\nu$ of E_{d+1} (which is invariant under G) induces on X_d a Lorentzian pseudo-Riemannian metric also invariant under G. Geodesics in this metric lie in 2-planes containing 0. Light-like geodesics are straight lines. G_0 (resp. $G_0^{(c)}$) acts transitively on X_d (resp. $X_d^{(c)}$). The point e_d belongs to X_d. Its stability subgroup is the connected Lorentz group acting in the d-Minkowski space $\{y \in E_{d+1}\ :\ (y, e_d) = 0\}$, parallel to the tangent hyperplane to X_d at e_d. We identify the future cone within this d-Minkowski space to the "future cone at e_d", denoted $V_+(e_d)$, i.e. $V_+(e_d) = \{y \in E_{d+1}\ :\ y^d = 0,\ \ y^0 - |\vec{y}| > 0\}$. If $x \in X_d$, it can be written (non-uniquely) as Λe_d with $\Lambda \in G_0$. We then define $V_+(x) = \Lambda V_+(e_d)$, a subset of $\{y \in E_{d+1}\ :\ (y, x) = 0\}$ which does not depend on the choice of Λ. It is easy to see that, for any $x \in X_d$,

$$V_+(x) = \mathcal{C}_+ x = \{y \in E_{d+1}\ :\ (y, x) = 0,\ \ (y, y) > 0,\ \ y^0 x^d - y^d x^0 > 0\}\ . \tag{7.27}$$

As a consequence, if $M \in \widehat{\mathcal{C}}_+$, the orbits of $\exp(tM)$ in X_d are all time-like, and, at each $x \in X_d$, $(d/dt)\exp(tM)x \in V_+(\exp(tM)x)$. This motivates the spectral condition (see below) for QFT on AdS.

Two points x_1, x_2 in X_d are *space-like separated* if $(x_1 - x_2)^2 < 0$ or equivalently $(x_1, x_2) > 1$, *time-like separated* if $(x_1 - x_2)^2 > 0$ or equivalently $(x_1, x_2) < 1$, *light-like separated* if $(x_1 - x_2)^2 = 0$ or equivalently $(x_1, x_2) = 1$.

In AdS (X_d) two space-like separated points (as Fig. 7.2 shows) can still be connected by a time-like curve, but not by a time-like geodesic. Two points $x,\ y \in X_d$ are said to be *exotically space-like separated* if $(x + y)^2 < 0$, i.e. if x and $-y$ are space-like separated.

In $\widetilde{X}_d$, the future set $\Gamma^+(x)$ of a point x is the set of all points y which can be reached from x by a positively time-like or light-like curve $t \mapsto c(t)$, i.e. such that $c(0) = x$, $c(1) = y$, and $(d/dt)c(t) \in \overline{V_+(c(t))}$ for all $t \in [0,\ 1]$.

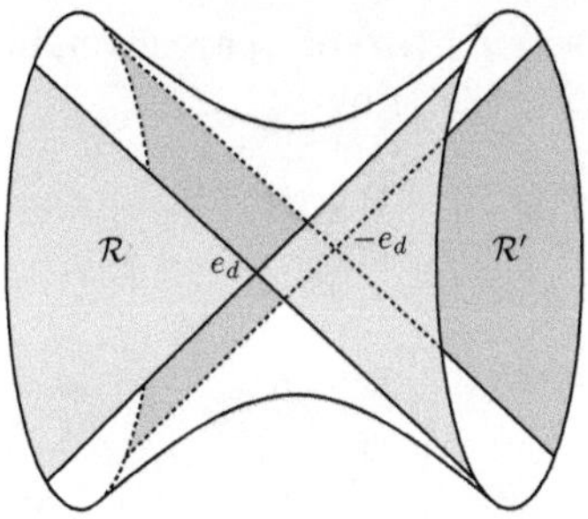

Fig. 7.1. $\mathcal{R} = \{x \in X_d \ : \ x^d > 1\}$, $\mathcal{R}' = \{x \in X_d \ : \ x^d < -1\}$.

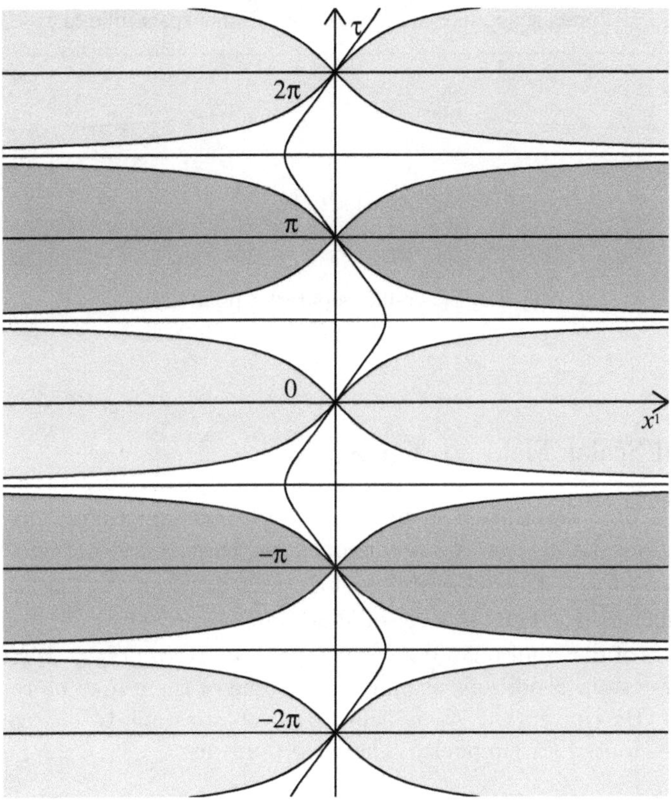

Fig. 7.2. X_d and $\widetilde{X}_d$ in angular coordinates. The sinuous line is a time-like geodesic.

The future set of an arbitrary $x = \Lambda e_d$, where $\Lambda \in \widetilde{G}_0$, is given by $\Lambda\Gamma^+(e_d)$. The relation $x \in \Gamma^+(y)$ defines a partial order on $\widetilde{X}_d$, denoted $x \geq y$ or $y \leq x$. The past set $\Gamma^-(x)$ of x is $\{y \in \widetilde{X}_d \ : \ x \geq y\}$. Two points $x, \ y \ \in \widetilde{X}_d$ are space-like separated if $x \not\geq y$ and $y \not\geq x$.

The future and past sets $\Gamma^{\pm}(e_d)$ of e_d are given in the angular coordinates, $(t,\ \vec{x}) = \widetilde{\chi}^{-1}(x)$ (see (7.25, 7.26)) by

$$\Gamma^{\pm}(e_d) = \{(t,\ \vec{x}) \in \mathbf{R} \times \mathbf{R}^{d-1}\ :\ \pm t \geq \mathrm{Arctg}\,|\vec{x}|\}\ . \tag{7.28}$$

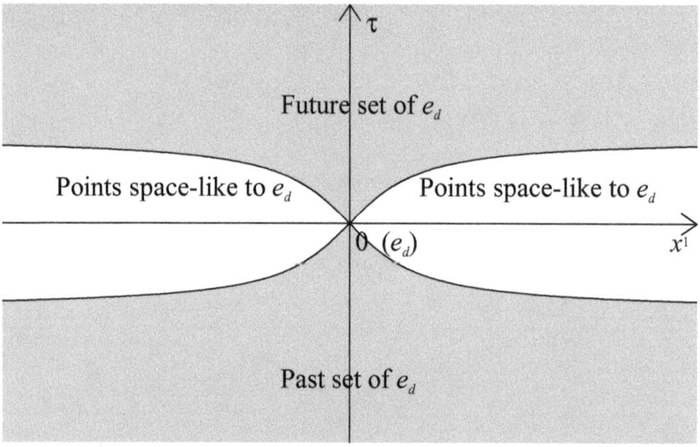

Fig. 7.3. Space-like separated points in $\widetilde{X}_d$.

7.3 Real Scalar Field on X_d, $\widetilde{X}_d$

We adopt a description in terms of Wightman functions. The Schwartz test-functions $\mathcal{S}(X_d)$ on X_d are defined as restrictions of Schwartz test-functions on E_{d+1}. The Schwartz test-functions $\mathcal{S}(\widetilde{X}_d)$ on $\widetilde{X}_d$ are the functions f such that $f \circ \widetilde{\chi} \in \mathcal{S}(\mathbf{R}^d)$. A set of Wightman functions on X_d (resp. $\widetilde{X}_d$) is a sequence of distributions $(\mathcal{W}_n \in \mathcal{D}'(X_d^n))_{n\in\mathbf{N}}$ (resp. $(\mathcal{W}_n \in \mathcal{D}'(\widetilde{X}_d^n))_{n\in\mathbf{N}}$) on which certain conditions are imposed. Some of them may be relaxed, e.g. positivity. By convention $\mathcal{W}_0$ is defined as the constant 1. The condition of reality is adopted for simplicity. The conditions are:

Temperedness: $\mathcal{W}_n \in \mathcal{S}'(X_d^n)$ (resp. $\mathcal{W}_n \in \mathcal{S}'(\widetilde{X}_d^n)$) for all $n \in \mathbf{N}$.

Covariance:

$$\mathcal{W}_n(x_1,\ \ldots,\ x_n) = \mathcal{W}_n(\Lambda x_1,\ \ldots,\ \Lambda x_n) \tag{7.29}$$

for all $\Lambda \in G_0$ (resp. $\widetilde{G}_0$).

Locality:

$$x_j \text{ and } x_{j+1} \text{ space-like separated} \Rightarrow$$
$$\mathcal{W}_n(x_1,\ \ldots,\ x_j,\ x_{j+1},\ \ldots,\ x_n) = \mathcal{W}_n(x_1,\ \ldots,\ x_{j+1},\ x_j,\ \ldots,\ x_n). \tag{7.30}$$

Reality (or hermiticity): For all $n \in \mathbf{N}$,

$$\mathcal{W}_n(x_1,\ldots,\ x_n) = \overline{\mathcal{W}_n(x_n,\ldots,\ x_1)}\ . \tag{7.31}$$

Positivity: for each terminating sequence $(f_n \in \mathcal{S}(X_d^n))_{n\in\mathbf{N}}$ (resp. $(f_n \in \mathcal{S}(\widetilde{X}_d^n))_{n\in\mathbf{N}}$) and each $f_0 \in \mathbf{C}$,

$$\sum_{0\le p,q\le N} \int \mathcal{W}_{q+p}(x'_q,\ldots,x'_1,\ x_1,\ldots,x_p)$$
$$\overline{f(x'_1,\ldots,x'_q)}\, f(x_1,\ldots,x_p)\, dx'_1\ldots dx_p \ge 0. \tag{7.32}$$

If this condition is satisfied, the reconstruction procedure can be carried out as in [3, 7, 11], and provides a Hilbert space $\mathcal{H}$, a unit vector $\Omega \in \mathcal{H}$, and an unbounded operator-valued distribution ϕ, such that

$$\mathcal{W}_n(x_1,\ldots,\ x_n) = (\Omega,\ \phi(x_1)\ldots\phi(x_n)\Omega). \tag{7.33}$$

If temperedness holds, ϕ is tempered. If locality holds,

$$x \text{ and } y \text{ space-like separated} \quad\Rightarrow\quad [\phi(x),\ \phi(y)] = 0. \tag{7.34}$$

If covariance holds, there is a continuous unitary representation U of G_0 (resp. $\widetilde{G}_0$) such that, for all Λ,

$$U(\Lambda)\Omega = \Omega, \quad U(\Lambda)\phi(x)U(\Lambda)^{-1} = \phi(\Lambda x). \tag{7.35}$$

Strong Spectral Condition: For any $M \in \mathcal{C}_+$ and $t \in \mathbf{R}$, $U(\exp tM) = \exp(it\widehat{M})$, where $\widehat{M}$ has positive spectrum. This makes sense only if covariance and positivity hold. This condition has been adopted by a large number of authors, in particular in the early paper of Fronsdal [5]. If positivity does not hold, the Weak Spectral Condition can be used:

Weak Spectral Condition: For all n, for any $M \in \mathcal{C}_+$, and for any test function $f \in \mathcal{D}(\mathbf{R})$ with support contained in $(-\infty,\ 0)$,

$$\int_{\mathbf{R}} \mathcal{W}(x_1,\ldots,\ \exp(tM)x_j,\ldots,\ x_n)\, dt \int_{\mathbf{R}} \exp(-it\omega)f(\omega)\, d\omega = 0\ . \tag{7.36}$$

This condition is equivalent to the Strong Spectral Condition if positivity holds. We will replace it by the "Tempered Spectral Condition", a property of analyticity of the Wightman functions. To formulate this condition, we have to describe the "n-point tuboids".

7.4 The n-point Tuboids

7.4.1 The Domains $G_0^+ \subset G_0^{(c)}$ and $\widetilde{G}_0^+ \subset \widetilde{G}_0^{(c)}$

We define

$$G_0^+ = \{\exp(\tau_1 M_1)\ldots\exp(\tau_N M_N)\ :\ N\in\mathbf{N},\ \mathrm{Im}\,\tau_j>0,\ \ M_j\in\mathcal{C}_+\ \ \forall j=1,\ldots,N\}. \tag{7.37}$$

In this formula $\mathcal{G}$ is identified with a set of matrices as in (7.18) and the exponentials are the matrix exponentials. We define $\widetilde{G}_0^+ \subset \widetilde{G}_0^{(c)}$ by the same formula, but interpreting the exponentials as exponential maps from the complexified $\mathcal{G}^{(c)}$ of $\mathcal{G}$ into $\widetilde{G}_0^{(c)}$. We also define

$$G_0^- = \{\Lambda\in G_0^{(c)}\ :\ \Lambda^{-1}\in G_0^+\},\quad \widetilde{G}_0^- = \{\Lambda\in \widetilde{G}_0^{(c)}\ :\ \Lambda^{-1}\in \widetilde{G}_0^+\}. \tag{7.38}$$

Note that G_0^- (resp. $\widetilde{G}_0^-$) is the complex-conjugate of G_0^+ (resp. $\widetilde{G}_0^+$). It is obvious that $G_0^\pm$ and $\widetilde{G}_0^\pm$ are semi-groups (without unit). Moreover (see [4])

Lemma 1.

(i) G_0^+ *is open in* $G_0^{(c)}$.
(ii) $G_0^+ = G_0 G_0^+ = G_0^+ G_0$.

Similar properties hold for $\widetilde{G}_0^+$.

7.4.2 The n-point Tuboids $\mathcal{T}_{n\pm}$ and $\widetilde{\mathcal{T}}_{n\pm}$

The tuboids $\mathcal{T}_{n\pm}$ are subsets of $X_d^{(c)}$ defined by

$$\begin{aligned}\mathcal{T}_{n+} &= \{(z_1,\ \ldots,\ z_n)\in X_d^{(c)n}\ :\ \forall j=1,\ldots,n,\\ &\qquad z_j=\Lambda_1\ldots\Lambda_j\,c_j\ ,\quad \Lambda_j\in G_0^+\ ,\quad c_j\in X_d\ \}\ ,\\ \mathcal{T}_{n-} &= \mathcal{T}_{n+}^*\ .\end{aligned} \tag{7.39}$$

By Lemma 1 (ii),

$$\begin{aligned}\mathcal{T}_{n+} = \{(z_1,\ \ldots,\ z_n)\in X_d^{(c)n}\ &:\ \forall j=1,\ldots,n,\\ z_j=\Lambda_1\ldots\Lambda_j\,e_d\ ,\quad &\Lambda_j\in G_0^+\}\ .\end{aligned} \tag{7.40}$$

Indeed suppose that for all $j=1,\ldots,\ n,\ z_j=\Lambda_1\ldots\Lambda_j\,c_j$ with $\Lambda_j\in G_0^+$ and $c_j=A_j e_d$ with $A_j\in G_0$. Denote $\Lambda_1'=\Lambda_1 A_1$, and $\Lambda_j'=A_{j-1}^{-1}\Lambda_j A_j$ for $j>1$. Then $z_j=\Lambda_1'\ldots\Lambda_j'\,e_d$ for all j. It is obvious that $G_0\mathcal{T}_{n+}=\mathcal{T}_{n+}$ and that $G_0^+\mathcal{T}_{n+}=\mathcal{T}_{n+}$ (and similarly with tildes). Note also that $\mathcal{T}_{n+}\subset\mathcal{T}_{1+}^n$.

Lemma 2. *For any* $n\in\mathbf{N}$, $\mathcal{T}_{n+}$ *is open.*

Proof. For a fixed $c \in X_d^{(c)}$ the map $h_c : G_0^{(c)} \to X_d^{(c)}$ given by $\Lambda \mapsto \Lambda c$ is holomorphic and open. Therefore so is the map $h_n : (G_0^{(c)})^n \to X_d^{(c)n}$ given by $(\Lambda_1, \dots, \Lambda_n) \mapsto (\Lambda_1 e_d, \dots, \Lambda_n e_d)$. The map $k_n : (\Lambda_1, \dots, \Lambda_n) \mapsto (\Lambda_1, \Lambda_1\Lambda_2, \dots, \Lambda_1 \dots \Lambda_n)$ is a biholomorphic map of $(G_0^{(c)})^n$ onto itself. Equation (7.40) can be rewritten as

$$\mathcal{T}_{n+} = h_n \circ k_n(G_0^{+n}), \tag{7.41}$$

hence $\mathcal{T}_{n+}$ is open. □

The tuboids $\widetilde{\mathcal{T}}_{n\pm}$ are defined as the universal covering spaces of $\mathcal{T}_{n\pm}$.

7.4.3 The special case of $\mathcal{T}_{1+}$

It is not hard to verify that

$$\mathcal{T}_{1+} = \{z = x + iy \in X_d^{(c)} \ : \ (y, y) > 0, \quad y^0 x^d - y^d x^0 > 0\}, \tag{7.42}$$

$$\mathcal{T}_{1+} = \{\Lambda \exp(itM_{0d})\, e_d \ : \ \Lambda \in G_0, \quad t > 0\}. \tag{7.43}$$

It follows from (7.42) that $\mathcal{T}_{1+} = -\mathcal{T}_{1+}$, $\mathcal{T}_{1+} \cap \mathcal{T}_{1-} = \emptyset$ and hence $G_0^+ \cap G_0^- = \emptyset$ and $\mathcal{T}_{n+} \cap \mathcal{T}_{n-} = \emptyset$. As shown in [4], there is a biholomorphic map of $\mathcal{T}_{1+}$ onto the subset of the complex d-Minkowski space $\mathrm{T}_+ \setminus \{(z^0, \dots, z^{d-1}) \ : \ z^{d-1} = 0\}$. In particular $\mathcal{T}_{1+}$ is a domain of holomorphy. The image of $\mathcal{T}_{1+}$ under the map $z \mapsto z_d$ is the cut-plane $\Delta = \mathbf{C} \setminus [-1, 1]$. This map lifts into a map of $\widetilde{\mathcal{T}}_{1+}$ onto the universal cover $\widetilde{\Delta}$ of Δ. The map $(z_1, z_2) \mapsto z_1 \cdot z_2$ maps $\mathcal{T}_{1-} \times \mathcal{T}_{1+}$ (resp. $\mathcal{T}_{2+}$) onto the cut-plane Δ, and can be lifted to a map of $\widetilde{\mathcal{T}}_{1-} \times \widetilde{\mathcal{T}}_{1+}$ (resp $\widetilde{\mathcal{T}}_{2+}$) onto $\widetilde{\Delta}$.

7.5 Spectral Condition for AdS

It is now possible to formulate the following condition.

Tempered Spectral Condition: For each pair of integers $m \geq 0$ and $n \geq 0$, with $m + n > 0$,
$\mathcal{W}_{m+n}(w_m, \dots, w_1, z_1, \dots, z_n)$ is the boundary value, in the sense of tempered distributions, of a function $W_{m,n}$ of (w, z) holomorphic and of tempered growth (both at infinity and at the boundaries) in $\mathcal{T}_{m+}^* \times \mathcal{T}_{n+} = \mathcal{T}_{m-} \times \mathcal{T}_{n+}$.

In particular $\mathcal{W}_n(z_1, \dots, z_n)$ is the boundary value of a function W_n holomorphic and of tempered growth in $\mathcal{T}_{n+}$.

We always suppose this condition satisfied in the sequel.

7.6 Two-Point Functions, Generalized Free Fields

In X_d, according to the tempered spectral condition, in the sense of tempered distributions,

$$\mathcal{W}_2(x_1,\ x_2) = \lim_{\substack{z_1\in\mathcal{T}_{1-},\ z_2\in\mathcal{T}_{1+} \\ z_1\to x_1,\ z_2\to x_2}} W_+(z_1,\ z_2),$$

$$\mathcal{W}_2(x_2,\ x_1) = \lim_{\substack{z_1\in\mathcal{T}_{1+},\ z_2\in\mathcal{T}_{1-} \\ z_1\to x_1,\ z_2\to x_2}} W_-(z_1,\ z_2), \tag{7.44}$$

where W_+ is holomorphic in $\mathcal{T}_{1-} \times \mathcal{T}_{1+}$, and W_- is holomorphic in $\mathcal{T}_{1+} \times \mathcal{T}_{1-}$. If either locality or covariance holds, $W_\pm$ are branches of the same function W, holomorphic in

$$\{(z_1,\ z_2) \in X_d^{(c)2}\ :\ z_1 \cdot z_2 \in \Delta\}. \tag{7.45}$$

If covariance holds, there is a function w holomorphic on Δ such that $W(z_1,\ z_2) = w(z_1 \cdot z_2)$.

In $\widetilde{X}_d$, (7.44) holds with $\widetilde{\mathcal{T}}_{1\pm}$ replacing $\mathcal{T}_{1\pm}$, and if locality or covariance holds, $W_\pm$ are branches of a function W holomorphic on the universal cover of (7.45). If covariance holds, there is a function w holomorphic on the universal cover $\widetilde{\Delta}$ of Δ such that $W(z_1,\ z_2) = w(z_1 \cdot z_2)$ on the "first sheet" of $\widetilde{\mathcal{T}}_{1+} \times \widetilde{\mathcal{T}}_{1-}$.

For both X_d and $\widetilde{X}_d$, in the case of a real field,

$$\mathcal{W}_2(x_1,\ x_2) = \overline{\mathcal{W}_2(x_2,\ x_1)}\quad \forall (x_1,\ x_2) \in X_d^2,\quad W(z_1,\ z_2) = \overline{W(\bar{z}_2,\ \bar{z}_1)}. \tag{7.46}$$

In the case of X_d, if locality or covariance holds, this implies that $\mathcal{W}_2(x_1,\ x_2)$ and $\mathcal{W}_2(x_2,\ x_1)$ coincide on the two regions

$$\mathcal{R}_2 = \{x \in X_d^2\ :\ (x_1 - x_2)^2 < 0\} = \{x \in X_d^2\ :\ (x_1, x_2) > 1\}, \tag{7.47}$$

and

$$\mathcal{R}_2' = \{x \in X_d^2\ :\ (x_1 + x_2)^2 < 0\} = \{x \in X_d^2\ :\ (x_1, x_2) < -1\}. \tag{7.48}$$

This raises the question whether our assumptions imply "exotic locality", i.e. whether $[\phi(x_1),\ \phi(x_2)]$ vanishes when $(x_1 + x_2)^2 < 0$, or equivalently whether $\phi(x)$ and $\phi(-x)$ are mutually local. Note that

$$(z_1, \ldots,\ z_j, \ldots,\ z_n) \in \mathcal{T}_{n+} \Rightarrow (z_1, \ldots,\ -z_j, \ldots,\ z_n) \in \mathcal{T}_{n+}\ , \tag{7.49}$$

Hence mixed Wightman functions of $\phi(x)$ and $\phi(-x)$ satisfy the tempered spectral condition. However it is possible to give an example of a field on X_d which satisfies locality, covariance, positivity and the strong spectral condition, but not exotic locality. This is sketched in Appendix 7.9 and will appear in detail elsewhere.

Suppose given a function $\mathcal{W}_2$ on X_d^2 (resp. $\widetilde{X}_d^2$) verifying (7.44) (resp. (7.44) with $\widetilde{\mathcal{T}}_{1\pm}$ instead of $\mathcal{T}_{1\pm}$), with $W_\pm$ holomorphic as above with tempered growth. One can define a generalized free-field on X_d (resp. $\widetilde{X}_d$) by defining its sequence of Wightman functions as built from $\mathcal{W}_2$ with the same formulae as in Minkowski space. They will satisfy the tempered spectral condition, and the other conditions if and only if they are satisfied by $\mathcal{W}_2$. Furthermore all Wick powers of this field are well-defined through their Wightman functions, which also satisfy the tempered spectral condition, and the other conditions if and only if they are satisfied by $\mathcal{W}_2$. In X_d, if locality holds, then so does exotic locality.

7.7 *n*-Point Permuted Tuboids

For $n > 1$ and for any permutation π of $(1, \dots, n)$ we define, as usual, the permuted Wightman function

$$\mathcal{W}_{n,\pi}(x_1, \dots, x_n) = \mathcal{W}_n(x_{\pi(1)}, \dots, x_{\pi(n)}), \tag{7.50}$$

the permuted tuboid (for X_d)

$$\mathcal{T}_{n,\pi} = \{z \in X_d^{(c)n} \ : \ (z_{\pi(1)}, \dots, z_{\pi(n)}) \in \mathcal{T}_{n+}\}, \tag{7.51}$$

and $\widetilde{\mathcal{T}}_{n,\pi}$ as the universal cover of $\mathcal{T}_{n,\pi}$. If locality holds, the various permuted Wightman functions are boundary values, from their respective permuted tuboids, of a single holomorphic function. An interesting feature of AdS geometry is that, in contrast to the Minkowskian situation, two adjacent permuted tuboids have a non-empty intersection. Let us choose π as the transposition $(j,\ j+1)$, where $1 \le j < n$. Let $c \in X_d^n$ belong to $\mathcal{R}_2 \cup \mathcal{R}_2'$, with

$$\mathcal{R}_2 = \{x \in X_d^n \ : \ (x_j - x_{j+1})^2 < 0\}, \quad \mathcal{R}_2' = \{x \in X_d^n \ : \ (x_j + x_{j+1})^2 < 0\}. \tag{7.52}$$

It is proved in [4] that, for all $(\Lambda_1, \dots, \Lambda_{j-1}, \Lambda, \Lambda_{j+2}, \dots, \Lambda_n) \in (G_0^+)^{n-1}$, the point $z \in X_d^{(c)n}$ given by

$$\begin{aligned} z_k &= \Lambda_1 \dots \Lambda_k\, c_k \ \text{ for } k < j, \\ z_j &= \Lambda_1 \dots \Lambda_{j-1} \Lambda\, c_j, \quad z_{j+1} \ = \ \Lambda_1 \dots \Lambda_{j-1} \Lambda\, c_{j+1}, \\ z_k &= \Lambda_1 \dots \Lambda_{j-1} \Lambda \Lambda_{j+2} \dots \Lambda_k\, c_k \ \text{ for } k > j+1 \end{aligned} \tag{7.53}$$

belongs to $\mathcal{T}_{n+}$, and therefore also to $\mathcal{T}_{n,\pi}$. This implies that $\mathcal{T}_{n+} \cap \mathcal{T}_{n,\pi}$ contains two non-empty open tuboids bordered by $\mathcal{R}_2$ and $\mathcal{R}_2'$ (see Fig.7.4). Each of these real regions is connected if $d > 2$, or has two connected components if $d = 2$. As a consequence, if locality (resp. exotic locality) holds in some open set it holds everywhere.

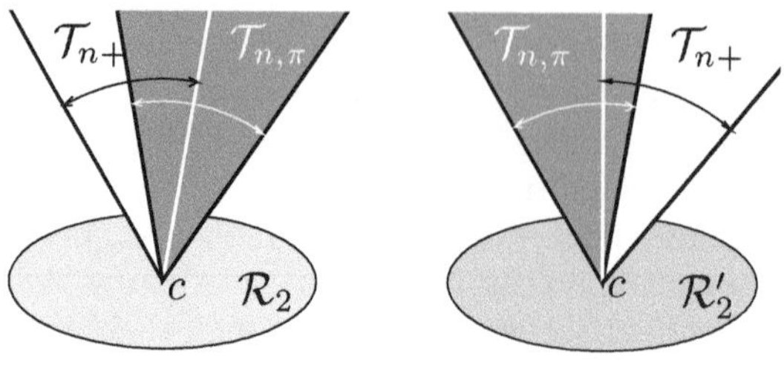

Fig. 7.4. Schematic picture of $\mathcal{T}_{n+} \cap \mathcal{T}_{n,\pi}$.

7.8 Bisognano-Wichmann-KMS Property, CTP

The Bisognano-Wichmann analyticity property can be proved (as for the dS and Minkowski spaces) without using positivity. For $\lambda \in \mathbf{C} \setminus \{0\}$, we define $[\lambda] \in G_0^{(c)}$ by

$$([\lambda]x)^0 = \frac{\lambda + \lambda^{-1}}{2} x^0 + \frac{\lambda - \lambda^{-1}}{2} x^1, \quad ([\lambda]x)^1 = \frac{\lambda - \lambda^{-1}}{2} x^0 + \frac{\lambda + \lambda^{-1}}{2} x^1 , \tag{7.54}$$

and $([\lambda]x)^\mu = x^\mu$ for $1 < \mu \le d$. For all $\zeta \in \mathbf{C}$, $[e^\zeta] = \exp(\zeta M_{10})$. For any test function $f_n \in \mathcal{S}(X_d^n)$ we denote

$$f_{n\lambda}(x_1, \dots, x_n) = f_n([\lambda^{-1}]x_1, \dots, [\lambda^{-1}]x_n). \tag{7.55}$$

Let

$$W_R = \{x \in X_d \ : \ x^1 > |x^0|, \ x^d > 1\} = [-1]W_L \ . \tag{7.56}$$

Theorem 1. *If a set of Wightman functions satisfies the locality and tempered spectral conditions, then for all $m, \ n \in \mathbf{N}$, $f_m \in \mathcal{D}(W_R^m)$ and $g_n \in \mathcal{D}(W_R^n)$, there is a function $G_{(f_m,g_n)}$ holomorphic on $\mathbf{C} \setminus \mathbf{R}_+$ with continuous boundary values $G^{\pm}_{(f_m,g_n)}$ on $\mathbf{R}_+ \setminus \{0\}$ from the upper and lower half-planes such that, for all $\lambda \in \mathbf{R}_+ \setminus \{0\}$,*

$$G^+_{(f_m,g_n)}(\lambda) = \langle \mathcal{W}_{m+n}, \ f_m \otimes g_{n\lambda} \rangle, \qquad G^-_{(f_m,g_n)}(\lambda) = \langle \mathcal{W}_{m+n}, \ g_{n\lambda} \otimes f_m \rangle \ . \tag{7.57}$$

If we also assume positivity, this gives rise to the Bisognano-Wichmann-KMS property [2]. Denote $\mathcal{P}(\mathcal{O})$ the polynomial algebra of fields smeared with test functions with supports in the open set $\mathcal{O}$. Let $A, \ B \in \mathcal{P}(W_R)$, and, for any real t,

$$A_t = U(e^{tM_{10}})\, A\, U(e^{-tM_{10}}). \tag{7.58}$$

Then

$$(\Omega, \ A_t\, B\, \Omega) = (\Omega, \ B\, A_{t+2i\pi}\, \Omega) \tag{7.59}$$

in the sense that $t \mapsto (\Omega, \ B\, A_t\, \Omega)$ extends to a function holomorphic in the strip $\{t + is \ : \ 0 < s < 2\pi\}$ and continuous in its closure, and the last expression in (7.59) is the boundary value of this function. This leads to an AdS-Unruh effect similar to those in the dS space.

The same process of analytic continuation which leads to Theorem 1 shows that

Lemma 3. *If a set of Wightman functions satisfies the locality, covariance, and tempered spectral conditions, then for all integer* $n \geq 1$, *and all* $z \in \mathcal{T}_{n+}$,

$$\mathcal{W}_n(z_1, \dots, z_n) = \mathcal{W}_n([-1]z_n, \dots, [-1]z_1). \tag{7.60}$$

If positivity holds this implies the existence of an anti-unitary operator θ such that $\theta\phi(x)\theta^{-1} = \phi([-1]x)^*$.

The contents of this section hold without change in $\widetilde{X}_d$.

7.9 Wick Rotations for $\widetilde{X}_d$

Here we assume that locality and covariance hold. We use the global chart $\widetilde{\chi}^{-1}$, i.e. a point of $\widetilde{X}_d$ is identified with an element $(t, \vec{x}) \in \mathbf{R}^d$. (see (7.26). The map $\widetilde{\chi}$ (see (7.26) can be extended to a holomorphic diffeomorphism of

$$\{\tau \in \mathbf{C}\} \times \{\vec{z} \in \mathbf{C}^{d-1} \ : \ \vec{z}^2 + 1 \notin \mathbf{R}_-\} \tag{7.61}$$

onto a holomorphic extension of $\widetilde{X}_d$ denoted $[\widetilde{X}_d]^{(c)}$. This contains in particular the image of $\mathbf{C}\times\mathbf{R}^{d-1}$, which in turn contains the image $\widetilde{X}_d^{(\mathcal{E})}$ of $i\mathbf{R}\times\mathbf{R}^{d-1}$. By using covariance and locality, it can be shown that the Schwinger function

$$S_n(s_1, \vec{x}_1, \dots, s_n, \vec{x}_n) = \mathcal{W}_n(\widetilde{\chi}(is_1, \vec{x}_1), \dots, \widetilde{\chi}(is_n, \vec{x}_n)) \tag{7.62}$$

(where $s_1, \dots, \vec{x}_n$ are real) extends to a function analytic in $(s_1, \dots, s_n)$ whenever $(s_j, \vec{x}_j) \neq (s_k, \vec{x}_k)$ for all $j \neq k$. As in the Minkowski case, positivity implies Osterwalder-Schrader positivity, and vice-versa if certain growth conditions are satisfied (see [6, 9, 10]).

Appendix. q-Sheeted AdS Covers and Exotic Locality

For any integer $q > 0$, let $H_q = \{\iota^{nq} \ : \ n \in \mathbf{Z}\}$ (see (7.23, 7.24)). The q-sheeted AdS cover ${}_q\widetilde{X}_d$ is obtained by identifying z and $\iota^q z$ for all $z \in \widetilde{X}_d$, i.e. ${}_q\widetilde{X}_d = \widetilde{X}_d/H_q$. We denote ${}_q\widetilde{G}_0 = \widetilde{G}_0/H_q$ and ${}_q\widetilde{G}_0^{(c)} = \widetilde{G}_0^{(c)}/H_q$.

A field defined by its Wightman functions $(\mathcal{W}_n)$ on $\widetilde{X}_d$ is actually a field on ${}_q\widetilde{X}_d$ if, for all $n \geq 1$, for all $j = 1, 2, \dots, n$, in the sense of distributions,

$$\mathcal{W}_n(x_1, \dots, \iota^q x_j, \dots, x_n) = \mathcal{W}_n(x_1, \dots, x_j, \dots, x_n), \tag{7.63}$$

or equivalently

$$W_n(z_1, \dots, \iota^q z_j, \dots, z_n) = W_n(z_1, \dots, z_j, \dots, z_n) \tag{7.64}$$

for all $(z_1, \dots, z_n) \in \widetilde{\mathcal{T}}_{n+}$, all $j = 1, 2, \dots, n$. Consider in particular a free Klein-Gordon field ϕ on $\widetilde{X}_d$ satisfying the covariance, tempered spectral and locality conditions. It is characterized by its two-point Wightman function $\mathcal{W}_2$ and (see Subsec. 7.6 and [1,4]), there exists a function w, holomorphic in the universal cover $\widetilde{\Delta}$ of $\Delta = \mathbf{C} \setminus [-1, 1]$, such that $W(z_1, z_2) = w(z_1 \cdot z_2)$ on the first sheet of $\widetilde{\mathcal{T}}_{1-} \times \widetilde{\mathcal{T}}_{1+}$ (or of $\widetilde{\mathcal{T}}_{2+}$), and

$$w(\zeta) = w_{\lambda + \frac{d-1}{2}}(\zeta) = C(\lambda, d) \int_1^\infty (\zeta + t\sqrt{\zeta^2 - 1})^{-\lambda - d + 1} (t^2 - 1)^{\frac{d-3}{2}} \, dt. \tag{7.65}$$

Here $C(\lambda, d) > 0$, and λ is related to the mass m by $m^2 = \lambda(\lambda + d - 1)$. For integer λ, w is a holomorphic function on Δ. If $\lambda = p/q$, with $p, q \in \mathbf{N}$ relatively prime, then w is q-sheeted and the free field is a free field on ${}_q\widetilde{X}_d$. We consider the case $\lambda = k + 1/2$, $k \in \mathbf{N}$ (this, together with $d \geq 2$, implies $\nu = \lambda + (d-1)/2 > 1$ so that the free field ϕ satisfies the positivity condition). In the integral on the rhs of (7.65), the function $\zeta \mapsto \sqrt{\zeta^2 - 1}$ is understood as holomorphic in Δ, and behaving like ζ at infinity. This function, as well as $\zeta \mapsto (\zeta + t\sqrt{\zeta^2 - 1})$, is Herglotz and positive on $(1, \infty)$, negative on $(-\infty, -1)$. It follows that $\zeta \mapsto w(\zeta)$ is holomorphic in the 2-sheeted covering of Δ and can be represented as the glueing together of w_{I} and $w_{\mathrm{II}} = -w_{\mathrm{I}}$, both holomorphic in $\mathbf{C} \setminus 1 + \mathbf{R}_-$ and continuing each other across $(-\infty, -1)$, $w_{\mathrm{I}}(\zeta) > 0$ for $\zeta \in (1, \infty)$. For $\zeta \in (-\infty, -1)$, $-iw_{\mathrm{I}}(\zeta + i0)$ is real $\neq 0$ and has the sign of $(-1)^{k+d}$, and $w_{\mathrm{I}}(\zeta + i0) = -w_{\mathrm{I}}(\zeta - i0)$. It follows that

$$\mathcal{W}_2(\iota^{-n} x_1, x_2) = \mathcal{W}_2(x_1, \iota^n x_2) = (-1)^n \, \mathcal{W}_2(x_1, x_2) \tag{7.66}$$

(in the sense of distributions) for all $n \in \mathbf{Z}$. The field ϕ satisfies (7.63) with $q = 2$ and is therefore a free field on ${}_2\widetilde{X}_d$. It is representable as an operator-valued tempered distribution acting in a Hilbert Fock space $\mathcal{H}$ as usual, and we claim that

$$U(\iota)\, \phi(x)\, U(\iota)^{-1} = -\phi(x) \quad \forall x \in \widetilde{X}_d\,. \tag{7.67}$$

Indeed

$$\mathcal{W}_{n+m+1}(y_1, \dots, y_n, x, y'_1, \dots, y'_m) \tag{7.68}$$

$$= \sum_{j=1}^n \mathcal{W}_{n+m-1}(y_1, \dots, \widehat{y_j}, \dots, \widehat{x}, \dots, y'_m) \mathcal{W}_2(y_j, x)$$

$$+ \sum_{k=1}^m \mathcal{W}_{n+m-1}(y_1, \dots, \widehat{x}, \dots, \widehat{y'_k}, \dots, y'_m) \mathcal{W}_2(x, y'_k) \tag{7.69}$$

where $\widehat{x}$, $\widehat{y_j}$ or $\widehat{y'_k}$ denotes the omission of the corresponding variable. Therefore, by (7.66), changing x to ιx in (7.68) is equivalent to changing the sign of (7.68). Let $A(x) = \,:\phi(x)^2:$. Then, by (7.67),

$$U(\iota)\, A(x)\, U(\iota)^{-1} = A(x) \quad \forall x \in {}_2\widetilde{X}_d\,. \tag{7.70}$$

Therefore A is a field on X_d. In the subspace $\mathcal{H}_2$ it generates from the vacuum, it satisfies locality, covariance, and the strong spectral condition. Let $(x_1,\ x_2) \in \widetilde{X}_d^2$ satisfy $(x_1 + x_2)^2 < 0$, i.e. $(x_1, x_2) < -1$. All boundary values of w at $\zeta = (x_1, x_2)$ are pure imaginary, so that, by (7.46), $\mathcal{W}_2(x_1,\ x_2) = -\mathcal{W}_2(x_2,\ x_1)$. In particular

$$\begin{aligned}&(\Omega, [A(x_1),\ A(x_2)]A(x_3)\Omega) = 2(\Omega, A(x_1)A(x_2)A(x_3)\Omega)\\ &= 16\,\mathcal{W}_2(x_1,\ x_2)\mathcal{W}_2(x_1,\ x_3)\mathcal{W}_2(x_2,\ x_3).\end{aligned} \tag{7.71}$$

The last expression cannot vanish in a real open set, therefore A does not satisfy exotic locality.

References

1. M. Bertola, J. Bros, U. Moschella and R. Schaeffer: A general construction of conformal field theories from scalar anti-de Sitter quantum field theories. *Nucl. Phys. B* 587:619–644 (2000). [hep-th/9908140].
2. J.J. Bisognano and E.H. Wichmann: On the duality condition for a Hermitian scalar field. *J. Math. Phys.* 16:985 (1975).
3. H.-J. Borchers: On the structure of the algebra of field observables. *Nuovo Cimento* 24:214 (1962).
4. J. Bros, H. Epstein and U. Moschella: Towards a General Theory of Quantized Fields on the Anti-de Sitter Space-Time. *Commun. Math. Phys.* 231:481-528 (2002). [hep-th/0111255].
5. C. Fronsdal: Elementary Particles In A Curved Space. II. *Phys. Rev. D* 10:589-598 (1974).
6. V. Glaser: On the equivalence of the Euclidean and Wightman formulations of field theory. *Commun. Math. Phys.* 37:257–272 (1974).
7. R. Jost: *The general theory of quantized fields.* Providence. A.M.S., Rhode Island, 1965.
8. J. Maldacena: The large N limit of superconformal field theories and supergravity. *Adv. Theor. Math. Phys.* 2:231–252 (1998). [hep-th/9711200].
9. K. Osterwalder and R. Schrader: Axioms for Euclidean Green's functions. *Commun. Math. Phys.* 33:83–112 (1973).
10. K. Osterwalder and R. Schrader: Axioms for Euclidean Green's functions, II. *Commun. Math. Phys.* 42:281–305 (1975).
11. R.F. Streater and A.S. Wightman: *PCT, Spin and Statistics, and all that.* W.A. Benjamin, New Yoek, 1964.

8

Quantum Energy Inequalities and Stability Conditions in Quantum Field Theory

Christopher J. Fewster

Department of Mathematics, University of York, Heslington, York YO10 5DD, United Kingdom
cjf3@york.ac.uk

Summary. Quantum Energy Inequalities (QEIs) are constraints on the extent to which quantum fields can violate the energy conditions of classical general relativity. As such they are closely related to the gravitational stability of quantised matter. In this contribution we discuss links between QEIs and other stability conditions in quantum field theory: the microlocal spectrum condition, passivity and nuclearity. The fist two links suggest an interconnection between stability conditions at three different length scales, while the third hints at a deeper origin of QEIs.

8.1 Introduction

The stress-energy tensor T_{ab} of the real scalar field, in common with those corresponding to most models of classical matter,[1] obeys the dominant energy condition (DEC): for any future-directed timelike vector u^a, the contraction $T^a{}_b u^b$ is itself timelike and future-directed. This may also be stated as the inequality $T_{ab}u^a v^b \geq 0$ for all pairs of future-directed timelike vectors u^a and v^a. In the special case $v^a = u^a$, we recover the weak energy condition (WEC), $T_{ab}u^a u^b \geq 0$, i.e., the energy density is nonnegative according to any observer.

In classical general relativity, energy conditions of this type play a key role, guaranteeing the stability of gravitational collapse (singularity theorems [29]), the stability of Minkowski space as a 'ground state' of the theory (positive mass theorems [35,48]) and also excluding certain exotic causal structures (see e.g., Hawking's discussion of chronology protection [28]). However, as has been known for a long time [6], the WEC (and hence DEC) are violated in quantum field theory. It is easy to give a simple proof in Minkowski space: suppose that $\varrho = T_{ab}u^a u^b$, where u^a is now a smooth timelike vector field, and let f be a nonnegative smooth function of compact support. We need assume only that the smeared field $\varrho(f)$ is (essentially) self-adjoint on the Hilbert space of the theory and that there is a vacuum state Ω in the operator domain

[1]The main exception is the nonminimally coupled scalar field, satisfying the field equation $(\Box + m^2 + \xi R)\phi = 0$ with $\xi \neq 0$.

of $\varrho(f)$ such that $\langle \Omega \,|\, \varrho(f)\Omega \rangle = 0$ but which is not annihilated by $\varrho(f)$, i.e., $\varrho(f)\Omega \neq 0$. Writing the spectral measure of $\varrho(f)$ as $\Delta E(\lambda)$, these last two properties tell us that the probability measure $\langle \Omega \,|\, \Delta E(\lambda)\Omega \rangle$ on $\mathbb{R}$ has zero expectation, but that its support is not simply the set $\{0\}$. Accordingly $\varrho(f)$ must have some negative spectrum.[2] Clearly the same argument applies in many circumstances, and for observables other than energy density.

Thus, the classical pointwise energy conditions are simply incompatible with the structures of quantum field theory. Further analysis of particular models shows that the pointwise energy density is typically unbounded from below as a function of the quantum state, and this can be proved for all theories with a suitable scaling limit [9].

This fact raises questions concerning the applicability of the singularity, positive mass and chronology protection results where quantised matter is concerned. Many authors have also sought to exploit quantum fields to support metrics (including wormhole or warp drive models) which require WEC-violating matter distributions. It is therefore important to understand whether the classical energy conditions are irretrievably lost, or whether one can identify some remnant in the quantum theory. This contribution will discuss a promising candidate: a group of results known as Quantum Energy Inequalities (QEIs), and will in particular focus on their emerging connections with other well-known stability conditions in quantum field theory, namely the microlocal spectrum condition, passivity and nuclearity. The hope is that, by unravelling these connections, further insight is provided into the nature of quantised matter and its (gravitational) stability.

It is a particular pleasure to dedicate this contribution to Jacques Bros, in view of his influential contributions to both microlocal analysis and the description of thermal behaviour in quantum field theory.

8.2 Quantum Energy Inequalities

As mentioned above, the pointwise energy conditions are unavoidably and severely violated in quantum field theory. However, observations at individual spacetime points are not physically achievable in any case (owing to the uncertainty principle), so it is more natural to consider weighted averages of the stress-energy tensor over a spacetime volume.

Definition 1. *Let $\mathscr{W}$ be a class of second-rank tensors on spacetime, and $\mathcal{S}$ a class of states of the theory. If, for each $\mathsf{f} \in \mathscr{W}$, the averaged expectation values $\int \Delta\mathrm{vol}(x)\langle T_{ab}(x)\rangle_\omega f^{ab}(x)$ are bounded from below as ω runs over $\mathcal{S}$, we say that the theory obeys a* Quantum Energy Inequality *(QEI) with respect to $\mathscr{W}$ and $\mathcal{S}$.*

---

[2] The same conclusion is easily drawn by examining the expectation values of $\Omega + \lambda\varrho(f)\Omega$ for small λ.

One generally aims to find an explicit lower bound $-\mathcal{Q}[\mathsf{f}]$ so that the QEI can be written as an inequality

$$\int \Delta\mathrm{vol}(x)\langle T_{ab}(x)\rangle_\omega f^{ab}(x) \geq -\mathcal{Q}[\mathsf{f}] \qquad \forall\omega\in\mathcal{S}\,.$$

Where $\mathscr{W}$ consists of tensors of a particular form e.g., $f^{ab} = u^a u^b$ or $f^{ab} = u^a v^b$ for timelike vector fields u^a, v^a, we use more specific terms, e.g., Quantum Weak Energy Inequality (QWEI) or Quantum Dominated Energy Inequality (QDEI). Of course a similar approach could be adopted for other quantities of interest.

For the most part, QEIs have been developed for averages along timelike curves, rather than over spacetime volumes, in which case the weights may be thought of as being singularly supported on a curve. By threading a spacetime volume by worldlines, these bounds imply the existence of spacetime-averaged QEIs, which may also be obtained directly, as sketched below. It is known that compactly supported weighted averages over spacelike hypersurfaces [23] or null lines [15] are not generally bounded from below, except for two-dimensional conformal fields [11,19].

QEIs were first proposed by Ford [21], who realised that suitable bounds of this type would be sufficient to prevent macroscopic violations of the second law of thermodynamics arising from negative energy phenomena in quantum field theory. They have since been established for the free Klein–Gordon [8,10, 16,19,20,22,24,26,39,47], Dirac [12,17,47], Maxwell [14,26,37] and Proca [14] quantum fields in both flat and curved spacetimes, the Rarita–Schwinger field in Minkowski space [49], and also for general unitary positive-energy conformal field theories in two-dimensional Minkowski space [11]. We will not give a full history of the development of the subject, referring the reader to the recent reviews [9, 42]. To give a flavour of the sort of results obtained, we give an example in which the energy density of a scalar field of mass m is averaged along the inertial trajectory $(t,0)$ in Minkowski space. It can be obtained by elementary means [10] or as a special case of the rigorous result [8]. Set $\rho = T_{ab}u^a u^b$, where $\mathsf{u} = \partial/\partial t$. Then the QWEI

$$\int \Delta t\,\langle \varrho(t,0)\rangle_\psi |g(t)|^2 \geq -\mathcal{Q}[g] := -\frac{1}{16\pi^3}\int_0^\infty \Delta u\, u^4\vartheta(u-m)|\widehat{g}(u)|^2\,, \quad (8.1)$$

holds for all Hadamard states ψ (see below) and smooth compactly supported g. Here $\widehat{g}$ denotes the Fourier transform[3] and ϑ is the Heaviside function. The bound is finite, owing to the rapid decay of $\widehat{g}$. In fact the bound given in [10] is slightly tighter than this, but (8.1) will suffice for our present purposes.

For later reference, let us note the scaling behaviour of the bound (8.1). Replacing g by $g_\tau(t) = \tau^{-1/2}g(t/\tau)$, so that τ controls the 'spread' of the weight, one may show that

[3]Our convention for the Fourier transform is $\widehat{g}(u) = \int \Delta t\, e^{iut}g(t)$ etc.

$$\mathcal{Q}[g_\tau] = \begin{cases} O(\tau^{-4}) & \text{as } \tau \to 0^+ \\ O(\tau^{-\infty}) & \text{as } \tau \to \infty \end{cases}$$

for $m > 0$, where the notation $O(\tau^{-\infty})$ indicates faster-than-inverse-polynomial decay. In the massless case, it turns out that $\mathcal{Q}(g_\tau) \propto \tau^{-4}$ for all $\tau > 0$. We note that the $\tau \to 0^+$ limit, which corresponds to sampling at a point, is consistent with the pointwise unboundedness below of the energy density. For intermediate scales, the QWEI allows for a limited violation of the classical WEC; bounds of this type therefore appear to be the natural remnant of the WEC in quantum field theory.

We mention briefly that related bounds appear elsewhere in quantum field theory [36] and quantum mechanics [7]; QEIs have also been used to place constraints on exotic spacetimes [25, 38, 42].

8.3 Stability at Three Scales

The work described in this section, conducted with Verch [18] and building on earlier work [8, 43], uncovers a circle of connections between stability conditions operating at three different scales: the microscopic (Hadamard condition/microlocal spectrum condition), mesoscopic (QEIs) and macroscopic (thermodynamic stability, expressed by the notion of passivity [40]). Each connection takes the form of a rigorous theorem; the reader should be cautioned, however, that the conclusions and hypotheses of successive links do not match perfectly. Moreover, two of the links (mesoscopic to macroscopic, and macroscopic to microscopic) are obtained in greater generality than the particular setting of quantum field theory on curved spacetimes, while the microscopic to mesoscopic link is currently known only for particular models of quantum field theory. Thus the existence of these connections should be regarded as indicative of a close relationship between these three stability conditions, rather than of proving their equivalence. In part, this work gives a precise expression to Ford's original insight [21], that bounds of QEI type would suffice to prevent macrosopic violations of the second law of thermodynamics.

8.3.1 Microscopic Stability: the Hadamard Condition

Stability of quantum field theory at the microscopic scale is (partly) expressed by the Hadamard condition, which requires that the singular structure of the two-point function takes a form determined for nearby points by the local geometry [33] of spacetime. As first shown by Radzikowski [41], this may be reformulated as a condition on the wave-front set [31] of the two-point function.[4] By passing to a Hilbert space representation, however, one obtains

[4]Appropriate conditions on higher n-point functions were given in [2]. For non-initiates: the wave-front set $\mathrm{WF}(S)$ of a distribution S on a manifold M is a subset of

a very simple formulation of the Hadamard condition [18, 46] (cf. also [1]): a state of the scalar field on (M, g) is Hadamard if and only if it may be represented by a vector ψ in some Hilbert space representation of the theory so that $f \mapsto \Phi(f)\psi$ is a vector-valued distribution whose wave-front set obeys

$$\mathrm{WF}(\Phi(\cdot)\psi) \subset \mathcal{V}^- , \tag{8.2}$$

where Φ is the field and

$$\mathcal{V}^- = \{(x, \mathsf{k}) \in T^*M : g^{ab}k_a k_b \geq 0,\ \mathsf{k}\text{ past directed}\}$$

is the bundle of past-pointing causal covectors (our signature convention is $+---$). This has the following practical upshot. Suppose f is smooth and compactly supported within some coordinate patch, with coordinates x^α so that $\partial/\partial x^0$ is future-pointing and timelike. Let V be any closed cone in $\mathbb{R}^4$ consisting of k such that the covector field $k_\alpha \Delta x^\alpha$ is nowhere causal and past-directed on the coordinate patch. In particular, V could be the half-space $V = \{k \in \mathbb{R}^4 : k_0 \geq 0\}$. Then

$$I(k) := \left\| \int \Delta^4 x\, \mathrm{e}^{ik_\alpha x^\alpha} f(x)\Phi(x)\psi \right\|$$

is of rapid decay in V; that is, it decays more rapidly than any inverse polynomial in the Euclidean norm of k as $k \to \infty$ in V. Moreover, the same is true if f is replaced by a partial differential operator with smooth coefficients compactly supported in the coordinate patch.

Microlocal formulations of the Hadamard condition are also known for the Dirac [30, 34, 44], Maxwell and Proca fields [14]. They may be regarded as local remnants of the spectrum condition, i.e., the Minkowski space requirement that the joint spectrum of the generators P_μ of spacetime translations should lie in the future causal cone.[5]

8.3.2 From Microscopic to Mesoscopic

We now show how QEIs may be derived from the Hadamard condition, using an argument based on that of [8]. The classical Klein–Gordon field ϕ obeying $(\Box + m^2)\phi = 0$ on spacetime (M, g) has stress-energy tensor

$$T_{ab} = \nabla_a\phi\nabla_b\phi - \frac{1}{2}g_{ab}g^{cd}\nabla_c\phi\nabla_d\phi + \frac{1}{2}g_{ab}m^2\phi^2 ,$$

---

the cotangent bundle T^*M which encodes the singular structure of S. Singularities are classified in terms of the (lack of) decay of local Fourier transforms of S in different directions.

[5]That the forward cone appears in the spectrum condition, but the backward cone in (8.2), is the result of an unfortunate clash of conventions.

and obeys the WEC and DEC because the relevant contractions of T_{ab} can be decomposed as sums of squares. Let us therefore consider — as representing the most general classical energy condition — any tensor field f^{ab} for which

$$T_{ab}f^{ab} = \frac{1}{2}\sum_j (P_j\phi)^2 , \tag{8.3}$$

where the P_j are finitely many linear partial differential operators (possibly of degree zero) with smooth real coefficients of compact support. Clearly $T_{ab}f^{ab} \geq 0$ for classical fields ϕ. For simplicity, assume that the supports of the P_j are contained within a single coordinate patch of (M, g) writing the coordinates as x^α and assuming as above that $\partial/\partial x^0$ is future-pointing and timelike. Write also $g(x) = |\det g_{\alpha\beta}(x)|$.

Let ψ_0 be a fixed Hadamard reference state. Defining the stress-energy tensor $:T_{ab}:$ by point-splitting and normal ordering with respect to ψ_0, we have

$$\langle :T_{ab}(x): \rangle_\psi f^{ab}(x) = \frac{1}{2\sqrt{g(x)}} F(x,x)$$

for any Hadamard state ψ, where

$$F(x,y) = (g(x)g(y))^{1/4} \sum_j \Big[\langle (P_j\Phi)(x)(P_j\Phi)(y)\rangle_\psi - \langle (P_j\Phi)(x)(P_j\Phi)(y)\rangle_{\psi_0} \Big]$$

is smooth (owing to the common singularity structure of Hadamard two-point functions) and symmetric (because two-point functions have a state-independent antisymmetric part). Thus we may write

$$\begin{aligned} \int \Delta\mathrm{vol}_{\mathbf{g}}(x) \langle :T_{ab}(x): \rangle_\psi f^{ab}(x) &= \frac{1}{2}\int \Delta^4 x\, \Delta^4 y\, F(x,y)\delta^{(4)}(x-y) \\ &= \int_{k_0 \geq 0} \frac{\Delta^4 k}{(2\pi)^4} \int \Delta^4 x\, \Delta^4 y\, \mathrm{e}^{-ik\cdot(x-y)} F(x,y) , \end{aligned}$$

where we have used the Fourier representation of the Dirac-δ and symmetry of F to restrict the outer domain of integration. Now the inner integral is $A(k;\psi) - A(k;\psi_0)$, where

$$A(k;\psi) := \sum_j \left\| \int \Delta^4 x\, \mathrm{e}^{ik\cdot x} g(x)^{1/4} P_j \Phi(x)\psi \right\|^2 \geq 0 ,$$

and so we obtain the QEI

$$\int \Delta\mathrm{vol}_{\mathbf{g}}(x)\, \langle :T_{ab}(x): \rangle_\psi f^{ab}(x) \geq -\int_{k^0 \geq 0} \frac{\Delta^4 k}{(2\pi)^4} A(k;\psi_0) , \tag{8.4}$$

the left-hand side of which depends on the reference state ψ_0, but not on ψ. The key point now is that the microlocal form of the Hadamard condition entails that $A(k;\psi_0)$ is of rapid decay in the half-space $k_0 \geq 0$. Thus the integral on the right-hand side of (8.4) exists and is finite. We conclude that the real linear scalar field obeys a QEI with respect to the class of weights delineated by (8.3) and the class of Hadamard states. The same argument would apply to a suitable class of adiabatic states [32] in which one replaces the smooth wave-front set by a wave-front set modulo Sobolev regularity.

Note that this QEI applies to the normal ordered stress-energy tensor, rather than the renormalised tensor.[6] By adding a term to both sides, which depends on the renormalised stress-energy tensor in state ψ_0 and certain other smooth local geometric terms, this defect can be remedied. (The bound is then typically not a 'closed form' expression.)

8.3.3 Macroscopic Stability: Passivity

Pusz and Woronowicz introduced the notion of passivity in the following way [40]. Let $(\mathcal{A}, \alpha_t)$ be a C^*-dynamical system; that is, $\mathcal{A}$ is a C^*-algebra, which we think of as the algebra of observables for some quantum system, while α_t is the map of evolution through time $t \in \mathbb{R}$ corresponding to the undisturbed evolution of the system, and has the group property $\alpha_t \circ \alpha_{t'} = \alpha_{t+t'}$. Provided α_t is strongly continuous (i.e., the map $\mathbb{R} \ni t \mapsto \alpha_t(A) \in \mathcal{A}$ is continuous for each $A \in \mathcal{A}$) we may define the generator δ of the evolution by

$$\delta(A) = \frac{\Delta}{\Delta t}\alpha_t(A)\bigg|_{t=0}$$

for the space of A for which the derivative exists (in $\mathcal{A}$), which we denote $D(\delta)$. For example, if $\mathcal{A}$ is the algebra of bounded operators on the Hilbert space of a quantum mechanical system with Hamiltonian H, then $\delta(A) = i[H, A]$. We also have $\delta(A) = \alpha_t^{-1}\left(\Delta/\Delta t\, \alpha_t(A)\right)$ for any t. The motivating idea of [40] is to understand thermodynamic stability of the dynamical system with respect to cyclical changes of external conditions. One might think of a box of gas which is compressed and then allowed to return to its initial volume. In the current setting, a cyclical process occuring during time interval $[0, T]$ may be modelled by a perturbed time evolution β_t satisfying

$$\beta_t^{-1}\left(\frac{\Delta}{\Delta t}\beta_t(A)\right) = \delta(A) + i[h_t, A] ,$$

and $\beta_0 = \mathrm{id}$ where $t \mapsto h_t$ is a differentiable assignment of a self-adjoint element $h_t \in \mathcal{A}$ to each time t, and $h_t = 0$ for $t \notin [0, T]$.

Suppose the system is initially in state ω. Then the work performed by the external agent driving the cyclical process is

---

[6] To form the renormalised tensor, we begin by splitting points as above, but then subtract appropriate derivatives of the locally determined Hadamard parametrix, rather than the two-point function of a reference state.

$$W_h = \int_0^T \Delta t\, \omega(\beta_t(\dot{h}_t)) ,$$

and the state ω is said to be *passive* if $W_h \geq 0$ for any h_t, i.e., if no cyclical process can extract energy from the system. Thus passivity isolates the property characteristic of the second law of thermodynamics in Kelvin's formulation, where we think of the system as a thermal reservoir from which we attempt to extract work.

Pusz and Woronowicz proved

Theorem 1. *A state ω is passive if and only if*

$$i^{-1}\omega(U^*\delta(U)) \geq 0$$

for all $U \in \mathcal{U}_1(\delta) := \mathcal{U}_1(\mathcal{A}) \cap D(\delta)$, where $\mathcal{U}_1(\mathcal{A})$ is the identity-connected component of the unitary elements of $\mathcal{A}$.

Particular examples of passive states are provided by ground and KMS states, or mixtures thereof. A key feature of passivity is that it introduces a definite thermodynamic 'arrow of time'.

8.3.4 From Mesoscopic to Macroscopic

Let us now see how passivity may be obtained from QEIs, giving a simplified and slightly modified version of the discussion in [18]. We begin by introducing an abstract formulation of QEIs for C^*-dynamical systems, to which end we must first provide a notion of the energy density. Accordingly, we assume that $\mathcal{A}$ is the algebra of observables of a system in a spacetime of the form $\mathbb{R} \times \Sigma$, for Σ compact and Riemannian, with volume measure $\Delta\mu(\underline{x})$. The evolution α_t corresponds to time-translations on spacetime and is assumed to be strongly continuous with generator δ.

As one would not expect the energy density to exist for all states, we must specify a smaller class of states and a class of unitary elements large enough to be dense in $\mathcal{U}_1(\delta)$, in a suitable sense, but which preserves the state space. Accordingly, let $\mathcal{O}$ be a $*$-subalgebra of $\mathcal{A}$ with $\mathbf{1} \in \mathcal{O} \subset \bigcap_n D(\delta^n)$, and be large enough that any element of $\mathcal{U}_1(\delta)$ may be approximated arbitrarily well by unitary elements of $\mathcal{O}$ with respect to the graph norm of δ. That is, to any $U \in \mathcal{U}_1(\delta)$ there is a sequence of unitaries $U_n \in \mathcal{O}$ with $U_n \to U$ and $\delta(U_n) \to \delta(U)$. In addition, let $\mathcal{S}$ be a convex set of states of $\mathcal{A}$ which is closed under operations in $\mathcal{O}$.[7]

The energy density $\varrho(t, \underline{x})$ is assumed to obey:

1. For each $A, B \in \mathcal{O}$ and $\varphi \in \mathcal{S}$, $\varphi(A\varrho(t, \underline{x})B)$ is a C^1 function on $\mathbb{R} \times \Sigma$.

[7]That is, for any $0 \neq A \in \mathcal{O}$ and $\omega \in \mathcal{S}$, we have $\omega(A^*A) > 0$ and $\omega^A(B) = \omega(A^*BA)/\omega(A^*A)$ defines a state $\omega^A \in \mathcal{S}$.

2. The energy density generates the dynamics, and energy is conserved, i.e,

$$\int_\Sigma \Delta\mu(\underline{x})\varphi(A[\varrho(t,\underline{x}),B]C) = \frac{1}{i}\varphi(A\delta(B)C) \tag{8.5}$$

   for arbitrary $A, B, C \in \mathcal{O}$, $\varphi \in \mathcal{S}$ and $t \in \mathbb{R}$.

Here, expressions of the form $\varphi(A\varrho(t,\underline{x})B)$ should be taken as a convenient shorthand: what is more precisely meant is the following. Let $\mathfrak{F}$ be the subspace of continuous linear functionals on $\mathcal{A}$ generated by functionals of the form $C \mapsto {}_A\varphi_B(C) := \varphi(ACB)$ (for $A, B \in \mathcal{O}$, $\varphi \in \mathcal{S}$). Then the energy density is a linear map $\boldsymbol{\varrho} : \mathfrak{F} \to C^1(\mathbb{R} \times \Sigma)$, and our shorthand notation $\varphi(A\varrho(t,\underline{x})B)$ means $(\boldsymbol{\varrho}({}_A\varphi_B))(t,\underline{x})$.

We are now in a position to define a general type of QWEI in this setting, by analogy with the result (8.1). Our definition differs slightly from that given in [18].

Definition 2. *Let $\mathscr{W}$ be a class of nonnegative integrable functions of compact support on $\mathbb{R}$. The system $(\mathcal{A}, \alpha_t, \mathcal{O}, \mathcal{S}, \varrho)$ obeys a static QWEI (SQWEI) with respect to $\mathscr{W}$ if, for some $\omega \in \mathcal{S}$, there exists a map $q_\omega : \mathscr{W} \to L^1(\Sigma)$ such that*

$$\int \Delta t\, f(t)\varphi(:\varrho(t,\underline{x}):) \geq -q_\omega(f)(\underline{x}) \qquad \mu\text{-a.e. in } \underline{x} \tag{8.6}$$

for all $\varphi \in \mathcal{S}$, where $:\varrho: = \varrho - \omega(\varrho)\mathbf{1}$. (In this case, the same is true for all $\omega' \in \mathcal{S}$, as we may take $q_{\omega'}(f)(\underline{x}) = q_\omega(f)(\underline{x}) + \int \Delta t\, f(t)\omega'(:\varrho(t,\underline{x}):)$.)

We now state and prove one of the main results of [18].

Theorem 2. *If $(\mathcal{A}, \alpha_t, \mathcal{O}, \mathcal{S}, \varrho)$ obeys a SQWEI, then $(\mathcal{A}, \alpha_t)$ admits at least one passive state.*

Proof. Fix a reference state $\omega \in \mathcal{S}$ and choose $f \in \mathscr{W}$ with $\int \Delta t\, f(t) = 1$ (we may assume $\mathscr{W}$ is conic without loss). For unitary $U \in \mathcal{O}$,

$$\begin{aligned}
\frac{1}{i}\omega(U^*\delta(U)) &= \int_\Sigma \Delta\mu(\underline{x})\, \omega(U^*[\varrho(t,\underline{x}),U]) \\
&= \int_\Sigma \Delta t\, f(t) \int \Delta\mu(\underline{x})\, \omega(U^*[\varrho(t,\underline{x}),U]) \\
&= \int_\Sigma \Delta\mu(\underline{x}) \int \Delta t\, f(t)\omega(U^*{:}\varrho(t,\underline{x}){:}U) \\
&\geq -\int_\Sigma \Delta\mu(\underline{x})\, q_\omega(f)(\underline{x})\,,
\end{aligned} \tag{8.7}$$

where we apply (8.6) with φ defined by $\varphi(A) = \omega(U^*AU)$. Because unitary elements of $\mathcal{O}$ provide arbitrarily good approximations to elements of $\mathcal{U}_1(\delta)$ we may choose unitaries $U_n \in \mathcal{O}$ such that

$$\frac{1}{i}\omega(U_n^*\delta(U_n)) \longrightarrow c_\omega := \inf_{U\in\mathfrak{U}_1(\delta)} \frac{1}{i}\omega(U^*\delta(U)) \,, \tag{8.8}$$

as $n \to \infty$, thereby deducing that

$$c_\omega \geq -\int_\Sigma \Delta\mu(\underline{x})\, q_\omega(f)(\underline{x}) > -\infty \,. \tag{8.9}$$

If $c_\omega \geq 0$ then ω is passive and we are done, so suppose instead that $c_\omega < 0$. By the Banach–Alaoglu Theorem there exists a state ω^p on $\mathcal{A}$ and a subnet $U_{n(\sigma)}$ of the U_n such that

$$\omega^p(A) = \lim_\sigma \omega(U_{n(\sigma)}^* A U_{n(\sigma)}) \qquad A \in \mathcal{A} \,. \tag{8.10}$$

To complete the proof, we calculate

$$\begin{aligned}\frac{1}{i}\omega^p(U^*\delta(U)) &= \lim_\sigma \frac{1}{i}\omega(U_{n(\sigma)}^* U^*\delta(U) U_{n(\sigma)}) \\ &= \lim_\sigma \big[\underbrace{i^{-1}\omega((UU_{n(\sigma)})^*\delta(UU_{n(\sigma)}))}_{\geq c_\omega} - \underbrace{i^{-1}\omega(U_{n(\sigma)}^*\delta(U_{n(\sigma)}))}_{\to c_\omega}\big] \\ &\geq 0 \,,\end{aligned} \tag{8.11}$$

so ω^p is passive. □

In [18] we also defined the notion of a state ω being *quiescent*, in terms of the behaviour of function $q_\omega(f_\lambda)$ in the limit $\lambda \to 0^+$, where $f_\lambda(t) = f(\lambda t)$. We showed that quiescent states are passive (and even ground states, under additional clustering assumptions).

Of course, we would like to see that this abstract set-up can be realised in practice, and in particular, that it applies to quantum field theory in static spacetimes with compact spatial section. Here, we encounter a problem with the scalar field because its C^*-algebraic description in terms of the Weyl algebra with generators $W(F)$ is not a C^*-dynamical system with respect to the time-translations

$$\alpha_t W(F) = W(F_t) \quad \text{where} \quad F_t(\tau, \underline{x}) = F(\tau - t, \underline{x}) \,. \tag{8.12}$$

(This problem would not occur with the Dirac field, but less was known about Dirac QEIs when [18] was written!) Instead one can generate $\mathcal{A}$ from objects of the form

$$\int \Delta t\, h(t) \alpha_t W(F) \qquad (h \in C_0^\infty(\mathbb{R})) \tag{8.13}$$

formed in quasifree Hadamard Hilbert space representations of the Weyl algebra; as shown in [18], all the requirements of the abstract setting are fulfilled with $\mathcal{S}$ equal to the set of finite convex combinations of Hadamard states occurring as vectors in quasifree Hadamard representations of the Weyl algebra. (Microlocal techniques turn out to be exactly the right tools for this nontrivial check.) The $*$-algebra $\mathcal{O}$ is generated by operators of the form $\exp iA$, where $A = A^*$ is a polynomial in objects of the type (8.13).

A further problem, however, is that the passive state obtained from the Banach–Alaoglu theorem lives on $\mathcal{A}$, rather than the Weyl algebra itself. Given sufficient regularity (e.g., energy compactness, believed to hold for this theory) we may reconstruct a passive state on the Weyl algebra [18]. Again, this problem would not arise for the Dirac field.

8.3.5 From Macroscopic to Microscopic

Finally, we briefly discuss the last link in our circle of stability conditions. In [43], Sahlmann and Verch considered general topological $*$-dynamical systems and defined a *strictly passive* state to be a mixture of ground and KMS states (at possibly different inverse temperatures). Note that this is a stronger requirement than the usual notion of passivity, as employed in [18, 40]. They also introduced the notion of an *asymptotic n-point correlation spectrum* which generalises the wave-front set to this setting, and formulated an appropriate generalisation of the microlocal spectrum condition. When applied to linear quantum field theory on stationary spacetimes, with respect to the stationary time evolution, the original microlocal spectrum condition is recovered. They then proved that strictly passive states obey the generalised microlocal spectrum condition: the key ingredient in their argument is that both (strict) passivity and the microlocal spectrum conditions share a common arrow of time.

8.4 Connections with Nuclearity

Quite recently, evidence has emerged to suggest the existence of a connection between QEIs and nuclearity criteria, with possibly far-reaching implications. We will consider the original nuclearity condition of Buchholz and Wichmann [5] (for other closely related criteria see, e.g., [4]). We work within the algebraic approach to quantum field theory [27], and consider a quantum field theory described by a Hilbert space $\mathscr{H}$, a strongly continuous unitary representation $g \mapsto U(g)$ on $\mathscr{H}$ of the universal cover of the proper orthochronous Poincaré group $\widetilde{\mathscr{P}_+^\uparrow}$, and a net of von Neumann algebras $\mathcal{R}(\mathcal{O})$, consisting of bounded operators on $\mathscr{H}$ and indexed by open bounded contractible spacetime regions $\mathcal{O}$. The following axioms are assumed to hold: *isotony* ($\mathcal{O}' \subset \mathcal{O}$ implies $\mathcal{R}(\mathcal{O}') \subset \mathcal{R}(\mathcal{O})$); *covariance* ($U(g)\mathcal{R}(\mathcal{O})U(g)^{-1} = \mathcal{R}(g\mathcal{O})$ for $g \in \widetilde{\mathscr{P}_+^\uparrow}$); *locality* ($\mathcal{R}(\mathcal{O})$ and $\mathcal{R}(\mathcal{O}')$ commute if $\mathcal{O}$ and $\mathcal{O}'$ are spacelike separated) and the *spectrum condition* (the generators of spacetime translations, P_μ, associated with the representation U, are self-adjoint operators such that P_0 and $P_0^2 - P_1^2 - P_2^2 - P_3^2$ are positive). Finally, we assume the existence of a unique vacuum state: namely, that the Hamiltonian $H = P_0$ has a simple eigenvalue at zero with normalised eigenvector Ω.

Given any double cone $\mathcal{O}_r$ based on a ball of radius r and any $\beta > 0$, let

$$\mathcal{N}_{\beta,r} = \{e^{-\beta H} W\Omega : W \in \mathcal{R}(\mathcal{O}_r) \text{ s.t. } W^*W = \mathbf{1}\} . \tag{8.14}$$

This set may be regarded as the set of local vacuum excitations associated with $\mathcal{O}_r$, damped exponentially in the energy. The theory is said to obey the *condition of nuclearity* if, firstly, each $\mathcal{N}_{\beta,r}$ is a *nuclear* subset of $\mathcal{H}$ [see below] and, secondly, there exist positive constants c, n, r_0 and β_0 so that the corresponding *nuclearity index* $\nu(\mathcal{N}_{\beta,r})$ obeys

$$\nu(\mathcal{N}_{\beta,r}) \leq \exp\left(cr^3\beta^{-n}\right) \tag{8.15}$$

for all $0 < \beta < \beta_0$ and $r > r_0$. This condition is therefore a restriction on the number of local degrees of freedom available to the theory.

In the above, a subset $\mathcal{L}$ of $\mathcal{H}$ has nuclearity index $\nu(\mathcal{L}) = \inf \operatorname{Tr}|T|$, where the infimum is taken over the set of trace-class operators T so that $\mathcal{L}$ is contained within the image of the unit ball $\mathcal{H}_{(1)}$ of $\mathcal{H}$ under T, and $\mathcal{L}$ is said to be nuclear if it has a finite nuclearity index.[8]

Despite its rather technical definition, the condition of nuclearity is well-motivated from a physical viewpoint as the discussion in [5] makes plain: the nuclearity index can be interpreted as a local partition function, and the form of the nuclearity bound (8.15) is suggested by the requirement that the associated pressure should remain finite in the thermodynamic limit and scale polynomially with temperature (as is the case, for example, in the Stefan–Boltzmann law).

Buchholz and Wichmann verified in [5] that the massive free scalar field satisfies the condition of nuclearity, and remark that the same is true of the system of countably many fields with masses m_j given suitable conditions on the density of states. Namely, the sets $\mathcal{N}_{\beta,r}$ are nuclear if [5] and only if [3] $\sum_j \exp(-\beta m_j) < \infty$ for all sufficiently small β; furthermore, the nuclearity index may be estimated from above by

$$\nu(\mathcal{N}_{\beta,r}) \leq \exp\left(c\left(\frac{r}{\beta}\right)^3 \sum_j \left|\log(1 - e^{-\beta m_j/2})\right|\right) \tag{8.16}$$

for all sufficiently large r and small β, and some constant c. It is convenient to introduce $N(u)$, the number of particle species with mass below u by

$$N(u) = \sum_j \vartheta(u - m_j) . \tag{8.17}$$

The assumption that $N(u)$ grows polynomially, $N(u) = O(u^p)$ as $u \to \infty$, is sufficient to show (using (8.16)) that (8.15) is satisfied, for any $n > 3 + p$.

[8] By convention, an infimum over an empty set is infinite, so this amounts to the assertion that there does exist a trace-class T with $\mathcal{L} \subset T\mathcal{H}_{(1)}$.

It is tempting to conjecture that this condition is also necessary, but this is currently an open question, and relies on finding better lower bounds on the nuclearity index than are currently known. We will return to this point below.

We now present some circumstantial evidence for a connection between nuclearity criteria and QEIs. Fix some inertial frame of reference in Minkowski space and let ϱ_j be the energy density of the free field of mass m_j with Hilbert space $\mathscr{H}_j$ and vacuum state Ω_j. Let $\mathsf{Had}_j \subset \mathscr{H}_j$ be the corresponding space of Hadamard vector states. The Hilbert space of the full theory is the tensor product

$$\mathscr{H} = \bigotimes_j {}^{\Omega_j} \mathscr{H}_j \, ; \tag{8.18}$$

that is, the completion with respect to the obvious inner product of the set of finite linear combinations of product states $\bigotimes_j \xi_j$ in which all but finitely many of the ξ_j are equal to Ω_j. We define the space of Hadamard states Had of the full theory to consist of finite linear combinations of product states $\bigotimes_j \xi_j$ in which each $\xi_j \in \mathsf{Had}_j$ and all but finitely many ξ_j equal Ω_j, and then define the total energy density as follows: for any $\eta = \bigotimes_j \eta_j$ and $\xi = \bigotimes_j \xi_j$ in Had we set

$$\langle \eta \,|\, \varrho(x)\, \xi \rangle = \sum_j \langle \eta_j \,|\, \varrho_j(x)\, \xi_j \rangle \prod_{k \neq j} \langle \eta_k \,|\, \xi_k \rangle \tag{8.19}$$

(noting that only finitely many terms contribute to the sum, and that each product involves only finitely many terms differing from unity) and then extend by linearity to all $\eta, \xi \in \mathsf{Had}$. The left-hand side should be regarded as a quadratic form on Had, taking values in the space of smooth functions on spacetime; clearly, any normal-ordered quantity could be treated in this way, and no constraints on the m_j have been imposed. Since the j'th component of the full theory obeys the QWEI (8.1) for each mass m_j,

$$\int \Delta t \, |g(t)|^2 \langle \psi_j \,|\, \varrho_j(t,0)\, \psi_j \rangle \geq -\frac{\|\psi_j\|^2_{\mathscr{H}_j}}{16\pi^3} \int_0^\infty \Delta u \, |\widehat{g}(u)|^2 u^4 \vartheta(u - m_j) \, , \tag{8.20}$$

for all Hadamard states $\psi_j \in \mathsf{Had}_j$, the full theory obeys

$$\int \Delta t \, |g(t)|^2 \langle \psi \,|\, \varrho(t,0)\, \psi \rangle \geq -\frac{1}{16\pi^3} \int_0^\infty \Delta u \, |\widehat{g}(u)|^2 u^4 N(u) \, , \tag{8.21}$$

for any normalised $\psi \in \mathsf{Had}$. Accordingly, polynomial growth of N is sufficient for the theory to admit a worldline QWEI with test-functions g drawn from $C_0^\infty(\mathbb{R})$, and it is possible to show that it is a necessary and sufficient condition if certain scaling behaviour is required:[9]

Theorem 3. *Consider a generalised free field with discrete mass spectrum described by $N(u)$. Let $p > 0$. Then the following are equivalent:*

[9] If $N(u)$ grows faster than polynomially, one may still formulate QWEIs, but for weight functions with sufficiently rapid decay in Fourier space. In particular, this would generally exclude compactly supported weights.

1) $N(u) = O(u^p)$ *as* $u \to \infty$;
2) *The generalised free field obeys the QWEI* (8.21) *for arbitrary* $g \in C_0^\infty(\mathbb{R})$, *and the bound has asymptotic behaviour of order* $O(\tau^{-(p+4)})$ *as* $\tau \to 0+$, *if we replace* g *by* $g_\tau(t) = \tau^{-1/2} g(t/\tau)$.

The proof of this result will be reported elsewhere. An immediate corollary is that the existence of a QWEI with polynomial scaling implies that the Buchholz–Wichmann nuclearity condition (8.15) is satisfied for any $n > p+3$.

All this raises two questions, which are being pursued in on-going work with Porrmann and Ojima. First, can we show that (8.15) implies that $N(u)$ is polynomially bounded? If so, we would have an equivalence between QWEIs and nuclearity for this model. This leads to the second question: Can we understand the link at a deeper level, or is it merely a coincidence, with no more significance than that both are manifestations of the uncertainty principle? A suitable understanding of this question might lead to a general framework for establishing QEIs in general quantum field theories. Part of the problem is to identify the right question, of course, and it may be that one or both of nuclearity or QEIs need to be carefully (re)phrased or even replaced. These questions also require consideration of lower bounds on nuclearity indices: here a potential stumbling block is the technical definition of many of the quantities appearing in discussions of nuclearity, which are therefore not easily amenable to direct calculation even in the simplest cases. Indeed this provides pitfalls for the unwary, one of which we have recently noted [13]: in the mathematical literature there is a notion of p-nuclear map, whose definitions for $p > 1$ and $p \leq 1$ take rather different forms. Although this difference has occasionally been noted in the physics literature [45], one often finds the $p \leq 1$ definition used for all p. However, as we show in [13], the corresponding nuclearity index would vanish identically for $p > 1$ according to this definition! Fortunately this confusion does not appear to have adverse consequences in the literature so far, but it serves as a warning.

8.5 Conclusion

Quantum Energy Inequalities are an expression of the uncertainty principle, and as such are deeply rooted within quantum theory. It is perhaps not surprising that they have connections with other fundamental properties: unravelling these interconnections has the potential to deepen our understanding of the structure of quantum field theory and the nature of quantised matter. Much remains to be done!

Acknowledgment. I am grateful to Lutz Osterbrink for a careful reading of the manuscript.

References

1. R. Brunetti and K. Fredenhagen: Microlocal analysis and interacting quantum field theories: Renormalization on physical backgrounds. *Commun. Math. Phys.* 208:623–661 (2000).
2. R. Brunetti, K. Fredenhagen and M. Köhler: The microlocal spectrum condition and Wick polynomials in curved spacetime. *Commun. Math. Phys.* 180:633–652 (1996).
3. D. Buchholz and P. Junglas: Local properties of equilibrium states and the particle spectrum in quantum field theory. *Lett. Math. Phys.* 11:51–58 (1986).
4. D. Buchholz and M. Porrmann: How small is the phase space in quantum field theory? *Ann. Inst. H. Poincaré* 52:237–257 (1990).
5. D. Buchholz and E.H. Wichmann: Causal independence and the energy-level density of states in local quantum field theory. *Commun. Math. Phys.* 106:321–344 (1986).
6. H. Epstein, V. Glaser and A. Jaffe: Nonpositivity of the energy density in quantized field theories. *Nuovo Cimento* 36:1016–1022 (1965).
7. S.P. Eveson, C.J. Fewster and R. Verch: Quantum Inequalities in Quantum Mechanics. To appear in *Ann. H. Poincaré.* Preprint arXiv:math-ph/0312046, 2003.
8. C.J. Fewster: A general worldline quantum inequality. Class. *Quantum Grav.* 17:1897–1911 (2000).
9. C.J. Fewster: Energy Inequalities in Quantum Field Theory (2003). Available at `http://www-users.york.ac.uk/~cjf3/QEIs.ps`
10. C.J. Fewster and S.P. Eveson: Bounds on negative energy densities in flat spacetime. *Phys. Rev. D* 58:084010 (1998).
11. C.J. Fewster and S. Hollands: Quantum Energy Inequalities in two-dimensional conformal field theory. Preprint arXiv:math-ph/0412028., 2004.
12. C.J. Fewster and B. Mistry: Quantum Weak Energy Inequalities for the Dirac field in Flat Spacetime. *Phys. Rev. D* 68:105010 (2003).
13. C.J. Fewster, I. Ojima and M. Porrmann: p-Nuclearity in a New Perspective. Preprint arXiv:math-ph/0412027, 2004.
14. C.J. Fewster and M.J. Pfenning: A Quantum Weak Energy Inequality for spin-one fields in curved spacetime. *J. Math. Phys.* 44:4480–4513 (2003).
15. C.J. Fewster and T.A. Roman: Null energy conditions in quantum field theory. *Phys. Rev. D.* 67:044003 (2003).
16. C.J. Fewster and E. Teo: Bounds on negative energy densities in static spacetimes. *Phys. Rev. D* 59:104016 (1999).
17. C.J. Fewster and R. Verch: A quantum weak energy inequality for Dirac fields in curved spacetime. *Commun. Math. Phys.* 225:331–359 (2002).
18. C.J. Fewster and R. Verch: Stability of quantum systems at three scales: passivity, quantum weak energy inequalities and the microlcal spectrum condition. *Commun. Math. Phys.* 240:329–375 (2003).
19. É.É. Flanagan: Quantum inequalities in two-dimensional Minkowski spacetime. *Phys. Rev. D* 56:4922–4926 (1997).
20. É.É. Flanagan: Quantum inequalities in two dimensional curved spacetimes. *Phys. Rev. D* 66:104007 (2002).
21. L.H. Ford: Quantum coherence effects and the second law of thermodynamics. *Proc. Roy. Soc. Lond. A* 364:227–236 (1978).

22. L.H. Ford: Constraints on negative-energy fluxes. *Phys. Rev. D* 43:3972–3978 (1991).
23. L.H. Ford, A. Helfer and T.A. Roman: Spatially averaged quantum inequalities do not exist in four-dimensional spacetime. *Phys. Rev. D* 66:124012 (2002).
24. L.H. Ford and T.A. Roman: Averaged energy conditions and quantum inequalities. *Phys. Rev. D* 51:4277–4286 (1995).
25. L.H. Ford and T.A. Roman: Quantum field theory constrains traversable wormhole geometries. *Phys. Rev. D* 53:5496–5507 (1996).
26. L.H. Ford and T.A. Roman: Restrictions on negative energy density in flat spacetime. *Phys. Rev. D* 55:2082–2089 (1997).
27. R. Haag: *Local quantum physics: Fields, particles, algebras.* Springer Verlag, Berlin, 1992.
28. S.W. Hawking: Chronology protection conjecture. *Phys. Rev. D* 46:603–611 (1992).
29. S.W. Hawking and G.F.R. Ellis: *The large scale structure of space-time.* Cambridge University Press, Cambridge, 1973.
30. S. Hollands: The Hadamard condition for Dirac fields and adiabatic states on Robertson–Walker spacetimes. *Commun. Math. Phys.* 216:635–661 (2001).
31. L. Hörmander: *The analysis of linear partial differential operators I.* Springer Verlag, Berlin, 1983.
32. W. Junker and E. Schrohe: Adiabatic vacuum states on general spacetime manifolds: Definition, construction, and physical properties. *Ann. H. Poincaré* 3:1113–1181 (2002).
33. B.S. Kay and R.M. Wald: Theorems on the uniqueness and thermal properties of stationary, nonsingular, quasifree states on spacetimes with a bifurcate Killing horizon. *Phys. Rep.* 207:49–136 (1991).
34. K. Kratzert: Singularity structure of the free Dirac field on a globally hyperbolic spacetime. *Annalen Phys.* 9:475–498 (2000).
35. M. Ludvigsen and J.A.G. Vickers: A simple proof of the positivity of the Bondi mass. *J. Phys. A Math. Gen.* 15:L67–L70 (1982).
36. P. Marecki: Application of quantum inequalities to quantum optics. *Phys. Rev. A* 66:053801 (2002).
37. M.J. Pfenning: Quantum inequalities for the electromagnetic field. *Phys. Rev. D* 65:024009 (2002).
38. M.J. Pfenning and L.H. Ford: The unphysical nature of 'warp drive'. *Class. Quantum Grav.* 14:1743–1751 (1997).
39. M.J. Pfenning and L.H. Ford: Scalar field quantum inequalities in static spacetimes. *Phys. Rev. D* 57:3489–3502 (1998).
40. W. Pusz and S.L. Woronowicz: Passive states and KMS states for general quantum systems. *Commun. Math. Phys.* 58:273–290 (1978).
41. M.J. Radzikowski: Micro-local approach to the Hadamard condition in quantum field theory in curved spacetime. *Commun. Math. Phys.* 179:529–553 (1996).
42. T.A. Roman: Some thoughts on energy conditions and wormholes. Preprint arXiv:gr-qc/0409090, 2004. To appear in: Proceedings of the Tenth Marcel Grossmann Meeting on General Relativity and Gravitation.
43. H. Sahlmann and R. Verch: Passivity and microlocal spectrum condition. *Commun. Math. Phys.* 214:705–731 (2000).
44. H. Sahlmann and R. Verch: Microlocal spectrum condition and Hadamard form for vector-valued quantum fields in curved spacetime. *Rev. Math. Phys.* 13:1203–1246 (2001).

45. R. Schumann: Operator ideals and the statistical independence in quantum field theory. *Lett. Math. Phys.* 37:249–271 (1996).
46. A. Strohmaier, R. Verch and M. Wollenberg: Microlocal analysis of quantum fields on curved spacetimes: Analytic Wavefront sets and Reeh–Schlieder theorems. *J. Math. Phys.* 43:5514–5530 (2002).
47. D.N. Vollick: Quantum inequalities in curved two-dimensional spacetimes. *Phys. Rev. D* 61:084022 (2000).
48. E. Witten: A new proof of the positive energy theorem. *Commun. Math. Phys.* 80:381–402 (1981).
49. H. Yu and P. Wu: Quantum inequalities for the free Rarita-Schwinger fields in flat spacetime. *Phys. Rev. D* 69:064008 (2004).

9

Action Ward Identity and the Stückelberg–Petermann Renormalization Group*

Michael Dütsch[1] and Klaus Fredenhagen[2]

[1] Institut für Theoretische Physik, Universität Zürich, CH-8057 Zürich, Switzerland
duetsch@physik.unizh.ch
[2] Institut für Theoretische Physik, Universität Hamburg, D-22761 Hamburg, Germany
klaus.fredenhagen@desy.de

Summary. A fresh look at the renormalization group (in the sense of Stückelberg-Petermann) from the point of view of algebraic quantum field theory is given, and it is shown that a consistent definition of local algebras of observables and of interacting fields in renormalized perturbative quantum field theory can be given in terms of retarded products. The dependence on the Lagrangian enters this construction only through the classical action. This amounts to the commutativity of retarded products with derivatives, a property named Action Ward Identity by Stora.

9.1 Introduction

Modern perturbative quantum field theory is mainly based on path integrals. Basic objects are the correlation functions

$$G(x_1,\ldots,x_n) = (\Omega, T\varphi(x_1)\cdots\varphi(x_n)\Omega) \tag{9.1}$$

which are calculated as moments of the Feynman path integral, i.e.

$$G(x_1,\ldots,x_n) = \frac{1}{Z}\int \mathcal{D}\varphi\,\varphi(x_1)\cdots\varphi(x_n)e^{iS(\varphi)/\hbar}\,. \tag{9.2}$$

Relations to Wightman fields or even to local algebras of observables are indirect and are not elaborated in typical cases. This fact leads to some severe problems. On the one hand, concepts from the algebraic approach to quantum field theory cannot be easily applied to perturbatively constructed models. This leads to the prejudice that these concepts are irrelevant for physically interesting models as long as they can be constructed only at the level of

*Talk given at the Symposium in honour of the 70th anniversary of Jacques Bros, Paris 2004, based on [9] and on private communication with R. Stora

formal perturbation theory. On the other hand, the path integral approach is intrinsically nonlocal; therefore difficulties arise for the treatment of infrared problems and of finite temperatures, and theories on curved back grounds or with external fields cannot properly be formulated. In algebraic quantum field theory, concepts for dealing with these problems have been developed [2,4,5,13,14,18].

We therefore started a program for a perturbative construction of the net of local algebras of observables [4,7]. There already exists a local formulation of renormalization due to Epstein and Glaser [11] and based on older ideas of Stückelberg and Bogoliubov [1]. The advantages of this method compared to other schemes of renormalization are that it can be formulated entirely in position space, that it is mathematically well elaborated and that it gives a direct construction of operators. It is, however, not equally well developed from the computational point of view, the rôle of the renormalization group is not yet fully established and, even worse, the application to non-Abelian gauge theories is not evident.

In particular the last problem was partially resolved by the Zürich school [20,21], who also developed many new tools for doing concrete calculations in the Epstein-Glaser framework.

The starting point for our approach is Bogoliubov's definition of interacting fields

$$\varphi_{\int \mathcal{L}g}(x) = (Te^{i\int \mathcal{L}g})^{-1} \frac{\delta}{\delta h(x)} Te^{i(\int \mathcal{L}g + \int \varphi h)}|_{h=0} . \tag{9.3}$$

In this contribution, we will first formulate axioms for interacting fields, will then construct solutions, thereafter discuss the renormalization group and will finally describe the arising local nets and local fields.

9.2 Basic properties required for interacting fields

We consider polynomial functionals of a classical scalar ($\mathcal{C}^\infty$) field φ on d-dimensional Minkowski space,

$$F(\varphi) = \sum_n \langle f_n, \varphi^{\otimes n} \rangle . \tag{9.4}$$

Here the smearing functions f_n are assumed to be distributions with compact support in n variables with wave front set

$$WF(f_n) \subset \{(x,k) \,,\, \sum_i k_i = 0\}, \tag{9.5}$$

e.g., $f_2(x_1,x_2) = g(x_1)\delta(x_1 - x_2)$ with $g \in \mathcal{D}$, so $\langle f_2, \varphi^{\otimes 2}\rangle = \int dx\, g(x)\varphi(x)^2$.

A functional F is called local, if $\mathrm{supp}(f_n) \subset D_n$, with the total diagonal

$$D_n = \{(x_1, \ldots, x_n) \,,\, x_1 = \cdots = x_n \in \mathbf{R}^d\} . \tag{9.6}$$

Let us first describe the classical free theory: There the field equation $(\Box + m^2)\varphi = 0$ generates an ideal in the algebra of functionals F (with respect to pointwise multiplication). The Poisson bracket of the classical model is

$$\{F, G\}(\varphi) = \int dx dy \frac{\delta F}{\delta\varphi(x)} \Delta(x-y) \frac{\delta G}{\delta\varphi(y)} \tag{9.7}$$

where Δ is the commutator function of the free scalar field with mass m.

For quantization we follow the recipes of deformation quantization and define an associative product $*$ by

$$F * G(\varphi) = e^{\hbar \int dx dy \Delta_+(x-y) \frac{\delta}{\delta\varphi(x)} \frac{\delta}{\delta\varphi'(y)}} F(\varphi) G(\varphi')|_{\varphi'=\varphi} \tag{9.8}$$

where Δ_+ is the 2-point function of the free scalar field.

The Poisson bracket as well as the $*$-product vanish on the ideal generated by the field equation and can therefore be defined also on the quotient algebra. In the classical case one obtains the Poisson algebra of the classical field theory, in the quantum case one obtains the algebra of Wick polynomials on Fock space, and the formula for the $*$-product translates into Wick's Theorem. For the treatment of interactions, it is however preferable not to go to the quotient but to formulate everything on the original space (off-shell formalism). This introduces some redundancy which will be very useful for the solution of cohomological problems, as for instance in the determination of counter terms in the Lagrangian which compensate changes in the renormalization prescription.

We now turn to the characterization of retarded interacting field functionals. Let F, S_n, $n \in \mathbf{N}$ be local functionals of φ. Let $S(\lambda) = \sum_{n=1}^{\infty} \lambda^n S_n$ be a formal power series with vanishing term of zeroth order. We associate to the pair $F, S(\lambda)$ a formal power series of functionals F_S which we interpret as the functional F of the interacting retarded field under the influence of the interaction S where λ is the expansion parameter of the formal power series.

We require the following properties:

Initial condition:

$$F_{S=0} = F. \tag{9.9}$$

Causality:

$$F_{S_1+S_2} = F_{S_1} \tag{9.10}$$

if S_2 takes place later than F.

Glaser-Lehmann-Zimmermann: [12]

$$\frac{i}{\hbar}[F_S, G_S]_\star = \frac{d}{d\mu}(F_{S+\mu G} - G_{S+\mu F})|_{\mu=0} \; . \tag{9.11}$$

In addition, we require

Unitarity:

$$(F_S)^* = F^*_{S^*}. \tag{9.12}$$

Covariance: Let β denote the natural action of the Poincaré group on the space of functionals F. Then

$$\beta_L(F_S) = \beta_L(F)_{\beta_L(S)} \tag{9.13}$$

for all Poincaré transformations L.

Field independence: The association of local functionals to their interacting counterpart should not explicitly depend on φ,

$$F_S(\varphi + \psi) = F(\varphi + \psi)_{S(\varphi+\psi)} \tag{9.14}$$

for test functions ψ.

Scaling: There is a natural scaling transformation, $x \mapsto \rho x$, on the space of functionals which scales also the mass in the $*$-product from m to $\rho^{-1}m$. But the limit $\rho \to \infty$ is singular because of scaling anomalies. This holds already for the free theory (in even dimensions).
Instead: Introduce an auxiliary mass parameter $\mu > 0$ and transform the $*$-product to an equivalent one which is smooth in the mass m. Require a smooth m dependence in the prescription $(F, S) \to F_S^{(m,\mu)}$ (in particular at $m = 0$ (one-sided)) such that $F_S^{(\rho^{-1}m,\mu)}$, scaled by ρ, is in every order of perturbation theory a polynomial of $\log \rho$.

9.3 Construction of solutions

We make the ansatz that the interacting field is a formal power series in the interaction,

$$F_{\lambda S} = \sum_{n=0}^{\infty} \frac{\lambda^n}{n!} R_{n,1}(S^{\otimes n} \otimes F) , \tag{9.15}$$

where $R_{n,1}$ is an $(n+1)$-linear functional, symmetric in the first n entries (called 'retarded product').

The inductive construction of retarded products can be done in an analogous way as the construction of time ordered products in the Epstein-Glaser method and was elaborated already (in a slightly different form) by Steinmann [22]. In a first step we represent the local functionals as fields smeared with a test function,

$$F(\varphi) = \int d^d x \, A(x) f(x) \tag{9.16}$$

where A is an element of the space $\mathcal{P}$ of all polynomials of φ and its derivatives,

$$A(x) = A(\varphi(x), \partial\varphi(x), \dots) , \tag{9.17}$$

and f is a test function. Using this representation, the retarded products can be represented as distributions $R_{n,1}(A_1(x_1), \dots, A_{n+1}(x_{n+1}))$ with values in the space of functionals of φ. With that the inductive construction can be done by first constructing the retarded products outside of the total diagonal D_{n+1} (9.6) in terms of retarded products with fewer arguments and then by extending them to the full space (in the sense of distributions). Perturbative renormalization is precisely this extension problem.

A difficulty with this procedure is that the representation of local functionals by fields is non-unique, e.g. one finds by partial integration

$$\int dx\, \partial A(x) f(x) = -\int dx\, A(x) \partial f(x) \ . \tag{9.18}$$

Therefore it was suggested by Stora [24] that one should impose an additional requirement, termed the **Action Ward Identity** (AWI), which guarantees that the retarded products depend only on the functionals, but not on the way they are represented as smeared fields. This amounts to the requirement, that the retarded products commute with derivatives,

$$\partial R_{n,1}(\ldots, A(x), \ldots) = R_{n,1}(\ldots, \partial A(x), \ldots) \ . \tag{9.19}$$

Such a relation cannot hold if the arguments of the retarded products are on-shell fields (i.e. satisfy the free field equations). In the off-shell formalism adopted here, however, one can actually show that there are no anomalies for the AWI. Namely one may introduce **balanced** fields, as was done recently for the purpose of non-equilibrium quantum field theory by Buchholz, Ojima and Roos [6]: the balanced fields form that subspace $\mathcal{P}_{\mathrm{bal}}$ of all local fields

$$A(x) = P(\partial_1, \ldots, \partial_n) \varphi(x_1) \cdots \varphi(x_n)|_{x_1 = \cdots = x_n = x} \in \mathcal{P} \tag{9.20}$$

(where P is a polynomial) which arises when P is restricted to depend only on the differences of variables $(\partial_i - \partial_j)$.

The argument relies on

Lemma 1. *Every local functional F is of the form*

$$F = \int d^d x\, f(x) \tag{9.21}$$

with a unique test function f with values in the space of balanced fields.

Proof. Existence: Every symmetric polynomial $P(p_1, \ldots, p_n)$ in n variables $p_1, \ldots, p_n$, $p_i \in \mathbf{R}^d$ may be written in the form

$$P(p_1, \ldots, p_n) = \sum_j a_j(p) b_j(p_{\mathrm{rel}}) \tag{9.22}$$

with polynomials a_j in the center of mass momentum $p = \sum p_i$ and symmetric polynomials b_j in the relative momentum $p_{\mathrm{rel}} = (p_i - p_j, i < j)$. Every local functional of n-th order in φ is of the form

$$F(\varphi) = \int d^d x \sum_j \lambda_j(x) \tag{9.23}$$

$$\times \int d^{d(n-1)} x_{\rm rel}\, \delta(x_{\rm rel}) a_j(i\partial)\, b_j(i\partial_{rel}) \varphi^{\otimes n}(x, x_{\rm rel}) \tag{9.24}$$

with test functions λ_j. By partial integration we find

$$F(\varphi) = \int dx \sum_j \mu_j(x) B_j(x) \tag{9.25}$$

with $\mu_j = a_j(-i\partial)\lambda_j$ and $B_j = b_j(i\partial_{\rm rel})\varphi^{\otimes n}|_{x_{\rm rel}=0} \in \mathcal{P}_{\rm bal}$.
Uniqueness: To show uniqueness let

$$F(\varphi) = \int dx \sum_j \mu_j(x) B_j(x) \tag{9.26}$$

with the balanced fields $B_j = b_j(i\partial_{\rm rel})\varphi^{\otimes n}|_{x_{\rm rel}=0}$, where b_j are *symmetrical* polynomials (with respect to permutations of $\partial_1, ..., \partial_n$). Then $F(\varphi) = 0$ implies that

$$\sum_j \mu_j(x) b_j(-i\partial_{\rm rel})\delta(x_{\rm rel}) = 0 \ . \tag{9.27}$$

Since the center of mass coordinate x and the relative coordinate $x_{\rm rel}$ are independent, this implies

$$\sum_j \mu_j \otimes b_j = 0 \ , \tag{9.28}$$

hence $\sum_j \mu_j B_j = 0$. □

By using this lemma the AWI can be fulfilled by the following procedure. One first extends $R(A_1(x_1), ..., A_{n+1}(x_{n+1}))$ only for $A_1, ..., A_{n+1} \in \mathcal{P}_{\rm bal}$. Since, by induction, the AWI holds outside of D_{n+1}, one may then define the extension for general fields $A_1, ..., A_{n+1}$ by using the AWI and linearity in the fields.

9.4 Renormalization group

The extension of $R_{n,1}$ from $\mathcal{D}(\mathbf{R}^{d(n+1)} \setminus D_{n+1})$ to $\mathcal{D}(\mathbf{R}^{d(n+1)})$ is generically non-unique. Hence, there is an ambiguity in the construction of interacting fields, which is well understood: it can be described in terms of the Stückelberg-Petermann renormalization group $\mathcal{R}$ [25].

The elements of $\mathcal{R}$ are analytic invertible maps $Z : S \mapsto Z(S)$ which map the space of formal power series of local functionals, which start with the first term, into itself such that

$$Z(0) = 0 \,, \tag{9.29}$$

$$Z'(0) = \mathrm{id} \qquad (\text{where } Z'(S)F := \frac{d}{d\tau} Z(S + \tau F)|_{\tau=0}) \,, \tag{9.30}$$

$$Z(S^\star) = Z(S)^\star \tag{9.31}$$

and

$$Z(S) = \int d^d x \, z(\mathcal{L}(x), \partial\mathcal{L}(x), \ldots) \tag{9.32}$$

if $S = \int dx \, \mathcal{L}(x)$. Here the Lagrangians are of the form $\mathcal{L}(x) = \sum_j A_j(x) g_j(x)$ with $A_j \in \mathcal{P}_{\text{bal}}$ and $g_j \in \mathcal{D}(\mathbf{R}^d)$. Derivatives are defined by $\partial\mathcal{L} := \sum_j A_j(x) \partial g_j$. z is of the form

$$z(\mathcal{L}(x), \partial\mathcal{L}(x), \ldots) = \sum_{n,a} \frac{1}{n!} d_{n,a}(A_{j_1} \otimes \cdots \otimes A_{j_n})(x) \prod_{i=1}^{n} \partial^{a_i} g_{j_i}(x) \tag{9.33}$$

with linear maps $d_{n,a} : \mathcal{P}_{\text{bal}}^{\otimes n} \to \mathcal{P}_{\text{bal}}$ which are Lorentz invariant, maintain homogeneous scaling of the fields and do not explicitly depend on φ.

The Main Theorem of Renormalization [9,19,23] amounts to the following relation between interacting field functionals for two renormalization prescriptions $(F, S) \to F_S$ and $(F, S) \to \hat{F}_S$ which both satisfy the mentioned axioms:

Theorem 1. *There is a unique element Zof the renormalization group $\mathcal{R}$ such that*

$$\hat{F}_S = (Z'(S)F)_{Z(S)} \,. \tag{9.34}$$

Conversely, given a renormalization prescription F_S satisfying the axioms and an arbitrary $Z \in \mathcal{R}$, equation (9.34) *gives a new renormalization prescription $\hat{F}_S$ which fulfills also the axioms.*

9.5 Local nets and local fields

We first define the algebra $\mathcal{A}_\mathcal{L}(\mathcal{O})$ of observables within the region $\mathcal{O}$ for a (fixed) interaction $\mathcal{L}$ with compact support. This algebra is generated by elements F_S, $S = \int dx \, \mathcal{L}(x)$ with local functionals F fulfilling supp $\frac{\delta F}{\delta\varphi} \subset \mathcal{O}$. In [4] it has been found that the algebraic structure of $\mathcal{A}_\mathcal{L}(\mathcal{O})$ is independent of the values of $\mathcal{L}$ outside of $\mathcal{O}$:

Theorem 2. *If the interactions $\mathcal{L}_1$ and $\mathcal{L}_2$ coincide within $\mathcal{O}$, there exist isomorphisms $\alpha : \mathcal{A}_{\mathcal{L}_1}(\mathcal{O}) \to \mathcal{A}_{\mathcal{L}_2}(\mathcal{O})$ such that*

$$\alpha(F_{S_1}) = F_{S_2} \tag{9.35}$$

for all F localized in $\mathcal{O}$ (where $S_j = \int dx \, \mathcal{L}_j(x)$).

We now want to construct the algebras for a not necessarily compactly supported Lagrangian $\mathcal{L}$ (i.e. we perform the so-called algebraic adiabatic limit). For this purpose we consider the bundle of algebras $\mathcal{A}_{\mathcal{L}_1}(\mathcal{O})$ over the space of compactly supported Lagrangians $\mathcal{L}_1$ which coincide within $\mathcal{O}$ with the given Lagrangian $\mathcal{L}$.

A section $B = (B_{\mathcal{L}_1})$ of this bundle is called covariantly constant if for all automorphisms α satisfying (9.35) the following relation holds:

$$\alpha(B_{\mathcal{L}_1}) = B_{\mathcal{L}_2} \; . \tag{9.36}$$

So in particular the interacting field functionals F_S are (by definition) covariantly constant sections. The covariantly constant sections now generate the local algebras associated to the interaction $\mathcal{L}$.

To define the net of local algebras one has to specify, for $\mathcal{O}_1 \subset \mathcal{O}_2$, the injections

$$i_{\mathcal{O}_2\mathcal{O}_1} : \mathcal{A}_{\mathcal{L}}(\mathcal{O}_1) \to \mathcal{A}_{\mathcal{L}}(\mathcal{O}_2) \; . \tag{9.37}$$

Let $B \in \mathcal{A}_{\mathcal{L}}(\mathcal{O}_1)$. Then $i_{\mathcal{O}_2\mathcal{O}_1}(B)$ is the section which is obtained from the section B by restriction to Lagrangians which coincide with $\mathcal{L}$ on the larger region. Clearly these injections satisfy the compatibility relation required for nets:

$$i_{\mathcal{O}_3\mathcal{O}_2} \circ i_{\mathcal{O}_2\mathcal{O}_1} = i_{\mathcal{O}_3\mathcal{O}_1} \tag{9.38}$$

for $\mathcal{O}_1 \subset \mathcal{O}_2 \subset \mathcal{O}_3$. Moreover, the net satisfies local commutativity and, if $\mathcal{L}$ is Lorentz invariant, it is covariant under the Poincaré group [7].

The next step is the construction of local fields associated to the local net. Following [5], a local field associated to the net is defined as a family of distributions $(A_{\mathcal{O}})$ with values in $\mathcal{A}_{\mathcal{L}}(\mathcal{O})$ such that

$$i_{\mathcal{O}_2\mathcal{O}_1}(A_{\mathcal{O}_1}(h)) = A_{\mathcal{O}_2}(h) \tag{9.39}$$

if the test function h has support contained in $\mathcal{O}_1$. In particular all classical fields $A \in \mathcal{P}$ induce local fields $A^{\mathcal{L}} \equiv (A^{\mathcal{L}}_{\mathcal{O}})_{\mathcal{O}}$ by the sections

$$A^{\mathcal{L}}_{\mathcal{O}}(h) \; : \; \mathcal{L}_1 \mapsto (A_{\mathcal{O}}(h))_{\mathcal{L}_1} = (\int dx\, A(x)h(x))_{S_1} \tag{9.40}$$

where $\mathcal{L}_1$ coincides with $\mathcal{L}$ within $\mathcal{O}$, $h \in \mathcal{D}(\mathcal{O})$ and $S_1 = \int dx\, \mathcal{L}_1$. It is an open question whether there are other local fields. The answer would amount to the determination of the Borchers class for perturbative quantum field theories.

We may now restrict the renormalization group flow to constant Lagrangians $\mathcal{L} \in \mathcal{P}_{\mathrm{bal}}$.

Theorem 3. *Let $\mathcal{L} \mapsto \mathcal{A}_{\mathcal{L}}$ and $\mathcal{L} \mapsto \hat{\mathcal{A}}_{\mathcal{L}}$ ($\mathcal{L} \in \mathcal{P}_{\mathrm{bal}}$) be associations of local nets which are defined by the two renormalization prescriptions F_S and $\hat{F}_S$, respectively.*

1. *Then there exists a unique map ('renormalization of the interaction')*

$$z_0 : \lambda\mathcal{P}_{\mathrm{bal}}[[\lambda]] \longrightarrow \lambda\mathcal{P}_{\mathrm{bal}}[[\lambda]] \ : \ z_0(\lambda\mathcal{L}) = \lambda\mathcal{L} + O(\lambda^2) \tag{9.41}$$

such that the nets $\hat{\mathcal{A}}_{\mathcal{L}}$ and $\mathcal{A}_{z_0(\mathcal{L})}$ are equivalent for all $\mathcal{L}$.

2. *Furthermore there exists a unique map ('field renormalization')*

$$z^{(1)} : \lambda\mathcal{P}_{\mathrm{bal}}[[\lambda]] \times \mathcal{P}[[\lambda]] \longrightarrow \mathcal{P}[[\lambda]] \ : \ (\lambda\mathcal{L}, A) \mapsto z^{(1)}(\lambda\mathcal{L})A = A + O(\lambda) \tag{9.42}$$

such that $z^{(1)}(\lambda\mathcal{L}) : \mathcal{P}[[\lambda]] \longrightarrow \mathcal{P}[[\lambda]]$ is a linear invertible map which commutes with partial derivatives and such that for the local fields (9.40) *the identification of $\hat{\mathcal{A}}_{\mathcal{L}}$ with $\mathcal{A}_{z_0(\mathcal{L})}$ is given by*

$$\hat{A}^{\mathcal{L}} = (z^{(1)}(\mathcal{L})A)^{z_0(\mathcal{L})} \ , \quad \forall \mathcal{L} \in \mathcal{P}_{\mathrm{bal}}, \ A \in \mathcal{P} \ . \tag{9.43}$$

We point out that z_0 and $z^{(1)}$ are independent of $\mathcal{O}$.

Proof. By applying the algebraic adiabatic limit to Theorem 2 the formula (9.34) goes over into (9.43) with

$$z_0(\mathcal{L}) = z(\mathcal{L}, 0) = \sum_n \frac{1}{n!} d_{n,0}(\mathcal{L}^{\otimes n}) \tag{9.44}$$

and

$$z^{(1)}(\mathcal{L})A = \sum_{n,a \in \mathbf{N}_0^d} \frac{1}{n!} (-1)^{|a|} \, \partial^a d_{n+1,(0,\dots,0,a)}(\mathcal{L}^{\otimes n} \otimes A) \ . \tag{9.45}$$

For the equivalence of the nets we refer to Theorem 5.1 of [9]. □

Example. $\mathcal{L} = \varphi^4$ in $d = 4$ dimensions. Then

$$z_0\mathcal{L} = (1 + a)\varphi^4 + b((\partial\varphi)^2 - \varphi\Box\varphi) + m^2 c\varphi^2 + m^4 e \ , \tag{9.46}$$

where $a, b, c, e \in \mathbf{C}[[\lambda]]$. In the case $m = 0$ the coupling constant renormalization a and the field renormalization b are explicitly computed to lowest non-trivial order in [9].

Remarks. (1) It may happen that for a certain interaction $\mathcal{L}_0$ the renormalization is trivial, $z_0(\mathcal{L}_0) = \mathcal{L}_0$, but the corresponding field renormalization $z^{(1)}(\mathcal{L}_0)$ is non-trivial. This corresponds to the Zimmermann relations.

(2) The scaling transformations on a given renormalization prescription induce a one-parameter subgroup of the renormalization group, which may be called a Gell-Mann-Low Renormalization Group. Its generator is related to the β-function. Gell-Mann-Low subgroups belonging to different renormalization prescriptions are conjugate to each other. The generator starts with a term of second order which is universal [3].

(3) An analysis of the perturbative renormalization group in the algebraic adiabatic limit was already given by Hollands and Wald in the more general framework of QFT on curved space times [16]. But the formalism presented here is not yet fully adapted to general Lorentzian space-times.

(4) Hollands and Wald generalized the Action Ward Identity to curved space times [17] (and called it 'Leibniz rule'). In that framework it is a non-trivial condition already for time ordered (or retarded) products of *one* factor.

(5) Ward identities, in particular the Master Ward Identity [8, 10] remain to be analyzed.

References

1. N.N. Bogoliubov and D.V. Shirkov: *Introduction to the Theory of Quantized Fields.* Interscience, New York, 1959.
2. J. Bros and D. Buchholz: Towards a Relativistic KMS Condition. *Nucl. Phys. B* 429:291-318 (1994).
3. R. Brunetti, M. Dütsch and K. Fredenhagen: work in progress.
4. R. Brunetti and K. Fredenhagen: Microlocal analysis and interacting quantum field theories: Renormalization on physical backgrounds. *Commun. Math. Phys.* 208:623 (2000).
5. R. Brunetti, K. Fredenhagen and R. Verch: The generally covariant locality principle – A new paradigm for local quantum physics. *Commun. Math. Phys.* 237:31-68.
6. D. Buchholz, I. Ojima and H. Roos: Thermodynamic Properties of Non-Equilibrium States in Quantum Field Theory. *Annals Phys.* 297:219-242 (2002).
7. M. Dütsch and K. Fredenhagen: Algebraic Quantum Field Theory, Perturbation Theory, and the Loop Expansion. *Commun. Math. Phys.* 219:5 (2001).
8. M. Dütsch and K. Fredenhagen: The Master Ward Identity and Generalized Schwinger-Dyson Equation in Classical Field Theory. *Commun. Math. Phys.* 243:275-314 (2003).
9. M. Dütsch and K. Fredenhagen: Causal perturbation theory in terms of retarded products, and a proof of the Action Ward Identity. To appear in *Rev. Math. Phys.*
10. M. Dütsch and F.-M. Boas: The Master Ward Identity. *Rev. Math. Phys.* 14:977-1049 (2002).
11. H. Epstein and V. Glaser: The role of locality in perturbation theory. *Ann. Inst. H. Poincaré A* 19:211 (1973).
12. V. Glaser, H. Lehmann and W. Zimmermann: Field Operators and Retarded Functions. *Nuovo Cimen.* 6:1122 (1957).
13. R. Haag: *Local Quantum Physics: Fields, particles and algebras.* Springer-Verlag, Berlin, 2nd ed., 1996.
14. S. Hollands and R.M. Wald: Local Wick Polynomials and Time-Ordered-Products of Quantum Fields in Curved Spacetime. *Commun. Math. Phys.* 223:289 (2001).
15. S. Hollands and R.M. Wald: Existence of Local Covariant Time-Ordered-Products of Quantum Fields in Curved Spacetime. *Commun. Math. Phys.* 231:309-345 (2002).

16. S. Hollands and R.M. Wald: On the Renormalization Group in Curved Spacetime. *Commun. Math. Phys.* 237:123-160 (2003).
17. S. Hollands and R.M. Wald: Conservation of the stress tensor in interacting quantum field theory in curved spacetimes. [arXiv gr-qc/0404074].
18. P. Marecki: Quantum Electrodynamics on background external fields. [arXiv hep-th/0312304].
19. G. Pinter: Finite Renormalizations in the Epstein-Glaser Framework and Renormalization of the S-Matrix of Φ^4-Theory. *Ann. Phys. (Leipzig)* 10:333 (2001).
20. G. Scharf: *Finite Quantum Electrodynamics. The causal approach.* Springer-Verlag, 2nd ed., 1995.
21. G. Scharf: *Quantum Gauge Theories – A True Ghost Story.* John Wiley and Sons, 2001.
22. O. Steinmann: Perturbation expansions in axiomatic field theory. *Lecture Notes in Physics* 11, Springer-Verlag, Berlin, Heidelberg New York, 1971.
23. G. Popineau and R. Stora: A pedagogical remark on the main theorem of perturbative renormalization theory. Unpublished preprint, 1982.
24. R. Stora: Pedagogical Experiments in Renormalized Perturbation Theory. Contribution to the conference *Theory of Renormalization and Regularization*, Hesselberg, Germany (2002).
http://wwwthep.physik.uni-mainz.de/~scheck/Hessbg02.html
and private communication.
25. E.C.G. Stueckelberg and A. Petermann: La normalisation des constantes dans la theorie des quanta. *Helv. Phys. Acta* 26:499-520 (1953).

10

On the Relativistic KMS Condition for the $P(\phi)_2$ Model

Christian Gérard[1] and Christian D. Jäkel[2]

[1] Université Paris Sud XI, F-91405 Orsay, France
christian.gerard@math.u-psud.fr
[2] ETH Zürich, Hönggerberg, CH-8093 Zürich, Switzerland
christian.jaekel@itp.phys.ethz.ch

Summary. The relativistic KMS condition introduced by Bros and Buchholz provides a link between quantum statistical mechanics and quantum field theory. We show that for the $P(\phi)_2$ model at positive temperature, the two-point function for fields satisfies the relativistic KMS condition.

10.1 Introduction

The operator algebraic framework allows us to characterize the equilibrium states of a quantum system by first principles: when the dynamical law is changed by a local perturbation, which is slowly switched on and slowly switched off again, an equilibrium state returns to its original form at the end of this procedure. In a pioniering work, Haag, Kastler and Trych-Pohlmeyer [15] showed that this characterization of an equilibrium state leads to a sharp mathematical criterion, the so-called *KMS condition*, first encountered by Haag, Hugenholtz and Winnink [14] and more implicitly by Kubo [15], Martin and Schwinger [24].

On the other hand, the vacuum states of a relativistic QFT are characterized by Poincaré invariance and the spectrum condition [27]. Both the KMS and the spectrum condition can be formulated in terms of analyticity properties of the correlation functions, but for almost 40 years the connection between these two conditions was not investigated in depth, and algebraic quantum statistical mechanics and algebraic quantum field theory were treated in this respect as disjoint subjects. But finally increased interest in cosmology and heavy-ion collisions led to a need to combine QFT and quantum statistical mechanics.

It was first recognized by Bros and Buchholz [6] that the thermal equilibrium states of a relativistic QFT should have stronger analyticity properties in configuration space than those imposed by the traditional KMS condition. The result of their careful analysis is a *relativistic KMS condition* which can be understood as a remnant of (and, in applications, as a substitute for) the relativistic spectrum condition in the vacuum sector.

10.1.1 Content of the Paper

The content of this short paper is as follows. In Section 10.2 we recall the construction of the $P(\phi)_2$ model at positive temperature presented in [12]. In Section 10.3 we review the main ingredients of the Euclidean approach to thermal field theory. In Section 10.4 we show that the Wightman two-point function for fields satisfies the relativistic KMS condition. Section 10.5 is devoted to a brief discussion of the implications of the relativistic KMS condition.

10.2 The spatially-cutoff $P(\phi)_2$ model at positive temperature

The construction of interacting thermal quantum fields in [12] includes several of the original ideas of Höegh-Krohn [17], but instead of starting from the interacting system in a box the authors started from the Araki-Woods representation for the free thermal system in infinite volume. We briefly recall the main steps of this construction.

10.2.1 Preliminary Material

The KMS condition

A state ω_β over a C^*-algebra $\mathfrak{A}$ equipped with a C^*-dynamics τ_t is called a (τ, β)-KMS state for some $\beta \in \mathbb{R} \cup \{\pm\infty\}$, if for all $A, B \in \mathfrak{A}$ there exists a function $F_{A,B}$ which is continuous in the strip $\{z \in \mathbb{C} \mid 0 \leq \operatorname{Im} z \leq \beta\}$ and analytic and bounded in the open strip $\{z \in \mathbb{C} \mid 0 < \operatorname{Im} z < \beta\}$, with boundary values given by

$$F_{A,B}(t) = \omega_\beta\big(A\tau_t(B)\big),\ F_{A,B}(t + i\beta) = \omega_\beta\big(\tau_t(B)A\big),\ \forall\, t \in \mathbb{R}.$$

Thus KMS states are characterized by a real parameter β, which has the meaning of inverse temperature.

The relativistic KMS condition

Lorentz invariance is always broken by a KMS state. A KMS state might also break spatial translation or rotation invariance, but the maximal propagation velocity of signals is not affected by such a lack of symmetry.

Definition 1. *A state ω_β over a C^*-algebra $\mathfrak{A}$ satisfies the* relativistic KMS condition *at inverse temperature $\beta > 0$ if there exists some positive timelike vector $e \in V_+$, $e^2 = 1$, such that for every pair of elements A, B of $\mathfrak{A}$ there exists a function $F_{A,B}$ which is analytic in the tube domain*

$$\mathcal{T}_\beta := \{z \in \mathbb{C}^2 \mid \operatorname{Im} z \in V_+ \cap (\beta e + V_-)\},$$

and continuous at the boundary sets $\mathbb{R}^2$ and $\mathbb{R}^2 + \mathrm{i}\beta e$ with boundary values given by

$$F_{A,B}(t,x) = \omega_\beta\big(A\alpha_{t,x}(B)\big) \quad \text{and} \quad F_{A,B}(t + \mathrm{i}\beta e, x) = \omega_\beta\big(\alpha_{t,x}(B)\big)$$

for all $(t,x) \in \mathbb{R}^2$. Here $V_\pm = \{(t,x) \in \mathbb{R}^2 \mid |t| < |x|, \pm t \geq 0\}$ denote the forward/backward light-cones.

Remark 1. Note that in the limit $\beta \to \infty$ the tube $\mathcal{T}_\beta$ tends toward the forward tube $\mathbb{R}^2 + iV_+$.

As we will discuss in Section 10.5, the relativistic KMS condition has numerous applications. For instance, the famous Reeh-Schlieder property (in the thermal representation) is a direct consequence of the relativistic KMS condition (and weak additivity), no matter if the spatial translation or rotation invariance is broken by the KMS state or not [19].

10.2.2 The free neutral scalar field at temperature β^{-1}

Let $\mathfrak{h} = H^{-\frac{1}{2}}(\mathbb{R})$ be the complex Sobolev space of order $-\frac{1}{2}$ equipped with the norm

$$\|h\|^2 = (h, (2\nu)^{-1}h)_{L^2(\mathbb{R},dx)}$$

for $\nu := (D_x^2 + m^2)^{\frac{1}{2}}$.

Let $\mathfrak{W}(\mathfrak{h})$ be the abstract Weyl C^*-algebra over $\mathfrak{h}$. On $\mathfrak{W}(\mathfrak{h})$ we define the free time evolution $\{\tau_t^\circ\}_{t\in\mathbb{R}}$ by

$$\tau_t^\circ(W(h)) = W(e^{it\nu}h), \quad h \in \mathfrak{h},\, t \in \mathbb{R}.$$

For $m > 0$ the unique (τ°, β)-KMS state on the Weyl algebra $\mathfrak{W}(\mathfrak{h})$ is given by

$$\omega_\beta^\circ(W(h)) := e^{-\frac{1}{4}(h,(1+2\rho)h)}, \quad h \in \mathfrak{h}, \tag{10.1}$$

where $\rho := (e^{\beta\nu} - 1)^{-1}$, $\beta > 0$.

The Araki-Woods representation

A realization of the GNS representation associated to the pair $(\mathfrak{W}(\mathfrak{h}), \omega_\beta^\circ)$ is provided by the Araki-Woods representation. We set:

$$\mathcal{H}_{AW} := \Gamma(\mathfrak{h} \oplus \overline{\mathfrak{h}}),$$

$$\Omega_{AW} := \Omega_F,$$

$$\pi_{AW}(W(h)) = W_{AW}(h) := W_F\big((1+\rho)^{\frac{1}{2}}h \oplus \overline{\rho}^{\frac{1}{2}}\overline{h}\big), \quad h \in \mathfrak{h}.$$

Here $\overline{\mathfrak{h}}$ is the conjugate Hilbert space to $\mathfrak{h}$, $W_F(.)$ denotes the Fock Weyl operator on the bosonic Fock space $\Gamma(\mathfrak{h} \oplus \overline{\mathfrak{h}})$ and $\Omega_F \in \Gamma(\mathfrak{h} \oplus \overline{\mathfrak{h}})$ is the Fock vacuum.

The generator of the time evolution

Since ω_β° is τ°-invariant, the time evolution can be unitarily implemented in the representation π_{AW}:

$$e^{itL_{AW}}\pi_{AW}(A)\Omega_{AW} := \pi_{AW}\big(\tau_t^\circ(A)\big)\Omega_{AW}, \quad A \in \mathfrak{W}(\mathfrak{h}). \tag{10.2}$$

The generator L_{AW} of the free time evolution, uniquely fixed by (10.2) and the request $L_{AW}\Omega_{AW} = 0$, is called the (free) Liouvillean. It is equal to $d\Gamma(\nu \oplus -\overline{\nu})$.

Thermal fields

The Araki-Woods representation π_{AW} is a regular representation, i.e., the map $\mathbb{R} \ni \lambda \mapsto W_\pi(\lambda h)$ is strongly continuous for any $h \in \mathfrak{h}$. Thus one can define *field operators*

$$\phi_{AW}(h) := -\mathrm{i}\frac{\mathrm{d}}{\mathrm{d}\lambda} W_{AW}(\lambda h)\Big|_{\lambda=0}, \quad h \in \mathfrak{h},$$

which satisfy in the sense of quadratic forms on $\mathcal{D}(\phi_{AW}(h_1)) \cap \mathcal{D}(\phi_{AW}(h_2))$ the commutation relations

$$[\phi_{AW}(h_1), \phi_{AW}(h_2)] = \mathrm{i}\mathrm{Im}\,(h_1, h_2), \quad h_1,\, h_2 \in \mathfrak{h}.$$

The net of local von Neumann algebras

The von Neumann algebra generated by $\{\pi_{AW}(W(h)) \mid h \in \mathfrak{h}\}$ is denoted by $\mathfrak{R}_{AW}$.

To define the net of local von Neumann algebras, we introduce the $\mathbb{R}$-linear map

$$\begin{array}{rll} U\colon \ \mathfrak{h} = H^{-\frac{1}{2}}(\mathbb{R}) & \to & \mathcal{S}'(\mathbb{R}) \\ h & \mapsto & \mathrm{Re}\, h + \mathrm{i}\nu^{-1}\mathrm{Im}\, h. \end{array}$$

For $I \subset \mathbb{R}$ a bounded open interval, we define the following real vector subspace of $\mathfrak{h}$,

$$\mathfrak{h}_I := \{h \in \mathfrak{h} \mid \mathrm{supp}\, Uh \subset I\}. \tag{10.3}$$

We denote by $\mathfrak{R}_{AW}(I)$ the von Neumann algebra generated by $\{W_{AW}(h) \mid h \in \mathfrak{h}_I\}$. The algebra

$$\mathfrak{A} := \overline{\bigcup_{I \subset \mathbb{R}} \mathfrak{R}_{AW}(I)}^{(C^*)}$$

is called the *algebra of local observables*. The union is over all open bounded intervals $I \subset \mathbb{R}$ and the symbol $\overline{\bigcup_{I \subset \mathbb{R}} \mathfrak{R}_{AW}(I)}^{(C^*)}$ denotes the C^*-inductive limit (see e.g. [20, Proposition 11.4.1.]).

Remark 2. We note that the Araki-Woods representation is locally normal w.r.t. the vacuum representation of the free field. Consequently, the C^*-algebra $\mathfrak{A}$ is identical (up to $*$-isomorphisms) to the algebra

$$\mathfrak{A}_\mathrm{F} := \overline{\bigcup_{I \subset \mathbb{R}} \pi_\mathrm{F}(\mathfrak{W}(\mathfrak{h}_I))''}^{(C^*)},$$

where π_F is the Fock representation.

10.2.3 The spatially-cutoff $P(\phi)_2$ model at temperature β^{-1}

Let $P(\lambda) = \sum_{j=0}^{2n} a_j \lambda^j$ be a real-valued polynomial, which is bounded from below, and let $l \in \mathbb{R}^+$ be a spatial cutoff parameter. The spatially cutoff $P(\phi)_2$ model on the real line $\mathbb{R}$ at temperature β^{-1} is then defined by the formal interaction term

$$V_l = \int_{-l}^{l} {:}P(\phi_{AW}(x)){:}\, \mathrm{d}x, \quad l > 0.$$

Here : : denotes the Wick ordering (see e.g. [13]) with respect to the *temperature* β^{-1} *covariance on* $\mathbb{R}$:

$$C_0(h_1, h_2) := \left(h_1, \frac{(1+\mathrm{e}^{-\beta\nu})}{2\nu(1-\mathrm{e}^{-\beta\nu})} h_2\right)_{L^2(\mathbb{R})}, \quad h_1, h_2 \in \mathcal{S}(\mathbb{R}). \tag{10.4}$$

Using a sequence of functions $\{\delta_k\}$ approximating the delta-function, the limit

$$V_l := \lim_{k\to\infty} \int_{-l}^{l} :P(\phi_{AW}(\delta_k(.-x)):\mathrm{d}x$$

exists on a dense set of vectors in $\Gamma(\mathfrak{h}\oplus\overline{\mathfrak{h}})$. An approximation of the Dirac δ function can be fixed by setting $\delta_k(x) := k\chi(kx)$, where χ is a function in $C^\infty_{0\,\mathbb{R}}(\mathbb{R}^d)$ with $\int \chi(x)\mathrm{d}x = 1$.

The perturbed KMS system

It can be shown (see [12]) that the operator sum $L_{AW}+V_l$ is essentially selfadjoint on $\mathcal{D}(L_{AW})\cap\mathcal{D}(V_l)$ and if we set $H_l := \overline{L_{AW}+V_l}$, then the perturbed time-evolution on $\mathfrak{A}$ is given by

$$\tau^l_t(A) := \mathrm{e}^{\mathrm{i}tH_l} A \mathrm{e}^{-\mathrm{i}tH_l}, \quad A \in \mathfrak{A}.$$

The perturbed KMS state ω_l on $\mathfrak{A}$ is normal w.r.t. the Araki-Woods representation π_{AW}. In fact, the GNS vector $\Omega_{AW} \in \Gamma(\mathfrak{h}\oplus\overline{\mathfrak{h}})$ belongs to $\mathcal{D}\big(\mathrm{e}^{-\frac{\beta}{2}H_l}\big)$ and the perturbed KMS state ω_l is the vector state induced by the state vector

$$\Omega_l := \frac{\mathrm{e}^{-\frac{\beta}{2}H_l}\Omega_{AW}}{\|\mathrm{e}^{-\frac{\beta}{2}H_l}\Omega_{AW}\|}.$$

These results are in complete analogy to the analytic perturbation theory for bounded perturbations due to Araki (see e.g. [5]). Identical formulas, valid for a certain class of unbounded perturbations, have recently been derived in [8].

10.2.4 The translation invariant $P(\phi)_2$ model at temperature β^{-1}

Existence of the dynamics

Let $I \subset \mathbb{R}$ be a bounded open interval. For $t \in \mathbb{R}$ fixed, the norm limit

$$\lim_{l\to\infty} \tau^l_t(A) =: \tau_t(A)$$

exists for all $A \in \mathfrak{R}_{AW}(I)$. In fact, for A and t fixed, $\tau^l_t(A)$ is independent of l for l sufficiently large.

By construction the elements of the local von Neumann algebras $\mathfrak{R}_{AW}(I)$, I open and bounded, are norm dense in $\mathfrak{A}$. Thus the map $\tau\colon t \mapsto \tau_t$ extends to a group of $*$-automorphisms of $\mathfrak{A}$. Moreover, if $\{\alpha_x\}_{x\in\mathbb{R}}$ denotes the group of space translations over $\mathfrak{A}$, defined by

$$\alpha_x\big(W_{AW}(h)\big) := W_{AW}(\mathrm{e}^{\mathrm{i}x.k}h), \quad x \in \mathbb{R},$$

where k is the momentum operator acting on $\mathfrak{h}$, then

$$\tau_t \circ \alpha_x = \alpha_x \circ \tau_t$$

for all $t, x \in \mathbb{R}$. Consequently the time evolution is translation invariant.

Existence of the thermodynamic limit

Let $\{\omega_l\}_{l>0}$ be the family of (τ^l, β)-KMS states for the spatially cutoff $P(\phi)_2$ models constructed in Subsection 10.2.3. Then it has been shown in [12] that

$$\mathrm{w} - \lim_{l \to +\infty} \omega_l =: \omega_\beta \text{ exists over } \mathfrak{A}. \tag{10.5}$$

Moreover, the state ω_β has the following properties:

(i) ω_β is a (τ, β)-KMS state over $\mathfrak{A}$;
(ii) ω_β is *locally normal*, i.e., if $I \subset \mathbb{R}$ is an open and bounded interval, then $\omega_{\beta|\mathfrak{R}_{AW}(I)}$ is normal w.r.t. the Araki-Woods representation;
(iii) ω_β is invariant under spatial translations, i.e.,

$$\omega_\beta(\alpha_x(A)) = \omega_\beta(A), \quad x \in \mathbb{R},\ A \in \mathfrak{A};$$

(iv) ω_β has the *spatial clustering property*, i.e.,

$$\lim_{x \to \infty} \omega_\beta(A\alpha_x(B)) = \omega_\beta(A)\omega_\beta(B) \quad \forall A, B \in \mathfrak{A}.$$

The rate of the spatial clustering is related to the infrared properties of the Liouvillean [18].

10.3 Euclidean approach

As far as the formulation of the spatially cut-off thermal $P(\phi)_2$ model is concerned, the Euclidean approach is only used to show that the sum of the operators L_{AW} and V_l (which are both unbounded from below) is essentially selfadjoint on the intersection of their domains.

However, for the existence of the thermodynamic limit (10.5) the Euclidean approach is used in a more sophisticated manner. The key argument in the proof of (10.5), *Nelson symmetry*, will be crucial for the proof of the relativistic KMS condition too. In order to formulate it, we briefly recall the Euclidean approach, in a framework adapted to the thermal $P(\phi)_2$ model (see [11], [21] for a more general abstract framework).

10.3.1 Euclidean reconstruction theorem

Euclidean measures on the cylinder

Let $S_\beta = [-\beta/2, \beta/2]$ be the circle of length β. The points in the cylinder $S_\beta \times \mathbb{R}$ are denoted by (t, x). Let $Q := \mathcal{S}'_{\mathbb{R}}(S_\beta \times \mathbb{R})$ be the space of real-valued, β-periodic Schwartz distributions on $S_\beta \times \mathbb{R}$ and let Σ be the Borel σ-algebra on Q. Let μ be a Borel probability measure on (Q, Σ).

For $f \in \mathcal{S}_{\mathbb{R}}(S_\beta \times \mathbb{R})$, we denote by $\phi(f)$ the function

$$\begin{aligned} \phi(f)\colon\ Q &\to \mathbb{R} \\ q &\mapsto \langle q, f \rangle. \end{aligned}$$

For $T \geq 0$, we denote by $\Sigma_{[0,T]}$, the sub σ-algebra of Σ generated by the functions $\mathrm{e}^{\mathrm{i}\phi(f)}$ for $\operatorname{supp} f \subset [0,T[\times \mathbb{R}$. Let $r\colon Q \to Q$ be the time-reflection around $t = 0$ defined by $r\phi(t,x) = \phi(-t,x)$ and let $\tau_t^E : Q \to Q$ be the group of euclidean time translations defined by $\tau_t^E\phi(s,x) = \phi(s-t,x)$. The map r lifts to a $*$-automorphism R of $L^\infty(Q,\Sigma)$ defined by $RF(q) := F(rq)$, and the group τ_t^E lifts to a group $U(t)$ of $*$-automorphisms of $L^\infty(Q,\Sigma)$. This group is unitary on $L^2(Q,\Sigma,\mu)$ if μ is invariant under τ_t^E.

Reconstruction theorem

Let $\mathcal{H}_{\mathrm{OS}} := L^2(Q, \Sigma_{[0,\beta/2]}, \mathrm{d}\mu)$. We assume that the measure μ satisfies the *Osterwalder-Schrader positivity*

$$(F,F) := \int_Q R(\overline{F})F\mathrm{d}\mu \geq 0 \quad \forall F \in \mathcal{H}_{\mathrm{OS}}.$$

Let $\mathcal{N} \subset \mathcal{H}_{\mathrm{OS}}$ be the kernel of $(.,.)$. We set

$$\mathcal{H}_{\mathrm{phys}} := \overline{\mathcal{H}_{\mathrm{OS}}/\mathcal{N}},$$

where the completion of $\mathcal{H}_{\mathrm{OS}}/\mathcal{N}$ is done w.r.t. the norm $(.,.)^{\frac{1}{2}}$. The canonical projection $\mathcal{H}_{\mathrm{OS}}$ to $\mathcal{H}_{\mathrm{phys}}$ is denoted by $\mathcal{V}$.

In $\mathcal{H}_{\mathrm{phys}}$ we have the distinguished vector

$$\Omega_{\mathrm{phys}} := \mathcal{V}(1),$$

where 1 is the constant function equal to 1 on Q.

The unitary group $U(t)$ for $t \geq 0$ does *not* preserve $\mathcal{H}_{\mathrm{OS}}$ (contrary to 0-temperature theories), because distributions supported in the 'future' $[0,\beta/2[\times\mathbb{R}$ come back in the 'past' $]-\beta/2,0]\times\mathbb{R}$ by time translations. Nevertheless $U(s)$ for $0 \leq s \leq t$ sends $L^2(Q,\Sigma_{[0,\beta/2-t[}, \mathrm{d}\mu)$ into $\mathcal{H}_{\mathrm{OS}}$.

Using the theory of *local symmetric semigroups* (see [10], [22]), it is possible to define a selfadjoint operator L_{phys} on $\mathcal{H}_{\mathrm{phys}}$ such that for $F \in L^2(Q,\Sigma_{[0,\beta/2-t[},\mathrm{d}\mu)$ and $0 \leq s \leq t$ one has

$$\mathcal{V}(U(s)F) = \mathrm{e}^{-sL_{\mathrm{phys}}}\mathcal{V}(F).$$

Finally if $A \in L^\infty(Q,\Sigma_{\{0\}})$, then multiplication by A preserves $\mathcal{H}_{\mathrm{OS}}$ and $\mathcal{N}$, and one obtains a representation π_{phys} of the *algebra of time-zero fields* $L^\infty(Q,\Sigma_{\{0\}})$:

$$\pi_{\mathrm{phys}}(A)\mathcal{V}(F) := \mathcal{V}(AF), \quad F \in \mathcal{H}_{\mathrm{OS}}.$$

From this reconstruction procedure one obtains a β-KMS system defined as follows:

(i) the C^*-algebra $\mathfrak{A}_{\mathrm{phys}}$ is the von Neumann algebra generated by the operators $\mathrm{e}^{\mathrm{i}tL_{\mathrm{phys}}}\pi_{\mathrm{phys}}(A)\mathrm{e}^{-\mathrm{i}tL_{\mathrm{phys}}}$, $t \in \mathbb{R}$, $A \in L^\infty(Q,\Sigma_{\{0\}})$;
(ii) the dynamics τ_t on $\mathfrak{A}_{\mathrm{phys}}$ is the dynamics generated by the unitary group $\mathrm{e}^{-\mathrm{i}tL_{\mathrm{phys}}}$;
(iii) the β-KMS state on $\mathfrak{A}_{\mathrm{phys}}$ is the state generated by the vector Ω_{phys}.

10.3.2 Euclidean measure for the translation invariant $P(\phi)_2$ model

The spatially-cutoff $P(\phi)_2$ model at positive temperature allows a Euclidean formulation: the measure $\mathrm{d}\mu_l$ for the spatially cutoff $P(\phi)_2$ model is given by

$$\mathrm{d}\mu_l := \frac{1}{Z_l}\,\mathrm{e}^{-\int_{-\beta/2}^{\beta/2}\int_{-l}^{l}:P(\phi(t,x)):_C \mathrm{d}t\mathrm{d}x}\,\mathrm{d}\phi_C, \tag{10.6}$$

where $\mathrm{d}\phi_C$ denotes the Gaussian measure on (Q,Σ) with covariance

$$C(u,u) = (u, (D_t^2 + D_x^2 + m^2)^{-1}u)$$

(with β-periodic b.c.) defined by

$$\int_Q \mathrm{e}^{\mathrm{i}\phi(f)}\mathrm{d}\phi_C = \mathrm{e}^{-C(f,f)/2}, \quad f \in \mathcal{S}_{\mathbb{R}}(S_\beta \times \mathbb{R}). \tag{10.7}$$

The partition function Z_l is chosen such that $\int_Q d\mu_l = 1$.

Existence of limiting measure

In order to show that one can remove the spatial cutoff one has to show that

$$\lim_{l\to+\infty}\int_Q F(q)\mathrm{d}\mu_l =: \int_Q F(q)\mathrm{d}\mu_\infty \tag{10.8}$$

exists and defines a Borel probability measure on $\mathcal{S}'_{\mathbb{R}}(S_\beta \times \mathbb{R})$.

Nelson Symmetry

Formally exchanging the role of t and x in (10.6) one notices that $\mathrm{d}\mu_\infty$ is the Euclidean measure of the $P(\phi)_2$ model on the circle at temperature zero. This formal argument can be made rigorous (see [12], [17]). In particular one has:

$$\mathrm{e}^{-\int_{-\beta/2}^{\beta/2}(\int_{-l}^{l}:P(\phi(t,x)):_{C_0}\mathrm{d}x)\mathrm{d}t} = \mathrm{e}^{-\int_{-l}^{l}(\int_{-\beta/2}^{\beta/2}:P(\phi(t,x)):_{C_\beta}\mathrm{d}t)\mathrm{d}x}. \tag{10.9}$$

Note that on the r.h.s. Wick ordering is done w.r.t. the covariance

$$C_\beta(g_1,g_2) := \left(g_1, \frac{1}{2\nu}g_2\right)_{L^2(S_\beta,\mathrm{d}t)}, \; g_1,g_2 \in \mathcal{S}(S_\beta).$$

The analog of (10.9) in the zero temperature case is called *Nelson symmetry* (see e.g. [26]).

It was first noticed by Höegh-Krohn [17] that the existence of the limit (10.8) is equivalent to the uniqueness of the vacuum state for the $P(\phi)_2$ model on the circle.

Using Nelson symmetry, the existence of the limit (10.8) is proved in [12]. Moreover it is shown in [12] that μ_∞ is OS positive, invariant under space-time translations, and that *sharp-time fields* are well defined: this means that if $\delta_k \in C_0^\infty(S_\beta)$ is a sequence of functions tending to the Dirac distribution δ_0, then the limits

$$\phi(t,h) := \lim_{k\to\infty}\phi(\delta_k(.-t)\otimes h)$$

exist in $\bigcap_{1\le p<\infty} L^p(Q,\Sigma,\mathrm{d}\mu_\infty)$ for any $h \in C_0^\infty(\mathbb{R})$.

10.4 The relativistic KMS condition

In this section we show that two-point functions for fields in the thermal $P(\phi)_2$ model satisfy the relativistic KMS condition.

Identification of physical objects

Let $(\mathcal{H}_\beta, \pi_\beta, \Omega_\beta)$ be the GNS triple associated to the KMS state ω_β over the C^*-algebra $\mathfrak{A}$. Let also P_β, L_β be the unique generators of space-time translations such that $L_\beta \Omega_\beta = P_\beta \Omega_\beta = 0$.

As we saw in Subsect. 10.3.1, the Euclidean reconstruction theorem provides a Hilbert space $\mathcal{H}_{\mathrm{phys}}$, a unit vector Ω_{phys}, a representation π_{phys} of the abelian von Neumann algebra $\mathcal{U}_{\mathrm{AW}}$ generated by $\{W_{\mathrm{AW}}(h) \mid h \in \mathfrak{h}, h \text{ real-valued }\}$, and a selfadjoint operator L_{phys}. Since the Euclidean measure μ_∞ is invariant under space translations, we obtain also a selfadjoint operator P_{phys} implementing the space translations.

Let us briefly check that these two families of objects are identical (up to unitary equivalence). In the sequel we will freely identify them.

Let $\mathfrak{U}(I)$ be the abelian von Neumann algebra generated by $\{W_{\mathrm{AW}}(h) \mid h \in \mathfrak{h}, h \text{ real-valued}, \operatorname{supp} h \subset I\}$. It was shown in [12] that for $A_j \in \mathfrak{U}(I)$, $1 \leq j \leq n$,

$$\begin{aligned}&(\Omega_{\mathrm{phys}}, \textstyle\prod_{j=1}^n \mathrm{e}^{\mathrm{i}t_j L_{\mathrm{phys}}} \pi_{\mathrm{phys}}(A_j) \mathrm{e}^{-\mathrm{i}t_j L_{\mathrm{phys}}} \Omega_{\mathrm{phys}})\\ &= \omega_\beta(\textstyle\prod_{j=1}^n \tau_{t_j}(A_j))\\ &= (\Omega_\beta, \textstyle\prod_{j=1}^n \mathrm{e}^{\mathrm{i}t_j L_\beta} \pi_\beta(A_j) \mathrm{e}^{-\mathrm{i}t_j L_\beta} \Omega_\beta).\end{aligned} \tag{10.10}$$

Thus we can define a map $U \colon \mathcal{H}_{\mathrm{phys}} \to \mathcal{H}_\beta$ by

$$U \mathrm{e}^{\mathrm{i}tL_{\mathrm{phys}}} \pi_{\mathrm{phys}}(A) \Omega_{\mathrm{phys}} := \mathrm{e}^{\mathrm{i}tL_\beta} \pi_\beta(A) \Omega_\beta, \quad t \in \mathbb{R},\ A \in \mathfrak{U}(I),\ I \subset \mathbb{R}. \tag{10.11}$$

From (10.10) we see that U preserves the scalar product, hence it can be uniquely extended by linearity to

$$\mathcal{E} = \mathrm{Vect}\{\mathrm{e}^{\mathrm{i}tL_{\mathrm{phys}}} \pi_{\mathrm{phys}}(A) \Omega_{\mathrm{phys}} \mid t \in \mathbb{R}, A \in \mathfrak{U}(I),\ I \subset \mathbb{R}\}$$

as an isometry. It follows from the Euclidean reconstruction theorem, that $\mathcal{E}$ is dense in $\mathcal{H}_{\mathrm{phys}}$. Moreover it is clear from (10.11) that U intertwines π_{phys} and π_β restricted to $\mathfrak{U}$ and also intertwines L_{phys} and L_β. Finally U also intertwines P_{phys} and P_β (note that the algebra of time-zero fields is clearly invariant under space translations).

To check that U is unitary, we use a result in [12]: let $\mathcal{B}_\alpha(I)$ be the von Neumann algebra generated by $\{\tau_t(A) \mid A \in \mathfrak{U}(I), |t| < \alpha\}$. Then it was shown in [12, Prop. 6.5] that

$$\mathfrak{R}_{AW}(I) = \bigcap_{\alpha > 0} \mathcal{B}_\alpha(I).$$

Since by construction Ω_β is cyclic for $\pi_\beta(\mathfrak{A})$, this implies that the range of U is dense in $\mathcal{H}_\beta$. Therefore U is unitary.

10.4.1 Wightman two-point function for the thermal $P(\phi)_2$ model

Let $I \subset \mathbb{R}$ be a bounded open interval and $\mathfrak{h}_I$ the real subspace of $\mathfrak{h}$ defined in (10.3).

By restriction to the local algebra $\mathcal{R}_{\mathrm{AW}}(I)$, π_β defines a CCR representation of the real symplectic space $\mathfrak{h}_I$.

Lemma 1. i) *The representation π_β restricted to $\mathcal{R}_{\mathrm{AW}}(I)$ is quasi-equivalent to the concrete representation of $\mathcal{R}_{\mathrm{AW}}(I)$;*
ii) *the CCR representation $\mathfrak{h}_I \ni h \mapsto \pi_\beta(W_{\mathrm{AW}}(h)) \in \mathcal{B}(\mathcal{H}_\beta)$ is regular.*

Proof. It is well known (see e.g. [12, Lemma 6.2]) that $\mathcal{R}_{\mathrm{AW}}(I)$ is a factor. Now it is shown in [20, Prop. 10.3.14] that if $\mathcal{R}$ is a C^*-algebra and π is a factor representation of $\mathcal{R}$, then π is quasi-equivalent to the GNS representation of any π-normal state ω. Applying this fact to $\mathcal{R}_{\mathrm{AW}}(I)$ (with its concrete representation) and to ω_β, we obtain that π_β is quasi-equivalent to the concrete representation of $\mathcal{R}_{\mathrm{AW}}(I)$. This proves *i)*. We know then that there exists a $*$-isomorphism γ from $\mathcal{R}_{\mathrm{AW}}(I)$ into $\pi_\beta(\mathcal{R}_{\mathrm{AW}}(I))''$ extending π_β. This isomorphism is automatically weakly continuous. Since the Araki-Woods representation is regular, the same holds true for the GNS representation π_β restricted to $\mathfrak{h}_I$. □

Since π_β is a regular CCR representation, we can define for $h \in \mathfrak{h}_I$ the *Segal field operators*

$$\phi_\beta(h) := -\mathrm{i}\frac{\mathrm{d}}{\mathrm{d}s}\pi_\beta(W_{\mathrm{AW}}(sh))\Big|_{s=0}.$$

In the sequel we will consider only the *time-zero fields* φ_β:

$$\varphi_\beta(h) := \phi_\beta(h), \text{ for } h \in \mathfrak{h}_I,\ h \text{ real.}$$

Remark 3. If we restrict ourselves to time-zero fields, it is possible to give a direct proof that the CCR representation is locally regular, avoiding the use of the local normality of the state ω_β. In fact, let us show that

$$\text{the map } \mathbb{R} \ni s \mapsto \pi_\beta\big(W_{\mathrm{AW}}(sh)\big) \text{ is strongly continuous for } h \in \mathfrak{h}_I,\ h \text{ real,} \tag{10.12}$$

using the Euclidean approach.

It suffices to prove the weak continuity on a dense subspace of $\mathcal{H}_\beta$. From the reconstruction theorem, we see that we may take as a dense subspace of $\mathcal{H}_\beta$ the linear span of the vectors $\mathcal{V}(\prod_1^k \mathrm{e}^{\mathrm{i}\phi(t_j,h_j)})$ for $h_j \in C_{0\mathbb{R}}^\infty(\mathbb{R})$, $0 \le t_j < \beta/2$.

We see that it suffices to prove the continuity of the map

$$\mathbb{R} \ni s \mapsto \int_Q (\prod_1^n \mathrm{e}^{\mathrm{i}\phi(t_j,h_j)})\mathrm{e}^{\mathrm{i}s\phi(0,h)}\mathrm{d}\mu_\infty$$

for $h_j \in C_{0\mathbb{R}}^\infty(\mathbb{R})$, $t_j \in S_\beta$. But this follows from the fact that $\phi(0,h) \in L^1(Q, \mathrm{d}\mu_\infty)$, shown in [12].

Lemma 2. *$\Omega_\beta \in \mathcal{D}(\varphi_\beta(h))$, $\forall h \in \mathfrak{h}_I$, h real-valued.*

Proof. Clearly it suffices to prove that

$$2 - (\Omega_\beta, \mathrm{e}^{\mathrm{i}s\varphi_\beta(h)}\Omega_\beta) - (\Omega_\beta, \mathrm{e}^{-\mathrm{i}s\varphi_\beta(h)}\Omega_\beta) \leq C|s|^2, \quad 0 \leq s \leq 1. \tag{10.13}$$

By the reconstruction theorem, the r.h.s. is equal to

$$\int_Q (2 - \mathrm{e}^{\mathrm{i}s\phi(0,h)} - \mathrm{e}^{-\mathrm{i}s\phi(0,h)})\mathrm{d}\mu_\infty.$$

Since $\phi(0,h) \in L^2(Q, \mathrm{d}\mu_\infty)$, we obtain (10.13). □

We now define Wightman two-point functions. For a function $h \in C_0^\infty(\mathrm{I\!R})$ we denote by h^- the function $h^-(x) = h(-x)$.

Proposition 1. *There exists a unique $\mathcal{W}_\beta(t,.) \in C^0(\mathrm{I\!R}_t, \mathcal{D}'(\mathrm{I\!R}))$ such that for $h_1, h_2 \in C^\infty_{0\mathrm{I\!R}}(\mathrm{I\!R})$:*

$$\big(\varphi_\beta(h_1)\Omega_\beta, \mathrm{e}^{\mathrm{i}tL_\beta}\varphi_\beta(\alpha_x h_2)\Omega_\beta\big) = h_1 \star h_2^- \star \mathcal{W}_\beta(t,x), \quad (t,x) \in \mathrm{I\!R}^2. \tag{10.14}$$

Proof. For fixed t the l.h.s. is a bilinear form Q_t w.r.t. h_1 and h_2. Moreover it is shown in [12] that $\|\varphi_\beta(h)\Omega_\beta\| \leq C\|h\|_S$, where $\|.\|_S$ is a Schwartz seminorm. Therefore Q_t is continuous for the topology of $C_0^\infty(\mathrm{I\!R})$, which by translation invariance, implies the existence of $\mathcal{W}_\beta(t,.)$. The continuity w.r.t. the variable t of $\mathcal{W}(t,.)$ follows from the obvious continuity in t of the l.h.s. of (10.14). □

10.4.2 Relativistic KMS condition

The rest of this section is devoted to the proof of the following theorem:

Theorem 1. *The distribution $W_\beta(t,x)$ extends holomorphically to $\mathrm{I\!R}^2 + \mathrm{i}V_\beta$, where $V_\beta := \{(t,y) \mid |y| < \inf(t, \beta - t)\}$. Therefore for $A_i = \varphi_\beta(h_i)$, $h_i \in C^\infty_{0\mathrm{I\!R}}(\mathrm{I\!R})$ the two-point function $F_{A_1,A_2}(t,x)$ is holomorphic in $\mathrm{I\!R}^2 + \mathrm{i}V_\beta$.*

10.4.3 Proof of Theorem 1

Let us briefly recall a few facts concerning the 0-temperature $P(\phi)_2$ model on the circle S_β. The Hilbert space is the Fock space $\Gamma(H^{-\frac{1}{2}}(S_\beta))$, where $H^{-\frac{1}{2}}(S_\beta)$ is the Sobolev space of order $-\frac{1}{2}$ on S_β with norm

$$\|g\|^2 = (g, (2b)^{-1}g)_{L^2(S_\beta, \mathrm{d}t)}$$

with $b = (D_t^2 + m^2)^{\frac{1}{2}}$. For $g \in H^{-\frac{1}{2}}(S_\beta)$, we denote by $\phi_C(g)$ the (Fock) field operator acting on $\Gamma(H^{-\frac{1}{2}}(S_\beta))$.

The operator sum

$$\mathrm{d}\Gamma(b) + \int_{S_\beta} :P(\phi_C(t)):_{C_\beta} \mathrm{d}t$$

is essentially selfadjoint and bounded below, and the Hamiltonian of the $P(\phi)_2$ model on S_β is

$$H_C := \mathrm{d}\Gamma(b) + \int_{S_\beta} :P(\phi(t)):_{C_\beta} \mathrm{d}t - E_C,$$

where E_C is an additive constant such that $\inf \sigma(H_C) = 0$. The Hamiltonian H_C has a unique (up to a phase) ground state (i.e., vacuum state) induced by the state vector Ω_C. Another fact we shall need is the following: if g_1, g_2 are *real* elements of $H^{-\frac{1}{2}}(S_\beta)$, then $(\phi_C(g_1)\Omega_C, \mathrm{e}^{-yH_C}\phi_C(g_2)\Omega_C)$ is real for $y \in \mathbb{R}^+$. This follows from the representation of e^{-yH_C} using the Feynman-Kac-Nelson (FKN) formula.

Finally we note that

$$P_C := \mathrm{d}\Gamma(D_t)$$

is the momentum operator on the circle S_β.

The two-point function for the $P(\phi)_2$ model on the circle

Now consider the two-point function $\mathcal{W}_C$ for the $P(\phi)_2$ model on the circle

$$\mathcal{W}_C(t,y) = \big(\Omega_C, \phi_C(\delta_0)\mathrm{e}^{\mathrm{i}yH_C}\mathrm{e}^{\mathrm{i}tP_C}\phi_C(\delta_0)\Omega_C\big), \quad t \in S_\beta,\ y \in \mathbb{R}.$$

$\mathcal{W}_C$ is a tempered distribution on $S_\beta \times \mathbb{R}$ rigorously defined by

$$\langle \mathcal{W}_C, f \otimes g\rangle := (2\pi)^{\frac{1}{2}}\big(\phi_C(\delta_0)\Omega_C, \tilde{g}(-H_C)\phi_C(f)\Omega_C\big),$$

for $f \in C^\infty(S_\beta)$, $g \in \mathcal{S}(\mathbb{R})$, and $\tilde{g}$ the Fourier transform of g.

To check that $\mathcal{W}_C$ is well defined as a tempered distribution, we use the bound

$$\|(H_C+1)^{-\frac{1}{2}}\phi_C(h)(H_C+1)^{-\frac{1}{2}}\| \le C\|h\|_{H^{-1}(S_\beta)}, \tag{10.15}$$

which using that $\delta_0 \in H^{-1}(S_\beta)$ yields

$$|\langle \mathcal{W}_C, f \otimes g\rangle| \le C\|(H_C+1)\tilde{g}(-H_C)\|\|f\|_{H^{-1}(S_\beta)}\|(H_C+1)^{\frac{1}{2}}\Omega_C\|^2.$$

The r.h.s. can clearly be estimated in terms of Schwartz seminorms of f and g.

Analytic continuation

We first recall the *spectrum condition on the circle* [14]:

$$|P_C| \le H_C.$$

Set

$$V_\pm := \{(t,y) \in \mathbb{R}^2 \mid |t| < |y|,\ \pm y \ge 0\}.$$

Using $\|\mathrm{e}^{-\epsilon H_C}(H_C+1)\| \le C\epsilon^{-1}$ and the bound (10.15) we conclude that

$$F(\tau,z) := \big(\Omega_C, \phi_C(\delta_0)\mathrm{e}^{\mathrm{i}zH_C}\mathrm{e}^{\mathrm{i}\tau P_C}\phi_C(\delta_0)\Omega_C\big)$$

is holomorphic in $S_\beta \times \mathbb{R} + \mathrm{i}V_+$, has a moderate growth when $\mathrm{Im}\,(\tau,z) \to 0$ and

$$\begin{aligned}
&\textstyle\int F(t+\mathrm{i}\epsilon, y+\mathrm{i}\epsilon)f(t)g(y)\mathrm{d}t\mathrm{d}y \\
&= (2\pi)^{\frac{1}{2}}\big(\phi_C(\delta_0)\Omega_C, \tilde{g}(-H_C)\mathrm{e}^{-\epsilon(H_C+P_C)}\phi_C(f)\Omega_C\big) \\
&\to \langle \mathcal{W}_C, f \otimes g\rangle,
\end{aligned}$$

when $\epsilon \to 0$, i.e.,

$$\lim_{\epsilon \to 0} F(.+\mathrm{i}\epsilon, .+\mathrm{i}\epsilon) = \mathcal{W}_C(.,.) \text{ in } \mathcal{S}'(S_\beta \times \mathbb{R}).$$

Locality on the circle

Clearly

$$\mathcal{W}_C = \lim_{k\to+\infty} G_k \text{ in } \mathcal{S}'(S_\beta \times \mathbb{R}),$$

where

$$G_k(t,y) := \left(\Omega_C, \phi_C(\delta_k)\mathrm{e}^{\mathrm{i}yH_C}\mathrm{e}^{\mathrm{i}tP_C}\phi_C(\delta_k)\Omega_C\right),$$

and δ_k is a sequence in $C^\infty(S_\beta)$ with support in $\{t \in \mathbb{R} \mid |t| \le k^{-1}\}$ and tending to δ_0 when $k \to \infty$.

Now using locality (i.e., finite speed of light) on the circle, we see that if $(t,y) \in V_{\beta,k}$,

$$V_{\beta,k} = \{(t,y) \mid |y| < \inf(t, \beta - t) - 2k^{-1}\},$$

then

$$[\phi_C(\delta_k), \mathrm{e}^{\mathrm{i}yH_C}\mathrm{e}^{\mathrm{i}tP_C}\phi_C(\delta_k)\mathrm{e}^{-\mathrm{i}tP_C}\mathrm{e}^{-\mathrm{i}yH_C}] = 0,$$

because no signal can go from supp δ_k to supp $(\delta_k + t)$ in time y if $(t,y) \in V_{\beta,k}$. This fact can be shown by exactly the same arguments as those used for the $P(\phi)_2$ model on $\mathbb{R}$. Thus for $(t,y) \in V_{\beta,k}$, the function G_k is real-valued.

Edge of the wedge

According to the Schwarz reflection principle, $\mathcal{W}_C$ can now be viewed as the boundary value of a function holomorphic in $V_\beta - iV_+$:

$$\mathcal{W}_C(t,x) = H(t,x)$$

where

$$H(\tau,z) := \left(\phi_C(\delta_0)\Omega_C, \mathrm{e}^{-\mathrm{i}zH_C}\mathrm{e}^{-\mathrm{i}\tau P_C}\phi_C(\delta_0)\Omega_C\right).$$

Thus

$$\overline{\mathcal{W}_C(t,y)} = \mathcal{W}_C(\bar{t},\bar{y}), \quad (t,y) \in V_\beta + i(V_+ \cup V_-). \tag{10.16}$$

We can now apply the edge of the wedge theorem [27]. It implies that there exists an open ball $B(0,d) := \{z \in \mathbb{C}^2 \mid |z| < d\}$ such that $\mathcal{W}_C(t,y)$ is holomorphic in $V_\beta + i\Gamma$, where

$$\Gamma := V_+ \cup V_- \cup B(0,d).$$

Moreover, $\mathcal{W}_C(t,iy)$ is real for $t \in S_\beta$, $y > 0$ (by using the representation of e^{-yH_C} as a Feynman-Kac-Nelson (FKN) kernel). Thus $\mathcal{W}_C(t,iy)$ is real for $t \in S_\beta$, $y \in \mathbb{R}$, by (10.16). Applying the Schwarz reflection principle one more time, we conclude that

$$\mathcal{W}_C(t,y) = \mathcal{W}_C(t,-y) \quad \forall (t,y) \in V_\beta + i\Gamma.$$

Schwinger two-point function for the thermal $P(\phi)_2$ model

Let $h \in C^\infty_{0\mathbb{R}}(\mathbb{R})$ and set

$$I(t,x) := \int_{\mathcal{S}'(S_\beta\times\mathbb{R})} \phi(0,h)\phi(t,\alpha_x h)\mathrm{d}\mu_\infty, \quad t \in S_\beta, x \in \mathbb{R},$$

where $\alpha_x h(y) = h(y-x)$. By [13, Thm. 7.2], we get:

$$I = 2\int_{-\infty<x_1\leq x_2<\infty} h(x_1)\big(\Omega_C, \phi_C(\delta_0)\mathrm{e}^{-(x_2-x_1)H_C}\phi_C(\delta_t)\Omega_C\big)h(x_2-x)\mathrm{d}x_1\mathrm{d}x_2$$
$$= 2\int_{-\infty<x_1\leq x_2<\infty} h(x_1-x)\big(\Omega_C, \phi_C(\delta_t)\mathrm{e}^{-(x_2-x_1)H_C}\phi_C(\delta_0)\Omega_C\big)h(x_2)\mathrm{d}x_1\mathrm{d}x_2.$$

But $\big(\Omega_C, \phi_C(\delta_0)\mathrm{e}^{-(x_2-x_1)H_C}\phi_C(\delta_t)\Omega_C\big)$ is real by the FKN formula and hence

$$\big(\Omega_C, \phi_C(\delta_0)\mathrm{e}^{-(x_2-x_1)H_C}\phi_C(\delta_t)\Omega_C\big) = \big(\Omega_C, \phi_C(\delta_t)\mathrm{e}^{-(x_2-x_1)H_C}\phi_C(\delta_0)\Omega_C\big),$$

which yields:

$$I(t,x) = 2\int_{-\infty<x_1\leq x_2<\infty}\big(h(x_1)r(t,x_2-x_1)h(x_2-x) \\ +h(x_1-x)r(t,x_2-x_1)h(x_2)\big)\mathrm{d}x_1\mathrm{d}x_2, \tag{10.17}$$

for

$$r(t,x) = \big(\Omega_C, \phi_C(\delta_0)\mathrm{e}^{-xH_C}\phi_C(\delta_t)\Omega_C\big) = \mathcal{W}_C(t,\mathrm{i}x), \quad x\geq 0.$$

We have seen that $\mathcal{W}_C(t,y) = \mathcal{W}_C(t,-y)$ for $(t,y)\in V_\beta + \mathrm{i}\Gamma$, which when restricted to $t\in S_\beta$, $\mathrm{Re}\, y = 0$, yields $\mathcal{W}_C(t,\mathrm{i}x) = \mathcal{W}_C(t,-\mathrm{i}x)$.

Therefore exchanging the variables x_1 and x_2 in the r.h.s. of (10.17), we obtain:

$$2I(t,x) = \int_{\mathbb{R}^2} h(x_1)\mathcal{W}_C(t,x_2-x_1)h(x_2-x)\mathrm{d}x_1\mathrm{d}x_2 = h\star h^- \star \mathcal{W}_C(t,x), \tag{10.18}$$

where $h^-(x) := h(-x)$.

Wightman two-point function for the thermal $P(\phi)_2$ model

Let Ω_β be the GNS vector for the thermal state ω_β, and let P_β and L_β be the generators of space-time translations such that $P_\beta\Omega_\beta = L_\beta\Omega_\beta = 0$. We recall that $\varphi_\beta(h)$ for $h\in C^\infty_{0\mathbb{R}}(\mathbb{R})$ is the field operator in the GNS representation.

Let $\mathcal{W}_\beta(t,x)$ be the two-point function for the thermal $P(\phi)_2$ model, defined by

$$\big(\varphi_\beta(h_1)\Omega_\beta, \mathrm{e}^{\mathrm{i}tL_\beta}\varphi_\beta(\alpha_x h_2)\Omega_\beta\big) = h_1\star h_2^- \star \mathcal{W}_\beta(t,x), \tag{10.19}$$

for $h_i\in C^\infty_{0\mathbb{R}}(\mathbb{R})$. Since $\mathrm{d}\mu_\infty$ is the Euclidean measure for the thermal $P(\phi)_2$ model on the line, we have

$$\mathcal{W}_\beta(\mathrm{i}t,x) = \mathcal{W}_C(t,\mathrm{i}x), \quad 0<t<\beta/2,\ x\in\mathbb{R}. \tag{10.20}$$

As we have seen, $\mathcal{W}_C(t,y)$ is holomorphic in $V_\beta + \mathrm{i}\Gamma$. Thus we deduce from (10.20) that $\mathcal{W}_\beta(t,x)$ is holomorphic in $\Gamma + \mathrm{i}V_\beta$. Applying (10.19) with $h_1 = h_2 =: h$, we deduce from this fact that

$$\varphi_\beta(h)\Omega_\beta \in \mathcal{D}(\mathrm{e}^{-(sL+yP)/2}) \quad \forall (s,y)\in V_\beta.$$

But by the spectral theorem this clearly implies that $\mathcal{W}_\beta(t,x)$ is holomorphic in $\mathbb{R}^2 + \mathrm{i}V_\beta$. □

10.5 Outlook

The condition of locality leads to strong constraints on the general form of the thermal two-point functions which allow one to apply the techniques of the Jost–Lehmann–Dyson representation. As has been shown by Bros and Buchholz [6] [7], the interacting two-point function $\mathcal{W}_\beta$ can be represented in the form

$$\mathcal{W}_\beta(t,x) = \int_0^\infty \mathrm{d}m\, \mathcal{D}_\beta(x,m)\, \mathcal{W}_\beta^{(0)}(t,x,m).$$

Here $\mathcal{D}_\beta(x,m)$ is a distribution in x, m which is symmetric in x, and

$$\mathcal{W}_\beta^{(0)}(x,m) = (2\pi)^{-1} \int \mathrm{d}\nu \mathrm{d}p\, \varepsilon(\nu)\delta(\nu^2 - p^2 - m^2)(1 - \mathrm{e}^{-\beta\nu})^{-1} \mathrm{e}^{i(\nu t - px)}$$

is the two-point correlation function of the free field of mass m in a thermal equilibrium state at inverse temperature β. In contrast to the vacuum case, the damping factors $\mathcal{D}_\beta(x,m)$ depend in general in a non-trivial way on the spatial variables x. They describe the dissipative effects of the thermal background on the propagation of sharply localized excitations.

If the underlying equilibrium state satisfies the relativistic KMS condition, the function $D_\beta(x,m)$ is regular in x and admits an analytic continuation into the domain $\{z \in \mathbb{C} \mid |\mathrm{Im}\, z| < \beta/2\}$.

References

1. H. Araki: A lattice of von Neumann algebras associated with the quantum theory of a free Bose field. *J. Math. Phys.* 4:1343–1362 (1963).
2. H. Araki: Relative Hamiltonian for faithful normal states of a von Neumann algebra. *Publ. Res. Int. Math. Soc. (RIMS)* 9:165–209 (1973).
3. H. Araki: Positive cone, Radon-Nikodym theorems, relative Hamiltonian and the Gibbs condition in statistical mechanics. An application of Tomita-Takesaki theory. In *C^*-algebras and their applications to Statistical Mechanics and Quantum Field Theory* (ed.: D. Kastler). North Holland, 1976.
4. L. Birke and J. Fröhlich: KMS, etc. *Rev. Math. Phys.* 14:829–871 (2002).
5. O. Bratteli and D.W. Robinson: *Operator Algebras and Quantum Statistical Mechanics I, II.* Springer-Verlag, New York, Heidelberg, Berlin, 1981.
6. J. Bros and D. Buchholz: Towards a relativistic KMS condition. *Nucl. Phys. B* 429:291–318 (1994).
7. J. Bros and D. Buchholz: Axiomatic analyticity properties and representations of particles in thermal quantum field theory, New problems in the general theory of fields and particles. *Ann. l'Inst. H. Poincaré* 64:495–521 (1996).
8. J. Derezinski, V. Jaksic and C.-A. Pillet: Perturbation theory of W^*-dynamics, Liouvilleans and KMS states. *Rev. Math. Phys.* 15:447–48. (2003).
9. J. Fröhlich: The reconstruction of quantum fields from Euclidean Green's functions at arbitrary temperatures. *Helv. Phys. Acta* 48:355–363 (1975).
10. J. Fröhlich: Unbounded, symmetric semigroups on a separable Hilbert space are essentially selfadjoint. *Adv. in Appl. Math.* 1:237–256 (1980).

11. C. Gérard and C.D. Jäkel: Thermal quantum fields with spatially cut-off interactions in 1+1 space-time dimensions. To appear in *Journal of Funct. Anal.*, 2005.
12. C. Gérard and C.D. Jäkel: Thermal quantum fields without cutoffs in 1+1 space-time dimensions. arXiv math-ph/0403047. To appear in *Rev. Math. Phys.*
13. J. Glimm and A. Jaffe: *Quantum Physics, A Functional Point of View.* Springer, 1981.
14. R. Haag, N.M. Hugenholtz and M. Winnink: On the equilibrium states in quantum statistical mechanics. *Comm. Math. Phys.* 5:215–236 (1967).
15. R. Haag, D. Kastler and E.B. Trych-Pohlmeyer: Stability and equilibrium states. *Comm. Math. Phys.* 38:173–193 (1974).
16. E.P. Heifets and E.P. Osipov: The energy momentum spectrum in the $P(\varphi)_2$ quantum field theory. *Comm. Math. Phys.* 56:161–172 (1977).
17. R. Höegh-Krohn: Relativistic quantum statistical mechanics in two-dimensional space-time. *Comm. Math. Phys.* 38:195–224 (1974).
18. C.D. Jäkel: Decay of spatial correlations in thermal states. *Ann. l'Inst. H. Poincaré* 69:425–440 (1998).
19. C.D. Jäkel: The Reeh-Schlieder property for thermal field theories. *J. Math. Phys.* 41:1745–1754 (2000).
20. R.V. Kadison and J.R. Ringrose: *Fundamentals of the Theory of Operator Algebras II.* Academic Press, New York, 1986.
21. A. Klein and L. Landau: Stochastic processes associated with KMS states. *J. Funct. Anal.* 42:368–428 (1981).
22. A. Klein and L. Landau: Construction of a unique selfadjoint generator for a symmetric local semigroup. *J. Funct. Anal.* 44:121–137 (1981).
23. R. Kubo: Statistical mechanical theory of irreversible processes I. *J. Math. Soc. Jap.* 12:570–586 (1957).
24. P.C. Martin and J. Schwinger: Theory of many-particle systems. I. *Phys. Rev.* 115:1342–1373 (1959).
25. M. Reed and B. Simon: *Methods of Modern Mathematical Physics, Vol. II: Fourier Analysis, Self-adjointness.* Academic Press, 1975.
26. B. Simon: *The $P(\varphi)_2$ Euclidean (Quantum) Field Theory.* Princeton University Press, 1974.
27. R.F. Streater and A.S. Wightman: *PCT, Spin and Statistics and all that.* Benjamin, New York, 1964.

11

The Analyticity Program in Axiomatic Quantum Field Theory

Daniel Iagolnitzer

Service de Physique Theorique CEA – Saclay, 91191 Gif-sur-Yvette Cedex, France
daniel.iagolnitzer@cea.fr

Summary. This paper reviews some topics and results of the linear and the nonlinear programs of Axiomatic Quantum Field Theory. Applications to Constructive Field Theory are also discussed.

11.1 Introduction

This review treats analyticity in energy-momentum variables (dual to time and space variables respectively) in traditional Quantum Field Theory (QFT). It roughly covers the period 1960–1990. In 1960, analyticity is clearly recognized as an important property of collision amplitudes in relativistic quantum physics, a fact which leads to a number of rigorous works and results, in various approaches, in successive steps in the 1960s through 1980s. A more recent result related to this analyticity program concerns analyticity in angular momentum (presented in this book by G.A. Viano).

Viewed from a current perspective, the analysis includes a number of basic results on the general structure of 2-body and multiparticle collision amplitudes. It has also been related to, and at the origin of some, deep mathematical problems and results in the domain of analytic functions of several complex variables and of what is called today "microlocal analysis". On the negative side, some of the results so far apply only to "massive" theories (particles with strictly positive masses only), whose existence in space-time dimension 4 is very doubtful. Nevertheless, it is hoped that the analysis will keep some value in more general theories.

The paper will mainly treat the *analyticity program in the axiomatic QFT approach*, in which Jacques Bros has played a major role. The related approach of Constructive QFT, in which his ideas also played a direct role, will be briefly presented at the end. Other approaches (analysis of Feynman integrals in perturbative QFT, "axiomatic" S-matrix theory) will be briefly mentioned in the course of the talk. Concerning the axiomatic analyticity program, two types of methods and results can be distinguished; they lead respectively to what we shall call *microlocal* and *global analyticity properties* of the N-point functions of the fields and $m \to n$ particle collision amplitudes. As a general rule, our survey will provide more information about the former than about the latter; in particular, such important and well-known global results of the case $N = 4$ as the dispersion relations for two-particle collision amplitudes and the rigorous derivation of the Froissart bounds (and other types of bounds) by A. Martin will only be briefly mentioned at the relevant place of our survey.

For conciseness, references to original works are omitted. For details and references, see the book by the author "Scattering in Quantum Field Theories: the Axiomatic and Constructive Approaches", Princeton University Press, 1991. Among contributions by many authors, most crucial works on the topics treated are due to A. S. Wightman et al in Sect. 1, J. Bros in Sect. 2, J. Bros, H. Epstein, V. Glaser in Sect. 3, H.P. Stapp in Sect. 4, J. Bros in Sect. 5 and J. Magnen in Sect. 6.

11.2 The axiomatic framework

11.2.1 Wightman axioms and Asymptotic Completeness

We denote by $x = (x_0, \vec{x})$ a (real) point in Minkowski space-time with respective time and space components x_0 and $\vec{x}$ (in a given Lorentz frame), the Minkowskian squared pseudo-norm of x being $x^2 = x_0^2 - \vec{x}^2$. Besides the usual physical space-time dimension $d = 4$, possible values 2 or 3 will also be considered. In all that follows, the unit system is such that the velocity c of light is equal to 1. Energy-momentum variables, dual (by Fourier transformation) to time and space variables respectively, are denoted by $p = (p_0, \vec{p})$; the corresponding squared mass variable is $p^2 = p_0^2 - \vec{p}^2$.

We describe below the Wightman axiomatic framework, though alternative frameworks such as the one of "Local Quantum Physics" based on the Araki-Haag-Kastler axioms, may be used similarly for present purposes with only minor changes. For simplicity, we consider a theory with only one basic (neutral, scalar) field A, defined on space-time as an operator-valued distribution: for each test function f, $A(f)$ (i.e. formally the integral $\int A(x)f(x)dx$) is an operator in a Hilbert space $\mathcal{H}$ of states (by abuse, we will also speak of $A(x)$, for a given point x, as an "operator in $\mathcal{H}$"). A physical state is more precisely represented by a (normalized) vector in $\mathcal{H}$ modulo scalar multiples. It has to be physically understood as "sub specie aeternitatis" (i.e. "with all its evolution", the Heisenberg picture of quantum mechanics being always adopted).

It is assumed that there exists in $\mathcal{H}$ a representation of the Poincaré group (semi-direct product of pure Lorentz transformations and space-time translations).

The Wightman axioms

– local commutativity : the operators $A(x)$ and $A(y)$ commute if $x - y$ is space-like (i.e. $(x-y)^2 < 0$).

– spectral condition (= positivity of the energy in relativistic form): the spectrum of the energy-momentum operators (infinitesimal generators of space-time translations) is contained in the future cone V_+ ($p^2 \geq 0, p_0 \geq 0$).

In a theory "with mass gap", the spectrum is more precisely assumed to be contained in the union of the origin (that will correspond to the vacuum vector introduced next), of one or more discrete mass-shell hyperboloids $H_+(m_i)$ ($p^2 = m_i^2$,

$p_0 > 0$) with strictly positive masses m_i, and of a continuum. For simplicity, we will mostly consider a theory with only one mass m and a continuum starting at $2m$, but this is not so in a theory "with bound states" (see Note ii) below).

– existence in $\mathcal{H}$ of a vacuum vector Ω, which is the only invariant vector under Poincare transformations up to scalar mutiples. It is moreover assumed that the vector space generated by the action of field operators on the vacuum is dense in $\mathcal{H}$.

– Poincaré covariance of the theory.

Haag-Ruelle-Hepp asymptotic theory

In a theory with mass gap, subspaces $\mathcal{H}_{in}$ and $\mathcal{H}_{out}$ of $\mathcal{H}$ can be defined by limiting procedures that we briefly explain (assuming only one mass m for simplicity). To that purpose, one considers test functions $f_{j,t}(x)$ with Fourier transforms of the form

$$\tilde{f}_{j,t}(p) = \tilde{f}_j(p) e^{i(p_o - [\vec{p}^2 + m^2]^{\frac{1}{2}})t},$$

where the functions $\tilde{f}_j$ have their supports in a neighborhhod of the mass-shell $H_+(m)$. It can then be shown that vectors of the form

$$\Psi_t = A(f_{1,t})A(f_{2,t}) \cdots A(f_{n,t})\Omega$$

converge to limits in $\mathcal{H}$ when t tends to minus or plus infinity respectively, and that these limits only depend on the mass-shell restrictions of the test functions $\tilde{f}_{j|H_+(m)}$.

$\mathcal{H}_{in}$ and $\mathcal{H}_{out}$ are interpreted physically as subspaces of states that are "asymptotically tangent" before, respectively after the interactions, to free-particle states, with particles of mass m. They are in fact both isomorphic to the free-particle Fock space $\mathcal{F}$,namely the direct sum of n-particle spaces of "wave functions" depending on n energy-momenta $p_1, p_2, \dots, p_n$ on the mass shell $H_+(m)$.

Asymptotic Completeness

Asymptotic Completeness (AC) is the assertion that $\mathcal{H} = \mathcal{H}_{in} = \mathcal{H}_{out}$, i.e. that each state in $\mathcal{H}$ is asymptotically tangent to a free-particle state, with particles of mass m, both before and after interactions (the two free-particle states are different if there are interactions). This condition can certainly NOT be expected to always hold in the general framework introduced above. Even if we restrict our attention to "physically reasonable" theories, in which states of $\mathcal{H}$ are assumed to be asymptotically tangent to free-particle states before and after interactions, the absence of other stable particles with different masses is not guaranteed. For instance, eventhough the field A as introduced above is "neutral", the action of field operators on the vacuum might generate neutral states corresponding to pairs of "charged" particles with opposite charges, whatever "charge" one might imagine. Individual charged particles cannot occur in $\mathcal{H}$ and their mass thus does not appear in the spectral condition for the field A. Hence such states of pairs of charged particles will not belong to $\mathcal{H}_{in}$ or $\mathcal{H}_{out}$ although they belong to $\mathcal{H}$.

If the set of charged particles is known, it can be shown that the above axiomatic framework might be enlarged by defining charged fields, in such a way that AC might still be valid in the enlarged framework. (For the analysis of charged fields in the general context of Local Quantum Physics, see the works by Doplicher, Haag , Roberts and by Buchholz and collaborators).

For simplicity, we will restrict below our attention to the simplest theories in which AC holds in the way stated above.

Notes

i) We are speaking so far of "stable" particles. Unstable particles may occur in the theory (see Sect. 4) but do not occur in the spectrum.

ii) In a theory with one basic field A, it may happen that one-particle states of several stable particles corresponding to different masses in the discrete spectrum, will be created by the action of field operators on the vacuum. This includes the case of the 2-particle bound states mentioned in Sect. 5 and 6, which then contribute to $\mathcal{H}_{in}$ and $\mathcal{H}_{out}$.

Theoretical models

Do there exist theoretical models satisfying axioms of the type described above? Besides free fields (corresponding to theories without interactions) or "generalized free fields", together with the class of "Wick-polynomials" of the latter, works in Constructive QFT (see Sect. 6) provide a positive answer in space-time dimension $d = 2$ or 3: non-trivial models of interest satisfying the Wightman axioms, with mass gap, have been defined rigorously at $d = 2$ or 3 in the 1970s and 1980s, such as ϕ^4 at $d = 2$ or 3 or the "renormalisable" but not "super-renormalisable", "asymptotically free", Gross-Neveu model at $d = 2$. Asymptotic Completeness has also been partly established for these models. At $d = 4$, no similar result has been achieved and the existence of such models is very doubtful for various basic reasons not discussed here.

11.2.2 Chronological N-point functions and collision amplitudes

Collision operator S, or S-matrix, and collision amplitudes

For simplicity, we consider an axiomatic framework including AC, though the latter is not needed to define the collision operator S. It is shown that there does exist a linear operator S from $\mathcal{H}$ to $\mathcal{H}$, called "collision operator" or "S-matrix", that relates the "initial" and "final" free-particle states to which a state in $\mathcal{H}$ is tangent before and after interactions respectively.

Collision amplitudes are the energy-momentum kernels of S for given numbers m and n of initial and final particles. As easily seen, they are well-defined distributions on the space of all initial and final on-shell energy-momenta. For convenience, we will denote by p_k the physical energy-momentum of a final particle with index k ($p_k \in H_+(m)$), and by $-p_k$ the physical energy-momentum of an initial particle ($-p_k \in H_+(m)$).

Wightman functions and chronological functions of the fields

The N-point Wightman functions W_N, which are as a matter of fact distributions, are defined as the vacuum expectation values (VEV) of the products of N field operators, namely:

$$W_N(x_1, x_2, \ldots, x_N) = < \Omega, A(x_1)A(x_2) \cdots A(x_N)\Omega > .$$

The chronological functions T_N are the VEV of the chronological products $\mathcal{T}_N(x_1, \ldots, x_N)$ of N field operators $A(x_1), \ldots, A(x_N)$: in the latter, fields are ordered according to decreasing values of the time components of the points x_k. T_N is essentially well-defined due to local commutativity, with, however, problems not treated here at coinciding points.

$\tilde{T}_N(p_1, \ldots, p_N)$ will denote the Fourier transform of T_N. It is shown that it has poles for each energy-momentum variable p_k on the mass-shell. The amputated chronological function $\tilde{T}_N^{amp}$ is defined by multiplying $\tilde{T}_N$ by the product of all factors $p_k^2 - m^2$ that cancel these poles.

L.S.Z.-Hepp "Reduction formulae"

A basic connection between chronological functions and collision amplitudes is then established: $\tilde{T}_N^{amp}$ can be restricted *as a distribution* to the mass-shell of any physical process with m initial and n final particles, with $m + n = N$, and this restriction coincides with the collision amplitude of the process. A process is here characterized by fixing the initial and final indices.

We note that, in view of the invariance of the theory under space-time translations, Wightman and chronological functions in space-time are invariant under global space-time translation of all points x_k together. As a consequence their Fourier transforms contain an energy-momentum conservation (e.m.c.) delta function. $\delta(p_1 + p_2 + \cdots + p_N)$.

Finally, *connected* (also called *truncated*) *N-point functions* $\tilde{T}_N^{amp,c}$ are defined by induction (over N) via a formula expressing each initial (non-connected) function as the sum of the corresponding connected function and of products of connected functions depending on subsets of points. In contrast to non-connected functions, the analysis shows that connected functions in energy-momentum space do not contain in general e.m.c. delta functions involving *subsets* of energy-momenta.

The analyticity properties of interest (described below) will apply to the connected functions after factoring out their global e.m.c. delta functions.

11.2.3 Linear and nonlinear programs

The Wightman axioms allow one to establish analyticity properties in complex energy-momentum space for the so-called *Analytic N-point functions* or *Green functions* that will be introduced in Sect. 11.4. The introduction of these functions makes only use of *linear* relations (either in x-space or in p-space) between the Wightman functions $W_N(x_{i_1}, x_{i_2}, \ldots, x_{i_N})$ corresponding for each fixed N to various permutations of the variables $x_1, \ldots, x_N$. The exploitation of these relations for obtaining the best analyticity properties of the Analytic N-point functions has been called *"the linear program"* of axiomatic QFT. In order to establish further results on the analytic structure of N-point functions and collision amplitudes, one then makes recourse to the exploitation of positivity conditions (positivity of the Hilbertian norm) and of relations (namely discontinuity formulae) arising from Asymptotic Completeness. These conditions and relations appear respectively as *nonlinear* integral inequations or equations involving various boundary values of N-point functions for different values of N. Their exploitation has been called *the "nonlinear program"* of axiomatic QFT.

11.3 Causality and Analyticity: mathematical results

11.3.1 The standard Fourier-Laplace transform theorem

Up to minor technical conditions, this theorem states the equivalence between the following properties of a tempered distribution f defined in $\mathbf{R}^n_{(x)}$ and of its Fourier transform $\tilde{f}$ defined in $\mathbf{R}^n_{(p)}$:

i) the support of f is contained in a closed convex salient cone C with apex at the origin,

ii) $\tilde{f}$ is the boundary value (b.v.), in the sense of distributions, of a function $\hat{f}$ analytic in a domain of the complexified p-space. More precisely, $\hat{f}$, which is called the Fourier-Laplace transform of f, is analytic in the *tube* Im $p \in \tilde{C}$, Re p arbitrary, where $\tilde{C}$ denotes the (open) dual cone of C with respect to the scalar product $< p.x >= p_1 x_1 + p_2 x_2 \cdots + p_n x_n$. The corresponding b.v. $\tilde{f}$ of $\hat{f}$ is obtained for Im p tending to zero from the directions of the cone $\tilde{C}$.

A similar result holds if C is not a strict support but is an *"essential support"* in the sense of the *exponential decay of f in all directions outside* C: the only difference is that $\hat{f}$ is then analytic only for Im p sufficiently small in the cone $\tilde{C}$. If there is exponential decay in all directions, $\tilde{f}(p)$ is analytic at all real points. The results rely on the real factor Im $p.x$ in the exponent of the kernel $e^{\pm ip.x}$ of the Fourier-Laplace transformation. The support or essential support of f in C provides absolute convergence of the integral, hence analyticity in the tube. In the converse direction, the integration contour $\mathbf{R}^n_{(p)}$ is replaced by $\mathbf{R}^n_{(p)} + ib$ for a convenient choice of b in $\tilde{C}$.

The following result is also obtained similarly: if the support or essential support of f is contained in a finite union of closed convex salient cones C_k, $\tilde{f}$ is a corresponding sum of b.v. of analytic functions from imaginary directions of the dual cones $\tilde{C}_k$. Since $\mathbf{R}^n_{(x)}$ itself is always contained in many ways in the union of at least $n+1$ convex salient cones, any tempered distribution can thus always be written in many ways as a sum of $n+1$ (or more) b.v. of analytic functions. On the other hand, if f satisfies particular support or essential support conditions, more refined decompositions of f with possibly less than $n+1$ terms and with specified directions of analyticity are obtained. The result may be of interest even when the support or essential support of f is contained in a convex salient cone C, but is smaller than C: it gives decompositions of $\tilde{f}$ for which each cone $\tilde{C}_k$ is larger than $\tilde{C}$ (which is the intersection of all cones $\tilde{C}_k$).

11.3.2 The essential support, or microsupport, in x-space for given p

A crucial limitation of the previous results comes from the fact that the analyticity properties of $\tilde{f}$ *do not depend* on the real point p considered. More refined analyticity properties depending on p (e.g. analyticity at certain real points but not all...) thus cannot be characterized by support or essential support properties of f in the sense specified above.

More refined results will be needed in applications to the chronological functions of Quantum Field Theory. To that purpose, one introduces the notion of essential support $ES_P(\tilde{f})$ associated with $\tilde{f}$ *for each given (real) value* P of p: for each P, $ES_P(\tilde{f})$ is defined as the cone C_P with apex at the origin in $\mathbf{R}^n_{(x)}$ which is composed of all directions along which an adequate *generalized Fourier transform of* $\tilde{f}$ *"at P"* does *not* decay exponentially. A typical form of such transform is

$$f_{P,\gamma}(x) = \int \tilde{f}(p) e^{-ip.x - \gamma|x||p-P|^2} dp.$$

More precisely, a point u in $\mathbf{R}^n_{(x)}$ is by definition outside $ES_P(\tilde{f})$ if there exists a positive constant α such that $f_{P,\gamma}(x)$ decays exponentially as $e^{-\alpha\gamma|x|}$ in the direction of u, for sufficiently small γ.

The function $f_{P,\gamma}(x)$ is the prototype of what mathematicians have later called "FBI transform". It can be shown that the definition of $ES_P(\tilde{f})$ is not changed if any given function $\chi(p)$ with compact support around P and locally analytic at P is inserted as an additional factor in the integrand. For a distribution $\tilde{f}$ defined on an analytic manifold $\mathcal{M}$, the essential support at a point P can then be defined similarly by using local coordinates; the set thus defined is however shown to be independent of the choice of local coordinates and therefore characterized as a well-defined cone C_P in the cotangent space $T^*_P\mathcal{M}$ at P to $\mathcal{M}$.

Results of Sect.11.3.1 can then be adapted by similar, but more refined methods which provide:

Theorem 1.

i) *If* $ES_P(\tilde{f})$ *is contained in a closed convex salient cone* $C(P)$ *for each point* P *in a given real region, then* $\tilde{f}$ *is equivalently the b.v. of an analytic function* $\hat{f}$ *from the imaginary directions of the dual cone* $\tilde{C}(P)$ *for each point* P *in the real region considered.*
ii) *If, for a given point* P, $ES_P(\tilde{f})$ *is contained in a (finite) union of closed convex salient cones* $C_k(P)$, *then* $\tilde{f}$ *is equivalently a sum of b.v. of analytic functions whose directions of analyticity at* P *are those of the respective dual cones* $\tilde{C}_k(P)$.

More generally, the data of all essential supports $ES_P(\tilde{f})$ *at all points* P *in a real region characterizes the possible decompositions of* $\tilde{f}$ *in that region as a sum of b.v. of analytic functions obtained from cones of imaginary directions that in general depend on* P.

Remark. The notion of essential support defined above (due to J.Bros and the author) is a refinement of a similar notion in which exponential fall-off in x-space is replaced by rapid fall-off (i.e. faster than any inverse power). The latter was used implicitly in various works by theoretical physicists in the 1960s, and was introduced independently in a more explicit way by L. Hörmander under the name of "wave front set". This notion is defined by introducing infinitely differentiable functions χ with compact support around P in the Fourier transform (without gaussian), and taking the limit when the support of χ gets smaller and smaller around P. However, analyticity properties cannot be characterized in that way (this is linked to the fact that there exists no function with compact support and analytic at all real points).

Besides the notion of essential support that we have presented, two different approaches have been proposed independently around the same period (1970): the "analytic wave front" of Hörmander and the "singular spectrum" of M. Sato, T. Kawai and M. Kashiwara in hyperfunction theory. It has been shown by J.M. Bony that the three notions coincide for distributions; since then, one often uses the terminology of "microsupport", proposed by M. Sato.

More precise notions of essential support at a point P in which rates of exponential fall-off are specified have also been given by J.Bros and the author, with corresponding more precise local analyticity domains ("local tubes").

11.3.3 Complements

Many further developments have been achieved under the general name of *microlocal analysis*. For our later purposes, only the following simple results are needed:

i) Lemmas 1 and 2 are straightforward consequences of the previous definitions and results:

Lemma 1. *If f has its support (or essential support in the sense specified above) in a cone C with apex at the origin, the essential support $ES_P(\tilde{f})$ of $\tilde{f}$ at any point P is contained in C.*

Lemma 2. *If $\tilde{f}$ and $\tilde{g}$ coincide in the neighborhood of a point P, their essential supports at P also coincide.*

ii) In the forthcoming sections, results on restrictions of distributions to submanifolds of $\mathbf{R}^n_{(p)}$, and on products of distributions, will be useful. Under some conditions, they give a simple and natural definition of the restriction or of the product, with information on their microsupports. In Lemma 4, $N(P)$ denotes the conormal space to the manifold $\mathcal{M}$ at P. and we recall that the cotangent space $T^*_P\mathcal{M}$ at P to $\mathcal{M}$ is the quotient by $N(P)$ of the dual $\mathbf{R}^n_{(x)}$ of $\mathbf{R}^n_{(p)}$.

Lemma 3. *(Products) Given two distributions $\tilde{f}_1$ and $\tilde{f}_2$, their product $\tilde{f}$ is a well-defined distribution in a given real region if, for any point P in that region, there is no common direction in $ES_P(\tilde{f}_1)$ and $-ES_P(\tilde{f}_2)$. In that case, $ES_P(\tilde{f})$ is contained in the set of points x of the form $x = x_1 + x_2$ with $x_j \in ES_P(\tilde{f}_j)$, $j = 1, 2$.*

Lemma 4. *(Restrictions) Given a distribution $\tilde{f}$ on $\mathbf{R}^n_{(p)}$ and an analytic submanifold $\mathcal{M}$ of $\mathbf{R}^n_{(p)}$, the restriction $\tilde{f}_{|\mathcal{M}}$ of $\tilde{f}$ to $\mathcal{M}$ is a well-defined distribution if, given any point P of $\mathcal{M}$, there is no common direction in $ES_P(\tilde{f})$ and in the conormal space $N(P)$ at P to $\mathcal{M}$. The essential support, at any point P of $\tilde{f}_{|\mathcal{M}}$, is then the set of points in $ES_P(\tilde{f})$ defined modulo $N(P)$.*

11.4 The linear axiomatic program

We consider a theory with mass gap and, for simplicity, with only one hyperboloid $H_+(m)$ in the spectrum and a continuum $V_+(2m)$. The aim is ultimately to establish results on (on-shell) collision amplitudes. We thus start with chronological functions. Some of the results on the latter apply also to theories without mass gap.

The analysis is essentially based on the interplay of support properties in x-space arising from the definition of the chronological operators (involving as we have seen the role of local commutativity) and support properties in p-space due to the spectral condition. As a matter of fact, all these support properties do not apply to the chronological functions themselves, but to related distributions. They will provide both general microsupport and related analyticity properties (near real points) of chronological functions and more global analyticity properties of the latter in complex p-space. In view of the reduction formulae, related analyticity properties of the collision amplitudes will in turn be obtained by restriction of the previous analyticity domains to the complex mass shell. These analyticity properties will provide excellent global results at $N = 4$ for 2-body collision amplitudes such as "crossing" and "dispersion relations", but only local results at $N > 4$.

Support properties in x-space

Chronological functions do not themselves have support properties in x-space nor exponential fall-off properties of interest (it can be shown that connected functions do decay exponentially for large *spacelike* separations of subgroups of points, in a theory with mass gap, but this is not a true causality property since it does not distinguish past and future and yields no interesting result on analyticity).

Actual support properties apply to related "paracell functions". The simplest ones are the functions R_I associated with subsets I of indices among $(1, \ldots, N)$. They are defined as the following VEV:

$$R_I(x_1, \ldots, x_N) = < \Omega, [\mathcal{T}(x_1, \ldots, x_N) - \mathcal{T}(x(J))\mathcal{T}(x(I))]\Omega >,$$

where J denotes the complement of I in the set $\{1, \ldots, N\}$ and $x(I)$ (resp. $x(J)$) denotes the multiplet of points x_k for k in I (resp. J). In view of its definition, the chronological operator $\mathcal{T}$ factorizes into the product $\mathcal{T}(x(J))\mathcal{T}(x(I))$ if $x(J) \leq x(I)$, where this notation means that each point x_i in $x(I)$ is either space-like to, or in the future cone of each point x_j in $x(J)$ $(x_i - x_j \in V_+)$. As a consequence R_I has its support in the closed cone C_I with apex at the origin in x-space composed of all points x that do not satisfy this condition.

General paracell functions $R_{\mathcal{S}}$ are defined similarly from more general combinations of products of "partial" chronological operators. Each $R_{\mathcal{S}}$ is also shown to have its support in a closed cone $C_{\mathcal{S}}$ with apex at the origin. Besides the class of functions R_I introduced above, another particular class of paracell functions (distinct from the class of functions R_I for $N > 2$) are the so-called "cell" functions which satisfy the following property:

Lemma 5. i) *If $R_{\mathcal{S}}$ is a cell function, the cone $C_{\mathcal{S}}$ is* convex *and salient.*

ii) *The Fourier transform $\tilde{R}_{\mathcal{S}}$ of $R_{\mathcal{S}}$ is the boundary value on the reals of a function analytic in the tube*

$$\mathcal{T}_{\mathcal{S}} = \{p;\ \text{Im}\ p \in \tilde{C}_{\mathcal{S}},\ \ \text{Re}\ p\ \ \text{arbitrary}\},$$

where $\tilde{C}_{\mathcal{S}}$ *denotes the open dual cone of* $C_{\mathcal{S}}$.

We omit the proof of Part i). Part ii) follows from the standard Laplace transform theorem.

Support properties in p-space

It is shown on the other hand that, given any real point $P = (P_1, \ldots, P_N)$, the (momentum space) chronological function $\tilde{T}_N$ coincides with many (but not all!) $\tilde{R}_{\mathcal{S}}$ in the neighborhood of P, as a consequence of the spectral condition. For instance, the latter entails that $\tilde{T}_N$ and $\tilde{R}_I$ coincide locally if the sum $P(I)$ of the energy-momenta P_k for k in I does not belong to the spectrum: to check this result, one notes that $\tilde{T}_N - \tilde{R}_I$ is the VEV of the product $\mathcal{T}(x(J))\mathcal{T}(x(I))$. For connected functions $\tilde{T}_N^c$, a similar result holds if $P(I)$ does not belong to $H_+(m) \cup V_+(2m)$. Hence, given any point P and any partition of the set of indices $\{1, \ldots, N\}$ into two subsets I,J, $\tilde{T}_N^c$ coincides locally with either one of the corresponding connected functions $\tilde{R}_I^c$, $\tilde{R}_J^c$ (since $P(I) = -P(J)$), or with both.

Finally, given a point $P = (P_1, \ldots, P_N)$ on the mass shell of a certain physical process, it is also shown that the coincidence between $\tilde{T}_N^{amp,c}$ and corresponding functions $\tilde{R}_I^{amp,c}$ also holds if $|I| = 1$ or $N - 1$, when $P(I)$ does not belong to $V_+(2m)$ (but may belong to $H_+(m)$).

On the other hand the following basic result can also be established.

Lemma 6. i) *Given any point* $P = (P_1, \ldots, P_N)$, *the connected function* $\tilde{T}_N^c$ *coincides with at least one (and possibly several) cell functions in a neighborhood of* P.

ii) *Any cell function does coincide with* $\tilde{T}_N^c$ *in a well-specified real region.*

In view of this information, the following result holds:

Theorem 2.

i) **Microsupport of chronological functions:** *Given any real point* P, *the microsupport of* $\tilde{T}_N^{amp,c}$ *at* P *is contained in the intersection of the cones* $C_{\mathcal{S}}$ *associated with all paracell functions such that* $\tilde{T}_N^{amp,c}$ *and* $\tilde{R}_{\mathcal{S}}^{amp,c}$ *coincide locally. It is in particular contained in each one of the following two sets:*
  a) *the intersection of the cones* C_I, *for all subsets* I *such that* $P(I)$ *does not belong to* $H_+(m) \cup V_+(2m)$ *(resp. to* $V_+(2m)$ *for* $|I| = 1$ *or* $N - 1$*),*
  b) *the closed convex salient cone* C_P *with apex at the origin, which is the intersection of the cones* $C_{\mathcal{S}}$ *associated with the cell functions that coincide with* $\tilde{T}_N^{amp,c}$ *locally.*

ii) **The Analytic** N**-point Functions:**
  a) *Given any* N, *there exists a unique analytic function* F_N *(called the "analytic* N*-point function"), whose domain contains all the tubes* $\mathcal{T}_{\mathcal{S}}$ *associated with the cell functions (see Lemma 3.1). For any given tube* $\mathcal{T}_{\mathcal{S}}$, *the boundary value of* F_N *on the reals from the directions of* $\tilde{C}_{\mathcal{S}}$ *is the corresponding (momentum space) cell function* $\tilde{R}_{\mathcal{S}}^{amp,c}$.

b) *The chronological function $\tilde{T}_N^{amp,c}$ is the boundary value of F_N at all real points P. This boundary value is obtained from imaginary directions which are, at each P, those of the dual cone $\tilde{C}_P$ of C_P. $\tilde{C}_P$ contains the convex envelope of the cones $\tilde{C}_{\mathcal{S}}$ associated with cell functions that coincide locally with $\tilde{T}_N^{amp,c}$.*

iii) **Microsupport of collision amplitudes:** *For each N, the distribution $\tilde{T}_N^{amp,c}$ can be restricted to the physical region of any process with N initial and final particles and yields the corresponding (connected) collision amplitude. For each such process and each physical point P of that process, the microsupport of the collision amplitude is obtained from that of $\tilde{T}_N^{amp,c}$ by replacing each point u_k ($k = 1, 2, \ldots, N$) by the space-time trajectory passing through u_k and parallel to P_k.*

Proof. By Lemma 2, $\tilde{T}_N^{amp,c}$ has the same microsupport at P as $\tilde{R}_{\mathcal{S}}^{amp,c}$ if both coincide locally. On the other hand, by Lemma 1, the microsupport of $\tilde{R}_{\mathcal{S}}^{amp,c}$ is contained in $C_{\mathcal{S}}$. Result i) follows by considering all relevant paracell functions at P, including the functions $\tilde{R}_I^{amp,c}$, or alternatively the cell functions $\tilde{R}_{\mathcal{S}}^{amp,c}$. The existence of F_N and its analyticity properties near the reals described in Part ii) follows from Theorem 1. The analytic continuation of F_N in each tube $\mathcal{T}_{\mathcal{S}}$ follows from the fact that each cell function $\tilde{R}_{\mathcal{S}}^{amp,c}$ is the boundary value of a function analytic in $\mathcal{T}_{\mathcal{S}}$ (in view of Lemma 5) and coincides with $\tilde{T}_N^{amp,c}$ in some real region (in view of Lemma 6). Hence the analytic functions from which $\tilde{T}_N^{amp,c}$ and $\tilde{R}_{\mathcal{S}}^{amp,c}$ are boundary values (respectively from the directions of $\tilde{C}_P$ and $\tilde{C}_{\mathcal{S}}$, with $\tilde{C}_{\mathcal{S}} \subset \tilde{C}_P$), also coincide. Finally, the existence of the restriction in Part iii) follows from Lemma 4: the condition needed to apply this lemma can be checked e.g. by using Part i)b) of the theorem. Note that the conormal space at P to the mass shell is indeed the set of points $u = (\lambda_1 P_1, \lambda_2 P_2, \ldots, \lambda_N P_N)$, in view of the mass shell conditions $p_k^2 = m^2$, $k = 1, 2, \ldots, N$. □

Remarks. 1) Rates of exponential fall-off outside the microsupport are also easily determined (and may provide more detailed information on corresponding analyticity domains).

2) More detailed physical explanations on the content of Parts i) and iii) of Theorem 2 in terms of (asymptotic) causality properties will be given in Section 11.5.

We end this section with a brief account of further analyticity properties whose derivation makes use of Theorem 2 as a starting point.

A further use of all relations of coincidence between the cell functions $\tilde{R}_{\mathcal{S}}^{amp,c}$ (in well-specified real regions) allows one to obtain the so-called "primitive domain of analyticity" of the analytic N-point function F_N (which improves the result stated in Part ii) of Theorem 2). However, it can be checked that this domain has an *empty* intersection with the complex mass-shell. It therefore gives *no* result by itself on the analyticity properties of collision amplitudes on the (real or complex) mass-shell. One important approach to improve the situation is based on the following fact: the primitive domain is *not* a domain of holomorphy; this means that any function analytic in that domain can be analytically continued in a larger domain, called its holomorphy envelope. It is this holomorphy envelope which has a non-empty intersection with the complex mass shell; the results which one obtains by computing part of this holomorphy envelope are however substantial only in the case $N = 4$ (two-particle collision amplitudes). Further improvements of these analyticity domains of the functions F_N (substantial again at $N = 4$) can still be obtained by exploiting a class of linear relations between the various cell functions $\tilde{R}_{\mathrm{S}}^{amp,c}$ called "Steinmann relations".

The case $N = 4$

We can distinguish results of local and global types:

i) The following local result can also be seen as implied by a result on the microsupport of $\tilde{T}_4^{amp,c}$ (see Sect. 4): the 4-point function F_4 can be restricted to the complex mass-shell in an open set Δ that admits each (real) physical region on its boundary (there is here one physical region for each choice of the two initial and the two final indices, the corresponding physical regions being disconnected from each other). In each physical region, the collision amplitude is the boundary value of the mass shell restriction of F_4, from the corresponding half-space of "$+i\varepsilon$" directions Im $s_{ij} > 0$, where $s_{ij} = (p_i + p_j)^2 = (p_k + p_\ell)^2$ for the collision process $(i, j) \to (k, \ell)$.

ii) The set Δ is connected, namely a domain of the complex mass-shell. That means that it contains paths of analytic continuation *on the complex mass shell* between the various physical regions ("crossing property"). As a matter of fact, analyticity on the complex mass-shell is obtained in a one-sheeted domain, called the "physical sheet", admitting cuts s_{ij} real $\geq (2m)^2$ covering the various physical regions. This property also holds for *all* possibly unequal values of the masses of the stable particles involved in the collision process. In the most favourable cases, there exist sets of sections of the domain Δ obtained by fixing a momentum transfer variable $s_{ik} = t$, with $t_0 < t \leq 0$, which consist in *a full cut-plane (with two cuts) of the complex energy variable* $s = s_{ij}$ (or $u = s_{i\ell} = m_i^2 + m_j^2 + m_k^2 + m_\ell^2 - s - t$). Such a cut-plane produces by itself a "crossing-domain of maximal type" which borders the physical regions of the s and u-collision processes along the corresponding s and u-cuts. By applying the Cauchy formula to the restriction of F_4 to such type of cut-plane domain, one obtains the famous "dispersion relations", which relate the real and imaginary parts of the collision amplitudes. All these global results have been obtained by methods of analytic continuation in several complex variables which will not be described in the present review.

The case $N > 4$

No similar result has been achieved at $N > 4$, and as a matter of fact, no similar result is expected. The best information achieved so far in the present framework is derived from the microsupport properties of Theorem 2 (and as a matter of fact from Part i)a) of the latter). Given any physical process, the collision amplitude cannot be shown to be, in its physical region, the boundary value of a function analytic on the complex mass-shell. Instead, decompositions of interest as sums of boundary values of functions analytic in domains of the complex mass-shell are obtained in various parts of the physical region. In contrast to the case $N = 4$, the sum does not reduce in general to one term. However, it has been shown that it does reduce to one term (which thus coincides locally with the analytic N-point function) in a certain subset of the physical region. Otherwise, the N-point analytic function cannot be restricted locally to the complex mass-shell, but it can be decomposed as a sum of terms which, individually, are locally analytic in a larger domain that intersects the complex mass-shell.

It remains an open question whether a crossing property can be established or not in this framework, namely the existence of an analyticity domain on the complex mass-shell of the N-point function which relates the local analyticity domains neighbouring two physical regions (in the parts of the latter where one term only is needed in the decompositions mentioned above). A partial result in that direction has been achieved by J. Bros for the 5-point function, but by using also some ingredients of the non linear program (see Section 11.6).

The fact that collision amplitudes are not shown to be everywhere boundary values of functions analytic on the complex mass-shell was a surprise. The discussion of Section 11.5 will show that, even under stronger assumptions, one cannot expect such a result at all real points.

11.5 Causality and local analyticity of chronological functions: physical discussion

By definition, a configuration $u = (u_1, \ldots, u_N)$ of N spacetime points u_k will be called "causal" (resp. "non-causal") at $P = (P_1, \ldots, P_N)$ if u belongs to (resp. does not belong to) the essential support of $\tilde{T}_N^{amp,c}$ at P.

Theorem 2 states in particular that, if $u(J) \leq u(I)$, the configuration u is "non-causal" at P if $P(I)$ is not "outgoing", namely if it does *not* belong to the spectral set $H_+(m) \cup V_+(2m)$, or to $V_+(2m)$ if $|I| = 1$ or $N-1$. For definiteness, let us now consider the case when P is a *physical* point of a given collision process. With each initial (resp. final) point u_k, can be associated the incoming energy-momentum $-P_k$, (resp. the outgoing energy-momentum P_k). The basic physical content of the latter statement can then be described as follows: the only possible causal configurations u at P are those for which energy-momentum can be transferred from the initial to the final points *in future cones*. It is also easily seen, from the case $|I| = 1$ or $N-1$, that if u is causal, at least two "extremal" initial points must coincide, as also two extremal final points.

Simple examples of possible causal situations are:

i) $N = 4$, 2-body process, $1, 2$ initial, $3, 4$ final: $u_1 = u_2$, $u_3 = u_4$, with u_3 in the future cone of u_1. By using Lorentz invariance, one shows in the linear program that moreover $u_3 - u_1$ must be parallel to $-(P_1 + P_2) = (P_3 + P_4)$.

ii) $N = 6$, $3 \to 3$ process, $1, 2, 3$ initial, $4, 5, 6$ final: two initial points coincide at a point a, two final points coincide at a point b, b is in the future cone of a, and the remaining initial and final points lie both in the future cone of a and in the past cone of b (no other condition at this stage).

In more general cases, the possible causal configurations u depend on P.

A better causality property "in terms of particles" — which is the best possible one — is expected for "physically reasonable" theories if the (stable) particles of the theory are known. By physically reasonable, we mean the absence of "a la Martin" pathologies such as the occurrence of an infinite number of unstable particles with arbitrary long life-time. That property expresses the idea that the only causal configurations u at P are those for which the energy-momentum can be transferred from the initial to the final points via intermediate stable particles in accordance with classical laws: there should exist a classical connected multiple scattering diagram in space-time joining the initial and final points with physical on-shell energy-momenta K_ℓ attributed to each intermediate particle ℓ and energy-momentum conservation at each interaction vertex (between all particles involved, including possibly some of the external P_k); moreover, each intermediate particle ℓ must move in the direction of its energy-momentum (if this particle joins two interaction vertices a and b, $b - a = \alpha_\ell K_\ell$ for some positive α_ℓ).

This microsupport property, if it holds, yields in turn improved analyticity of the analytic N-point function near real physical regions, and it is in turn shown that the (on-shell) collision amplitude is then the b.v. of a *unique* analytic function in its physical region, at least away from some "exceptional points". This b.v. (namely the collision amplitude) is moreover analytic outside $+\alpha$-Landau surfaces $L_+(G)$ of connected multiple scattering graphs G; along these surfaces, it is in general obtained from well-specified "$+i\varepsilon$" directions. Note that analyticity outside $+\alpha$-Landau surfaces entails the existence of analytic continuations in "unphysical sheets".

By definition P belongs to $L_+(G)$ if there is a space-time diagram with the structure of the graph G and external energy-momenta P_k. The "surfaces" $L_+(G)$ have codimension 1 or more; those of codimension 1 are in general smooth manifolds. At a point P of such a smooth surface $L_+(G)$, there is in general only one causal direction which is (co)normal to $L_+(G)$ at P. The $+i\varepsilon$-directions are those of the dual half-space.

The existence of exceptional points in the neighborhood of which the collision amplitude is *not* the b.v. of this analytic function was a surprise. They are those that lie at the intersection of *two* (or several) surfaces $L_+(G_1)$, $L_+(G_2)\dots$, with opposite causal directions, and hence having no $+i\varepsilon$-directions in common. Such points do not occur at $N = 4$ for 2-body processes, in which case the $+\alpha$-Landau surfaces are the n-particle thresholds $s = (nm)^2$,with $n \geq 2$, $s = (p_1 + p_2)^2$. They do occur more generally: in a $3 \to 3$-process, $1, 2, 3$ initial, $4, 5, 6$ final, this is the case of all points P such that $-P_1 = P_4$, $-P_2 = P_5, -P_3 = P_6$ which all belong to the surfaces of the two graphs G_1, G_2, with only one internal line joining two interaction vertices: in the case of G_1, (resp. G_2), the first vertex involves the external particles $1, 2, 4$ (resp. $1, 3, 5$), while the second one involves $3, 5, 6$ (resp. $2, 4, 6$). The configurations of external trajectories that define the respective causal configurations at P are obviously opposite. (This is of course not the case for the respective causal configurations of points u in the off-shell framework). If moreover P_1, P_2, P_3 lie in a common plane, previous points P also lie on $+\alpha$-Landau surfaces of "triangle" graphs with again opposite causal directions at P.

Remark. The above "exceptional" points are no longer exceptional in space-time dimension 2. In fact, all $+\alpha$-Landau surfaces mentioned then coincide with the (on-shell) codimension 1 surface $-p_1 = p_4$,$-p_2 = p_5$,$-p_3 = p_6$, with two opposite causal directions. The previous causality property (exponential fall-off for non-causal configurations), together with a further "causal factorization" property for causal configurations, then yields along that surface an actual *factorization* of the 3-body (non-connected) S matrix into a product of 2-body scattering functions *modulo an analytic background.* The latter vanishes outside the surface, hence is identically zero, for some special two-dimensional models. The analysis can also be extended to n-body processes.

If we come back to the linear program, one clearly sees why the above refined causality in terms of particles cannot be established: as we have seen in Sec.1, there is a priori no control in the linear program on the stable particles of the theory and on their masses, and pathologies such as those mentioned above cannot be excluded. Hopefully the first problem should be solved if Asymptotic Completeness is assumed, and the second one should be removed from adequate regularity assumptions to be added to AC in the nonlinear program. Results obtained will be described in Sect. 11.6.

Besides the previous causality properties and related analyticity, one wishes to examine many further problems in the nonlinear program, e.g. analytic continuation into unphysical sheets, with the occurence of possible unstable particles, poles and other singularities, nature of physical-region and other singularities, possible multiparticle dispersion relations,..., to cite only a few. Results so far remain limited but provide a first insight into such problems.

11.6 The nonlinear program: some results

Results described below are based on discontinuity formulae arising from — and essentially equivalent in adequate energy regions to — Asymptotic Completeness, together with some regularity conditions. They can be established either by "direct" methods presented below or through the introduction of adequate "irreducible" kernels. The latter method, due to J. Bros, relies on some general preliminary results on Fredholm theory in complex space (and with complex parameters). Irreducible kernels are defined through integral (Fredholm-type) equations, first in the euclidean region (imaginary energies) and then by local distortions of integration contours allowing one to reach the Minkowskian region. From discontinuity formulae and algebraic arguments, these irreducible kernels are shown to have analyticity (or meromorphy) properties associated with the physical idea of irreducibility (see examples below).

Results obtained so far by direct methods or through the use of irreducible kernels are comparable in the simplest cases: see Sec 11.6.1 and 11.6.2 below. However, the method based on irreducible kernels gives more refined results and seems best adapted to "extricate" the analytic structure of N-point functions for $N > 4$. Indications in that direction are mentioned in Sec. 11.6.2 and 11.6.3.

11.6.1 N=4, 2-body processes in the low-energy region

Standard results on 2-body processes with initial (resp. final) energy-momenta p_1, p_2 (resp. p_1', p_2') in the low-energy region $(2m)^2 \leq s < (3m)^2$ ($s = (p_1 + p_2)^2 = (p_1' + p_2')^2$) are based on the "off-shell unitarity equation"

$$F_+ - F_- = F_+ \star F_-,$$

where $F_+(p_1, p_2; p_1', p_2')$ and $F_-(p_1, p_2; p_1', p_2')$ denote respectively the $+i\varepsilon$ and $-i\varepsilon$ boundary values of the 4-point function F_4 from above or below the cut $s \geq 4m^2$ in the physical sheet, and $\star$ denotes *on-shell convolution* over two intermediate energy-momenta. This relation is a direct consequence of Asymptotic Completeness for s less than $(3m)^2$, in a theory with only one mass m. When the four external energy-momentum vectors p_1, p_2, p_1', p_2' are put on the mass shell (on both sides of that relation) one recovers the usual elastic unitarity relation for the collision amplitude T_+ and its complex conjugate T_-:

$$T_+ - T_- = T_+ \star T_- .$$

In fact, in view of the reduction formulae, T_+ and T_- appear respectively as the restrictions of F_+ and F_- to the mass shell ($p_i \in H_+(m)$, $p_i' \in H_+(m)$)). In the exploitation of these relations outlined below, a regularity condition is moreover needed, e.g. the continuity of F_+ in the low-energy region.

By considering the unitarity equation as a Fredholm equation for T_+ at fixed s (in the complex mass shell) one obtains the following result: T_+ can be analytically continued as a meromorphic function of s through the cut (in the low-energy-region) in a 2-sheeted (d even) or multisheeted (d odd) domain around the 2-particle threshold. Possible poles in the second sheet (generated by Fredholm theory) will correspond physically to unstable particles. The singularity at the 2-particle threshold is of the square-root type in s for d even, or in $\frac{1}{\log s}$ for d odd. The difference between the two cases is due to the power $\frac{(d-1)}{2}$ of s, integer or half-integer, in the kinematical factor arising from on-shell convolution. This result can also be extended to the off-shell function F_4 by applying a further argument of analytic continuation making use of the off-shell unitarity equation.

A similar result also follows from the introduction of a "*2-particle irreducible*" (in short "2 p.i.") Bethe-Salpeter type kernel G satisfying (and here defined from F through) a regularized B.S. equation of the form

$$F = G + F \circ G,$$

where $\circ$ denotes convolution over two intermediate energy-momenta with 2-point functions on the internal lines and a regularization factor in order to avoid convergence problems at infinity (G then depends on the choice of this factor but its properties and the subsequent analysis do not). Alternatively, one may also introduce a kernel satisfying a renormalized B.S. equation, but this is not directly useful for present purposes.

Starting from the above discontinuity formula (arising from AC in the low-energy region), one shows in turn that G is indeed "2 p.i." in the analytic sense:

$$G_+ = G_-$$

in the low-energy region (below the 3-particle threhold). More precisely, G is analytic or meromorphic (with poles that may arise from Fredholm theory) in a domain that includes the 2-particle threshold $s = 4m^2$, in contrast to F itself.

The proof of this property of G is based on the relation

$$\circ_+ - \circ_- = \star$$

for the operation $\circ$ (which is a non-trivial adaptation of the decomposition of a mass-shell delta-function as a sum of plus and minus $i\varepsilon$ poles), and on a simple algebraic argument which then shows essentially the equivalence between the discontinuity formula $F_+ - F_- = F_+ \star F_-$ and the relation $G_+ = G_-$.

In turn, this analyticity of G allows one to recover the two-sheetedness (d even) or multisheetedness (d odd, singularity in $\frac{1}{\log}$) of F, in view of the B.S. type equation.

11.6.2 N = 6, 3-3 process in the low-energy region (even theory)

The result, in the neighborhood of the 3-3 physical region, is here a "structure equation" expressing the 3-3 function F in the low-energy region as a sum of "à la Feynman contributions" associated with graphs with one internal line and with triangle graphs, with 2-point functions on internal lines and 4-point functions at each vertex, plus a remainder R. The latter is shown to be a b.v. from $+i\varepsilon$-directions Im s positive, where $s = (p_1 + p_2 + p_3)^2$, p_1, p_2, p_3 denoting the energy-momentum vectors of the initial particles. Further regularity conditions are needed to recover its local physical-region analyticity. The various explicit contributions that we have just mentioned yield the actual physical-region singularities expected in the low-energy 3-3 physical region.

A more refined result, in the approach based on irreducible kernels outlined below, applies in a larger region and then includes further *à la Feynman* contributions associated with 2-loop and 3-loop diagrams (the latter do not contribute to "effective" singularities in the neighborhood of the physical region).

The first result can be established by "direct" methods using discontinuity formulae for the 3-point function around 2-particle thresholds, arising from AC, and microsupport analysis of all terms involved. In the approach based on irreducible kernels, it is useful to introduce further kernels, and in particular a 3-p.i. kernel L that, in contrast to the 3-3 function, will be analytic or meromorphic in a domain including the 3-particle threshold. To that purpose an adequate set of integral equations is introduced and the 3-particle irreducibility of L in "the analytic sense" is then established. In turn it provides the complete structure equation mentioned above.

11.6.3 More general analysis

There are so far only preliminary steps in more general situations in view of the technical problems involved. As already mentioned, the approach based on irreducible kernels seems best adapted when expected singularities can be associated with graphs with more than two lines between some vertices. The analysis will clearly involve more general irreducible kernels with various irreducibility properties with respect to various channels (and not only with respect to the basic channel considered such as the 3-3 channel in Sect. 11.6.2. From a heuristic viewpoint, one may first consider to that purpose adequate formal expansions into (infinite) sums of "à la Feynman contributions" adapted to the energy regions under investigation. These à la Feynman contributions will involve adequate irreducible kernels in the graphical sense at each vertex, and the above expansions correspond formally to the best possible regroupings of Feynman integrals with respect to the energy region considered. From such expansions, one might determine adequate sets of integral equations allowing one, together with regularity assumptions, to carry an analysis similar to above.

11.6.4 Special applications of positivity conditions on the four-point functions

The present review of the nonlinear program would not be complete without mentioning an important and completely different type of analytic continuation which concerns an enlargement *in the physical sheet* of the analyticity domain of the two-particle collision amplitudes obtained in the linear program (see the paragraph entitled "The case $N = 4$" at the end of Sec. 11.4).

By making use of Hilbertian positivity conditions on the discontinuity $T_+ - T_-$ of the collision amplitude (i.e. of F_4) across the s-cut together with the local and global analyticity properties of the restriction T of F_4 to the complex mass shell, A. Martin obtained the following results, which are closely linked:

a) The set of values of the momentum transfer variable t for which T is holomorphic in a cut plane of s (or a "quasi-cut-plane" excluding only a bounded region from the full cut-plane) can be extended from its initial range $t < 0$ to a large complex domain of the t-plane including a certain positive interval $[0, t_{max}[$.

b) As a function of s, the function T admits a very restrictive bound at infinity in its (quasi-)cut-plane, obtained previously by M.Froissart under much stronger assumptions. This type of bound shows the validity of dispersion relations with at most two subtractions, which can be considered as a beautiful application of rigorous quantum field theory. What must also be stressed as remarkable is the fact (proved by Epstein, Glaser and Martin) that these results still hold true when the Wightman axioms of quantum field theory are replaced by the more general framework of *local quantum physics* in which the pointlike fields are replaced by fields generated by Araki-Haag-Kastler local observables.

11.7 The Analyticity Program in Constructive QFT

Models in Constructive QFT are first rigorously defined in Euclidean space-time (i.e., at imaginary times) through cluster and, more generally, phase-space expansions which are convergent at small coupling (and replace the non-convergent expansions, in terms of Feynman integrals, of perturbative QFT). Examples of such models in space-time dimension $d = 2$ or 3 have been mentioned in Sect. 11.1. For later purposes, we note that, in view of these expansions, the N-point functions of these models can also be shown to have exponential fall-off in Euclidean space-time, of the form

$$\exp[-(m - \varepsilon(\lambda))L(x_1, x_2, \ldots, x_N)],$$

for some strictly positive mass m (which will be the basic physical mass as outlined below); in the latter, $L(x_1, x_2, \ldots, x_N)]$ denotes an appropriate positive function depending on the mutual distances between the various points $x_1, x_2, \ldots, x_N$. By the usual Fourier-Laplace transform theorem (Sect. 11.3), one obtains in turn analyticity properties in corresponding regions around the Euclidean energy-momentum space.

On the other hand, *à la Osterwalder-Schrader* properties can be established in Euclidean space-time. By analytic continuation from imaginary to real times, it is in turn shown that a corresponding non-trivial theory satisfying the Wightman axioms is recovered on the Minkowskian side. However, note that no information is obtained in that way on the mass spectrum, Asymptotic Completeness, energy-momentum space analyticity. Such results can be obtained through the use of *irreducible kernels.* The latter can here be defined through "higher order" phase-space expansions which are again convergent at sufficiently small coupling. Irreducible kernels involved then satisfy exponential fall-off in Euclidean space-time with rates better than those of the N-point functions, and hence corresponding analyticity in larger regions around Euclidean energy-momentum space. As in Sect. 11.6, they depend on a regularization factor at infinity (in energy-momentum space) but the corresponding analyticity property does not.

The 1 p.i. two-point kernel is analytic up to $s = (2m)^2 - \varepsilon(\lambda)$. A simple rigorous argument (extending partial heuristic arguments of perturbative QFT) allows one to show in turn analyticity of the actual two-point function in the same region up to a pole at $k^2 = m^2$: this in turn will show the existence of a mass gap with a first basic physical mass m.

The 2 p.i. four-point kernel G is shown to be analytic up to $s = (4m)^2 - \varepsilon(\lambda)$ in an even theory. On the other hand, in view of its definition, it also satisfies a (regularized) B.S. equation. From the arguments presented in Sect. 11.6 (used here in the converse direction), the actual four-point function F is in turn analytic or meromorphic in that region up to the cut at $s \geq 4m^2$, and the analyticity of G together with the relation $\circ_+ - \circ_- = \star$ imply in turn the discontinuity formula associated with AC in the low-energy region.

For some (though not all) of the models (depending on some signs), it can be rigorously shown that there may be more precisely *a pole in the physical sheet,* below the 2-particle threshold (at a distance which tends to zero as the coupling itself tends to zero). Their possible occurrence is due to specific kinematical factors at $d = 2$ or 3 (as can also be seen in a heuristic way from a partial perturbative argument). This pole will then correspond to the occurrence of a further stable particle in the mass spectrum, with mass close to (and below) $2m$, which is referred to as a *"2-particle bound state".*

More generally, and up to technical problems only partly solved so far, the (here direct) introduction of general irreducible kernels leads one to structure equations, in higher and higher energy regions, of the type already mentioned in Sect. 11.6 and with here convergence properties (under some conditions). They allow one in turn to derive various discontinuity formulae, including those associated with AC in the corresponding energy region, and others (see Sect. 11.5 in this connection). However, convergence in this approach may be obtained only for smaller and smaller couplings as the energy region considered increases.

12

Renormalization Theory Based on Flow Equations*

Christoph Kopper

Centre de Physique Théorique de l'Ecole Polytechnique, F-91128 Palaiseau, France
`kopper@cpht.polytechnique.fr`

Summary. I give an overview over some work on rigorous renormalization theory based on the differential flow equations of the Wilson renormalization group. I first consider massive Euclidean φ_4^4-theory. The renormalization proofs are achieved through inductive bounds on regularized Schwinger functions. I present relatively crude bounds which are easily proven, and sharpened versions (which seem to be optimal as regards large momentum behaviour). Then renormalizability statements in Minkowski space are presented together with analyticity properties of the Schwinger functions. Finally I give a short description of further results.

12.1 Introduction

In this lecture I would like to give a short overview of part of the work on renormalization theory based on flow equations I have been involved in since 1990. The differential flow equations of the Wilson renormalization group [1] appear for the first time in a paper of Wegner and Houghton [2] in 1972. In a seminal paper by Polchinski [3] in 1984 a renormalization proof for scalar φ_4^4-theory was performed. It is based on the observation that the flow equations give access to a tight inductive scheme wherefrom bounds on the regularized Schwinger functions implying renormalizability may be deduced. The Schwinger functions are regularized by an UV-cutoff Λ_0 and by an infrared cutoff Λ, the flow parameter. The bounds on these Schwinger functions obtained in [4] are *uniform in the UV cutoff* and finite for $\Lambda \to 0$. This basically solves the renormalization problem. Later on renormalization theory with flow equations was extended in various directions, for a recent review covering some of the rigorous work see [5]. Among the issues treated are the renormalization of composite operators and the short distance expansion, sharp bounds on the Schwinger functions, massless theories, the transition to Minkowski space, the treatment of abelian and nonabelian gauge theories and field theories at finite temperatures.

Lecture given at the symposion in honour of Jacques Bros

I will address subsequently massive Euclidean φ_4^4-theory. I first outline the simplest version of the renormalization proof based on relatively crude bounds on the regularized Schwinger functions. Then I present sharpened versions of these bounds in momentum and position space. Hereafter analyticity properties of the Schwinger functions and statements on renormalizability in Minkowski space are presented. These notes end with a short overview on further rigorous results obtained in the flow equation framework.

Colleagues I had the pleasure to work with on renormalization theory with flow equations are Frédéric Meunier, Walter Pedra, Thomas Reisz, Manfred Salmhofer, Clemens Schophaus and Vladimir Smirnov. In particular I am grateful for the fruitful long term collaborations with Georg Keller and Volkhard Müller.

12.2 Renormalization of φ_4^4-Theory

Massive Euclidean φ_4^4-theory is the simplest theory on which the general issues of renormalization theory can be studied. When one analyzes the divergence structure of its bare Feynman amplitudes one faces the problem to be solved by general renormalization theory, namely the so-called overlapping divergences which had paved the way to a rigorous theory of renormalization with so many difficulties during the first few decades of existence of perturbative quantum field theory. The most astonishing message from the flow equation framework is that the focus on overlapping divergences and the heavy combinatorial and analytical machinery employed for its solution are not intrinsic to the problem[2]. In the flow equation approach overlapping divergences do not leave any trace.

12.2.1 The basic tools

The bare propagator of the theory is replaced by a regularized *flowing* propagator

$$C^{\Lambda,\Lambda_0}(p) = \frac{1}{p^2+m^2}\,\{e^{-\frac{p^2+m^2}{\Lambda_0^2}} - e^{-\frac{p^2+m^2}{\Lambda^2}}\}\,, \quad 0 \le \Lambda \le \Lambda_0 \le \infty\,.$$

The full propagator is recovered by taking the regulator Λ_0 to ∞ and the flow parameter Λ to 0. We calculate the derivative of C^{Λ,Λ_0} as

$$\dot{C}^{\Lambda}(p) = \partial_\Lambda C^{\Lambda,\Lambda_0}(p) = -\frac{2}{\Lambda^3}\,e^{-\frac{p^2+m^2}{\Lambda^2}}\,.$$

Subsequently we will denote by $d\mu_{\Lambda,\Lambda_0}$ the Gaussian measure with covariance $\hbar\, C^{\Lambda,\Lambda_0}$. The parameter $\hbar$ is introduced as usual to obtain a systematic expansion in the number of loops.

The theory we want to study is massive Euclidean φ_4^4-theory. This means that we start from the *bare action*

$$L_0(\phi) = \int d^4x\,\{\frac{g}{4!}\,\varphi^4 + a_0\,\varphi^2 + b_0\,(\partial_\mu\,\varphi)^2 + c_0\,\varphi^4\},$$

[2]This is of course of no consequence as for the merit attached to the hard and profound work done in early rigorous renormalization theory.

$$a_0\,,\; c_0 = O(\hbar)\,, \quad b_0 = O(\hbar^2)\,.$$

From the bare action and the flowing propagator we may define Wilson's *flowing effective action* L^{Λ,Λ_0} by integrating out momenta in the region $\Lambda^2 \le p^2 \le \Lambda_0^2$. It is defined through

$$e^{-\frac{1}{\hbar}L^{\Lambda,\Lambda_0}(\varphi)} := \mathcal{N} \int d\mu_{\Lambda,\Lambda_0}(\phi)\, e^{-\frac{1}{\hbar}L_0(\phi+\varphi)}$$

and can be recognized to be the generating functional of the connected free propagator amputated Schwinger functions of the theory with propagator C^{Λ,Λ_0} and bare action L_0. For the normalization factor $\mathcal{N}$ to be finite we have to restrict the theory to finite volume. All subsequent formulae are valid also in the thermodynamic limit since they involve no longer the vacuum functional or partition function.

The fundamental tool for our study of the renormalization problem is then the functional *Flow Equation*

$$\partial_\Lambda L^{\Lambda,\Lambda_0} = \frac{\hbar}{2}\langle \frac{\delta}{\delta\varphi}, \dot{C}^\Lambda \frac{\delta}{\delta\varphi}\rangle L^{\Lambda,\Lambda_0} - \frac{1}{2}\langle \frac{\delta L^{\Lambda,\Lambda_0}}{\delta\varphi}, \dot{C}^\Lambda \frac{\delta L^{\Lambda,\Lambda_0}}{\delta\varphi}\rangle\,.$$

It is obtained by deriving both sides of the previous equation w.r.t. Λ and performing an integration by parts in the functional integral on the r.h.s., and finally rearranging the powers of $\hbar$ coming from $L/\hbar$ and from $\hbar\dot{C}$. We then expand L^{Λ,Λ_0} in moments w.r.t. φ,

$$(2\pi)^{4(n-1)}\,\delta_{\varphi(p_1)}\cdots\delta_{\varphi(p_n)}L^{\Lambda,\Lambda_0}|_{\varphi\equiv 0} = \delta^{(4)}(p_1+\cdots+p_n)\,\mathcal{L}_n^{\Lambda,\Lambda_0}(p_1,\ldots,p_n)$$

and also in a formal powers series w.r.t. $\hbar$ to select the loop order l,

$$\mathcal{L}_n^{\Lambda,\Lambda_0} = \sum_{l=0}^{\infty}\hbar^l\,\mathcal{L}_{l,n}^{\Lambda,\Lambda_0}\,.$$

From the functional flow equation we then obtain the perturbative flow equations for the (connected free propagator amputated) n-point functions by identifying coefficients

$$\partial_\Lambda\partial^w\,\mathcal{L}_{l,n}^{\Lambda,\Lambda_0} = \frac{1}{2}\int_k \partial^w\mathcal{L}_{l-1,n+2}^{\Lambda,\Lambda_0}(k,-k,\ldots)\,\dot{C}^\Lambda$$
$$-\sum_{l_i,n_i,w_i} c_{\{w_i\}}\left[\partial^{w_1}\mathcal{L}_{l_1,n_1}^{\Lambda,\Lambda_0}\,(\partial^{w_3}\dot{C}^\Lambda)\,\partial^{w_2}\mathcal{L}_{l_2,n_2}^{\Lambda,\Lambda_0}\right]_{sym}\,,$$
$$l_1+l_2=l\,,\quad n_1+n_2=n+2\,,\quad w_1+w_2+w_3=w\,.$$

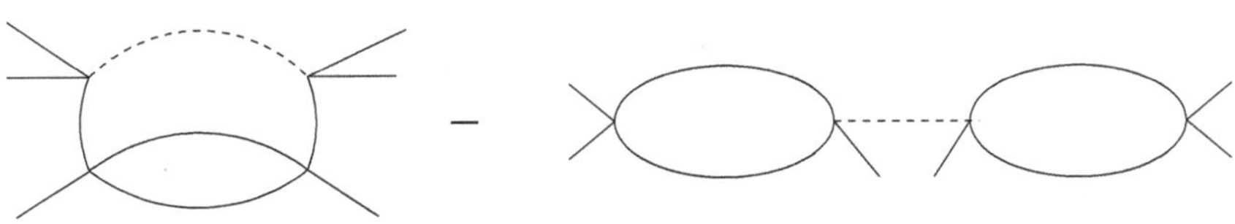

Fig. 12.1. A contribution to the r.h.s. of the flow equation for $l=2$, $n=6$. The dashed line represents the derived propagator $\dot{C}^\Lambda$.

Here we wrote the equation directly in a form where a number $|w|$ of momentum derivatives, characterized by a multi-index w, act on both sides. Derived Schwinger functions are needed to make close the inductive scheme (see below). The $c_{\{w_i\}}$ are combinatoric constants The subscript *sym* indicates a symmetrization procedure w.r.t. external momenta.

12.2.2 Renormalizability

Before bounding the solutions of the system of flow equations we first have to specify the boundary conditions :

At $\Lambda = \Lambda_0$ we find as a consequence of our choice of the bare action $L_0 = L^{\Lambda_0,\Lambda_0}$,

$$\partial^w \mathcal{L}_{l,n}^{\Lambda_0,\Lambda_0} \equiv 0 \quad \text{for} \quad n + |w| \geq 5 \ .$$

The so-called relevant parameters of the theory or renormalization constants are explicitly fixed by renormalization conditions imposed for the fully integrated theory at $\Lambda = 0$:

$$\mathcal{L}_4^{0,\Lambda_0}(0) = g \, , \quad \mathcal{L}_2^{0,\Lambda_0}(0) = 0 \, , \quad \partial_{p^2} \mathcal{L}_2^{0,\Lambda_0}(0) = 0$$

(where we chose for simplicity BPHZ renormalization conditions).

Once the boundary conditions are specified the renormalization problem can be solved *inductively* as follows :

The *inductive scheme* may for example be chosen as in the subsequent figure, namely by ascending in $n + 2l$ and for fixed $n + 2l$ ascending in l. For this scheme to work it is important to note that by definition there is no 0-loop two-point function in L^{Λ,Λ_0}.

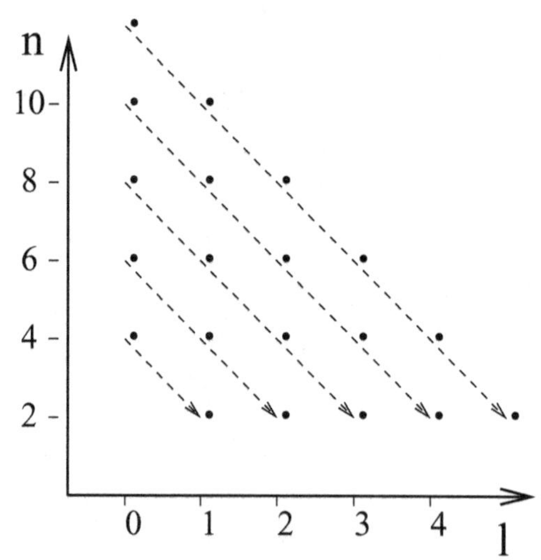

Fig. 12.2. The inductive scheme starts from the point (l=0, n=4) and follows the arrows, along the lines of fixed value of $n + 2l$ in increasing order.

We then may verify the following

Induction hypothesis:

$$|\partial^w \mathcal{L}_{l,n}^{\Lambda,\Lambda_0}(\vec{p})| \le (\Lambda_m)^{4-n-|w|}\, \mathcal{P}_1(\ln \frac{\Lambda_m}{m})\, \mathcal{P}_2(\frac{|\vec{p}|}{\Lambda_m})\,.$$

Here we note $\vec{p} = (p_1, \ldots, p_n)$, $|\vec{p}| = \sup\{|p_1|, \ldots, |p_n|\}$, $\Lambda_m = \sup(\Lambda, m)$, and the $\mathcal{P}_i$ are (each time they appear possibly new) suitable polynomials with *coefficients,* which *do not depend on* Λ_0. The important point to note on this induction hypothesis is that it is *independent of (and thus uniform in) the UV cutoff* Λ_0. This means that its verification implies essentially the solution of the renormalization problem. We shortly describe the

Method of proof:

The induction hypothesis is proven by first verifying it for the boundary conditions specified previously and then by inserting it on the right-hand side of the flow equation bounding the integrals in an elementary way. One has to distinguish *irrelevant terms with* $n + |w| \ge 5$ *which are integrated downwards from* Λ_0 *to* Λ *and relevant terms which are integrated upwards from* 0 *to* Λ. We note that the relevant terms are imposed by the renormalization conditions at *fixed* external momentum. To move away from the renormalization point one uses the Schlömilch interpolation formula, e.g. for the two-point function

$$\mathcal{L}_{l,2}^{\Lambda,\Lambda_0}(p) = \mathcal{L}_{l,2}^{\Lambda,\Lambda_0}(0) + \sum_\mu p_\mu \int_0^1 d\lambda\, (\frac{\partial}{\partial p_\mu}\mathcal{L}_{l,2}^{\Lambda,\Lambda_0})\,(\lambda p)\,,$$

and similar formulas where $\mathcal{L}_{l,2}^{\Lambda,\Lambda_0}(p)$ is replaced by $\frac{\partial}{\partial p_\mu}\mathcal{L}_{l,2}^{\Lambda,\Lambda_0}(p)$ or $\frac{\partial^2}{\partial p_\mu \partial p_\nu}\mathcal{L}_{l,2}^{\Lambda,\Lambda_0}(p)$. When applying three derivatives to the two-point function or one derivative to the four-point function the contribution becomes irrelevant and is integrated downwards from Λ_0. In this way the analogue of the previous equation written for $\frac{\partial^2}{\partial p_\mu \partial p_\nu}\mathcal{L}_{l,2}^{\Lambda,\Lambda_0}(p)$ gives full control of the twice derived two-point function. Then the equation for $\frac{\partial}{\partial p_\mu}\mathcal{L}_{l,2}^{\Lambda,\Lambda_0}(p)$ gives control of the once derived two-point function, and finally the one for $\mathcal{L}_{l,2}^{\Lambda,\Lambda_0}(p)$ gives control of the two-point function itself. By Euclidean invariance one realizes that no renormalization conditions are needed for terms which are not scalars w.r.t. this symmetry. For the four-point function only one step is required.

As an example we now bound by induction the first term on the r.h.s. of the flow equation in the irrelevant case, i.e. for $n + |w| \ge 5$,

$$\int_\Lambda^{\Lambda_0} d\lambda \int_k \frac{2}{\lambda^3}\, e^{-\frac{k^2+m^2}{\lambda^2}}\, \lambda_m^{4-(n+2)-|w|}\, \mathcal{P}_1(\ln \frac{\lambda_m}{m})\, \mathcal{P}_2(\frac{|(\vec{p},k,-k)|}{\lambda_m})$$

$$\le \int_\Lambda^{\Lambda_0} d\lambda\, \lambda_m^{4+1-(n+2)-|w|}\, \mathcal{P}_1(\ln \frac{\lambda_m}{m})\, \tilde{\mathcal{P}}_2(\frac{|\vec{p}|}{\lambda_m}) \le (\Lambda_m)^{4-n-|w|}\, \tilde{\mathcal{P}}_1(\ln \frac{\Lambda_m}{m})\, \tilde{\mathcal{P}}_2(\frac{|\vec{p}|}{\Lambda_m})$$

which verifies the induction hypothesis.

To complete the proof of renormalizability one also proves

$$|\partial_{\Lambda_0} \partial^w \mathcal{L}_{l,n}^{\Lambda,\Lambda_0}(\vec{p})| \le \frac{1}{\Lambda_0^2}(\Lambda_m)^{5-|n|-|w|}\, \mathcal{P}_1(\ln(\Lambda_0/m))\, \mathcal{P}_2(\frac{|\vec{p}|}{\Lambda_m})\,.$$

This bound is obtained applying the same inductive scheme on the flow equation derived once w.r.t. Λ_0 and using the previous bounds on $\partial^w \mathcal{L}_{l,n}^{\Lambda,\Lambda_0}(\vec{p})$. It implies convergence of the $\partial^w \mathcal{L}_{l,n}^{\Lambda,\Lambda_0}(\vec{p})$ for $\Lambda_0 \to \infty$, whereas the previous induction hypothesis still allows for solutions of the flow equations which are uniformly bounded but oscillating in terms of Λ_0.

12.2.3 Generalizations and Improvements

The previous statement of renormalizability may be improved and generalized in various ways. A first short remark is that we may generalize the boundary conditions by enlarging the class of bare actions : we also admit *nonvanishing irrelevant terms in the bare action* as long as they satisfy the bound

$$|\partial^w \mathcal{L}_{l,n}^{\Lambda_0,\Lambda_0}(\vec{p})| \le \Lambda_0^{4-n-|w|}\, \mathcal{P}_1(\ln \frac{\Lambda_0}{m})\, \mathcal{P}_2(\frac{|\vec{p}|}{\Lambda_0}) \quad \text{for } n+|w| \ge 5\,.$$

Remembering the previous method of proof we only have to note that these boundary conditions still satisfy the induction hypothesis (for $\Lambda = \Lambda_0$) so that the proof goes through without change. One might note that this generalization is of practical importance on one hand, since such irrelevant terms typically will appear, e.g. in effective actions used to describe critical behaviour in statistical mechanics [6]. On can calculate subdominant corrections to scaling near second order phase transitions due to such terms.

On the other hand these terms pose considerable problems in the BPHZ type renormalization proofs of which we are not sure whether they have been solved in full rigour up to the present day [7].

It is also possible to give much *sharper bounds* on the Schwinger functions with the aid of the flow equations which do not only imply renormalizability but also restrict the high momentum or large distance behaviour of the Schwinger functions in a basically optimal way (in the sense that they are saturated by individual Feynman amplitudes). To phrase those bounds we need the following concept of

Weighted trees:
We regard trees with n external lines and $\mathcal{V}$ vertices of coordination number 3 or 4. To the n external lines of a tree T we associate n external incoming momenta $\vec{p} = (p_1, \ldots, p_n)$ respectively positions $\vec{x} = (x_1, \ldots, x_n)$. We then define weight factors in momentum and position space.

Weight factor in momentum space:
To each internal line $I \in \mathcal{I}$ of the tree is attached a weight $\mu(I) \in \{1,2\}$ such that $\sum_{I\in\mathcal{I}} \mu(I) = n-4$. Let $p(I)$ be the momentum flowing through the internal line $I \in \mathcal{I}$. For given Λ and tree T we define

$$g^{\Lambda}(T) = \prod_{I\in\mathcal{I}} \frac{1}{(\sup(\Lambda_m, |p(I)|)^{\mu(I)}}\,.$$

Weight factor in position space:
For $\alpha = 1-\varepsilon$ fixed and letting $|I|$ be the distance of the points joined by I in position space we define

$$\mathcal{F}_{\Lambda}(T) := \Lambda_m^{4-n} \prod_{I\in\mathcal{I}} e^{-[\Lambda_m\, |I|]^{\alpha}}\,.$$

In terms of these weight factors we may now state

Sharp bounds in momentum space: [8]

$$|\mathcal{L}_{l,n}^{\Lambda,\Lambda_0}(\vec{p})| \leq \sup_T g^{\Lambda}(T)\ \mathcal{P}_l\big(\ln \sup(\frac{|\vec{p}|}{\Lambda_m}, \frac{\Lambda_m}{m})\big)\ ,\ n \geq 4\ ,$$

$$|\mathcal{L}_{l,2}^{\Lambda,\Lambda_0}(p)| \leq \sup(|p|, \Lambda_m)^2\ \mathcal{P}_{l-1}\big(\ln \sup(\frac{|p|}{\Lambda_m}, \frac{\Lambda_m}{m})\big)\ ,$$

$$\deg \mathcal{P}_l \leq l\ .$$

The most important ingredient to obtain such sharpened bounds is to implement in the inductive procedure the fact that momentum derivatives improve the large momentum behaviour. These derivatives necessarily appear due to the use of the Schlömilch interpolation formula. Here the main problem to solve is to find optimal paths in the interpolation formula for the four point function so as to avoid that derivatives only lead to a net gain of a small external momentum, whereas subsequent application of the interpolation formula then may lead to multiplication by a large one. Thus the problem is related to the exceptional momentum problem.

Sharp bounds in position space:
On smearing out the positions of $n-1$ external vertices with smooth bounded test functions $\varphi := \varphi_2 \cdot \dots \cdot \varphi_n$, φ_i being supported in a square of side length $1/\Lambda_m$ around x_i, we have for suitable (maximizing) choice of the positions of the internal vertices of the trees for $n \geq 4$,

$$|\mathcal{L}_{l,n}^{\Lambda,\Lambda_0}(\varphi, x_1)| \leq \sup_T \mathcal{F}_{\Lambda}(T)\ \mathcal{P}_l(\ln(\varphi, \Lambda_m))\ ||\varphi||\ ,$$

$$|\mathcal{L}_{l,2}^{\Lambda,\Lambda_0}(\varphi, x_1)|$$
$$\leq \Big(||\varphi|| + \Lambda_m^{-2}\,||\varphi^{(2)}|| + \Lambda_m^{-3}\,||\varphi^{(3)}||\Big)\ e^{-[\Lambda_m\, dist(x_1,\, supp\varphi)]^{\alpha}}\ \mathcal{P}_{l-1}(\ln(\varphi, \Lambda_m))\ .$$

Here the definiton of $\mathcal{P}_{l-1}(\ln(\varphi, \Lambda_m))$ is given through

$$\mathcal{P}_l(\ln(\varphi, \Lambda_m)) := \mathcal{P}_l(\ln \sup(\frac{||\varphi'||}{||\varphi||\Lambda_m}, \frac{\Lambda_m}{m}))\ ,$$

with the definition $||\varphi^{(k)}|| := \sup_{w,|w|\leq k} ||\varphi^{(w)}||_{\infty}$. This means that the polynomial is in logarithms of the maximal momentum content of the test functions if the latter exceeds Λ_m/m . The proof of these bounds has not been published so far.

Sharp combinatoric bounds: [9]
Finally the flow equation method also permits us to recover the bounds on large orders of perturbation theory which were established by de Calan and Rivasseau using BPHZ type methods [10]. They can be stated now in a form somewhat more compact than the one given in the original paper [9]:

$$|\mathcal{L}_{l,n}^{\Lambda,\Lambda_0}(\vec{p})| \leq \Lambda_m^{4-n}\ K^{\frac{n-3}{2}+2l}\ (\frac{n}{2})!\ \sum_{s=0}^{l}(\frac{n}{2}+s-2)!\ \ln^{l-s}(\frac{\Lambda_m}{m})$$

for $|\vec{p}| \leq \sup(2\Lambda_m, k)$ with k fixed, and for K sufficiently large.

12.3 Relativistic Theory

We want to indicate how the flow equation method, which at first sight is restricted to euclidean theories, can be employed to prove renormalizability in Minkowski space [11]. In this respect we emphasize the beautiful framework for renormalization theory developed by Epstein and Glaser, which takes into account the structural locality and causality properties of relativistic theories from the beginning, and which was presented at the Saclay conference in the contributions of Dorothea Bahns and Klaus Fredenhagen. In our framework we address the relativistic renormalization problem in momentum space, and we want to recover the results from BPHZ renormalization theory saying that the perturbative relativistic Green functions are distributions in momentum space, and at the same time analytic functions in those domains where the external incoming energy stays below the physical thresholds. For a fully satisfactory renormalization theory in Minkowski space one would also like to have continuity properties of the Green functions above threshold, sufficient at least to define a physical renormalized coupling for physical values of the external momenta which necessarily lie above threshold. Satisfactory results in this respect do not seem to exist, and we hope to come back to this issue in the future.

12.3.1 The Flow Equation for one-particle irreducible Schwinger functions

To discuss analyticity properties it is preferable to work with one-particle irreducible (1PI) Schwinger functions, the generating functional of which is obtained from the one for connected Schwinger functions by a Legendre transform. Using this relation one can deduce the (still Euclidean) flow equations for the perturbative 1PI Schwinger functions $\Gamma_{l,n}^{\Lambda,\Lambda_0}$ which take the form

$$\partial_\Lambda \Gamma_{l,n}^{\Lambda,\Lambda_0} = \frac{1}{2}\int_k \hat{\Gamma}_{l-1,n+2}^{\Lambda,\Lambda_0}(k,-k,\ldots)\, \dot{C}^{\Lambda,\Lambda_0}(k)\,,$$

$$\hat{\Gamma}_{l,n}^{\Lambda,\Lambda_0} = \sum_{i\ge 1}(-1)^{i+1}\sum_{l_i,n_i}\Big[\Big(\prod_{j=1}^{i-1} C^{\Lambda,\Lambda_0}(k_j)\; \Gamma_{l_j,n_j+2}^{\Lambda,\Lambda_0}\Big)\, \Gamma_{l_i,n_i+2}^{\Lambda,\Lambda_0}\Big]_{sym}\,,$$

$$\sum_i l_i = l\,,\quad \sum_i n_i = n-2\,(1-\delta_{i,1})\,.$$

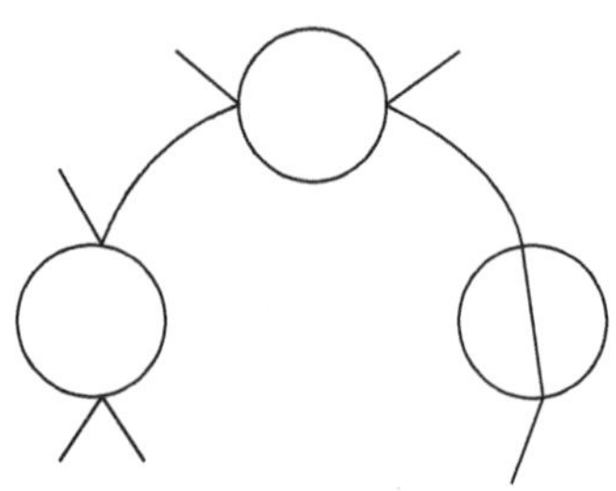

Fig. 12.3. A contribution to $\hat{\Gamma}_{l,n}$ for $l=4,\ n=6$.

The momentum arguments k_j are determined by momentum conservation. They are a sum of the loop momentum k and a subsum of incoming momenta p_i. One immediately realizes that the inductive scheme used for the renormalization proof of connected Schwinger functions is also viable in the 1PI case, and the same inductive bounds may also be proven for the euclidean 1PI functions $\Gamma^{\Lambda}_{l,n}(\vec{p})$.

12.3.2 Analyticity of the Euclidean theory

Using the 1PI flow equation it is possible to extend the inductive renormalizability proof to the *complex domain*

$$\mathcal{D} = \Big\{ (p_{01}, \underline{p}_1, \ldots, p_{0n-1}, \underline{p}_{n-1}) \mid \underline{p}_i \in \mathbf{R}^3 \ , \ p_{0i} \in \mathbf{C} \ , \\ |Im \sum_{i\in J} p_{0i}| < 2m - \eta \quad \forall J \subset \{1, \ldots, n\} \Big\}$$

(with $\eta > 0$ arbitrarily small).

That is to say:
The 1PI Schwinger functions $\Gamma^{\Lambda,\Lambda_0}_{l,n}(\vec{p})$ are analytic in $\mathcal{D}$ with respect to the arguments $p_{01}, \ldots, p_{0n}$ and still uniformly bounded with respect to Λ_0 by bounds analogous to those given for real external momenta (for $\eta > 0$ fixed).

The proof of this statement obviously uses the fact that we have chosen an analytic regulator in momentum space. It relies on the observation that the external momentum sets of the terms $\Gamma^{\Lambda,\Lambda_0}_{l_i,n_i+2}$ appearing on the r.h.s. of the 1PI flow equation also fulfill the conditions appearing in the definition of $\mathcal{D}$, if the full external momentum set does. Thus the inductive procedure is still valid. Then by displacing the integration contour of the first component of the integration variable k in the flow equation by an amount of modulus smaller than m in the imaginary direction — in fact one has to move towards the centre of the interval of width $2m - \eta$ appearing in the definition of $\mathcal{D}$ — one can show that all propagators $C^{\Lambda,\Lambda_0}(k_i)$ in the 1PI flow equation have still $|Im\, k_{0i}| < m - \eta$, so that the k-integral is still controlled by the exponential fall-off $\exp(-\frac{k_i^2+m^2}{\Lambda^2})$ of the regulating factors. We note again that the bounds thus obtained are (obviously) not uniform in η.

To obtain the statement on existence of the relativistic Green functions as distributions, we need more explicit information on the dependence of the 1PI Schwinger functions on the external momenta. This information is obtained from the following α-parametric integral representation for these Schwinger functions:

$$\Gamma^{\Lambda,\Lambda_0}_{l,n}(\vec{p}) = \int_{1/\Lambda_0^2}^{1/\Lambda^2} d\alpha_1 \ldots d\alpha_s \; G^{\Lambda,\Lambda_0}_{l,n}(\vec{\alpha}, \vec{p})$$

with

$$G^{\Lambda,\Lambda_0}_{l,n}(\vec{\alpha}, \vec{p}) = \sum_j V_j(\vec{\alpha})\, P_j(\vec{p})\, Q_j(\vec{\alpha})\, e^{-(\vec{p}, A_j\, \vec{p})_{eu} - m^2 \sum_{k=1}^{s} \alpha_k} \ .$$

By induction one verifies the following statements:

(i) P_j is a monomial in the $O(4)$-invariant scalar products of the external momenta.
(ii) Q_j is a rational function homogeneous of degree $d_j > -s$ in the α_i. Here $s = 2l - 2 + n/2$ is the number of internal lines of $\Gamma_{l,n}$.

(iii) A_j is a positive-semidefinite symmetric $(n-1\times n-1)$-matrix, homogeneous of degree 1 in $\vec{\alpha}$.
(iv) V_j is a product of θ-functions restricting the α-integration domain. Its origin traces back to previous integrations over the flow parameter, which either start from Λ_0 for the irrelevant terms or from 0 (corresponding to $\alpha \to \infty$) for the relevant terms.

On changing variables

$$\alpha_k = \tau\beta_k\,, \quad d\vec{\alpha} = \tau^{s-1}\delta(1-\sum \beta_k)d\vec{\beta}\,d\tau$$

and *using homogeneity of A_j and Q_j* we may write

$$\Gamma^{\Lambda,\Lambda_0}_{l,n}(\vec{p}) = \int_0^1 d\vec{\beta}\,\delta(1-\sum \beta_k)\sum_j V_j(\vec{\beta})\;P_j(\vec{p})\;Q_j(\vec{\beta})\;\frac{1}{[(\vec{p},A_j\,\vec{p})_{eu}+m^2]^{d_j+s}}\,.$$

The integrand is absolutely integrable with respect to $\vec{\beta}$ for $\vec{p}\in\mathcal{D}$. This statement follows from the previous one on the degree of Q_j. The previous integral representation is obtained quite naturally from the starting observation that

$$C^{\Lambda,\Lambda_0}(k) = \int_{1/\Lambda_0^2}^{1/\Lambda^2} e^{-\alpha(k^2+m^2)}\,d\alpha$$

and on using recursively (inductively) the flow equation where each successive Λ-integration is written as a new α-integration.

12.3.3 Transition to Minkowski space

We use the relativistic Feynman propagator

$$C_{rel}(p) = \frac{i}{p^2_{rel}-m^2+i\varepsilon(p^2_{eu}+m^2)}\,,$$

$$C^{\Lambda,\Lambda_0}_{rel}(p) = \int_{\Lambda_0^{-2}}^{\Lambda^{-2}} e^{i\alpha[p^2_{rel}-m^2+i\varepsilon(p^2_{eu}+m^2)]}\,d\alpha\,,$$

which for finite ε can be bounded in terms of the Euclidean one and $1/\varepsilon$ to make all integrals well-defined. Then following the steps which led to the integral representation for the Euclidean theory one obtains the same representation for the relativistic theory, replacing Euclidean by Minkowski scalar products and

$$\frac{1}{(\vec{p},A_j\,\vec{p})_{eu}+m^2}$$

by

$$\frac{1}{(\vec{p},A_j\,\vec{p})_{rel}-m^2+i\varepsilon((\vec{p},A_j\,\vec{p})_{eu}+m^2)}\,.$$

This representation together with results from distribution theory [12] implies the following

Results:

1) The relativistic 1PI Green functions are *Lorentz-invariant tempered distributions.*
2) For external momenta $(p_{01}, \underline{p}_1, \ldots, p_{0n}, \underline{p}_n)$ with $|\sum_{i\in J} p_{0i}| < 2m$ for all $J \subset \{1, \ldots, n\}$ they agree with the Euclidean ones for $(ip_{01}, \underline{p}_1, \ldots, ip_{0n}, \underline{p}_n)$ and are thus *smooth functions* in the (image of the) corresponding domain under the Lorentz group.

12.4 A short look at further results

Among the results of rigorous renormalization theory with flow equations not covered here we mention the following:

1. One of the earliest applications of Polchinski's method was by Mitter and Ramadas [13], who gave a renormalization proof for the two-dimensional nonlinear σ-model on the ultraviolet side.

2. Massless theories [14]
Massles theories or partially massless theories have been treated in momentum space. In this case the inductive bounds of course have to allow for the singularities present in the connected Schwinger functions at exceptional external momentum, i.e. when subsums of external momenta vanish. One then has to find and prove bounds which characterize these singularities in a basically optimal way so that the successive integration procedures of the flow equation reveal how the possible singularities disappear when passing from $\mathcal{L}_{l-1,n+2}$ to $\mathcal{L}_{l,n}$. We think that a better and more transparent way of treating this problem than that of [14] is by the more recent method of [8]. There we give sharp large momentum bounds for the massive theory. It should be possible to translate them to the massless theory, taking care of restrictions on the admissible renormalization conditions in this case.

3. Composite Operators, Zimmermann identities [15] and the Wilson short distance expansion [16]
The renormalization of Schwinger functions with composite operator insertions is an indispensable tool in many applications of renormalization theory. In particular in gauge theories these composite operator insertions appear as quantities which characterize the violation of the Ward identities due to the presence of cutoffs. Besides, inserted Green functions are of physical interest in their own right, e.g. current-current-correlators in relativistic field theory, or the $< \varphi^2\ \varphi^2 >$-correlator from which one reads the behaviour of the specific heat when applying Euclidean field theory to critical phenomena. The Zimmermann identities relate these composite operator insertions in different renormalization schemes with each other. They are obtained in a straightforward way from the linear structure of flow equations for inserted Green functions. Having at hand the general composite operator renormalization techniques one can then prove asymptotic expansions for perturbative Green functions of the Wilson type. This has been done in Euclidean space only.

3. Gauge theories [17, 18]
We mentioned that the Ward identities are violated in momentum space regulated gauge theories. In the flow equation framework the renormalization of gauge theories then requires us to prove that the Ward identities can be restored in the limit

where the cutoffs are taken away. To do so one disposes of the freedom of imposing renormalization conditions on all relevant terms of the theory. This programme has been achieved explicitly for QED and for massive SU(2) Yang-Mills-Higgs theory.

4. Finite temperature theory [19]
Using flow equations it is possible to study explicitly the difference between the (massive Euclidean scalar) finite temperature and zero temperature field theories and to show that this difference theory is completely irrelevant in the sense of the renormalization group. That is to say, both theories ($T > 0$ and $T = 0$) are renormalized with identical counter terms. This result is new (though expected by experts) and its proof is simple: our framework is particularly suited for this kind of analysis where one has to lay hands on the bare and renormalized actions at the same time, since both appear automatically as opposite boundary values of the flow to be controlled.

Two further topics of related nature have been treated within the flow equation framework. First the Symanzik programme of improved actions [20] has been carried out: it is possible to modify the bare action through irrelevant terms such that the convergence of the regularized towards the renormalized theory is accelerated, in terms of inverse powers of $1/\Lambda_0$. Secondly decoupling theorems [21] for heavy particles have been proven.

In conclusion we may say that

1. The Wilson-Wegner flow equation allows for a simple transparent rigorous solution of the perturbative renormalization problem, without introducing Feynman diagrams.

2. The method gives new results and perspectives on various aspects of the problem.

3. It places the problem in a more general, more physical and less technical context, relating perturbative and nonperturbative aspects.

We should also mention that flow equations in different forms and in various approximation schemes are presently applied to many problems in high energy and solid state physics, a topic not covered in this lecture. For a relatively recent review see [22].

Acknowledgement. This manuscript was written when the author was a visitor to DESY, Vienna. Support by ESI is gratefully acknowledged.

References

1. K. Wilson: Renormalization group and critical phenomena I. Renormalization group and the Kadanoff scaling picture. *Phys. Rev. B* 4:3174-3183 (1971).

   K. Wilson: Renormalization group and critical phenomena II. Phase cell analysis of critical behaviour. *Phys. Rev. B* 4:3184-3205 (1971).

   K. Wilson and J.B.Kogut: The Renormalization Group and the ε-Expansion. *Phys. Rep. C* 12:75-199 (1974).
2. F.J. Wegner and A. Houghton: Renormalization Group Equations for Critical Phenomena. *Phys. Rev. A* 8:401-412 (1973).

3. J. Polchinski: Renormalization and Effective Lagrangians. *Nucl. Phys. B* 231:269-295 (1984).
4. G. Keller, Ch. Kopper and M. Salmhofer: Perturbative Renormalization and effective Lagrangians in φ_4^4. *Helv. Phys. Acta* 65:32-52 (1991).

   See also:

   Ch. Kopper: *Renormierungstheorie mit Flußgleichungen.* Shaker-Verlag, Aachen, 1998.

   M. Salmhofer: *Renormalization, an Introduction.* Springer, Berlin-Heidelberg-New York, 1999.
5. V.F. Müller: Perturbative Renormalization by Flow Equations. *Rev. Math. Phys.* 15:491-558 (2003).
6. J. Zinn-Justin: *Quantum Field Theory and Critical Phenomena.* Clarendon Press, Oxford, 4th ed., 2002.
7. G. Chalmers: Extended BPH Renormalization of Cutoff Scalar Field Theories. *Phys. Rev. D* 53:7143-7156 (1996).
8. Ch. Kopper and F. Meunier: Large Momentum Bounds from Flow Equations. *Ann. Henri Poincaré* 3:435-449 (2002).
9. G. Keller: Local Borel summability of Euclidean Φ_4^4: A simple Proof via Differential Flow Equations. *Commun. Math. Phys.* 161:311-323 (1994).
10. C. de Calan and V. Rivasseau: Local existence of the Borel transform in Euclidean Φ_4^4. *Commun. Math. Phys.* 82:69-100 (1981).
11. G. Keller, Ch. Kopper and C. Schophaus: Perturbative Renormalization with Flow Equations in Minkowski Space. *Helv. Phys. Acta* 70:247-274 (1997).
12. E. Speer: Generalized Feynman Amplitudes. *Ann. of Math. Studies* 62, Princeton Univ. Press, 1969.
13. P.K. Mitter and T.R. Ramadas: The two-dimensional $O(N)$ nonlinear σ-model: renormalization and effective actions. *Commun. Math. Phys.* 122:575-596 (1989).
14. G. Keller and Ch. Kopper: Perturbative Renormalization of Massless Φ_4^4 with Flow Equations. *Commun. Math. Phys.* 161:515-532 (1994).
15. G. Keller and Ch. Kopper: Perturbative Renormalization of Composite Operators via Flow Equations I. *Commun. Math. Phys.* 148:445-467 (1992).
16. G. Keller and Ch. Kopper: Perturbative Renormalization of Composite Operators via Flow Equations II: Short distance expansion. *Commun. Math. Phys.* 153:245-276 (1993).
17. G. Keller and Ch. Kopper: Perturbative Renormalization of QED via flow equations. *Phys. Lett. B* 273:323-332 (1991).

    G. Keller and Ch. Kopper: Renormalizability Proof for QED Based on Flow Equations. *Commun. Math. Phys.* 176:193-226 (1996).
18. Ch. Kopper and V.F. Müller: Renormalization Proof for Spontaneously Broken Yang-Mills Theory with Flow Equations. *Commun. Math. Phys.* 209:477-516 (2000).

    See also [5], where the proof presented in the previous paper is simplified.
19. Ch. Kopper, V.F. Müller and Th. Reisz: Temperature Independent Renormalization of Finite Temperature Field Theory. *Ann. Henri Poincaré* 2:387-402 (2001).

20. Ch. Wieczerkowski: Symanzik's Improved actions from the viewpoint of the Renormalization Group. *Commun. Math. Phys.* 120:148-176 (1988).

G. Keller: The Perturbative Construction of Symanzik's improved Action for ϕ_4^4 and QED_4. *Helv. Phys. Acta* 66:453-470 (1993).

21. C. Kim: A Renormalization Group Flow Approach to Decoupling and Irrelevant Operators. *Ann. Phys. (N.Y.)* 243:117-143 (1995).
22. C. Bagnuls and C. Bervillier: Exact Renormalization Group Equations, An Introductory Review. *Phys. Rep.* 348:91 (2001).

13

Towards the Construction of Quantum Field Theories from a Factorizing S-Matrix

Gandalf Lechner

Institut für Theoretische Physik, Universität Göttingen,
37077 Göttingen, Germany
lechner@physik.uni-goe.de

Summary. Starting from a given factorizing S-matrix S in two space-time dimensions, we review a novel strategy to rigorously construct quantum field theories describing particles whose interaction is governed by S. The construction procedure is divided into two main steps: Firstly certain semi-local Wightman fields are introduced by means of Zamolodchikov's algebra. The second step consists in proving the existence of local observables in these models. As a new result, an intermediate step in the existence problem is taken by proving the modular compactness condition for wedge algebras.

13.1 Introduction

In quantum field theory, the rigorous construction of models with non-trivial interaction is one of the most challenging open problems. Although collision theory was established a long time ago and the calculation of the scattering matrix is well understood, little is known about the inverse problem, *i.e.* the reconstruction of interacting models from a given S-matrix. The only situation in which certain steps of such a reconstruction have been carried out is the class of factorizing S-matrices on two-dimensional Minkowski space, which correspond to scattering processes in which the particle number and momenta are conserved. This issue is usually taken up in the framework of the so-called formfactor program [2,3,35], which aims at the construction of local quantum field theories corresponding to factorizing S-matrices by determining expectation values of local operators in scattering states. In spite of the interesting results that have been obtained for many S-matrices, the final construction of interacting Wightman fields has not been achieved up to now.

In the present paper, we shall review a novel approach to this construction problem, which was initiated by Schroer in the last few years[1]. As it mainly uses the framework of local quantum physics [1,19] instead of Wightman theory, we will term this new approach "algebraic" as opposed to the more field theoretic concepts of the formfactor program. The starting point of the algebraic approach was Schroer's insight that for the family of factorizing S-matrices describing a single type of massive,

[1]See also the contribution of B. Schroer in the present volume.

scalar particles, certain field operators arising from Zamolodchikov's algebra [40] (in which the given factorizing S-matrix S is encoded) can be interpreted as being localized in wedge-shaped regions of Minkowski space [33]. Subsequently, the understanding of these wedge-local fields was deepened in [34] and [22], and connection to the algebraic formulation of quantum field theory was made by investigating the von Neumann algebras generated by them [14]. The construction of the wedge-local fields and their corresponding operator algebras will be reviewed in Section 13.2.

The second step of the algebraic program is devoted to exhibiting *local* observables. Compared to the formfactor program, where the aim is an explicit construction of local field operators, the algebraic approach focusses on the question of *existence* of local operators, which can be phrased in terms of the aforementioned wedge algebras. In [14], the modular nuclearity condition [10,11] for wedge algebras was put forward as a sufficient condition for the existence of local observables. This condition was then shown to be fulfilled in specific models in [14, 23]. These subjects will be discussed in Section 13.3.

Whereas the subsequent Sections 13.2 and 13.3 have the character of a review, we will prove a new result in Section 13.4, already announced in [14]. In a specific class of models with factorizing S-matrices, the modular compactness criterion for wedge algebras will be verified, thus taking a further step towards proving the existence of local observables. Regarding the occasion of this conference, we emphasize the relation of our compactness proof to the work of J. Bros [7] on the 'Haag-Swieca compactness property' [20]. Inspired by his strategy, we will also employ techniques of complex analysis in Section 13.4.

The article ends with a short summary in Section 13.5.

We conclude this introductory section by stating our assumptions: In the spirit of the inverse scattering approach, our construction begins with the specification of the particle content of the theory and the S-matrix. We deal here only with a single species of scalar particles of mass $m > 0$. It will be convenient to parametrize the upper mass shell by the rapidity θ via

$$p(\theta) = m \begin{pmatrix} \cosh\theta \\ \sinh\theta \end{pmatrix} . \tag{13.1}$$

In this variable, the physical sheet of the complex energy plane is transformed to the horizontal strip $S(0,\pi) := \{\zeta \in \mathbb{C} : 0 < \mathrm{Im}\zeta < \pi\}$.[2]

As mentioned before, the S-matrix is assumed to be of the factorizing type. This implies that it can be described by means of a single function S_2, called the *scattering function* in the following, which is related to two-particle S-matrix elements by

$$_{\mathrm{out}}\langle\theta_1,\theta_2|\theta_1,\theta_2\rangle_{\mathrm{in}} = S_2(|\theta_1-\theta_2|) , \tag{13.2}$$

and is required to satisfy the following conditions:

1. $S_2 : \overline{S(0,\pi)} \to \mathbb{C}$ is continuous and bounded, and analytic on $S(0,\pi)$.
2. For real θ one has

$$S_2(\theta)^{-1} = S_2(-\theta) = \overline{S_2(\theta)} = S_2(\theta+i\pi) . \tag{13.3}$$

---

[2]More generally, we will use the notation $S(a,b) = \{\zeta \in \mathbb{C} : a < \mathrm{Im}\zeta < b\}$ in the following.

By excluding poles of S_2 in the strip $S(0,\pi)$, the first condition characterizes models without bound states. The equations summarized in (13.3) arise from the requirements of unitarity, crossing symmetry and hermitian analyticity for the corresponding S-matrix (*cf.*, for example, the review in [16, p. 46] and the references cited there), and put strong constraints on the possible form of the function S_2. Indeed, the general solution of (13.3) is quite explicitely known [27]. We note here the existence of two particularly simple solutions, namely $S_2(\theta) = \pm 1$, which will be discussed in some detail. A more generic scattering function is given by

$$S_2(\theta) = \frac{1 + ig\sinh\theta}{1 - ig\sinh\theta}, \tag{13.4}$$

where $g > 0$ is some constant. It interpolates between the preceding solutions in the limit of small and large g, respectively. We also note that products of scattering functions again satisfy (13.3).

13.2 Wedge-local fields

13.2.1 Hilbert space and Zamolodchikov's algebra

The starting point of the construction of the wedge-local fields is Zamolodchikov's algebra[3] [40], which is a basic ingredient in the context of factorizing S-matrices. We do not deal with the abstract algebra here, but rather with a particular representation of it on a conveniently chosen Hilbert space $\mathcal{H}$, which we define first. For details we refer the reader to [22, 24].

In view of the assumptions on the particle spectrum made above, a reasonable choice for the one-particle Hilbert space is $\mathcal{H}_1 = L^2(\mathbb{R}, d\theta)$, the space of square integrable functions over the upper mass shell of mass m. The n-particle space $\mathcal{H}_n$ is defined as a particular subspace of the (unsymmetrized) n-fold tensor product $\mathcal{H}_1^{\otimes n}$; namely its elements are wavefunctions $\psi_n \in L^2(\mathbb{R}^n, d^n\theta)$ which satisfy the symmetry relations

$$\psi_n(\theta_1, \ldots, \theta_{k+1}, \theta_k, \ldots, \theta_n) = S_2(\theta_k - \theta_{k+1}) \cdot \psi_n(\theta_1, \ldots, \theta_k, \theta_{k+1}, \ldots, \theta_n) \tag{13.5}$$

for $k = 1, \ldots, n-1$. Here S_2 is the scattering function corresponding to the S-matrix we are considering. The full Hilbert space of the theory is

$$\mathcal{H} := \bigoplus_{n=0}^{\infty} \mathcal{H}_n, \tag{13.6}$$

where we have put $\mathcal{H}_0 := \mathbb{C} \cdot \Omega$ to denote the zero particle space containing the vacuum unit vector Ω. For the special scattering functions $S_2 = 1$ and $S_2 = -1$, $\mathcal{H}$ coincides with the symmetric and antisymmetric Fock space over $\mathcal{H}_1$, respectively. But in general we deal with a "twisted" Fock space with rapidity dependent symmetry structure.

On $\mathcal{H}$ we have a positive energy representation U of the proper orthochronous Poincaré group $\mathcal{P}_+^\uparrow$. If $(x, \lambda) \in \mathcal{P}_+^\uparrow$ denotes the transformation consisting of a boost

[3] Sometimes also called Zamolodchikov-Faddeev algebra.

with rapidity parameter $\lambda \in \mathbb{R}$ and a subsequent translation along $x \in \mathbb{R}^2$, then $U(x,\lambda)$ is defined as

$$(U(x,\lambda)\psi)_n(\theta_1,\ldots,\theta_n) = e^{i\sum_{k=1}^n p(\theta_k)x} \cdot \psi_n(\theta_1 - \lambda,\ldots,\theta_n - \lambda)\,. \qquad (13.7)$$

In the following, we will also use the shorthand notation $U(x) := U(x,0)$ for pure translations.

The creation and annihilation operators familiar from symmetric and antisymmetric Fock space have their counterparts on $\mathcal{H}$. These operator-valued distributions will be denoted $z(\theta)$ and $z^\dagger(\theta) = z(\theta)^*$, and are defined by

$$(z(\theta)\psi)_n(\theta_1,\ldots,\theta_n) = \sqrt{n+1} \cdot \psi_{n+1}(\theta,\theta_1,\ldots,\theta_n) \qquad (13.8)$$

and by taking the adjoint on $\mathcal{H}$. This definition yields a representation of Zamolodchikov's algebra on $\mathcal{H}$, i.e., $z(\theta)$, $z^\dagger(\theta)$ satisfy the exchange relations

$$z^\dagger(\theta_1)z^\dagger(\theta_2) = S_2(\theta_1 - \theta_2)z^\dagger(\theta_2)z^\dagger(\theta_1)\,, \qquad (13.9)$$

$$z(\theta_1)z^\dagger(\theta_2) = S_2(\theta_2 - \theta_1)z^\dagger(\theta_2)z(\theta_1) + \delta(\theta_1 - \theta_2)\cdot 1\,. \qquad (13.10)$$

We will also write $z(\psi) = \int d\theta\, z(\theta)\psi(\theta)$, $z^\dagger(\psi) = \int d\theta\, z^\dagger(\theta)\psi(\theta)$, for wave functions $\psi \in \mathcal{H}_1$. Note that with these conventions, $z(\psi)^* = z^\dagger(\overline{\psi})$.

13.2.2 Wedge locality

With the help of the creation and annihilation operators $z^\dagger(.)$ and $z(.)$, a scalar quantum field ϕ can be defined on (a dense domain in) $\mathcal{H}$ in a manner analogous to the definition of the free field on symmetric Fock space. For $f \in \mathscr{S}(\mathbb{R}^2)$, we consider the restrictions of the Fourier transform of this function to the upper and lower mass shell, parametrized by the rapidity:

$$f^\pm(\theta) := \frac{1}{2\pi}\int d^2x\, f(x)e^{\pm ip(\theta)x}\,, \qquad (13.11)$$

and set

$$\phi(f) := z^\dagger(f^+) + z(f^-)\,, \qquad (13.12)$$

which is a well-defined operator on the dense subspace $\mathcal{D} \subset \mathcal{H}$ of vectors of finite particle number. In the case of the scattering function $S_2 = 1$, this definition yields the well-known free scalar field. But also for different scattering functions, ϕ has many properties in common with a free field.

Proposition 1. [22] *The field operator $\phi(f)$ has the following properties:*

1. *$\phi(f)$ is defined on $\mathcal{D}$ and leaves this space invariant.*
2. *For $\psi \in \mathcal{D}$ one has*
$$\phi(f)^*\psi = \phi(\overline{f})\psi. \qquad (13.13)$$
All vectors in $\mathcal{D}$ are entire analytic for $\phi(f)$. If $f \in \mathscr{S}(\mathbb{R}^2)$ is real, $\phi(f)$ is essentially self-adjoint on $\mathcal{D}$.
3. *ϕ is a solution of the Klein-Gordon equation: For every $f \in \mathscr{S}(\mathbb{R}^2)$, $\psi \in \mathcal{D}$ one has*
$$\phi((\Box + m^2)f)\psi = 0\,. \qquad (13.14)$$

4. *$\phi(f)$ transforms covariantly under the representation U of $\mathcal{P}_+^\uparrow$, cf.* (13.7)*:*

$$U(g)\phi(f)U(g)^{-1} = \phi(f_g), \quad f_g(x) = f(g^{-1}x), \quad g \in \mathcal{P}_+^\uparrow . \tag{13.15}$$

5. *The vacuum Ω is locally cyclic for the field ϕ. More precisely, given any open subset $\mathcal{O} \subset \mathbb{R}^2$, the subspace*

$$D_{\mathcal{O}} := \text{span}\{\phi(f_1)\cdots\phi(f_n)\Omega \,:\, f_k \in \mathscr{S}(\mathcal{O}),\, n \in \mathbb{N}_0\} \tag{13.16}$$

is dense in $\mathcal{H}$.

In spite of these pleasant properties of the field operator, a simple calculation shows that ϕ is local if and only if $S_2 = 1$. As locality is one of the fundamental principles in quantum field theory, the generically non-local field operators $\phi(f)$ cannot be interpreted as the basic physical fields of our model, but rather as an auxiliary tool in the construction of the theory: They are polarization-free generators in the sense of [6].

To clarify the role of the field ϕ, we consider subsets W of $\mathbb{R}^2$ called *wedges*, which are the Poincaré transforms of the so-called left wedge

$$W_L := \{x \in \mathbb{R}^2 \,:\, |x_0| + x_1 < 0\} . \tag{13.17}$$

As W_L is invariant under boosts, any wedge has the form $W = W_L + x$ or $W = W_R + x$ for some $x \in \mathbb{R}^2$, where $W_R = -W_L = W_L'$ is the right wedge. The set of wedges will be denoted by $\mathcal{W}$.

Following Schroer and Wiesbrock [34], we address the question whether it is possible to consistently interpret the field ϕ as being localized in a wedge region, say in W_L for the sake of concreteness. Put differently, we take

$$\mathcal{A}(W_L) := \{e^{i\phi(f)} \,:\, f \in \mathscr{S}_{\mathbb{R}}(W_L)\}'' \tag{13.18}$$

as the von Neumann algebra generated by the observables in W_L and look for a map $\mathcal{W} \ni W \longmapsto \mathcal{A}(W)$ of wedge regions to von Neumann subalgebras of $\mathcal{B}(\mathcal{H})$ such that $\mathcal{A}(W_L)$ is given by (13.18) and the following standard properties [1, 19] hold: ($W, W_1, W_2 \in \mathcal{W}$):

1. $\mathcal{A}(W_1) \subset \mathcal{A}(W_2)$ for $W_1 \subset W_2$ (Isotony).
2. $U(g)\mathcal{A}(W)U(g)^* = \mathcal{A}(gW)$, $g \in \mathcal{P}_+^\uparrow$ (Covariance).
3. $\mathcal{A}(W') \subset \mathcal{A}(W)'$ (Wedge-Locality).
4. Ω is cyclic for each $\mathcal{A}(W)$ (Reeh-Schlieder property).

Such a map $W \longmapsto \mathcal{A}(W)$ will be called a *wedge-local covariant net.* Within the present context one obtains a net by setting

$$\mathcal{A}(W_R) := \mathcal{A}(W_L)', \tag{13.19}$$

$$\mathcal{A}(W + x) := U(x)\mathcal{A}(W)U(x)^*, \qquad x \in \mathbb{R}^2,\; W \in \mathcal{W}, \tag{13.20}$$

where the prime denotes taking the commutant in $\mathcal{B}(\mathcal{H})$. Whereas the first three properties (isotony, covariance, and wedge locality) follow immediately from the definitions (13.18-13.20), the cyclicity of Ω for $\mathcal{A}(W_R)$ is not so obvious – proving it is equivalent to showing that ϕ can be interpreted as being localized in W_L.

Proposition 2. [14,22]
The correspondence $W \longmapsto \mathcal{A}(W)$ *defined in* (13.18,13.19,13.20) *is a wedge-local covariant net. In particular,* Ω *is cyclic and seperating for* $\mathcal{A}(W)$, $W \in \mathcal{W}$. *Moreover,* $\mathcal{A}(W') = \mathcal{A}(W)'$, *i.e. wedge duality holds.*

The cyclicity of Ω can be proven by considering the antilinear operator J,

$$(J\psi)_n(\theta_1, \ldots, \theta_n) := \overline{\psi_n(\theta_n, \ldots, \theta_1)}\,, \tag{13.21}$$

which can be adjoined to the representation U to obtain a representation of the proper Poincaré group $\mathcal{P}_+$. More importantly, it gives rise to a second field ϕ',

$$\phi'(f) := J\phi(f^j)J, \qquad f^j(x) := \overline{f(-x)}\,. \tag{13.22}$$

The "reflected field" $\phi'(f)$, can be shown to commute with $\phi(g)$, in the sense that their associated unitary groups commute, whenever $\operatorname{supp} f + W_R$ is spacelike separated from $\operatorname{supp} g + W_L$. As the vacuum is cyclic for ϕ' as well, the cyclicity of Ω for all wedge algebras then follows.

In the next section, the modular data associated to $(\mathcal{A}(W_L), \Omega)$ will become important. As the J maps $\mathcal{A}(W_L)$ onto $\mathcal{A}(W_R)$, it can be shown to coincide with the modular conjugation of this couple. The modular group $\Delta_L^{i\lambda}$ of $(\mathcal{A}(W_L), \Omega)$ acts as expected from the Bisognano-Wichmann theorem [4,5,29]:

Proposition 3. [14]
The modular group and modular conjugation of $(\mathcal{A}(W_L), \Omega)$ *are given by* $\Delta_L^{i\lambda} = U(0, 2\pi\lambda)$ *and* J, *respectively.*

Before entering into the discussion of local observables in these models, we mention that it is possible to calculate the two-particle scattering states with the help of the wedge-local fields ϕ and ϕ', since left and right wedges can be causally separated by translation. Because of wedge-locality, it turns out that the particles are "Bosons". It is then possible to determine the two-particle S-matrix, which is the "right" one, *i.e.* the one associated to the scattering function S_2 we started with [22]. The construction of the wedge algebras thus leads to a (wedge-local) quantum field theory of particles whose interaction is described by S_2.

13.3 Existence of local observables

13.3.1 Observables localized in a double cone

Having constructed a wedge-local quantum theory with the correct two-particle scattering states, the next step is to exhibit observables localized in *bounded* spacetime regions. Typical examples of such regions are double cones, which in two dimensions can always be realized as intersections of two opposite wedges. To fix ideas, consider the double cone

$$\mathcal{O}_x := W_R \cap (W_L + x), \qquad x \in W_R\,. \tag{13.23}$$

An operator A representing an observable localized in $\mathcal{O}_x$ has to commute with any observable localized in $\mathcal{O}_x{}' = W_L \cup (W_R + x)$ because of Einstein causality. Any such

A is therefore an element of the algebra

$$\mathcal{A}(\mathcal{O}_x) := (\mathcal{A}(W_L) \vee \mathcal{A}(W_R + x))' = \mathcal{A}(W_R) \cap \mathcal{A}(W_L + x)\,, \tag{13.24}$$

the relative commutant of $\mathcal{A}(W_R + x)$ in $\mathcal{A}(W_R)$. We will adopt (13.24) as the definition of the algebra generated by the observables localized in $\mathcal{O}_x$ in our model. The algebras associated to translated double cones are then fixed by covariance.

The net $\mathcal{O} \longmapsto \mathcal{A}(\mathcal{O})$ of double cone algebras arising in this manner inherits the basic properties of isotony, covariance and locality from the corresponding features of the wedge net, as can be verified in a straightforward manner. But without further information on the structure of the wedge algebras, it is not clear whether the relative commutants (13.24) are non-trivial. As a physical theory should describe local observables, we would like to rule out the pathological cases in which $\mathcal{A}(\mathcal{O}_x) = \mathbb{C} \cdot 1$.

In [34], a method to construct explicitly non-trivial operators localized in $\mathcal{O}_x$ has been proposed. However, this procedure faces substantial difficulties related to the convergence of certain "perturbation" series. We will concentrate here on an existence proof without trying to give concrete expressions for local operators.

13.3.2 Split property and modular nuclearity

The basic idea in the approach to the existence problem proposed in [14] is the observation that the non-triviality of the relative commutant $\mathcal{A}(\mathcal{O}_x)$ (13.24) can be established if the net of wedge algebras $\mathcal{A}(W)$ has the *split property*, *i.e* if to each $x \in W_R$ there is a type I_∞ von Neumman factor $\mathcal{N}_x$ such that

$$\mathcal{A}(W_R + x) \subset \mathcal{N}_x \subset \mathcal{A}(W_R)\,. \tag{13.25}$$

In this case, the observables localized in W_L and $W_R + x$ satisfy a strong form of statistical independence (for a review, see [39]), and the algebraic structure of $\mathcal{A}(\mathcal{O}_x)$ is completely fixed. According to a result of Longo [25], the wedge algebras $\mathcal{A}(W)$ are type III_1 von Neumann factors in the present context. Using this information and the split assumption (13.25), one can establish the unitary equivalence [14,17]

$$\mathcal{A}(\mathcal{O}_x) \cong \mathcal{A}(W_R) \otimes \mathcal{A}(W_L + x). \tag{13.26}$$

Thus the split property for the wedge algebras implies that the local algebras $\mathcal{A}(\mathcal{O}_x)$ are of type III as well, and in particular non-trivial.

In the following, the split property will be used as a sufficient condition for the non-triviality of the relative commutants (13.24)[4]. However, as the existence of an interpolating type I factor (13.25) is difficult to establish directly, one needs another condition, implying the split property, which is better manageable in concrete models. In the literature different "nuclearity" criteria for the split property have been discussed, the one which is relevant in our context being introduced in [10]. As these criteria involve the notions of nuclear maps, we recall that a bounded operator between two Banach spaces is called *nuclear* if it can be expanded into a norm

[4]Note that in two space-time dimensions the split property for wedges is a reasonable assumption, as Araki's argument [9] to the effect that inclusions of wedge algebras cannot be split does not apply here because of the missing translation invariance along the edge of the wedge.

convergent series of rank 1 operators [30]. Note that a nuclear map is in particular compact.

To formulate the relevant condition for the split property, we denote by J, Δ the modular involution and modular operator of the pair $(\mathcal{A}(W_R), \Omega)$, respectively[5], and introduce the maps

$$\begin{aligned} \Xi(x) : \mathcal{A}(W_R) &\longrightarrow \mathcal{H}, \\ A &\longmapsto \Delta^{1/4} U(x) A\Omega . \end{aligned} \tag{13.27}$$

Using modular theory, one easily finds

$$\|\Xi(x)A\|^2 = \langle U(x)A\Omega, \Delta^{1/2}U(x)A\Omega\rangle = \langle U(x)A\Omega, JU(x)A^*\Omega\rangle \leq \|A\|^2,$$

i.e. $\Xi(x)$ is a bounded map with $\|\Xi(x)\| \leq 1$ for any $x \in \overline{W_R}$. Based on results of [10], the following "modular nuclearity condition" has been discussed in [14].

Proposition 4. [14]
If $\Xi(x)$ (13.27) is nuclear, the inclusion $\mathcal{A}(W_R + x) \subset \mathcal{A}(W_R)$ is split and the local algebra $\mathcal{A}(\mathcal{O}_x)$ (13.24) is isomorphic to the unique hyperfinite type III_1 factor.

Proposition 4 is used as a sufficient condition for the non-triviality of the local algebras in the model theories presented in Section 13.2. If the scattering function is constant, $S_2(\theta) = \pm 1$, the structure of Zamolodchikov's algebra simplifies to the CCR- and CAR-algebra, respectively, and the estimates needed for the nuclearity proof of $\Xi(x)$ are fully under control. We have

Proposition 5. [14, 23]
In the models corresponding to the constant scattering functions $S_2(\theta) = 1$ and $S_2(\theta) = -1$, the maps $\Xi(x)$ are nuclear, $x \in W_R$. In particular, the split property for wedges holds and all double cone algebras (13.24) contain non-trivial observables.

The proof for the case $S_2 = 1$ can be found in [14], where previous results obtained in [13, 15] have been used. For the case $S_2 = -1$, see [23]. In these articles, one also finds explicit bounds on the nuclear norms $\|\Xi(x)\|_1$. The case $S_2 = 1$ gives just the free scalar field in two dimensions, and the model corresponding to the scattering function $S_2 = -1$ is related to the scaling limit of the Ising model (see [3] and the references cited there).

Although the existence of local observables is well known in free field theory, the check of the modular nuclearity condition in the case $S_2 = 1$ was an important test for its value in the discussion of models with non-constant scattering functions. In view of Proposition 5 and the earlier mentioned fact that $S_2 = \pm 1$ may be considered as the "limiting cases" of typical non-constant scattering functions, we conjecture that $\Xi(x)$ is nuclear in the family of models considered in Section 13.2.

It was shown in [10] that whereas the nuclearity of $\Xi(x)$ is sufficient for the split property, the *compactness* of $\Xi(x)$ is a necessary condition for split. In the next section, our conjecture about the nuclearity of $\Xi(x)$ will be further substantiated by proving the compactness of this map in a wide class of models with certain typical scattering functions.

[5] Δ is connected to the earlier discussed modular operator of the left wedge by $\Delta = \Delta_L^{-1}$.

13.4 Modular Compactness for Wedge Algebras

In this section we concentrate on the model corresponding to the scattering function (13.4) with arbitrary constant $g > 0$, or a finite product of such functions with different values of g. In view of the general solution [27] of the constraining equations (13.3) for S_2, this is a typical example of a non-constant scattering function. The aim of this section is the proof of the following proposition.

Proposition 6. *Consider the model theory corresponding to the scattering function*

$$S_2(\theta) := \prod_{r=1}^{R} \frac{1 + ig_r \sinh\theta}{1 - ig_r \sinh\theta}, \tag{13.28}$$

where $R < \infty$ and $g_1, \ldots, g_R > 0$. The maps $\Xi(x)$ are compact, $x \in W_R$.

Before explaining our strategy of the proof, we make a few remarks about the maps $\Xi(x)$ and introduce some notation.

First note that it is sufficient to consider the maps $\Xi(0,s)$ corresponding to wedge inclusions of the type

$$W_R + (0,s) \subset W_R, \qquad s > 0. \tag{13.29}$$

As W_R is stable under boosts, a more general inclusion $W_R + x \subset W_R$, $x \in W_R$, of right wedges can be transformed to (13.29) by a velocity transformation with appropriately chosen rapidity parameter λ. Using the fact that the boosts commute with the modular operator, one easily shows $\Xi(x) = U(0,-\lambda)\Xi(0,s)\mathrm{Ad}U(0,\lambda)$, where $s = (x_1^2 - x_0^2)^{1/2} > 0$. Hence $\Xi(x)$, $x \in W_R$, is nuclear (compact) if and only if $\Xi(0,s)$, $s > 0$, is nuclear (compact). In the case of nuclear maps, $\|\Xi(x)\|_1 = \|\Xi(0,(x_1^2 - x_0^2)^{1/2})\|_1$. For this reason, we will consider in the following only inclusions of the form (13.29), and use the shorthand notation $\Xi(s) := \Xi(0,s)$.

It will be useful to consider, as a generalization of (13.27), the maps

$$\Xi^\alpha(s)\mathcal{A}(W_R) \longrightarrow \mathcal{H}\,, \quad \Xi^\alpha(s)A := \Delta^\alpha U(s)A\Omega\,, \qquad 0 \leq \alpha \leq \tfrac{1}{2}, \tag{13.30}$$

and we adopt the convention to suppress the upper index for the "canonical" value $\alpha = \frac{1}{4}$. Furthermore, we introduce the n-particle projections[6]

$$\Xi_n^\alpha(s) := P_n \Xi^\alpha(s)\,, \tag{13.31}$$

where $P_n \in \mathcal{B}(\mathcal{H})$ denotes the orthogonal projection onto the n-particle subspace $\mathcal{H}_n$.

In the proof of Proposition 6, the maps $\Xi_n(s)$ will be shown to be nuclear by estimating the "size" of their images in $\mathcal{H}_n$. This is achieved by exploiting certain analytic properties of the n-particle rapidity wavefunctions

$$\psi_n^s := P_n U(0,s)A\Omega\,, \qquad A \in \mathcal{A}(W_R), \tag{13.32}$$

which are considered as elements of $L^2(\mathbb{R}^n, d^n\vec{\theta})$ and constitute our main objects of interest in the following. (For a different compactness proof based on the techniques of complex analysis, see [7].)

[6] Note that the modular operator Δ can be restricted to the n-particle space $\mathcal{H}_n$.

From modular theory and the Bisognano-Wichmann property (as specified in Proposition 3) it is known that

$$\lambda \longmapsto (\Delta^{-i\lambda/2\pi}\psi_n^s)(\theta_1,\ldots,\theta_n) = \psi_n^s(\theta_1 - \lambda,\ldots,\theta_n - \lambda) \tag{13.33}$$

is a strongly analytic function in the strip $S(0,\pi)$, and $\|\Delta^{\lambda/2\pi}\psi_n^s\| \le \|A\|, 0 \le \lambda \le \pi$. In particular, the vectors in the image of $\Xi_n(s)$ are given by the functions

$$(\theta_1,\ldots,\theta_n) \longmapsto (\Delta^{1/4}\psi_n^s)(\theta_1,\ldots,\theta_n) = \psi_n^s(\theta_1 - \tfrac{i\pi}{2},\ldots,\theta_n - \tfrac{i\pi}{2}), \tag{13.34}$$

which have an analytic continuation (in the sense of distributions) in the "center of mass rapidity" $n^{-1}\cdot(\theta_1+\cdots+\theta_n)$ to the strip $S(-\frac{\pi}{2},\frac{\pi}{2})$. Furthermore, the L^2-bound of the continuation is uniformly bounded over this strip and the convergence to the boundary values is valid in the norm topology of $\mathcal{H}_n$.

The main idea in the proof of Proposition 6 consists in the observation that in the models at hand, the wavefunctions (13.32) enjoy considerably stronger analytic properties, namely they are holomorphic, as functions of n complex variables, in a certain tube domain. More precisely, we find:

Lemma 1. *Let $A \in \mathcal{A}(W_R)$, $s > 0$, $n \in \mathbb{N}_0$ and ψ_n^s as in (13.32). There exists a constant $\alpha > 0$ (independent of A and s, but dependent on n) such that*

a) *$\Delta^\alpha\psi_n^s$ is analytic in the tube*

$$\mathcal{T}_n(\alpha) := \{\vec{\zeta} \in \mathbb{C}^n \,:\, -2\pi\alpha < \mathrm{Im}(\zeta_k) < 2\pi\alpha,\; k = 1,\ldots,n\}. \tag{13.35}$$

b) *For any $\vec{\lambda} \in]-2\pi\alpha, 2\pi\alpha[^{\times n}$,*

$$\mathbb{R}^n \ni \vec{\theta} \longmapsto (\Delta^\alpha\psi_n^s)(\vec{\theta} + i\vec{\lambda}) \tag{13.36}$$

is in $L^2(\mathbb{R}^n, d^n\vec{\theta})$, with norm bounded by $K\cdot\|A\|$, where K depends on α, s and n, but is independent of A and $\vec{\lambda}$. Moreover, $\lim\limits_{|\vec{\theta}|\to\infty} |\psi_n^s(\vec{\theta}+i\vec{\lambda})| = 0$.

c) *$(\Delta^\alpha\psi_n^s)(\vec{\zeta})$ converges strongly to its boundary values at $\mathrm{Im}(\zeta_k) = \pm 2\pi\alpha$.*

Here we introduced the convention to denote vectors in $\mathbb{C}^n$ or $\mathbb{R}^n$ by bold face letters $\vec{\zeta}, \vec{\lambda}, \vec{\theta}$, and their components by $\zeta_k, \lambda_k, \theta_k$, $k = 1,\ldots,n$. Note in particular that by considering the limit $\mathrm{Im}(\zeta_k) \to 2\pi\alpha$, $k = 1,\ldots,n$, the wavefunctions ψ_n^s (13.32) are recovered as a (strong) boundary value of the analytic function $\Delta^\alpha\psi_n^s$.

The constants α and K appearing in Lemma 1 specify the size of the tube $\mathcal{T}_n(\alpha)$ in which $\Delta^\alpha\psi_n^s$ is analytic and its bound in that region, respectively. They depend on the scattering function S_2 (13.28) under consideration, *i.e.* on the parameters $g_1,\ldots,g_R$. This dependence will be made explicit in the proof of Lemma 1, which is based on wedge locality and the symmetry properties (13.5) of n-particle functions. However, it will require the discussion of some technical points. We therefore postpone it and and rather begin by showing how Lemma 1 can be used to derive estimates on the nuclear norms of $\Xi_n(s)$.

Lemma 2. *The maps $\Xi_n(s)$ are nuclear, $s > 0$.*

Proof. In view of the definition of the translations (13.7), we have, $\vec{\theta} \in \mathbb{R}^n$,

$$\begin{aligned}(\Delta^\alpha \psi_n^s)(\vec{\theta}) &= (\Delta^\alpha U(0, \tfrac{s}{2})\psi_n^{\frac{s}{2}})(\vec{\theta}) \\ &= e^{-im\frac{s}{2}\sum_{k=1}^n \mathrm{sh}(\theta_k - 2\pi\alpha\, i)}(\Delta^\alpha \psi_n^{\frac{s}{2}})(\vec{\theta}) \\ &= \prod_{k=1}^n e^{-i\frac{ms}{2}\cos(2\pi\alpha)\,\mathrm{sh}\theta_k} e^{-\frac{ms}{2}\sin(2\pi\alpha)\,\mathrm{ch}\theta_k} \cdot (\Delta^\alpha \psi_n^{\frac{s}{2}})(\vec{\theta}).\end{aligned} \tag{13.37}$$

The strongly decreasing factor appearing here allows us to conclude the nuclearity of $\Xi_n(s)$ by application of Cauchy's integral formula: Let $\vec{\theta} \in \mathbb{R}^n$ and consider a polydisc $\mathscr{D} \subset \mathcal{T}_n(\alpha)$ with $\vec{\theta} \in \mathscr{D}$. As $\Delta^\alpha \psi_n^s$ is holomorphic in $\mathcal{T}_n(\alpha)$,

$$(\Delta^\alpha \psi_n^s)(\vec{\theta}) = \frac{1}{(2\pi i)^n} \oint_{\partial\mathscr{D}} d^n\vec{\zeta}\, \frac{(\Delta^\alpha \psi_n^s)(\vec{\zeta'})}{\prod_{k=1}^n (\zeta_k' - \theta_k)}\,.$$

Because of the decrease properties of $\Delta^\alpha \psi_n^s$ in real directions, and the good convergence to its boundary values, as specified in Lemma 1 b),c), we can deform the contour of integration to the boundary of $\mathcal{T}_n(\alpha)$ and get

$$\begin{aligned}(\Delta^\alpha \psi_n^s)(\vec{\theta}) = \frac{1}{(2\pi i)^n} \sum_{\vec{\varepsilon}} \int_{\mathbb{R}^n} d^n\vec{\theta'} \prod_{k=1}^n \frac{\varepsilon_k}{(\theta_k' - \theta_k - i\cdot 2\pi\alpha\varepsilon_k)} \\ \times\ (\Delta^\alpha \psi_n^s)(\theta_1' - i\cdot 2\pi\alpha\varepsilon_1, \ldots, \theta_n' - i\cdot 2\pi\alpha\varepsilon_n).\end{aligned} \tag{13.38}$$

The summation extends over the 2^n terms parametrized by $\varepsilon_1, \ldots, \varepsilon_n = \pm 1$. Consider the integral operators $T_{s,\alpha}^{\pm} \in \mathcal{B}(L^2(\mathbb{R}, d\theta))$ which are defined by their kernels

$$T_{s,\alpha}^{\pm}(\theta, \theta') := \pm\frac{1}{i\pi}\frac{e^{-\frac{ms}{2}\sin(2\pi\alpha)\,\mathrm{ch}\theta}}{\theta' - \theta \mp 2\pi i\alpha}, \tag{13.39}$$

and the unitary operator $M_{s,\alpha}$ multiplying with $e^{-i\frac{ms}{2}\cos(2\pi\alpha)\mathrm{sh}\theta}$. With these definitions, inserting (13.37) into (13.38) yields

$$\Xi_n^\alpha(s)A = \Delta^\alpha \psi_n^s = 2^{-n}\sum_{\vec{\varepsilon}} M_{s,\alpha}{}^{\otimes n}(T_{s,\alpha}^{\varepsilon_1} \otimes \ldots \otimes T_{s,\alpha}^{\varepsilon_n})\Delta^\alpha \psi_{n,\vec{\varepsilon}}^{\frac{s}{2}}, \tag{13.40}$$

where we used the shorthand notation $\psi_{n,\vec{\varepsilon}}^{\frac{s}{2}}(\vec{\theta}) = \psi_n^{\frac{s}{2}}(\vec{\theta} - 2\pi\alpha i\cdot\vec{\varepsilon})$. It is now important to note that $T_{s,\alpha}^{\pm}$ are trace class operators on $L^2(\mathbb{R}, d\theta)$ [14]. (This can be seen by Fourier transforming the kernels (13.39) in the variable θ', and the applying standard techniques to split the resulting operators into a product of two Hilbert Schmidt maps [38, Thm. XI.21].) Note also that $T_{s,\alpha}^{+}$ and $T_{s,\alpha}^{-}$ are unitarily equivalent, the equivalence being given by the operator V, $(V\psi)(\theta) := \psi(-\theta)$. Hence they have the same trace norm $\|T_{s,\alpha}^{+}\|_1$.

According to Lemma 1 b), $\|\psi_{n,\vec{\varepsilon}}^{\frac{s}{2}}\| \leq K\|A\|$ for each $\vec{\varepsilon}$ occuring in the above summation. Put differently, $A \longmapsto \Delta^\alpha \psi_{n,\vec{\varepsilon}}^{\frac{s}{2}}$ is bounded as a linear map between the Banach spaces $\mathcal{A}(W_R)$ and $\mathcal{H}_n$, with norm dominated by K. As $M_{s,\alpha}$ is unitary, this implies that $\Xi_n^\alpha(s)$ is a nuclear map, and as a crude bound on its nuclear norm we have

$$\|\Xi_n^\alpha(s)\|_1 \leq K\,\|T_{s,\alpha}^{+}\|_1^n\,. \tag{13.41}$$

To proceed from this nuclearity result to the statement in Lemma 2, the power of the modular operator needs to be adjusted from α to $\frac{1}{4}$. This can be achieved as in [10, Cor. 3.4]: Let $A \in \mathcal{A}(W_R)$. From modular theory we know

$$\Xi_n^\alpha(s)A = P_n \Delta^\alpha U(0,s)A\Omega = P_n\Delta^\alpha J \Delta^{1/2} U(0,s)A^*\Omega \tag{13.42}$$

$$= JP_n \Delta^{1/2-\alpha} U(0,s)A^*\Omega = J\Xi_n^{1/2-\alpha}(s)A^*\,, \tag{13.43}$$

where we used the fact that Δ and J commute with P_n. Hence $\Xi_n^{1/2-\alpha}(s)$ is nuclear, too, and $\|\Xi_n^{1/2-\alpha}(s)\|_1 = \|\Xi_n^\alpha(s)\|_1$. Since $X := \Delta^{1/4}(\Delta^\alpha + \Delta^{1/2-\alpha})^{-1}$ is a bounded operator with norm $\|X\| \le \frac{1}{2}$, it follows from

$$\Xi_n(s) = X\Big(\Xi_n^\alpha(s) + \Xi_n^{1/2-\alpha}(s)\Big) \tag{13.44}$$

that $\Xi_n(s)$ is nuclear, with nuclear norm bounded by

$$\|\Xi_n(s)\|_1 \le \|X\|\left(\|\Xi_n^\alpha(s)\|_1 + \|\Xi_n^{1/2-\alpha}(s)\|_1\right) \le \|\Xi_n^\alpha(s)\|_1\,. \tag{13.45}$$

This completes the proof of the Lemma. □

The proof of Proposition 6 now follows as a short corollary of Lemma 2 [14]. Let $A \in \mathcal{A}(W_R)$. Since

$$(\Xi_n(s)A)(\vec{\theta}) = \prod_{k=1}^n e^{-ms\cosh\theta_k} \cdot (\Xi_n(0)A)(\vec{\theta}) \le e^{-msn} \cdot (\Xi_n(0)A)(\vec{\theta})$$

and $\|\Xi_n(0)\| \le 1$, we have $\|\Xi_n(s)\| \le e^{-msn}$, and hence the series $\sum_{n=0}^\infty \Xi_n(s)$ converges in the norm topology of $\mathcal{B}(\mathcal{A}(W_R), \mathcal{H})$ to $\Xi(s)$. But as nuclear maps, the $\Xi_n(s)$ are in particular compact, and the set of compact operators between two Banach spaces is norm closed [36, Thm. VI.12]. Thus $\Xi(s)$ is compact, too. □

To accomplish the existence proof for local observables, one has to show that the map $\Xi(s)$ is not only compact, but also nuclear. This amounts to proving that the series $\sum_n \Xi_n(s)$ converges not only in the operator norm $\|\cdot\|$, but also in the nuclear norm $\|\cdot\|_1$, i.e., better bounds than (13.41) on $\|\Xi_n(s)\|_1$ are required. Such a refined analysis will be presented elsewhere.

After having shown how analytic properties of the wavefunctions ψ_n^s lead to the nuclearity of $\Xi_n(s)$, it remains to derive these properties, *i.e.* prove Lemma 1. This proof will be devided into three steps: First, we consider the dependence of ψ_n^s on its first variable θ_1 only, and derive analytic properties of $\theta_1 \mapsto \psi_n^s(\theta_1, \dots, \theta_n)$ by using the localization of A in W_R (Lemmata 3 and 4). This is a kind of one-particle analysis, and the form of the scattering function S_2 does not matter here. In a second step, the symmetry (13.5) is exploited to transfer these results to the other variables $\theta_2, \dots, \theta_n$. Here two important properties (13.67, 13.68) of S_2 enter. Finally, the n-variable analyticity claimed in Lemma 1 is established by using the Malgrange-Zerner ("flat tube") theorem (*cf.*, for example, [18]).

To exploit the localization of A, it is useful to consider the time zero fields φ, π of the "left localized" field ϕ (13.12) and study expectation value functionals of the commutator of these fields with A. In rapidity space, these functionals give rise to certain analytic functions (Lemma 3). Their relations to the wavefunctions ψ_n^s are explained in Lemma 4.

The formal definitions $\varphi(x) = \phi(0,x)$, $\pi(x) = (\partial_0\phi)(0,x)$, $x \in \mathbb{R}$, can be rephrased as ($f \in \mathscr{S}(\mathbb{R})$)

$$\hat{f}(\theta) := \widetilde{f}(m\sinh\theta), \qquad \hat{f}_-(\theta) := \hat{f}(-\theta), \tag{13.46}$$
$$\varphi(f) := z^\dagger(\hat{f}) + z(\hat{f}_-), \tag{13.47}$$
$$\pi(f) := i(z^\dagger(\omega\hat{f}) - z(\omega\hat{f}_-)). \tag{13.48}$$

Here $\omega = m\cosh\theta$ is considered as an (unbounded) multiplication operator on its maximal domain in $L^2(\mathbb{R}, d\theta)$. Along the same lines as in [22, Prop. 2 (2)], one can show that these fields are localized on the left half line, *i.e.* $\varphi(f)$, $\pi(f)$ are affiliated with $\mathcal{A}(W_L)$ if $f \in \mathscr{S}(\mathbb{R}_-)$.

Choosing an $(n-1)$-particle vector $\xi_{n-1} \in \mathcal{H}_{n-1}$, an operator $A \in \mathcal{A}(W_R)$, and a translation $s > 0$, we define two linear functionals on $\mathscr{S}(\mathbb{R})$ as

$$\begin{aligned} C_s^-(f) &:= \langle \xi_{n-1}, [\varphi(f), A(0,s)]\Omega\rangle, \\ C_s^+(f) &:= \langle \xi_{n-1}, [\pi(f), A(0,s)]\Omega\rangle . \end{aligned} \tag{13.49}$$

The operator A is an arbitrary element of $\mathcal{A}(W_R)$, but fixed in the following. The (anti-) linear dependence of the above distributions on A and ξ_{n-1} will not be indicated in our notation.

We recall that the creation and annihilation operators $z^\dagger(.), z(.)$ satisfy the following bounds with respect to the particle number [22], familiar from free field theory ($\chi \in \mathcal{H}_1$):

$$\begin{aligned} \|z(\chi)\xi_{n-1}\| &\le (n-1)^{1/2}\|\chi\|\|\xi_{n-1}\|, \\ \|z^\dagger(\chi)\xi_{n-1}\| &\le n^{1/2}\|\chi\|\|\xi_{n-1}\| . \end{aligned} \tag{13.50}$$

Also note that for $f \in L^2(\mathbb{R}, dx)$ (we put $\|f\|_2 := (\int dx|f(x)|^2)^{1/2}$),

$$\|\omega^{1/2}\hat{f}\|^2 = \int d\theta\, m\mathrm{ch}\theta\, |\hat{f}(\theta)|^2 = \int dp\, |\widetilde{f}(p)|^2 = \|f\|_2^2 . \tag{13.51}$$

Combining (13.50) and (13.51) with the Cauchy-Schwarz inequality and using the annihilation property of $z(\cdot)$, we obtain the bounds

$$|C_s^\pm(\omega^{\mp 1/2}f)| \le c_n\, \|\xi_{n-1}\|\|A\| \cdot \|f\|_2, \quad c_n := \sqrt{n} + \sqrt{n-1} + 1. \tag{13.52}$$

By carrying out the same estimate for $|C_s^\pm(f)|$ and taking into account that both, $\|\hat{f}\|$ and $\|\omega\hat{f}\|$, can be dominated by certain linear combinations of Schwartz space seminorms of f, we first note $C_s^\pm \in \mathscr{S}'(\mathbb{R})$. Moreover, (13.52) shows that the distributions $f \mapsto C_s^\pm(\omega^{\mp 1/2}f)$ are regular in the sense that they are given by L^2-functions, whose norm is bounded by $c_n\|\xi_{n-1}\|\|A\|$.

In view of the localization properties of A, $\varphi(f)$ and $\pi(f)$, the support of $C_s^\pm$ is contained in the half line $[s, \infty[\; \subset \mathbb{R}_+$. Consequently, their Fourier transforms are boundary values (in the sense of distributions) of functions $\tilde{C}_s^\pm$ which are analytic in the lower half plane and polynomially bounded in imaginary direction [37, Thm. IX.16]. Taking into account supp $C_s^\pm \subset [s, \infty[$, it follows that $\tilde{C}_s^\pm$ actually decays exponentially in the imaginary direction.

We now proceed to the rapidity picture by setting

$$\hat{C}_s^-(\theta) := m\cosh(\theta)\cdot\tilde{C}_s^-(m\sinh\theta), \qquad \hat{C}_s^+(\theta) := \tilde{C}_s^+(m\sinh\theta)\,. \tag{13.53}$$

Note that because of the regularity of $\omega^{\mp1/2}\tilde{C}_s^\pm$, these are well-defined *functions.* Their properties are collected in the following lemma.

Lemma 3. *Let $A \in \mathcal{A}(W_R)$, $s > 0$, and $\xi_{n-1} \in \mathcal{H}_{n-1}$. The functions $\hat{C}_s^\pm$ (13.53) corresponding to the distributions (13.49) have the following properties:*

a) *$\hat{C}_s^\pm$ is the boundary value of a function analytic in the strip $S(-\pi,0)$.*
b) *$\hat{C}_s^\pm$ is (anti-) symmetric with respect to reflections about $\mathbb{R}-\frac{i\pi}{2}$:*

$$\hat{C}_s^\pm(\theta-\tfrac{i\pi}{2}+i\mu) = \pm\hat{C}_s^\pm(-\theta-\tfrac{i\pi}{2}-i\mu), \qquad -\tfrac{\pi}{2}<\mu<\tfrac{\pi}{2}\,. \tag{13.54}$$

c) *Let $0\le\lambda\le\pi$ and $c_n = \sqrt{n}+\sqrt{n-1}+1$. The functions*

$$\mathbb{R}\ni\theta\longmapsto \hat{C}_{s,\lambda}^\pm(\theta) := \hat{C}_s^\pm(\theta-i\lambda) \tag{13.55}$$

are elements of $L^2(\mathbb{R},d\theta)$, with norm bounded by $\|\hat{C}_{s,\lambda}^\pm\|\le c_n\|\xi_{n-1}\|\|A\|$.
d) *$\overline{S(-\pi,0)}\ni\zeta\longmapsto\hat{C}_{s,\zeta}^\pm$ is continuous in the norm topology of $L^2(\mathbb{R},d\theta)$.*
e) *For $\theta\in\mathbb{R}$, $0<\lambda<\pi$,*

$$|\hat{C}_s^\pm(\theta-i\lambda)| \le \sqrt{\frac{2}{\pi}}\,c_n\|\xi_{n-1}\|\|A\|\frac{e^{-\frac{ms}{2}\sin\lambda\cosh\theta}}{\min\{\lambda,\pi-\lambda\}^{1/2}}\,. \tag{13.56}$$

Proof. a) Recalling

$$\begin{aligned}\cosh(\theta-i\lambda) &= \cos\lambda\,\cosh\theta - i\sin\lambda\,\sinh\theta,\\ \sinh(\theta-i\lambda) &= \cos\lambda\,\sinh\theta - i\sin\lambda\,\cosh\theta,\end{aligned} \tag{13.57}$$

we see that sinh(.) maps the strip $S(-\pi,0)$ to the lower half plane. Hence $\hat{C}_s^\pm$ is analytic in $S(-\pi,0)$. b) is also a direct consequence of (13.57).

To prove c), note that on the real line the claimed bound holds in view of the former estimates on $\|\omega^{\mp1/2}\tilde{C}_s^\pm\|$:

$$\int d\theta\,|\hat{C}_s^\pm(\theta)|^2 = \int dp\,|\omega(p)^{\mp1/2}\tilde{C}_s^\pm(p)|^2 = \|\omega^{\mp1/2}\tilde{C}_s^\pm\|_2^2 \le c_n^2\|\xi_{n-1}\|^2\|A\|^2.$$

But b) implies in particular $\hat{C}_s^\pm(\theta-i\pi)=\pm\hat{C}_s^\pm(-\theta)$, and hence $\hat{C}_s^\pm$ is also square integrable over the lower boundary $\mathbb{R}-i\pi$, with the same bound on its norm. As $\tilde{C}_s^\pm$ decays exponentially in the imaginary direction, we have $\hat{C}_{s,\lambda}^\pm\in L^2(\mathbb{R},d\theta)$ for any fixed $0<\lambda<\pi$. Moreover, the limits $\lim_{\lambda\to0}\hat{C}_{s,\lambda}^\pm$ and $\lim_{\lambda\to\pi}\hat{C}_{s,\lambda}^\pm$ are known to hold in the sense of distributions. These facts allow for the application of a version of the three lines theorem adapted to L^2-bounds, which is proven in the appendix as Lemma 5. The results are: $\hat{C}_{s,\lambda}^\pm$ converges in the norm topology of $L^2(\mathbb{R},d\theta)$ as $\lambda\to0$ or $\lambda\to\pi$, and the bound calculated on the boundary holds also for $\hat{C}_{s,\lambda}^\pm$, $0<\lambda<\pi$. This proves the claims c) and d).

Finally, e) is a consequence of a) and c): Let $\theta\in\mathbb{R}$, $0<\lambda<\pi$, and put $\rho:=\min\{\lambda,\pi-\lambda\}$. Then the disc D_ρ with center $\theta-i\lambda$ and radius ρ is contained in the closed strip $\overline{S(-\pi,0)}$. By the mean value theorem for analytic functions, Hölder's inequality and the norm bound given in c), we get

$$
\begin{aligned}
|\hat{C}_s^{\pm}(\theta - i\lambda)| &\leq \frac{1}{\pi\rho^2}\int_{D_\rho} d\theta'\, d\lambda'\, |\hat{C}_s^{\pm}(\theta' + i\lambda')| \\
&\leq \frac{1}{\sqrt{\pi}\rho}\left(\int_{D_\rho} d\theta'\, d\lambda'\, |\hat{C}_s^{\pm}(\theta' + i\lambda')|^2\right)^{1/2} \\
&\leq \frac{1}{\sqrt{\pi}\rho}\left(\int_{-\lambda-\rho}^{-\lambda+\rho} d\lambda' \int_{-\infty}^{\infty} d\theta'\, |\hat{C}_s^{\pm}(\theta' + i\lambda')|^2\right)^{1/2} \\
&\leq \sqrt{\frac{2}{\pi\rho}} \cdot c_n \|\xi_{n-1}\| \|A\| .
\end{aligned} \tag{13.58}
$$

Taking into account the covariance of $\varphi(f)$ and the translation invariance of Ω,

$$
\begin{aligned}
C_s^-(f) &= \langle \xi_{n-1},\, [\varphi(f), U(0,\tfrac{s}{2})A(0,\tfrac{s}{2})U(0,-\tfrac{s}{2})]\Omega\rangle \\
&= \langle U(0,-\tfrac{s}{2})\xi_{n-1},\, [\varphi(f_{\frac{s}{2}}), A(0,\tfrac{s}{2})]\Omega\rangle =: C'(f_{\frac{s}{2}}),
\end{aligned}
$$

where $f_{\frac{s}{2}}(x) = f(x + \frac{s}{2})$, $x \in \mathbb{R}$. Hence $\hat{C}_s^-(\theta) = e^{-\frac{ims}{2}\operatorname{sh}\theta}\hat{C}'(\theta)$, and $\hat{C}'$ also fulfills the estimate (13.58) since $\|U(0, \pm\frac{s}{2})\| = 1$. Taking the absolute value of the exponential factor evaluated on points $\theta - i\lambda \in S(-\pi, 0)$ yields the claimed estimate (13.56) for $\hat{C}_s^-$. The argument for $\hat{C}_s^+$ is the same. □

The relation between the functions $\hat{C}_s^{\pm}$ and the wavefunctions ψ_n^s (13.32) is specified in the next lemma.

Lemma 4. *Consider the function*[7]

$$
h_\xi^s(\zeta) := \frac{1}{2}(\hat{C}_s^-(\zeta) + i\hat{C}_s^+(\zeta)), \qquad \zeta \in \overline{S(-\pi, 0)}\,, \tag{13.59}
$$

which depends on $\xi_{n-1} \in \mathcal{H}_{n-1}$ through the definitions (13.49) *and* (13.53). *h_ξ^s has the properties* a), c), d), e) *of the preceding lemma. Moreover,* $\theta_1 \in \mathbb{R}$,

$$
h_\xi^s(\theta_1) = \sqrt{n}\int d\theta_2 \cdots d\theta_n\, \overline{\xi_{n-1}(\theta_2, \ldots, \theta_n)} \cdot \psi_n^s(\theta_1, \ldots, \theta_n)\,. \tag{13.60}
$$

Proof. From the definition of h_ξ^s it follows immediately that properties a) and c)–e) of Lemma 3 hold. To show (13.60), let $f \in \mathscr{S}(\mathbb{R})$.

$$
\begin{aligned}
\langle \overline{\hat{f}_-}, h_\xi^s\rangle &= \frac{1}{2}\int d\theta \hat{f}(-\theta)\left(\hat{C}_s^-(\theta) + i\hat{C}_s^+(\theta)\right) \\
&= \frac{1}{2}\int dp \tilde{f}(-p)\left(\tilde{C}_s^-(p) + i\omega(p)^{-1}\tilde{C}_s^+(p)\right) \\
&= \frac{1}{2}\left(C_s^-(f) + iC_s^+(\omega^{-1}f)\right) \\
&= \frac{1}{2}\langle \xi_{n-1},\, [\varphi(f) + i\pi(\omega^{-1}f), A(0,s)]\Omega\rangle \\
&= \langle \xi_{n-1},\, [z(\hat{f}_-), A(0,s)]\Omega\rangle && (13.61) \\
&= \langle \xi_{n-1},\, z(\hat{f}_-)A(0,s)\Omega\rangle && (13.62) \\
&= \langle z^\dagger(\overline{\hat{f}_-})\xi_{n-1},\, P_n U(0,s)A\Omega\rangle && (13.63) \\
&= \sqrt{n}\,\langle \overline{\hat{f}_-} \otimes \xi_{n-1},\, \psi_n^s\rangle\,. && (13.64)
\end{aligned}
$$

[7] We write h_ξ^s instead of $h_{\xi_{n-1}}^s$ in order not to overburden our notation.

In the last steps, we wrote the annihilation operator as a linear combination of the time zero fields in (13.61), *cf.* (13.47,13.48), used the annihilation property of $z(.)$ in (13.62), the relations of Zamolodchikov's algebra in (13.63) and the definition of $z^\dagger(.)$ in (13.64). By continuity, the above calculated relation

$$\langle \overline{\hat{f}_-}, h_\xi^s \rangle = \sqrt{n}\, \langle \overline{\hat{f}_-} \otimes \xi_{n-1},\, \psi_n^s \rangle \tag{13.65}$$

holds also if $\hat{f}_-$ is replaced by an arbitrary function in $L^2(\mathbb{R}, d\theta)$. This implies (13.60). □

For $n = 1$, Lemma 4 states $\psi_1^s = h_\Omega^s$. Hence the one particle wavefunctions ψ_1^s have the properties a), c)–e), listed in Lemma 3. In particular the claims of Lemma 1 follow if the parameters α and K appearing there are chosen as $\alpha = \frac{1}{4}$ and $K = 1$. But for $n > 1$, only information about ψ_n^s, considered as a function of the first variable θ_1, has been obtained. To extend this to the other variables, two properties of the scattering function (13.28),

$$S_2(\theta) = \prod_{r=1}^{R} \frac{1 + ig_r \sinh\theta}{1 - ig_r \sinh\theta}, \qquad g_1, \ldots, g_R > 0, \tag{13.66}$$

are important, which we extract now. Firstly, we see that S_2 is a meromorphic function in the entire complex plane, with certain $(2\pi i)$-periodic sequences of poles. In particular, S_2 is analytic not only in the physical sheet $S(0, \pi)$, but also in the wider strip $S(-\kappa_g, \pi + \kappa_g)$, where $\kappa_g > 0$ is given by

$$\kappa_g := \begin{cases} \arcsin(g^{-1})\,; & g > 1 \\ \frac{\pi}{2}\,; & 0 < g \leq 1 \end{cases}, \qquad g := \max_{r=1,\ldots,R} g_r\,. \tag{13.67}$$

Furthermore, given arbitrary $\kappa \in [0, \kappa_g[$, we note that S_2 is uniformly bounded on $S(-\kappa, \pi + \kappa)$. This bound will be denoted by

$$\sigma(\kappa) := \sup\{|S_2(\zeta)| \,:\, \zeta \in \overline{S(-\kappa, \pi + \kappa)}\} < \infty, \qquad 0 < \kappa < \kappa_g\,. \tag{13.68}$$

These two features of S_2, the analyticity in the enlarged strip and the bound (13.68), will be essential in the following proof of Lemma 1. Furthermore, the parameters α and K appearing there, will be specified explicitly in terms of $\kappa, \sigma(\kappa)$ and n.

Proof of Lemma 1. Fix $\zeta_1 = \theta_1 - i\lambda_1 \in S(-\pi, 0)$. In view of the definitions of $C_s^\pm$ (13.49), $\hat{C}_s^\pm$ (13.53) and h_ξ^s (13.59), it is clear that $\xi \longmapsto h_\xi^s(\zeta_1)$ is an anti-linear functional on $\mathcal{H}_{n-1}$, which by Lemma 3 e) is norm-continuous. Hence, by application of Riesz' Theorem, $h_\xi^s(\zeta_1)$ can be written as

$$h_\xi^s(\zeta_1) =: \sqrt{n} \int d\theta_2 \cdots d\theta_n\, \overline{\xi_{n-1}(\theta_2, \ldots, \theta_n)} \cdot \psi_n^s(\zeta_1, \theta_2, \ldots, \theta_n)\,, \tag{13.69}$$

where $(\theta_2, \ldots, \theta_n) \mapsto \psi_n^s(\zeta_1, \theta_2, \ldots, \theta_n)$ is a function in $\mathcal{H}_{n-1} \subset L^2(\mathbb{R}^{n-1} d^{n-1}\vec{\theta})$ which is defined by this equation. Taking ξ_{n-1} to be the characteristic function of a compact set $\mathcal{C} \subset \mathbb{R}^{n-1}$, and χ_I to be the characteristic function of a compact interval $I \subset \mathbb{R}$, it follows from

$$\sqrt{n} \left| \int\limits_{I \times \mathcal{C}} d^n\vec{\theta}\, \psi_n^s(\theta_1 - i\lambda_1, \theta_2, \ldots, \theta_n) \right| = \left| \int d\theta_1 \overline{\chi_I(\theta_1)} h_\xi^s(\theta_1 - i\lambda_1) \right| < \infty$$

that $\vec{\theta} \longmapsto \psi_n^s(\theta_1 - i\lambda_1, \theta_2, \ldots, \theta_n) =: \psi_{n,\lambda_1}^s(\vec{\theta})$ is locally integrable and hence measurable. According to Lemma 3 e),

$$\int d\theta_2 \cdots d\theta_n \, |\psi_{n,\lambda_1}^s(\vec{\theta})|^2 \leq \frac{2c_n^2 \|A\|^2 \, e^{-ms \sin\lambda_1 \mathrm{ch}\theta_1}}{\pi n \min\{\lambda_1, \pi - \lambda_1\}} \,. \tag{13.70}$$

In view of the measurability of ψ_{n,λ_1}^s, it follows from the estimate (13.70) that for fixed $\lambda_1 \in (0,\pi)$, ψ_{n,λ_1}^s is an element of $L^2(\mathbb{R}^n, d^n\vec{\theta})$. Furthermore, note that $\{n^{-1}c_n^2 \,:\, n \in \mathbb{N}\}$ is bounded. Hence there is a function $b(\lambda_1, s)$, independent of n and A, such that

$$\|\psi_{n,\lambda_1}^s\| \leq b(\lambda_1, s) \cdot \|A\| \,, \qquad 0 < \lambda_1 < \pi, \; s > 0 \,. \tag{13.71}$$

Next we want to show that ψ_{n,λ_1}^s converges to ψ_n^s in the norm topology of $L^2(\mathbb{R}^n, d^n\vec{\theta})$ as $\lambda_1 \to 0$. To this end, we need to improve on the bound (13.71), because $b(\lambda_1, s) \to \infty$ for $\lambda_1 \to 0$ and for $\lambda_1 \to \pi$. Note first that as a consequence of its rapid decrease in the real direction, the Fourier transform of $(h_\xi^s)_{\lambda_1}$ is given by $\beta_1 \mapsto e^{\lambda_1\beta_1}\widetilde{h_\xi^s}(\beta_1)$, $\beta_1 \in \mathbb{R}$ (*cf.* the proof of Lemma 5 in the appendix). In view of (13.69), this relation implies that the Fourier transform of ψ_{n,λ_1}^s is $\mathbb{R}^n \ni \vec{\beta} \mapsto e^{\lambda_1\beta_1} \cdot \widetilde{\psi_n^s}(\vec{\beta})$. Now one can proceed as in the proof of Lemma 5 and apply Lebesgue's dominated convergence theorem to arrive at the norm limit $\psi_{n,\lambda_1}^s \to \psi_n^s$ as $\lambda_1 \to 0$. The limit $\psi_{n,\lambda_1}^s \to \psi_{n,\pi}^s$ as $\lambda_1 \to \pi$ can be established by exploiting the symmetry properties (Lemma 3 b)) of $\hat{C}_s^\pm$, but will not be needed here.

We now turn to the analyticity properties of the wavefunctions and consider a point $\zeta_0 \in S(-\pi, 0)$ with an appropriate curve $\mathcal{C}_0 \subset S(-\pi, 0)$ containing it in its interiour. As h_ξ^s is analytic, we have

$$0 = \oint_{\mathcal{C}_0} d\zeta \, h_\xi^s(\zeta) = \sqrt{n} \oint_{\mathcal{C}_0} d\zeta \int d\theta_2 \cdots d\theta_n \, \overline{\xi_{n-1}(\theta_2, \ldots, \theta_n)} \, \psi_n^s(\zeta, \theta_2, \ldots, \theta_n).$$

But since the integrand is integrable over $\mathcal{C}_0 \times \mathbb{R}^{n-1}$, we may reverse the order of the two integrals and conclude $\oint_{\mathcal{C}_0} d\zeta \, \psi_n^s(\zeta, \theta_2, \ldots, \theta_n) = 0$ for almost all $\theta_2, \ldots, \theta_n \in \mathbb{R}$, *i.e.* ψ_n^s is the boundary value of an analytic function in the first variable if the other variables $\theta_2, \ldots, \theta_n \in \mathbb{R}$ are held fixed.

Consider the symmetry condition (*cf.* (13.5))

$$\psi_n^s(\theta_2, \theta_1, \theta_3, \ldots, \theta_n) = S_2(\theta_1 - \theta_2) \cdot \psi_n^s(\theta_1, \theta_2, \theta_3, \ldots, \theta_n) \tag{13.72}$$

and let $\theta_2, \ldots, \theta_n \in \mathbb{R}$ be fixed. In view of the analyticity of the scattering function in the enlarged strip $S(-\kappa_g, \pi + \kappa_g)$, we see that the right-hand side of (13.72) is the boundary value of a function analytic in its first variable θ_1, in the region $S(-\kappa_g, \pi + \kappa_g) \cap S(-\pi, 0) = S(-\kappa_g, 0)$. Hence the left-hand side can be continued to $S(-\kappa_g, 0)$ as well, and we conclude that ψ_n^s has also an analytic extension in the second variable, to the strip $S(-\kappa_g, 0)$, if the other variables are held fixed. In the same way we see inductively from

$$\psi_n^s(\theta_1, \ldots, \theta_{k+1}, \theta_k, \ldots, \theta_n) = S_2(\theta_k - \theta_{k+1})\psi_n^s(\theta_1, \ldots, \theta_k, \theta_{k+1}, \ldots, \theta_n) \tag{13.73}$$

that ψ_n^s is analytic in each variable $\theta_k \in S(-\kappa_g, 0)$ if the remaining $(n-1)$ variables are held fixed and real. (We neglect here the even stronger analyticity in $\theta_1 \in S(-\pi, 0)$.) By application of the Malgrange-Zerner theorem (for a proof of this theorem see [18]) we conclude that ψ_n^s is analytic (as a function of n complex variables) in the tube region

$$\mathscr{T}_n(\kappa_g) := \mathbb{R}^n - i\mathscr{B}_n(\kappa_g), \tag{13.74}$$

$$\mathscr{B}_n(\kappa_g) := \left\{\vec{\lambda} \in \,]0, \kappa_g\,[^{\times n} :\ 0 < \sum_{k=1}^{n} \lambda_k < \kappa_g\right\}. \tag{13.75}$$

To avoid the divergences due to the poles of S_2 at the boundary of $\mathscr{T}_n(\kappa_g)$, we now fix some $\kappa \in \,]0, \kappa_g]$ and consider the smaller tube $\mathscr{T}_n(\kappa)$ instead of $\mathscr{T}_n(\kappa_g)$. Note that $\mathscr{B}_n(\kappa)$ contains the n-dimensional cube $]0, \frac{\kappa}{n}[^{\times n}$. As

$$(\Delta^\lambda \psi_n^s)(\vec{\theta}) = \psi_n^s(\theta_1 - 2\pi\lambda \cdot i, \ldots, \theta_n - 2\pi\lambda \cdot i)\,, \tag{13.76}$$

the analyticity of ψ_n^s in $\mathscr{T}_n(\kappa)$ implies the claim of Lemma 1 a), and the parameter α appearing there can be chosen as

$$\alpha = \frac{\kappa}{4\pi n}\,. \tag{13.77}$$

Now we use the uniform bound (13.68) to prove part b). The relations (13.72) and (13.71) imply $\|\psi_n^s(\,.\,, \,.\, - i\lambda_2, \ldots)\| \leq \sigma(\lambda_2) b(\lambda_2, s)\|A\|$, and inductively we get from (13.73)

$$\int d^n\vec{\theta}\, |\psi_n^s(\theta_1, \ldots, \theta_k - i\lambda_k, \ldots, \theta_n)|^2 \leq \sigma(\lambda_k)^{2(k-1)} b(\lambda_k, s)^2 \cdot \|A\|^2. \tag{13.78}$$

Thus the analytic continuations of ψ_n^s in each single variable have L^2-norm bounded by $\sigma(\kappa)^{n-1} b(\kappa, s)\|A\|$. This bound can be transported to the interiour of the tube $\mathscr{T}_n(\kappa)$ by using the flat tube theorem, see Lemma 6 in the appendix. We arrive at

$$\int d^n\vec{\theta}\, |\psi_n^s(\vec{\theta} - i\vec{\lambda})|^2 \leq \sigma(\kappa)^{2(n-1)} b(\kappa, s)^2 \cdot \|A\|^2\,, \quad \vec{\lambda} \in \overline{\mathscr{B}_n(\kappa)}\,. \tag{13.79}$$

In particular, the norm bound claimed in Lemma 1 b) follows, and the parameter K can be chosen as

$$K = \sigma(\kappa)^{n-1} \cdot b(\kappa, s)\,. \tag{13.80}$$

To establish the limit $\lim_{|\vec{\theta}| \to \infty} \psi_n^s(\vec{\theta} - i\vec{\lambda}) = 0$, let $\vec{\theta} \in \mathbb{R}^n$, $\vec{\lambda} \in \mathscr{B}_n(\kappa)$ and consider a polydisc $\mathscr{D}_\rho \subset \mathscr{T}_n(\kappa)$ with sufficiently small radius ρ and $\vec{\theta} - i\vec{\lambda} \in \mathscr{D}_\rho$. By the mean value property for analytic functions and Hölder's inequality,

$$\begin{aligned} |\psi_n^s(\vec{\theta} - i\vec{\lambda})|^2 &\leq (\pi\rho^2)^{-n} \int_{\mathscr{D}_\rho} d^n\vec{\zeta}\, |\psi_n^s(\vec{\zeta})|^2 \\ &\leq (\pi\rho^2)^{-n} \int_{[-\rho,\rho]^{\times n}} d^n\vec{\lambda}' \int_{[-\rho,\rho]^{\times n}} d^n\vec{\theta}' |\psi_n^s(\vec{\theta}' + \vec{\theta} + i(\vec{\lambda}' - \vec{\lambda}))|^2\,. \end{aligned}$$

Because of (13.79) the last integral is convergent, also when the integration in $\vec{\theta}'$ is carried out over $\mathbb{R}^n$ instead of $[-\rho, \rho]^{\times n}$. Hence it vanishes in the limit $|\vec{\theta}| \to \infty$.

Finally, c) is a consequence of the earlier discussed strong continuity (Lemma 3 d)) of $[0, \kappa] \ni \lambda_1 \mapsto \psi_{n,\lambda_1}^s$. □

13.5 Summary and Outlook

We have reviewed a novel approach to the construction of quantum field theories with a factorizing S-matrix on two-dimensional Minkowski space. Starting from a scattering function describing the interaction of one type of massive, scalar particles without bound states, in a first step certain auxiliary quantum fields ("polarization-free generators") were constructed. It was shown how to define a covariant net of wedge algebras from these fields. Furthermore, we mentioned that the "correct" two-particle scattering behaviour, namely the one expected from the input scattering function, can be recovered from the wedge-local fields.

In a second step, the local operator content of these wedge local theories has to be analyzed to ensure physically meaningful models. In this context, the modular nuclearity condition constitutes a sufficient criterion for the existence of local observables. In two particular examples, namely the models with constant scattering functions ± 1, this criterion has been verified already.

In the present paper, the modular compactness criterion, a necessary condition for the split property of the wedge net and thus providing an intermediate step in proving the existence of local operators, has been checked in a wide class of models with typical, non-constant scattering functions. In view of these results, it seems reasonable to conjecture that the question of the existence of local observables will have an affirmative answer in the family of models considered.

If this conjecture can be proven, the program reviewed here provides a possibility to rigorously construct interacting quantum field theories in two dimensions, without taking recourse to classical concepts. Although the family of S-matrices considered is limited, it seems possible to generalize the procedure to more complicated models with several kinds of massive particles, ultimately leading to an existence proof for quantum field theories with an arbitrary factorizing S-matrix.

Appendix

In this appendix we prove two lemmata in complex analysis which are used in the main text. The first one is an adaptation of the three lines Theorem [31, Thm 12.8] to the case of L^2-bounds, and the second one shows how to obtain such bounds in the situation of the Malgrange-Zerner theorem. Both statements seem to be well-known; but as we did not find them in this form in the literature, we give their proofs here.

Lemma 5. *Let* $a, b \in \mathbb{R}$, $a < b$,

$$S(a,b) := \{z \in \mathbb{C} \, : \, a < \operatorname{Im}(z) < b\}, \tag{A.1}$$

and let f *denote a function which is analytic in* $S(a,b)$*. Assume that for each* $y \in [a,b]$*, the function* $x \mapsto f(x+iy) =: f_y(x)$ *is an element of* $L^2(\mathbb{R}, dx)$ *and that* $f_y \to f_c$ *for* $y \to c$*, where* $c = a, b$*, in the sense of distributions.*
Then the limits $f_y \to f_c$ *are also valid in the norm topology of* $L^2(\mathbb{R}, dx)$*, and*

$$\|f_y\| \leq \max\{\|f_a\|, \|f_b\|\}, \qquad a \leq y \leq b. \tag{A.2}$$

Proof. Let $\widetilde{g} \in C_0^\infty(\mathbb{R})$. Then g is entire analytic and $x \mapsto g(x+iy) =: g_y(x)$ is of rapid decrease at infinity for fixed $y \in \mathbb{R}$. Let $0 < y < b - a$.

$$\int dp\, \widetilde{f}_{a+y}(-p)\widetilde{g}(p) = \int dx\, f_{a+y}(x)g(x) = \int dx\, f_{a+\varepsilon y}(x)g_{(\varepsilon-1)y}(x).$$

Here we used the rapid decrease of g_y and the analyticity of f and g to shift the integration from $\mathbb{R}$ to $\mathbb{R} + iy(\varepsilon - 1)$, where $0 < \varepsilon < 1$. As $\widetilde{g} \in C_0^\infty(\mathbb{R})$, the limit $\lim_{\varepsilon\to 0} g_{(\varepsilon-1)y} = g_{-y}$ holds in the topology of $\mathscr{S}(\mathbb{R})$. Together with the distributional convergence $f_{a+\varepsilon y} \to f_a$, this implies that the above integral is equal to

$$\int dx\, f_a(x)g(x-iy) = \int dp\, \widetilde{f}_a(-p)e^{yp}\widetilde{g}(p)\,.$$

Hence $\widetilde{f}_{a+y}(p) = e^{-yp}\widetilde{f}_a(p)$. This implies

$$\|f_{a+y} - f_a\|^2 = \int dp\, |\widetilde{f}_a(p)|^2(e^{-yp} - 1)^2\,,$$

and since we have the integrable bound

$$|\widetilde{f}_a(p)|^2(e^{-yp} - 1)^2 \leq |\widetilde{f}_a(p)|^2 \left(4\Theta(p) + \Theta(-p)(1 + e^{-p(b-a)})^2\right),$$

we may use Lebesgue's dominated convergence theorem to conclude $\lim_{y\searrow a} f_y = f_a$ in the norm toplogy of $L^2(\mathbb{R}, dx)$. The limit $\lim_{y\nearrow b} f_y = f_b$ is established in the same manner.

Now let $h \in \mathscr{S}(\mathbb{R})$ be a test function and consider the convolution $f * h$, which is an analytic function in $S(a,b)$. It satisfies the bound, $a < y < b$,

$$|(f * h)(x+iy)| \leq \int dx'\, |h(x')| \cdot |f_y(x-x')| \leq \|h\| \cdot \|f_y\| < \infty\,.$$

But in view of the above established continuity of $[a,b] \ni y \mapsto f_y$, the norm $\|f_y\|$ depends continuously on y, and hence we can find a uniform bound on $|(f*h)(z)|$, $z \in \overline{S(a,b)}$. By the three lines theorem, we conclude

$$\left|\int dx'\, h(x')f_y(x-x')\right| \leq \|h\| \cdot \max\{\|f_a\|, \|f_b\|\}, \qquad a \leq y \leq b\,.$$

As $h \in \mathscr{S}(\mathbb{R})$ was arbitrary and $\mathscr{S}(\mathbb{R})$ is dense in $L^2(\mathbb{R}, dx)$, the claim follows. □

Lemma 6. *Let*

$$\mathscr{B}_n := \left\{\vec{y} \in (0,1)^{\times n} \;:\; 0 < \sum_{j=1}^n y_j < 1\right\}, \qquad \mathscr{T}_n := \mathbb{R}^n + i\mathscr{B}_n \tag{A.3}$$

and consider an analytic function $f : \mathscr{T}_n \to \mathbb{C}$ of n complex variables. Setting $f_{\vec{y}} : \mathbb{R}^n \to \mathbb{C}, \vec{x} \mapsto f(\vec{x} + i\vec{y})$, assume that $f_{\vec{y}} \in L^2(\mathbb{R}^n, d^n\vec{x})$ for any $\vec{y} \in \mathscr{B}_n$, that the map $\vec{y} \mapsto f_{\vec{y}}$ can be extended norm continuously to $\overline{\mathscr{T}_n}$, and that

$$\|f_{(0,\dots,y_k,\dots,0)}\|_2 \leq 1, \qquad 0 < y_k < 1,\; k \in \{1,\dots,n\}, \tag{A.4}$$

holds. Then one has the bound

$$\|f_{\vec{y}}\|_2 \leq 1 \tag{A.5}$$

for any $\vec{y} \in \overline{\mathscr{B}_n}$.

Proof. Let $g \in \mathscr{S}(\mathbb{R}^n)$ be a test function and consider the convolution $f * g$. According to the hypothesis of the lemma, $f * g$ is analytic in $\mathscr{T}_n$ and $\vec{y} \longmapsto (f * g)_{\vec{y}}$ is continuous on $\overline{\mathscr{T}_n}$. On the boundary we have

$$|(f * g)(x_1, \ldots, x_k + iy_k, \ldots, x_n)| \leq \|g\|_2 \|f_{(0,\ldots,y_k,\ldots,0)}\|_2 \leq \|g\|_2. \qquad \text{(A.6)}$$

Now consider $h_g(\vec{z}) := ((f * g)(\vec{z}) - e^{i\alpha}(\|g\| + \varepsilon))^{-1}$, where $\varepsilon > 0$ and $\alpha \in \mathbb{R}$ are arbitrary. Due to the bound (A.6), h_g is, in each variable separately, analytic in the strip $S(0,1)$ if the remaining variables are held fixed and real. By the Malgrange-Zerner Theorem, we conclude that h_g has an analytic continuation, as a function of n complex variables, to the tube $\mathscr{T}_n$. Varying α and letting $\varepsilon \to 0$, we conclude

$$|(f * g)(\vec{x} + i\vec{y})| = |\langle \hat{g}_{\vec{x}}, f_{\vec{y}} \rangle| \leq \|g\|_2 = \|\hat{g}_{\vec{x}}\|_2 \,, \qquad \text{(A.7)}$$

where we have put $\hat{g}_{\vec{x}}(\vec{x}') := \overline{g(\vec{x} - \vec{x}')}$. But as $g \in \mathscr{S}(\mathbb{R}^n)$ was arbitrary and $\mathscr{S}(\mathbb{R}^n)$ is norm dense in $L^2(\mathbb{R}^n, d^n\vec{x})$, the claim (A.5) follows. □

Acknowledgements

Many thanks are due to D. Buchholz for supervision and advice. I also benefitted from conversations with K.-H. Rehren and B. Schroer. Regarding the theory of complex analysis, several discussions with J. Bros and H.-J. Borchers have been very helpful. Financial support by the DFG (Deutsche Forschungsgemeinschaft) is acknowledged.

References

1. H. Araki: *Mathematical Theory of Quantum Fields.* Oxford University Press, New York, 1999.
2. H. Babujian and M. Karowski: *The "Bootstrap Program" for Integrable Quantum Field Theories in* $1+1$ *Dim.* Preprint (2001). [arXiv: hep-th/0110261].
3. B. Berg, M. Karowski and P. Weisz: Construction of Green's functions from an exact S matrix. *Phys. Rev. D* 19:2477 (1979).
4. J.J. Bisognano and E.H. Wichmann: On the duality condition for a hermitian scalar field. *J. Math. Phys.* 16:985 (1975).
5. H.-J. Borchers: The CPT theorem in two-dimensional theories of local observables. *Commun. Math. Phys.* 143:315 (1992).
6. H.-J. Borchers, D. Buchholz and B. Schroer: Polarization-Free Generators and the S-Matrix. *Commun. Math. Phys.* 219:125 (2001). [arXiv: hep-th/0003243].
7. J. Bros: A Proof of Haag-Swieca's Compactness Property for Elastic Scattering States. *Commun. Math. Phys.* 237:289 (2003).
8. R. Brunetti, D. Guido and R. Longo: Modular localization and Wigner particles. *Rev. Math. Phys.* 14:759 (2002). [arXiv: math-ph/0203021].
9. D. Buchholz: Product States for Local Algebras. *Commun. Math. Phys.* 36:287 (1974).
10. D. Buchholz, C. D'Antoni and R. Longo: Nuclear Maps and Modular Structures I: General Properties. *J. Funct. Anal.* 88:233 (1990).

11. D. Buchholz, C. D'Antoni and R. Longo: Nuclear Maps and Modular Structures II: Applications to Quantum Field Theory. *Commun. Math. Phys.* 129:115 (1990).
12. D. Buchholz and P. Jacobi: On the nuclearity condition for massless fields. *Lett. Math. Phys.* 13:313 (1987).
13. D. Buchholz and P. Junglas: On The Existence of Equilibrium States in Local Quantum Field Theory. *Commun. Math. Phys.* 121:255 (1989).
14. D. Buchholz and G. Lechner: Modular Nuclearity and Localization. *Ann. H. Poincaré* 5:1065 (2004). [arXiv: math-ph/0402072].
15. D. Buchholz and E. H. Wichmann: Causal Independence and the Energy-Level Density of States in Local Quantum Field Theory. *Comm. Math. Phys.* 106:321 (1986).
16. O.A. Castro-Alvaredo: *Bootstrap Methods in 1+1-Dimensional Quantum Field Theories: The Homogeneous Sine-Gordon Models.* PhD thesis, 2001. [arXiv: hep-th/0109212].
17. S. Doplicher and R. Longo: Standard and split inclusions of von Neumann algebras. *Commun. Math. Phys* 75:493 (1984).
18. H. Epstein: Some analytic properties of scattering amplitudes in quantum field theory. In: *Particle Symmetries and Axiomatic Field Theory*, Brandeis Summer School 195, Gordon and Breach, New York, 1966.
19. R. Haag: *Local Quantum Physics.* Springer Verlag, Berlin, 2nd ed., 1996.
20. R. Haag and J.A. Swieca: When does a quantum field theory describe particles?. *Commun. Math. Phys.* 1:308 (1965).
21. R.V. Kadison and J.R. Ringrose: *Fundamentals of the Theory of Operator Algebras*, Vol. II. Academic Press, Orlando, 1986.
22. G. Lechner: Polarization-Free Quantum Fields and Interaction. *Lett. Math. Phys.* 64:137 (2003). [arXiv: hep-th/0303062].
23. G. Lechner: On the existence of local observables in theories with a factorizing S-matrix. To appear in *J. of Phys. A* (2005). [arXiv: math-ph/0405062].
24. A. Liguori and M. Mintchev: Fock spaces with generalized statistics. *Commun. Math. Phys.* 169:635 (1995). [arXiv: hep-th/9403039].
25. R. Longo: Notes on algebraic invariants for noncommutative dynamical systems. *Commun. Math. Phys.* 69:195 (1979).
26. B.M. McCoy, C.A. Tracy and T.T. Wu: Two Dimensional Ising Model as an Exactly Solvable Relativistic Quantum Field Theory: Explicit Formulas for n-Point Functions. *Phys. Rev. Lett.* 38:793 (1977).
27. P. Mitra: Elasticity, Factorization and S-Matrices in $(1+1)$-Dimensions. *Phys. Lett. B* 72:62 (1977).
28. M. Müger: Superselection structure of massive quantum field theories in $(1+1)$-dimensions. *Rev. Math. Phys.* 10:1147 (1998). [arXiv: hep-th/9705019].
29. J. Mund: The Bisognano-Wichmann theorem for massive theories. *Annales Henri Poincaré* 2:907 (2001). [arXiv: hep-th/0101227].
30. A. Pietsch: *Nuclear locally convex spaces.* Springer Verlag, Berlin, Heidelberg, New York, 1972.
31. W. Rudin: *Real and complex analysis.* McGraw-Hill Book Company, 1987.
32. S. Sakai: *C^*-Algebras and W^*-Algebras.* Springer Verlag, 1971.
33. B. Schroer: Modular localization and the bootstrap form-factor program. *Nucl. Phys. B* 499:547 (1997). [arXiv: hep-th/9702145].
34. B. Schroer and H.W. Wiesbrock: Modular constructions of quantum field theories with interactions. *Rev. Math. Phys* 12:301 (2000). [arXiv: hep-th/9812251].

35. F.A. Smirnov: *Formfactors in completely integrable models in quantum field theory.* Advanced Series in Mathematical Physics 14, World Scientific, 1992.
36. M. Reed and B. Simon: *Methods of modern mathematical physics I: Functional Analysis.* Revised and enlarged edition, Academic Press, 1980.
37. M. Reed and B. Simon: *Methods of modern mathematical physics II: Fourier Analysis, Self-Adjointness.* Academic Press, 1975.
38. M. Reed and B. Simon: *Methods of modern mathematical physics III: Scattering Theory.* Academic Press, 1979.
39. S.J. Summers: On the Independence of Local Algebras in Quantum Field Theory. *Rev. Math. Phys.* 2:201 (1990).
40. A. Zamolodchikov: Factorized S-matrices as the exact solutions of certain relativistic quantum field theory models. *Ann. Phys.* 120:253 (1979).

14

String-Localized Covariant Quantum Fields

Jens Mund

Instituto de Física, Universidade de São Paulo, CP 66 318, 05315-970 São Paulo, SP, Brazil
mund@fma.if.usp.br

Summary. We present a construction of string-localized covariant free quantum fields for a large class of irreducible (ray) representations of the Poincaré group. Among these are the representations of mass zero and infinite spin, which are known to be incompatible with point-like localized fields. (Based on joint work with B. Schroer and J. Yngvason [13].)

14.1 Introduction

The principles of relativistic quantum physics admit certain "exotic" particle types which do not allow for point-localized quantum fields, namely the massless "infinite spin" representations found by Wigner [17] and Wilczek [19]. However, it is known [8, 12] that all Wigner particle types[1] do allow for localization, in a certain sense, in spacetime regions which extend to infinity in some space-like direction.

In this contribution, we present the construction of free Wightman type fields for the massless "infinite spin" particles, which are localized in semi-infinite strings extending to space-like infinity. This result solves the old problem [6, 18, 20] of reconciling these representations with the principle of causality. It has been obtained in collaboration with B. Schroer and J. Yngvason and partly published in [13]. The details will be presented in [14]. Here, we emphasize the relation with the work of Bros et al. [2], as appropriate for the occasion.

The construction also works for the usual, "non-exotic", particle types. Our motivation to consider such fields, despite the fact that they generate the same algebras as the corresponding point-like localized free fields, is the hope that they may serve as a starting point for the construction of interacting string-localized quantum fields.

Let us make precise what we mean by a string-localized covariant free quantum field for a given particle type.[2] The "*string*" is a ray which extends from a point

[1] By Wigner particle type, we mean here an irreducible unitary ray representation of the Poincaré group with positive energy.

[2] Our notion of a string-localized covariant quantum field is a generalization of the generalized Wightman fields of Steinmann [16].

$x \in \mathbb{R}^d$ to infinity in a space-like direction. That is to say, it is of the form $x + \mathbb{R}^+ e$, where e is in the manifold of space-like directions

$$H^{d-1} := \{e \in \mathbb{R}^d : e \cdot e = -1\}. \tag{14.1}$$

Let now U be a unitary ray representation of the Poincaré group acting on a Hilbert space $\mathcal{H}$ with positive energy and a unique invariant vector Ω, which contains an irreducible ray representation $U^{(1)}$ acting on $\mathcal{H}^{(1)} \subset \mathcal{H}$.

Definition 1. *A* string-localized covariant quantum field *for $U^{(1)}$ is an operator valued distribution $\varphi(x, e)$ over $\mathbb{R}^d \times H^{d-1}$ acting on $\mathcal{H}$ such that the following requirements are satisfied.*

0) Reeh–Schlieder property: *Ω is cyclic for the polynomial algebra of fields $\varphi(f, h)$ with* $\mathrm{supp} f \times \mathrm{supp} h$ *in a fixed region in $\mathbb{R}^d \times H^{d-1}$.*
i) Covariance: *For all $(a, \Lambda) \in \mathcal{P}_+^\uparrow$ and $(x, e) \in \mathbb{R}^d \times H^{d-1}$,*

$$U(a, \Lambda)\varphi(x, e)\, U(a, \Lambda)^{-1} = \varphi(\Lambda x + a, \Lambda e). \tag{14.2}$$

ii) String-locality: *If the strings $x_1 + \mathbb{R}^+ e_1'$ and $x_2 + \mathbb{R}^+ e_2$ are space-like separated for all e_1' in some open neighborhood of e_1, then*

$$[\varphi(x_1, e_1), \varphi(x_2, e_2)] = 0. \tag{14.3}$$

The field is called free, *if it creates only single particle states from the vacuum vector, $\varphi(f, h)\Omega \in \mathcal{H}^{(1)}$.*

Our construction of such fields in [13,14] is reduced to a single particle problem. Namely, consider the single particle vector $\psi(x, e) := \varphi(x, e)\Omega$ if a free field $\varphi(x, e)$ as above is given. It enjoys certain specific properties reflecting the covariance and locality of the field. The crucial point is that these properties are intrinsic to the representation $U^{(1)}$ and can be formulated without reference to the field, using the concept of a modular localization structure [5,8,12] based on Tomita–Takesaki modular theory. We will call an $\mathcal{H}^{(1)}$-valued distribution satisfying the ensuing properties a *string-localized covariant wave function* for $U^{(1)}$, cf. Definition 2. Our strategy is to reverse the route, namely to construct such an $\mathcal{H}^{(1)}$-valued distribution $\psi(x, e)$ for given $U^{(1)}$ and then to obtain the field via second quantization.

The idea of the construction of $\psi(x, e)$ is as follows. Recall that an irreducible representation $U^{(1)}$ of the Poincaré group is induced by a representation D of a subgroup G of the Lorentz group. If V is an extension of D to the Lorentz group, then $U^{(1)}$ is contained in $U_0 \otimes V$, where U_0 is the scalar representation. Thus the problem can be separated. The U_0 part is solved by Fourier transformation. Now Bros et al. exhibit in [2] a suitable representation V, for which they (implicitly) construct a localized covariant wave function living on H^{d-1}. Consider then the tensor product of a wave function localized at x for U_0 and a wave function localized at $e \in H^{d-1}$ for V. Our basic result is that the projection onto $U^{(1)}$ of this vector turns out to be a vector which is localized for $U^{(1)}$ in the string with initial point x and direction e.

We recall the relevant representations $U^{(1)}$ of the Poincaré group and the concept of a modular localization structure in Sections 14.2 and 14.3, respectively. We will concentrate on the bosonic representations with positive mass and for those with zero mass and infinite spin, in dimension $d = 3$ and 4. In Section 14.4 we present the (definition and) construction of a string-localized covariant wave function, as sketched above. In Section 14.5 we summarize our results and give a brief outlook.

14.2 Wigner Particles

Following Wigner [17], the state space of an elementary relativistic particle corresponds to an irreducible ray representation of the Poincaré group with positive energy. We recall the relevant representations here for spacetime dimension $d = 3$ and 4, restricting to proper representations since we are at the moment only interested in bosons. We denote the proper orthochronous Poincaré and Lorentz groups by $\mathcal{P}_+^\uparrow$ and $\mathcal{L}_+^\uparrow$, respectively. Reflecting the semidirect product structure $\mathcal{P}_+^\uparrow = \mathbb{R}^d \rtimes \mathcal{L}_+^\uparrow$, elements of the Poincaré group will be denoted $g = (a, \Lambda)$.

An irreducible positive energy representation $U^{(1)}$ of $\mathcal{P}_+^\uparrow$ is characterized by two data. The first one is the mass value $m \geq 0$, determining the energy-momentum spectrum of the corresponding particle as the mass hyperboloid

$$H_m^+ := \{p \in \mathbb{R}^d : p \cdot p = m^2,\, p_0 > 0\}. \tag{14.4}$$

Given m, one fixes a base point $\bar{p} \in H_m^+$, and considers the stabilizer subgroup, within $\mathcal{L}_+^\uparrow$, of this point. This so-called "little group" will be denoted $G_{\bar{p}}$ in the sequel. Then the second characteristic of $U^{(1)}$ is a unitary irreducible representation V of $G_{\bar{p}}$, acting in a Hilbert space $\mathfrak{h}$. The representation $U^{(1)}$ fixed by these data is said to be induced from D. It acts on

$$\mathcal{H}^{(1)} := L^2(H_m^+, \Delta\mu) \otimes \mathfrak{h}, \tag{14.5}$$

which we identify with $L^2(H_m^+, \Delta\mu; \mathfrak{h})$, according to

$$\big(U^{(1)}(a, \Lambda)(\phi \otimes \varphi)\big)(p) = e^{ia \cdot p}\, \phi(\Lambda^{-1}p)\, D(R(\Lambda, p))\varphi. \tag{14.6}$$

Here, $R(\Lambda, p)$ is the so-called Wigner rotation, defined by

$$R(\Lambda, p) := A_p^{-1}\, \Lambda\, A_{\Lambda^{-1}p}, \tag{14.7}$$

where $(A_p,\, p \in H_m^+)$ is a section of the bundle $\mathcal{L}_+^\uparrow \to H_m^+$, i.e. A_p maps $\bar{p}$ to p.

The little groups $G_{\bar{p}}$ can be conveniently determined as follows. Let

$$\Gamma_{\bar{p}} := \{q \in H_0^+ : q \cdot \bar{p} = 1\}. \tag{14.8}$$

Then $G_{\bar{p}}$ is precisely the (unit component of the) isometry group of $\Gamma_{\bar{p}}$. But $\Gamma_{\bar{p}}$, with the induced metric from ambient Minkowski space, is isometric to the sphere S^{d-2} for $m > 0$, and to $\mathbb{R}^{d-2}$ for $m = 0$. (E.g. for $m = 0$ and $d = 4$, the map $\xi : \mathbb{R}^2 \to \Gamma_{\bar{p}}$, with $\bar{p} = \frac{1}{2}(1, 1, 0, 0)$, defined by

$$\xi(z) := \big(|z|^2 + 1, |z|^2 - 1, z_1, z_2\big) \tag{14.9}$$

can be checked to be an isometric diffeomorphism.) It follows that the little group $G_{\bar{p}}$ is for $m > 0$ isomorphic to $SO(d-1)$, and for $m = 0$ isomorphic to the euclidean group in $d-2$ dimensions, i.e. $G_{\bar{p}} \cong E(2)$ in $d = 4$ and $G_{\bar{p}} \cong \mathbb{R}$ in $d = 3$. Now faithful representations of $E(2)$ are infinite dimensional. Owing to this fact, a representation $U^{(1)}$ resulting from $m = 0$ and a faithful representation of $G_{\bar{p}}$ is called a *massless infinite spin representation.* The faithful representations D of $E(2)$ are labelled by a strictly positive number $\kappa > 0$, and $D = D_{(\kappa)}$ acts on $L^2(\mathbb{R}^2, \delta(|k|^2 - \kappa^2))$ according to

$$\big(D_{(\kappa)}(c, R)\tilde{\varphi}\big)(k) := \exp(ic \cdot k)\, \tilde{\varphi}(R^{-1}k), \quad (c, R) \in E(2). \tag{14.10}$$

14.3 Modular Localization Structure for $U^{(1)}$

As the first step in our construction, the single particle space is endowed with a family of so-called Tomita operators, labelled by a specific class of spacetime regions. This family will be called a modular localization structure for the single particle space.

The basic geometrical ingredient is the family of wedge regions. A *wedge* is a region in Minkowski space which arises by a Poincaré transformation from the "standard wedge"

$$W_0 := \{x \in \mathbb{R}^d : |x^0| < x^1\}.$$

Associated with each wedge W is the one-parameter group of Lorentz boosts $\Lambda_W(t)$ leaving W invariant, and the reflection j_W about the edge of W. More precisely, for the standard wedge W_0 the boosts $\Lambda_{W_0}(t)$ act on the coordinates x^0, x^1 as

$$\begin{pmatrix} \cosh(t) & \sinh(t) \\ \sinh(t) & \cosh(t) \end{pmatrix}, \tag{14.11}$$

and the reflection j_{W_0} inverts the sign of the coordinates x^0, x^1 and leaves the other coordinate(s) invariant. For a general wedge $W = g\,W_0$, $g \in \mathcal{P}_+^\uparrow$, the boosts and reflection are defined as[3]

$$\Lambda_{gW_0}(t) := g\,\Lambda_{W_0}(t)\,g^{-1}, \tag{14.12}$$

$$j_{gW_0} := \quad g\,j_{W_0}\,g^{-1}. \tag{14.13}$$

Let now U be an (anti-) unitary representation of the proper Poincaré group acting in some Hilbert space $\mathcal{H}$. Then there is, in particular, for each wedge W an anti-unitary representer $U(j_W)$ of the reflection j_W. Let further K_W denote the self-adjoint generator of the unitary group representing the corresponding boosts, i.e. K_W is defined by $\exp(itK_W) = U(\Lambda_W(t))$ for all $t \in \mathbb{R}$. Then we define an anti-linear operator associated with W by

$$S_U(W) := \; U(j_W)\exp(-\pi K_W). \tag{14.14}$$

Owing to the group relations, it is an antilinear involution, $S_U(W)^2 \subset 1$, i.e. a so-called Tomita operator.

We now consider the class of causally complete, convex spacetime regions, which we denote by $\mathcal{C}$. It is known [3] that each $C \in \mathcal{C}$ coincides with the intersection of all wedges which contain C. Typical regions belonging to this class are double cones, space-like cones, and wedges. For each $C \in \mathcal{C}$ we now define the subspace of vectors which are "localized in C" by

$$\mathcal{D}_U(C) := \{\psi \in \bigcap_{W \supset C} \mathrm{dom} S_U(W),\; S_U(W)\psi \text{ independent of } W\}. \tag{14.15}$$

Brunetti et al. have shown [3] that if U has positive energy, then $\mathcal{D}_U(C)$ is dense in $\mathcal{H}$ if C contains a space-like cone. On this domain we define a closed anti-linear

---

[3]This definition is consistent because every Poincaré transformation which leaves W_0 invariant commutes with $\Lambda_{W_0}(t)$ and j_{W_0}, cf. [3].

involution $S_U(C)$ by[4]

$$S_U(C)\psi := S_U(W)\psi, \quad W \supset C. \tag{14.16}$$

The family of these anti-linear involutions satisfies isotony [3], $S_U(C_1) \subset S_U(C_2)$ for $C_1 \subset C_2$, and covariance, $U(g)S_U(C)U(g)^{-1} = S_U(gC)$. It has further a property [3] which will soon turn out to correspond to locality:

Lemma 1. *If C_1 and C_2 are causally disjoint, then*

$$S_U(C_1) \subset S_U(C_2)^*. \tag{14.17}$$

Proof. Choose a wedge W which contains C_1 and whose causal complement W' contains C_2. The group relations $\Lambda_{W'}(t) = \Lambda_W(-t)$ and $j_{W'} = j_W$, cf. [3], imply that $S(W') = U(j_W)\exp(\pi K_W)$. On the other hand, $\Lambda_W(t)$ commutes with j_W, hence $U(j_W)\exp(\pi K_W) = \exp(-\pi K_W)U(j_W) \equiv S(W)^*$. Hence $S(W') = S(W)^*$. Therefore

$$S(C_1) \subset S(W) = S(W')^* \subset S(C_2)^*,$$

which proves the claim. □

All these properties motivate us to call the family $S_U(C)$, $C \in \mathcal{C}$, a *modular localization structure*[5] for the representation U. They allow the construction of a local and covariant theory for a given particle type from the single particle space via second quantization as follows. Given the corresponding irreducible representation $U^{(1)}$ of $\mathcal{P}_+^\uparrow$, extend it to $\mathcal{P}_+$ as e.g. in Appendix A, and define $\mathcal{D}(C) = \mathcal{D}_{U^{(1)}}(C)$ as above. Let $a^*(\psi)$ and $a(\psi)$, for $\psi \in \mathcal{H}^{(1)}$, denote the creation and annihilation operators acting on the symmetrized Fock space over $\mathcal{H}^{(1)}$. Then define, for $\psi \in \mathcal{D}(C)$,

$$\Phi(\psi) := a^*(\psi) + a(S(C)\psi). \tag{14.18}$$

These operators generate a covariant and local theory [3], the locality property coming about as follows. For $\psi_1 \in \mathcal{D}(C_1)$, $\psi_2 \in \mathcal{D}(C_2)$, the commutator $[\Phi(\psi_1), \Phi(\psi_2)]$ equals $(S(C_1)\psi_1, \psi_2) - (S(C_2)\psi_2, \psi_1)$. But if C_1 is causally disjoint from C_2, this expression vanishes by Lemma 1, hence

$$[\Phi(\psi_1), \Phi(\psi_2)] = 0 \tag{14.19}$$

in this case. Thus, the property (14.17) of our modular localization structure implies the locality property of the second quantization.

---

[4]We shall skip the index U when no confusion can arise.

[5]Note that our notion of a modular localization structure is equivalent to the usual one, as formulated e.g. in [5,8,12,13]. There, one considers for each $C \in \mathcal{C}$ the real subspace

$$K_U(C) := \bigcap_{W \supset C} \{\phi \in \mathrm{dom} S_U(W) : \; S_U(W)\phi = \phi\}.$$

These are precisely the +1 eigenspaces of our Tomita operators (14.15), (14.16). But these eigenspaces are well-known [8,15] to be in one-to-one correspondence with the latter.

The motivation to construct the modular localization structure from the representation U is the Bisognano-Wichmann theorem [4, 11]. This theorem states that for a large class of local relativistic quantum fields $\varphi(f)$ the so-called modular covariance property holds:

$$S(C)\varphi(f)\Omega = \varphi(f)^*\Omega \quad \text{if } C \supset \operatorname{supp} f,$$

where $S(C)$ is constructed as above, cf. (14.14) to (14.16), from the representation U under which the field is covariant. Thus, given a local quantum field $\varphi(f)$, the vectors $\varphi(f)\Omega$, $\operatorname{supp} f \subset C$, are the prototypes for elements of the subspace $\mathcal{D}_U(C)$.

14.4 String-Localized Covariant Wave Functions

In view of the above discussion, our task of constructing a string-localized covariant free quantum field for a given particle type reduces to the first-quantized version of the problem: Namely, the construction of the spaces $\mathcal{D}(C)$ in terms of "covariant string-localized wave functions" as mentioned in the introduction. These are defined as follows. Let $U^{(1)}$ be the corresponding representation, acting on $\mathcal{H}^{(1)}$.

Definition 2. *A* string-localized covariant wave function *for $U^{(1)}$ is a weak $\mathcal{H}^{(1)}$-valued distribution $\psi(x, e)$ on $\mathbb{R}^d \times H^{d-1}$ satisfying the following requirements.*

0) *The set of $\psi(f, h)$, with* $\operatorname{supp} f \times \operatorname{supp} h$ *in a fixed compact region in $\mathbb{R}^d \times H^{d-1}$, is dense in $\mathcal{H}^{(1)}$.*
i) Covariance: *For all $(a, \Lambda) \in \mathcal{P}_+^\uparrow$ and $(x, e) \in \mathbb{R}^d \times H^{d-1}$,*

$$U^{(1)}(a, \Lambda)\psi(x, e) = \psi(\Lambda x + a, \Lambda e). \tag{14.20}$$

ii) String-locality: *If* $\operatorname{supp} f + \mathbb{R}^+ \operatorname{supp} h \subset C \in \mathcal{C}$, *then $\psi(f, h)$ is in $\mathcal{D}_{U^{(1)}}(C)$.*

Given such $\psi(x, e)$, one verifies that

$$\varphi(x, e) := \Phi(\psi(x, e)), \tag{14.21}$$

with $\Phi(\psi)$ as in (14.18), is a string-localized covariant free quantum field in the sense of Definition 1. (Locality (14.3) follows from (14.19).)

Example 1. To illustrate the concept, we consider the scalar irreducible unitary representation U_0 with mass $m \geq 0$. (Scalar means that the little group is represented trivially.) For $f \in \mathcal{S}(\mathbb{R}^d)$, let Ff denote the restriction of the Fourier transform of f to the mass shell H_m^+. This map enjoys the covariance properties

$$U_0(g)\, Ff = F\, g_* f, \quad g \in \mathcal{P}_+^\uparrow, \tag{14.22}$$

$$U_0(j)\, Ff = F\, j_* \bar{f}, \quad j \in \mathcal{P}_+^\downarrow, \tag{14.23}$$

where $(g_* f)(x) := f(g^{-1}x)$. Further, if f has compact support contained in some wedge W, then

$$\exp(-\pi (K_0)_W) Ff = F (j_W)_* f, \tag{14.24}$$

where $(K_0)_W$ is the generator of $U_0(\Lambda_W(t))$. The basic fact underlying this identity is that for $x \in W$, the analytic function $t \mapsto \Lambda_W(t)x$ has imaginary part in the forward light cone for t in the strip $\mathbb{R} + i(0, \pi)$, and goes to $j_W x$ if t goes to $i\pi$. Lemma 5 in the appendix then implies (14.24). It follows that $S_{U_0}(W)Ff = F\bar{f}$, hence $Ff \in \mathcal{D}_{U_0}(\mathcal{O})$ if $\mathrm{supp} f \subset \mathcal{O}$. Consequently the map $f \mapsto Ff$ is, in analogy to the above definition, a (point-) localized covariant wave function for U_0. Note that the definition (14.21), namely $\varphi(f) := \Phi(Ff)$ with $\Phi(\cdot)$ as in (14.18), then coincides with the usual scalar free field.

We now turn to the construction of a string-localized covariant wave function for arbitrary $U^{(1)}$ with mass $m \geq 0$ and faithful (or scalar) inducing representation D of the little group $G_{\bar{p}}$. Bros et al. exhibit in [2] a family of unitary irreducible representations V^α of the Lorentz group $\mathcal{L}_+$, labelled by a complex number α with real part $-(d-2)/2$.[6] As we show in Lemma 3, the inducing representation D is contained in the restriction of V^α to $G_{\bar{p}}$, namely as a subrepresentation if $m > 0$ and in a direct integral decomposition if $m = 0$. This implies that $U^{(1)}$ is contained in the representation induced by $V^\alpha | G_{\bar{p}}$. But the latter is equivalent to the representation $U_0 \otimes V^\alpha$, hence $U^{(1)}$ is contained in $U_0 \otimes V^\alpha$. More precisely, there is a map R^α from (a dense domain in) the tensor product of the representation spaces of U_0 and V^α into the representation space of $U^{(1)}$ satisfying the intertwiner relation

$$U^{(1)}(a, \Lambda) \circ R^\alpha = R^\alpha \circ U_0(a, \Lambda) \otimes V^\alpha(\Lambda), \quad (a, \Lambda) \in \mathcal{P}_+^\uparrow, \tag{14.25}$$

on its domain. We write down a suitable intertwiner R^α in Lemma 3, which turns out to satisfy also

$$U^{(1)}(j) \circ R^\alpha = R^{\bar{\alpha}} \circ U_0(j) \otimes V^\alpha(j), \quad j \in \mathcal{L}_+^\downarrow. \tag{14.26}$$

Thus, the problem of finding a string-localized wave function can now be separated. For U_0 we already have a localized wave function, cf. Example 1. Now for V^α, Bros et al. [2] construct implicitly a "localized covariant wave function" on H^{d-1}, in the following sense:

Example 2. There is a continuous linear map F^α from $\mathcal{D}(H^{d-1})$ into the representation space of V^α with the following properties:

0) The set of $F^\alpha h$, with $\mathrm{supp} h$ in a fixed region in H^{d-1}, is dense.
i) For $h \in \mathcal{D}(H^{d-1})$,

$$V^\alpha(\Lambda)\, F^\alpha h = F^\alpha \Lambda_* h, \quad \Lambda \in \mathcal{L}_+^\uparrow, \tag{14.27}$$

$$V^\alpha(j)\, F^\alpha h = F^{\bar{\alpha}} j_* \bar{h}, \quad j \in \mathcal{L}_+^\downarrow. \tag{14.28}$$

ii) For a wedge W whose edge contains the origin, let K_W^α be the generator of $V^\alpha(\Lambda_W(t))$. Then for all $h \in \mathcal{D}(H^{d-1})$ with $\mathrm{supp} h \subset W \cap H^{d-1}$, the vector $F^\alpha h$ is in the domain of $\exp(-\pi K_W^\alpha)$, and

$$\exp(-\pi K_W^\alpha)\, F^\alpha h = F^\alpha\, (j_W)_* h. \tag{14.29}$$

(We recall the definition of F^α in the appendix, cf. (14.49), and show in Lemma 4 that the mentioned properties are implicitly contained in [2].)

[6] V^α is in fact equivalent to the irreducible principal series representation corresponding to the value $-|\alpha|^2$ of the Casimir operator.

All this implies that a good candidate for a covariant string-localized wave function in the sense of Definition 2 is given by

$$\psi^\alpha(f,h) := R^\alpha(Ff \otimes F^\alpha h), \tag{14.30}$$

with $f \in \mathcal{D}(\mathbb{R}^d)$ and $h \in \mathcal{D}(H^{d-1})$. We have in fact the following result.

Proposition 1. *Equation* (14.30) *defines a string-localized covariant wave function for* $U^{(1)}$ *in the sense of Definition* 2. *Moreover, for* $C \in \mathcal{C}$ *the Tomita operator* $S(C)$ *acts as follows. Let* $\mathcal{O} \subset \mathbb{R}^d$ *and* $\hat{\mathcal{O}} \subset H^{d-1}$ *be such that* $\mathcal{O} + \mathbb{R}^+\hat{\mathcal{O}} \subset C$. *Then for all* f *with* $\mathrm{supp} f \subset \mathcal{O}$ *and* h *with* $\mathrm{supp} h \subset \hat{\mathcal{O}}$,

$$S(C)\,\psi^\alpha(f,h) = \psi^{\bar{\alpha}}(\bar{f},\bar{h}). \tag{14.31}$$

For $m=0$, $\psi^\alpha(f,h)$ *has the explicit form*

$$\psi^\alpha(f,h)(p) = Ff(p)\,u^\alpha(h,p), \tag{14.32}$$

where $h \mapsto u^\alpha(h,p)$ *is the* $\mathfrak{h}$*-valued distribution on* H^{d-1} *with kernel*

$$u^\alpha(e,p)(k) = \mathrm{e}^{-\mathrm{i}\pi\alpha/2}\int_{\mathbb{R}^{d-2}} \Delta z e^{ikz}\left(\xi(z)\cdot A_p^{-1}e\right)^\alpha. \tag{14.33}$$

Here $z \mapsto \xi(z)$ *is the isometry from* $\mathbb{R}^{d-2}$ *onto* $\Gamma_{\bar{p}}$ *exhibited in* (14.9).

Proof. We first consider $m>0$, in which case the intertwiner R^α is a partial isometry defined on the whole Hilbert space, cf. Lemma 3. Then the covariance condition *i)* of Definition 2 is satisfied by construction, cf. (14.22), (14.25) and (14.27). The "Reeh-Schlieder" property *0)* of Definition 2 follows from the well-known Reeh–Schlieder property of Ff and that of $F^\alpha h$, cf. *0)* of Example 2. It remains to prove (14.31), which then implies the locality property *ii)*. As a first step, let f and h be such that $\mathrm{supp} f + \mathbb{R}^+\mathrm{supp} h$ is contained in the standard wedge W_0. It then follows that $\mathrm{supp} f$ is contained in W_0 and $\mathrm{supp} h$ in its *closure*. Suppose first that $\mathrm{supp} h \subset W_0$. Then from (14.24) and (14.29) we know that the vectors Ff and $F^\alpha h$ are in the domains of the corresponding "modular operators" $\exp(-\pi K_{W_0})$ and that the latter maps them to $F(j_0)_* f$ and $F^\alpha(j_0)_* h$, respectively. From the intertwining property (14.25) of R^α and its continuity it follows (e.g. using Lemma 5) that $\psi^\alpha(f,h)$ is in the domain of $\exp(-\pi K_{W_0})$ and that

$$\exp(-\pi K_{W_0})\,\psi^\alpha(f,h) = \psi^\alpha\big((j_0)_* f,(j_0)_* h\big). \tag{14.34}$$

Further, (14.23), (14.26) and (14.28) imply that $U^{(1)}(j)\,\psi^\alpha(f,h) = \psi^{\bar{\alpha}}(j_*\bar{f}, j_*\bar{h})$ for $j \in \mathcal{L}_+^\downarrow$. Now the last two equations imply that

$$S(W_0)\,\psi^\alpha(f,h) = \psi^{\bar{\alpha}}(\bar{f},\bar{h}). \tag{14.35}$$

If, on the other hand, $\mathrm{supp} h$ meets the boundary of W_0 (but is contained in its closure), then one finds a sequence $h_n \to h$ so that $\mathrm{supp} h_n \subset W_0$ for all n. Then $S(W_0)\psi^\alpha(f,h_n)$ goes to $\psi^{\bar{\alpha}}(\bar{f},\bar{h})$, and (14.35) also holds in this case because $S(W_0)$ is closed. By covariance, it follows that for any wedge W, the operator $S(W)$ acts as in (14.35) if $\mathrm{supp} f + \mathbb{R}^+\mathrm{supp} h \subset W$. This proves (14.31). The continuity property follows from that of F, F^α and R^α. The proof is complete for $m>0$.

For $m = 0$, we show in [14] the following facts. R^α is well-defined on vectors of the form $Ff \otimes F^\alpha h$, leading to the formula (14.32), (14.33), and the intertwining properties (14.25) and (14.26) hold on these vectors. Further, if h has support in a wedge W, then for almost all p the $\mathfrak{h}$-valued function $t \mapsto u^\alpha(\Lambda_W(t)_* h, p)$ is analytic on the strip $\mathbb{R} + i(0, \pi)$ and weakly continuous on its closure. It is uniformly bounded in p and t, for p in a dense set of H_0^+ and for t in any compact subset of the closure of the strip. As t goes to $i\pi$, it goes to $u((j_W)_* h, p)$. This implies (14.34), e.g. using Lemma 5 (details are spelled out in [14]). The proof of (14.31) is then completed as in the case $m > 0$. Finally, we show in [14] an analyticity and growth property in e of $u^\alpha(e, p)$ which implies continuity of $\psi^\alpha(f, h)$. □

14.5 Summary and Outlook

We have constructed, for each $\alpha \in \mathbb{C}$ with $\mathrm{Re}\,\alpha = -(d-2)/2$ and each massless "infinite spin" representation $U^{(1)}$, a $\mathcal{H}^{(1)}$-valued distribution on $\mathbb{R}^d \times H^{d-1}$ with certain specific properties, which motivate our name "string-localized covariant wave function". They guarantee that second quantization (14.21) of these objects leads to a string-localized covariant free quantum field, cf. Definition 2 and discussion thereafter. Summarizing, and using the explicit formula (14.32), we have as our main result:

Theorem 1. *Let $\varphi(x, e)$ be the operator-valued distribution given by*[7]

$$\varphi^\alpha(x, e) = \int_{H_m^+} \Delta\mu(p) \left\{ e^{ip\cdot x}\, u^\alpha(e, p) \circ a^*(p) + e^{-ip\cdot x}\, \overline{u^{\bar{\alpha}}(e, p)} \circ a(p) \right\}, \quad (14.36)$$

with u^α as in (14.33). Then $\varphi(x, e)$ is a string-localized covariant free quantum field for $U^{(1)}$ in the sense of Definition 1.

It turns out [14] that the formula works for all $\alpha \in \mathbb{C} \setminus \mathbb{N}_0$, and that for a certain range of values the fields need not be smeared in the directional variable e. It also works with u^α and $u^{\bar{\alpha}}$ replaced by $F(p\cdot e) u^\alpha(e, p)$ and $\overline{F(-p \cdot e)} u^{\bar{\alpha}}(e, p)$, respectively, where F is the distributional boundary value of a suitable function which is analytic on the upper half plane. The resulting fields are all relatively "string-local" to each other. It is shown in [14] that every string-localized covariant free field, in the sense of Definition 1, is of the above form.

An important open problem for our infinite spin fields is the existence of local observables. These are operators which are localized in bounded regions, in the sense that they commute with field operators localized causally disjoint from the respective region. In this sense, a local expression for the energy density is of particular interest, since it would be valuable for a discussion of the thermodynamic properties of the KMS states [13] of our fields.

We have perfomed the above construction also for massive bosons with arbitrary spin, and similar constructions work for fermions with half-integer spin and for photons [14]. Our photon field $A_\mu(x, e)$ is a string-localized covariant version of the "axial gauge", acting on the physical photon Hilbert space. The resulting fields in

[7]We use the symbolic notation $a^*(\psi) =: \int d\mu(p)\, \psi(p) \circ a^*(p)$ and $a(\psi) =: \int d\mu(p)\, \overline{\psi(p)} \circ a(p)$.

all these cases are strictly string-localized, but relatively local to the corresponding standard point-localized free fields. (In fact, they can be written as certain line integrals over the latter [14].)

The reason why these fields nevertheless have the potential for applications is that they might serve as ingredients for the construction of interacting models with string-like localization. Recall that the results of [4, 8] support the viewpoint that localization of charged quantum fields in space-like cones (the idealizations of which are our strings) is a natural concept, yet there is so far a lack of rigorous model realizations[8]. There are two reasons to believe that our free fields are good starting points for a construction of interacting fields with strict string-localization. Firstly, since the obstruction to point-like localization is due to the charge, which is already carried by the single particle states, one should expect that already the latter are strictly string-localized. That is to say, the single particle states $E^{(1)}\varphi\Omega$, where $E^{(1)}$ denotes the projection onto the single particle space and φ is an interacting field, are string-like (but not point-like) localized in the sense of (14.15). But then the LSZ relations imply that the corresponding incoming and outgoing free fields are also strictly string-localized. Therefore, our fields might represent the in- and out-fields of such a model, in contrast to the usual point-localized free fields. Secondly, the distributional character of our free fields is less singular than that of the point-localized free fields, as is made precise in [14], even more so in the direction of the localization string. This fact should lead to a larger class of admissable interactions in a perturbative construction, as compared to taking the standard point-localized free fields as starting point.

Acknowledgements

This work has been supported by FAPESP. I also wish to thank the organizers and the sponsors!

A Extension of the Representations to $\mathcal{P}_+$

The proper Poincaré group $\mathcal{P}_+$ is generated by $\mathcal{P}_+^\uparrow$ and any single element j_0 in $\mathcal{P}_+^\downarrow$. We choose

$$j_0 := jw_0 . \tag{14.37}$$

(Note that $-j_0$ is in $\mathcal{P}_+^\uparrow$ and hence leaves each mass shell H_m^+, $m \geq 0$, invariant.) As to the irreducible representation $U^{(1)}$, we choose the base point $\bar{p} \in H_m^+$ so that

$$-j_0\bar{p} = \bar{p}. \tag{14.38}$$

Then the section $p \mapsto A_p$ of the bundle $\mathcal{L}_+^\uparrow \to \mathcal{L}_+^\uparrow/G_{\bar{p}} = H_m^+$ can be chosen [14] so that it transforms under the adjoint action of j_0 as

$$j_0 A_p j_0 = A_{-j_0 p}. \tag{14.39}$$

[8] apart from non-Lorentz covariant infra-vacua models as in [7] and lattice models as in [9, 10].

Let $D(j_0)$ be the anti-unitary involution from Lemma 2 below. Then, by virtue of (14.39) and (14.41), the anti-unitary involution defined by

$$(U^{(1)}(j_0)\psi)(p) = D(j_0)\,\psi(-j_0 p) \tag{14.40}$$

extends $U^{(1)}$ to an (anti-) unitary representation of $\mathcal{P}_+$ within the same Hilbert space $\mathcal{H}^{(1)} = L^2(H_m^+, \Delta\mu) \otimes \mathfrak{h}$. Due to irreducibility, $U^{(1)}(j_0)$ is fixed up to a phase factor.

Lemma 2. *There is an anti-unitary involution $D(j_0)$ acting on $\mathfrak{h}$ satisfying the representation properties*

$$D(j_0)^2 = 1 \quad \textit{and} \quad D(j_0)D(\Lambda)D(j_0) = D(j_0\Lambda j_0), \quad \Lambda \in G_{\bar{p}}. \tag{14.41}$$

The existence of such a representer is established in Lemma 3. Note that the adjoint action of j_0 leaves $G_{\bar{p}}$ invariant due to (14.38), hence the lemma states that D extends to a representation of the subgroup of $\mathcal{P}_+$ generated by $G_{\bar{p}}$ and j_0.

B Intertwiners and Localization Structure for the Principal Series Representations

We recall the representation of the Lorentz group presented by Bros et al. in [2]. Fix a complex number α with real part $-(d-2)/2$. Let H_0^+ denote the mantle of the forward light cone in $\mathbb{R}^d$ as before, and let $C^\alpha(H_0^+)$ denote the space of continuous $\mathbb{C}$-valued functions on H_0^+ which are homogenous of degree α, i.e.

$$C^\alpha(H_0^+) := \{\psi \in C(H_0^+) : \psi(tp) = t^\alpha\,\psi(p),\ t > 0\}.$$

Consider the maps $V^\alpha(\Lambda)$, $\Lambda \in \mathcal{L}_+$, defined on $C(H_0^+)$ by

$$(V^\alpha(\Lambda)\psi)(p) := \psi(\Lambda^{-1}p)\,, \quad \Lambda \in \mathcal{L}_+^\uparrow\,, \tag{14.42}$$

$$(V^\alpha(j)\psi)(p) := \overline{\psi(-jp)}\,, \quad j \in \mathcal{L}_+^\downarrow. \tag{14.43}$$

Clearly, $V^\alpha|\mathcal{L}_+^\uparrow$ establishes a representation of $\mathcal{L}_+^\uparrow$ in $C^\alpha(H_0^+)$, while $V^\alpha|\mathcal{L}_+^\downarrow$ maps $C^\alpha(H_0^+)$ onto $C^{\bar{\alpha}}(H_0^+)$, and the pair V^α, $V^{\bar{\alpha}}$ satisfies the following representation property:

$$V^{\bar{\alpha}}(j_1)V^{\bar{\alpha}}(\Lambda)\,V^\alpha(j_2) = V^\alpha(j_1\Lambda j_2), \quad \Lambda \in \mathcal{L}_+^\uparrow, j_k \in \mathcal{L}_+^\downarrow. \tag{14.44}$$

Let now Γ be any $(d-2)$-dimensional cycle which encloses the origin. Then $C^\alpha(H_0^+)$ can (and will) be identified with $C(\Gamma)$. Let $\Delta\nu_\Gamma$ be the restriction of the Lorentz invariant measure $\Delta\nu$ on H_0^+ to Γ, and define a scalar product on $C(\Gamma)$ by

$$(\psi, \psi') := \int_\Gamma \overline{\psi(p)}\,\psi'(p)\, d\nu_\Gamma(p). \tag{14.45}$$

As Bros and Moschella point out [2], the representation V^α of the Lorentz group is unitary w.r.t. this scalar product. The corresponding Hilbert space completion of $C(\Gamma)$ will be denoted by $\mathfrak{h}'$, and the extension of V^α to this space will be denoted by the same symbol. It is equivalent to the irreducible principal series representation corresponding to the value $-|\alpha|^2$ of the Casimir operator [2].

Lemma 3. *i) Let D be a faithful irreducible representation of $G_{\bar{p}}$, and let $\mathrm{Re}\,\alpha = -(d-2)/2$. Then $V^\alpha|G_{\bar{p}}$ contains D, i.e. there is a map T from a dense domain in $\mathfrak{h}'$ onto a dense subspace of $\mathfrak{h}$ which intertwines the representations $V^\alpha|G_{\bar{p}}$ and D in the sense that*

$$D(\Lambda)\circ T = T\circ V^\alpha(\Lambda), \quad \Lambda\in G_{\bar{p}}, \tag{14.46}$$

holds on its domain. In the case $m>0$, D is a subrepresentation of V^α, while for $m=0$, D occurs in a direct integral decomposition of V^α. T also intertwines $V^\alpha(j_0)$, in the sense of (14.46), with an anti-unitary operator $D(j_0)$ satisfying the representation properties (14.41).

ii) Let R^α be the map from (a dense domain in) $L^2(H_m^+)\otimes\mathfrak{h}'$ into $\mathcal{H}^{(1)} = L^2(H_m^+)\otimes\mathfrak{h}$ defined by

$$\big(R^\alpha(\phi\otimes\varphi)\big)(p) := \phi(p)\, T V^\alpha(A_p^{-1})\varphi. \tag{14.47}$$

Then R^α satisfies on its domain the intertwiner relations (14.25) and (14.26).

Proof. Ad *i*). We choose the cycle Γ conveniently as $\Gamma := \Gamma_{\bar{p}}$ defined in (14.8).[9] As mentioned, the cycle $\Gamma = \Gamma_{\bar{p}}$ is isometric to the sphere S^{d-2} for $m>0$, and to $\mathbb{R}^{d-2}$ for $m=0$, and its isometry group coincides with $G_{\bar{p}}$. Hence the action of $G_{\bar{p}}$ on Γ corresponds to the natural action of $SO(d-1)$ on S^{d-2} for $m>0$, and to the natural action of $E(d-2)$ on $\mathbb{R}^{d-2}$ for $m=0$. It also follows that the invariant measure $\Delta\nu_\Gamma$ goes over into the $SO(d-2)$ invariant measure $\Delta\Omega$ on S^{d-2} or the Lebesgue measure Δz on $\mathbb{R}^{d-2}$, respectively. In the case $m>0$, it follows that $\mathfrak{h}'$ is naturally isomorphic to $L^2(S^{d-2},\Delta\Omega)$, and $V^\alpha|G_{\bar{p}}$ acts as the push-forward representation. As is well-known, this representation decomposes into the direct sum of all irreducible representations $D_{(s)}$ of $SO(d-1)$. (In $d=4$, s runs through $\mathbb{N}_0$ and the irreducible subspaces are spanned by the spherical harmonics $Y_{s,m}$, and in $d=3$, s runs through $\mathbb{Z}$ and the irreducible subspaces are spanned by $\theta\mapsto\exp(is\theta)$.) Hence, for each s there is a partial isometry $T=T_{(s)}$ with the claimed property (14.46). Further, under the mentioned equivalence $\Gamma\cong S^{d-2}$ the representer of j_0 acts as $(V^\alpha(j_0)\varphi)(n) = \overline{\varphi(I_0 n)}$, where I_0 corresponds to $-j_0$ and is hence in $O(d-1)$. Since the spherical harmonics $\{Y_{s,m}, m=-s,\dots,s\}$ for given $s\in\mathbb{N}$ are invariant under $O(3)$ as well as under complex conjugation, it follows that $V^\alpha(j_0)$ leaves each irreducible subrepresentation invariant in d=4. This implies that $V^\alpha(j_0)$ is intertwined by T with an (anti-unitary) operator $D(j_0)$ satisfying (14.41), as claimed. In d=3, the same conclusion follows from the facts that I_0 is an orientation reversing isometry of the circle, hence $SO(2)$-conjugate to $\theta\mapsto-\theta$, and that $\overline{\exp(is(-\theta))} = \exp(is\theta)$.

Similarly, in the case $m=0$, $\mathfrak{h}'$ is naturally isomorphic to $L^2(\mathbb{R}^{d-2},\Delta z)$, and $V^\alpha|G_{\bar{p}}$ acts as the push-forward representation. Via Fourier transformation, this representation decomposes into a direct integral of irreducible representations $D_{(\kappa)}$, where κ runs through $\mathbb{R}$ for $d=3$, and through $\mathbb{R}^+$ for $d=4$. Thus there is a densely defined intertwiner T satisfying (14.46) on its domain: $T\varphi$ is the restriction of the Fourier transform of φ to the circle with radius κ for $d=4$, respectively its value at κ for $d=3$. Further, under the mentioned equivalence $\Gamma\cong\mathbb{R}^{d-2}$ the representer of j_0 acts as $(V^\alpha(j_0)\varphi)(z) = \overline{\varphi(I_0 z)}$, where I_0 corresponds to $-j_0$. With

[9] Note that $\Gamma_{\bar{p}}$ is invariant under $G_{\bar{p}}$ and $-j_0$, which implies that the restriction of V^α to the subgroup of $\mathcal{L}_+$ generated by $G_{\bar{p}}$ and j_0 does not depend on α.

our explicit formula (14.9), I_0 coincides with the reflection $z \mapsto -z$.[10] The identity $TV^\alpha(j_0)\varphi = \overline{T\varphi}$ then implies that $V^\alpha(j_0)$ leaves the kernel of T invariant. Hence

$$D(j_0)T\varphi := TV^\alpha(j_0)\varphi,$$

defines an anti-unitary operator $D(j_0)$ on the image of T, which also has the representation property (14.41), as claimed.

Ad *ii*). The intertwiner relations (14.25) and (14.26) follow from part *i*), (14.39) and (14.44). □

We now discuss the map F^α defined by Bros et al. [2], which we used in Example 2. It is the Fourier-Helgason type transformation given by

$$F^\alpha : \mathcal{D}(H^{d-1}) \to C^\alpha(H_0^+) \subset \mathfrak{h}', \tag{14.48}$$

$$(F^\alpha h)(p) := \mathrm{e}^{-i\pi\alpha/2} \int_{H^{d-1}} \Delta\sigma(e)\, h(e)\, (e \cdot p)^\alpha. \tag{14.49}$$

Here, $e \cdot p$ denotes the scalar product in d-dimensional Minkowski space, of which H^{d-1} and H_0^+ are considered submanifolds. The power t^α is defined via the branch of the logarithm on $\mathbb{R} \setminus \mathbb{R}_0^-$ with $\ln 1 = 0$, and as $\lim_{\varepsilon \to 0+}(t + i\varepsilon)^\alpha$ for $t < 0$. Further, $\Delta\sigma$ denotes the Lorentz invariant measure on H^{d-1}. In our context, the upshot of this transformation is the following.

Lemma 4 (Bros et al.). *The map $h \mapsto F^\alpha h$ establishes a "localized covariant wave function" on H^{d-1} in the sense of the properties* 0) ... ii) *listed in Example* 2.

Proof. The transformation F^α has been taken over from [2] in such a way that $F^\alpha h = \phi(h)\Omega$, where $\phi(\cdot)$ is the free field of [2], cf. [2, eq. (4.30)]. In this context, property 0) of our Example 2 is the Reeh–Schlieder property, Proposition 5.4 of [2]. The covariance property i) corresponds to the covariance of the field $\phi(\cdot)$ (and also follows directly from the definitions). Finally, the geometrical KMS condition [2, Prop. 2.3] enjoyed by the two-point function of $\phi(\cdot)$ implies that $F^\alpha h$ is in the domain of the Tomita operator $\exp(-\pi K_W^\alpha)$. The antipodal condition [2, Prop. 2.4] then shows that this operator acts on $F^\alpha h$ as in (14.29). This proves ii). □

We finally mention a standard result, which we have used occasionally in the context of our modular operators.

Lemma 5. *Let U_t be a continuous unitary one-parameter group, with generator K. Then ψ is in the domain of* $\exp(-\pi K)$ *if, and only if, the vector-valued map*

$$t \mapsto U_t\psi$$

is analytic in the strip $\mathbb{R} + i(0,\pi)$ and weakly continuous on the closure of that strip. In this case, $\exp(-\pi K)\psi$ *coincides with the analytic continuation of $U_t\psi$ into $t = i\pi$.*

---

[10] Using another isometric diffeomorphism yields the same I_0 up to conjugation with a euclidean transformation, leading to the same conclusion.

References

1. J.J. Bisognano and E.H. Wichmann: On the duality condition for a Hermitean scalar field. *J. Math. Phys.* 16:985 (1975).
2. J. Bros and U. Moschella: Two-point functions and quantum fields in de Sitter universe. *Rev. Math. Phys.* 8:324 (1996).
3. R. Brunetti, D. Guido and R. Longo: Modular localization and Wigner particles. *Rev. Math. Phys.* 14:759–786 (2002).
4. D. Buchholz and K. Fredenhagen: Locality and the structure of particle states. *Commun. Math. Phys.* 84:1–54 (1982).
5. L. Fassarella and B. Schroer: Wigner particle theory and local quantum physics. *J. Phys. A* 35:9123–9164 (2002).
6. G.J. Iverson and G. Mack: Quantum fields and interactions of massless particles: The continuous spin case. *Ann. Phys.* 64:211–253 (1971).
7. Kunhardt: On infravacua and the localisation of sectors. *J. Math. Phys.* 39:6353–6363 (1998).
8. P. Leyland, J. Roberts and D. Testard: *Duality for quantum free fields.* Unpublished notes, 1978.
9. M. Lüscher: Bosonization in $2+1$ dimensions. *Nucl. Phys. B* 326:557–582 (1989).
10. V.F. Müller: Intermediate statistics in two space dimensions in a lattice-regularized Hamiltonian quantum field theory. *Z. Phys. C* 47:301–310 (1990).
11. J. Mund: The Bisognano-Wichmann theorem for massive theories. *Ann. H. Poincaré.* 2:907–926 (2001).
12. J. Mund: Modular localization of massive particles with "any" spin in $d = 2+1$. *J. Math. Phys.* 44:2037–2057 (2003).
13. J. Mund, B. Schroer and J. Yngvason: String-localized quantum fields from Wigner representations. *Phys. Lett. B* 596:156–162 (2004).
14. J. Mund, B. Schroer and J. Yngvason: *String-localized quantum fields and modular localization.* In preparation.
15. M.A. Rieffel and A. Van Daele: A bounded operator approach to Tomita-Takesaki theory. *Pacific J. Math.* 69 no. 1:187–221 (1977).
16. O. Steinmann: A Jost-Schroer Theorem for String Fields. *Commun. Math. Phys.* 87:259–264 (1982).
17. E.P. Wigner: On unitary representations of the inhomogeneous Lorentz group. *Ann. Math.* 40:149 (1939).
18. E.P. Wigner: Relativistische Wellengleichungen. *Z. Physik* 124:665–684 (1948).
19. F. Wilczek: Quantum mechanics of fractional-spin particles. *Phys. Rev. Lett.* 49:957–1149 (1982).
20. D. Yngvason: Zero-mass infinite spin representations of the Poincaré group and quantum field theory. *Commun. Math. Phys.* 18:195–203 (1970).

15

Quantum Anosov Systems

Heide Narnhofer

Institut für Theoretische Physik, Universität Wien
Boltzmanngasse 5, A-1090 Wien
narnh@ap.univie.ac.at

Summary. The concept of Anosov flows and of Kolmogorov systems can be translated from classical to quantum systems. It is shown that modifications of the concepts are necessary to keep the same clustering behavior as is typical for classical Anosov systems. With such modifications, Anosov structure appears rather naturally in a type III_1 algebra. Here Anosov structure and Kolmogorov structure with respect to modular evolution are even equivalent. The Rindler wedge of quantum field theory offers a typical example.

15.1 Introduction

The idea that thermodynamics and an approach to equilibrium has to be understood on the basis of statistics, as has been emphasized by Boltzmann, is now common belief. However we have not been successful in proving the concepts in mathematically rigorous and at the same time physically realistic models. Nevertheless it has motivated us to study and characterize in detail the long time behavior of evolutions. In fact ergodic theory has become a rich and fruitful discipline in the classical framework. Concepts such as dynamical entropy and Lyapunov exponents have been developed and can be used to describe approaches to an invariant state. They can be studied in various models and enable us to study long time behavior in an efficient way.

There have been many attempts to translate these concepts to quantum systems. One path is followed in quantum chaology : one starts with a classical evolution in a finite system and studies how the spectral behavior of its quantum counterpart, and especially correlations between eigenvalues, reflect ergodic properties of the classical system. Concepts such as dynamical entropy and Lyapunov exponents however cannot be translated. The other possibility is to start with these most powerful concepts of classical theory and to try to translate them to quantum systems. It turns out that it is essential to leave the type I algebras, that describe few nonrelativistic particles and move to the type II and III cases. Only here is it possible to obtain powerful results on convergence properties of correlation functions. This is of course promising in understanding how statistics and thermodynamics can be related, because thermodynamics and infinite systems offer us type III algebras in which we can hope that in realistic models ergodic concepts are realized.

But it also turns out that small variations in the starting point lead to strong structural differences. This makes it even more important that it has been possible to observe the relevant ergodic behavior in quantum field theory and therefore enables us to test how we have to modify relevant questions in the context of ergodic quantum theory.

15.2 The classical system

We start with a measure space on which time evolution is described as a measure preserving action that can be either discrete or continuous. Already in [16] it was pointed out that we can formulate the problem in the language of abelian algebras. On this algebra the invariant measure defines an invariant state and we are concerned with the behavior of $\omega(A\tau_t B)$ and its convergence to $\omega(A)\omega(B)$. As a starting point the convergence can be read off from the spectral properties of the unitary U_t that implements the time evolution in the Koopman formalism [11]. If U_t has absolutely continuous spectrum apart from the nondegenerate eigenvalue 1 that corresponds to the (unique) invariant state, then convergence is guaranteed, but this statement is still very rough. We want to control the rate of convergence and its dependence on the operators A and B. Here we have two descriptions:

A.) Kolmogorov systems: These are systems $(\mathcal{A}, \mathcal{A}_0, \tau_t, \omega)$, where τ_t is an automorphism on the algebra $\mathcal{A}$, ω a τ_t invariant state over $\mathcal{A}$ and $\mathcal{A}_0$ a subalgebra such that

$$\tau_t \mathcal{A}_0 = \mathcal{A}_t \subset \mathcal{A}_0 \quad \forall t < 0, \quad \{\bigcup_t \mathcal{A}_t\}'' = \mathcal{A}, \quad \bigcap_t \mathcal{A}_t = c1. \tag{15.1}$$

In a K-system the time evolution is K-clustering in the following sense: For a given B and given ϵ there exists a t_0 such that

$$|\omega(A\tau_t B) - \omega(A)\omega(B)| \leq \epsilon ||A|| \quad \forall t > t_0. \tag{15.2}$$

In a Kolmogorov system U_t has, apart from a simple point spectrum 1, infinitely degenerate absolutely continuous spectrum. The algebra $\mathcal{A}_0$ can be found by starting with a finite dimensional subalgebra (a finite partition of the measure space) $\mathcal{B}$ and constructing $\mathcal{A}_0 = (\bigcup_{t \in R^+} \tau_{-t}\mathcal{B})''$.

B.) Anosov systems [2]: The measure space has in addition the structure of a hyperbolic manifold. The geodesic flow defines a continuous time evolution τ_t. In addition there exist two other automorphism groups $\sigma^{\pm}$, the horocyclic actions, that satisfy

$$\tau_t \sigma_s^{\pm} \tau_{-t} = \sigma_{e^{\pm\lambda t} s}. \tag{15.3}$$

These automorphism groups $\sigma^{\pm}$ permit a foliation of the manifold: in one direction the time evolution acts as expansion, in the other as contraction, and the amount of expansion and contraction is uniform over the manifold. The foliation helps to control the refinement of a given partition (preferably a partition respecting the foliation) in such a way that it can be shown that every partition has a trivial tail. Therefore every Anosov system is also a Kolmogorov system. In addition we can control the clustering properties also via

$$\limsup_{t\to\infty} \frac{|\omega(A\tau_t B) - \omega(A)\omega(B)|}{e^{-\lambda t}} < \infty. \tag{15.4}$$

These clustering properties are used in numerical calculations to characterize the ergodic behavior of an evolution.

Evidently the assumption of two automorphism groups τ_t and σ_s that satisfy the Anosov commutation relations (15.3) is not restricted to abelian algebras. Also the definition of a Kolmogorov system is not restricted to the abelian situation [13]. What remains is the question whether also in the noncommutative case we can draw conclusions on the clustering properties and on a Kolmogorov structure.

15.3 Spectral properties and clustering properties

We do not make any assumptions on the underlying algebra but just assume that τ_t and σ_s are strongly continuous. Then the two spectra of $U_t, \tau_t = adU_t$, $V_s, \sigma_s = adV_s$ are determined by the Anosov relation. The Hilbert space can be written as $\mathcal{H} = |\Omega> \oplus (L^2(R^+, dx) \times L^2(y, d\mu(y)))$ Here $|\Omega>$ is the state invariant both under U_t and V_s. In the remaining space U_t, V_s act as

$$V_s f(x,y) = e^{ixs} f(x,y), \quad U_t f(x,y) = e^{t/2} f(e^t x, y). \tag{15.5}$$

We can use the Anosov structure in various ways to characterize the clustering behavior.

Theorem 1. [14] *Let* $A(f) = \int_0^\infty ds f(s)\sigma_s A$, $f = F'$, $F(0) = 0$, $||F||_1 < \infty$, $||F'||_1 < \infty$, $||F''||_1 < \infty$. *Then:*

a) *The set of these operators is dense in the algebra.*
b) $\delta A(f) = \frac{d}{ds}\sigma_s A(f)|_{s=0}$ *exists and is norm bounded by* $||F''||_1$.
c) *Let* $\omega \circ \sigma_s = \omega, \omega \circ \delta = 0$. *Then*

$$\lim |\omega(A_1(f)\tau_t A_2(f) A_3(f)) - \omega(A_1(f)A_3(f))\omega(A_2(f))| \le 2e^{-\lambda t}||F||_1||F''||_1. \tag{15.6}$$

Proof. We may assume that $\omega(A_2(f)) = 0$. By partial integration we obtain

$$|\omega(A_1(f)\tau_t A_2(f)A_3(f))| = \left|e^{-\lambda t} < \Omega|A_1(f) \int dsF(s)\sigma_{e^{-\lambda t}s}\delta\tau_t A_2 A_3(f)\right|\Omega > |$$

$$= e^{-\lambda t}\Big|< \Omega|\delta A_1(f) \int dsF(s)\sigma_{e^{-\lambda t}s}\tau_t A_2 A_3(f)$$

$$+ A_1(f) \int dsF(s)\sigma_{e^{-\lambda t}s}\tau_t A_2 \delta A_3(f)|\Omega > |$$

$$\le 2e^{-\lambda t}||F||_1||F''||_1.$$

□

Variations of the restrictions on the operators on which we want to study the convergence of the correlation functions are possible. We have chosen a way which is expressed only in the operators and the action of the automorphisms. Alternatives use the spectral representations of U_t, V_s, but always also some smearing of the operators [9].

15.4 Verification of the Anosov structure in crossed product constructions

The first attempt to construct non-abelian algebras other than type I was the crossed product construction. Already in [17] von Neumann used automorphisms that satisfy the Anosov relations to construct the first example of a type III algebra. However he had to apply some modifications:

If we start with the abelian algebra $L^\infty(R, dx)$, we have as automorphisms the shift $\sigma_s f(x) = f(x+s)$, and the dilation $\tau_t f(x) = f(e^{-t}x)$ such that

$$\tau_t \sigma_s \tau_{-t} f(x) = \tau_t \sigma_s f(e^t x) = \tau_t f(e^t(x+s)) = f(x + e^t s).$$

The crossed product with σ_s $L^\infty \times |_{\sigma_s} R$ acts on $L^2(R^2, dxdy)$ and is built by operators $\{f(x+y), p_y\}$. Therefore it is isomorphic to $\mathcal{B}(L^2(R, dx))$. The automorphism τ_t can be extended to the crossed product and on the type I algebra becomes an inner automorphism. As a consequence we have no invariant state, in correspondence to the fact that we started with an infinite measure. There exist operators invariant under time evolution (the functions of the unitary U_t) and therefore we do not have time clustering in general. As an analog of the Kolmogorov structure we can construct an ideal $\mathcal{A}_a = \Xi_{[-a,a]}(x)\mathcal{B}\Xi_{[-a,a]}(x)$ where $\Xi_{[-a,a]}(x)$ is the projection operator in x-space on the interval $[-a, a]$. This ideal has a trivial tail $\bigcap_{t<0} \tau_t \mathcal{A}_a = \Omega$.

To avoid type I situations where $L^\infty(R, dx) \times |_{\sigma_s} R$ is isomorphic to $\mathcal{B}(L^2(R, dx)$, von Neumann restricted the crossed product construction to the rationals C. With $L^\infty(R, dx) \times |_{\sigma_s} C$ we obtain a type II_∞ algebra with the trace $tr\hat{f}(x, s) = \int f(x, 0)dx$. The natural extension of the automorphism τ_t to an automorphism $\hat{\tau}_t$ on the crossed product $(\hat{\tau}_t \hat{f}(x, s) = \hat{f}(e^t x, e^t s))$ (where we have to restrict the permitted values of t such that $e^t \in C$) satisfies $tr \circ \hat{\tau}_t = e^t tr$. Therefore [17], we obtain a type III algebra, more precisely a type III_0 algebra, with automorphism groups with Anosov type, which however are not continuous. This has to be compared to the result of [19]: if the automorphism τ_t (now $t \in R$) of a type II_∞ algebra satisfies $tr \circ \hat{\tau}_t = e^t tr$ for all $t \in R$, then the crossed product is a type III_1 algebra.

Another example is provided by the irrational rotation algebra that can be obtained as a crossed product of the abelian algebra over the one-dimensional torus with the shift [18]. We can also think of this algebra as the subalgebra of $\mathcal{B}(L^2(R, dx))$, $\mathcal{A}_a = \{U^n = e^{2\pi i n x}, V^m = e^{iamp_x}\}$ [10]. The parameter a is called a rotation parameter. The algebra appears in this form in the quantum Hall effect [15] where the rotation parameter a is related to the magnetic field. On this algebra we have the automorphism groups

$$\begin{aligned} \sigma^+_\gamma e^{2\pi i x n} &= e^{2\pi i n(x+\gamma)}, \quad \sigma^+_\gamma e^{iap_x m} = e^{iamp_x}, \\ \sigma^-_\gamma e^{2\pi i x n} &= e^{2\pi i n x}, \quad \sigma^-_\gamma e^{iap_x m} = e^{iam(p_x+\gamma)}. \end{aligned} \tag{15.7}$$

They replace the horocyclic actions in the classical Anosov system. In addition the dilation in appropriate direction and appropriate step size acts as a discrete automorphism group:

$$W(\vec{n}) = U^{n_1} V^{n_2}, \quad \tau_T W(\vec{n}) = W(T\vec{n}) \tag{15.8}$$

where T is a 2×2 matrix built of integers and with determinant 1. The algebra is type II_1, if a is irrational, it is type I with nontrivial center, if a is rational. In both cases it permits the construction of a trace. If $a > 1$ this trace can be obtained as expectation value with $\Psi(x) = \Xi_{[0,1]}(x)$. In the corresponding GNS representation all unitaries that implement the above automorphisms are independent of the rotation parameter a. Therefore the system has the same clustering properties as the abelian analog. However multicorrelations become different, and it is an open problem, if an underlying Kolmogorov structure can be found.

As another example in this framework we start with the abelian algebra $L^\infty(R^+, dx)$. Here σ_s acts only as (not unity preserving) endomorphism:$\sigma_s f(x) = f(x-s), s > 0$. We consider $L^\infty(R^+, dx)$ as an ideal in $L^\infty(R, dx)$ and take f as a function in $L^\infty(R, dx)$ satisfying $f(x) = f(x)\Theta(x)$ with the Heaviside Θ-function ($\Theta(x) = 1$ iff $x \geq 0$), which enables us to extend σ_s to all $s \in R$:

$$\sigma_s f(x) = f(x-s)\Theta(x)\Theta(x-s). \tag{15.9}$$

The group property is changed to

$$\sigma_{-s}\sigma_s f(x) = f(x), s \geq 0, \quad \sigma_s\sigma_{-s} f(x) = f(x)\Theta(x-s) = \gamma_s f(x), s \geq 0 \tag{15.10}$$

where γ_s is the conditional expectation on the ideal $\mathcal{A}_s = \{f(x)\Theta(x-s)\}$. $\tau_t f(x) = f(e^{-x}x)$ remains as an automorphism that satisfies Anosov relations with the endomorphisms σ_s.

Also with the endomorphism we can construct a crossed product. It is the algebra $\mathcal{A}$ acting on $\mathcal{H} = L^2(R^+ \times R^+, dxds)$ and generated by operators $\widehat{f(x)}$ corresponding to $f(x) \in L^\infty(R^+, dx) = \mathcal{F}$ and $\widehat{V}(t), t \in R$ defined as

$$\widehat{f(x)}\Omega(x,s) = (\sigma_{-s}f(x))\Omega(x,s) = f(x+s)\Omega(x,s), \tag{15.11}$$

$$\widehat{V}(t)\Omega(x,s) = \Theta(s-t)\Theta(s)\Omega(x,s-t) \tag{15.12}$$

such that $\widehat{V}(t)\widehat{V}(-t) = \widehat{P}(t)$ is a nontrivial projector for $t > 0$. We collect the multiplication rules that follow from the definition of the operators:

$$\widehat{V}(t)\widehat{f(x)}\widehat{V}(-t) = (\widehat{f(x-t)\Theta(x)}\widehat{P}(t) = (\widehat{\sigma_t f(x)})\widehat{P}(t), \tag{15.13}$$

$$\widehat{f(x)}\widehat{P}(t) = \widehat{P}(t)\widehat{f(x)}, \tag{15.14}$$

$$\widehat{V}(t)\widehat{P}(z)\widehat{V}(-t) = \widehat{P}(z+t)\widehat{P}(t), \tag{15.15}$$

$$\widehat{P}(z)\widehat{P}(t) = \widehat{P}(t)\widehat{P}(z). \tag{15.16}$$

For $f(x,t,z)$ continuous and bounded functions the operators

$$\widehat{f(x,t,z)} = \int (f(x,t,z))^\wedge \widehat{V}(t)\widehat{P}(z)dtdz$$

are well defined and dense in the algebra. They act as

$$\widehat{f(x,t,z)}\Omega(x,s) = \int_{-\infty}^{\infty} dt \int_0^{\infty} dz f(x+s,t,z)\Theta(s-t-z)\Theta(x)\Theta(s-t)\Omega(x,s-t). \tag{15.17}$$

As subalgebra of $\mathcal{B}(L^2(R^+ \times R^+, dxds)$ together with the imbedding $L^2(R^+ \times R^+, dxds) \subset L^2(R \times R, dxds)$ it is constructed by operators $\{\Theta(x)\Theta(s)f(x+s), \Theta(x)\Theta(s)e^{itp_s}\Theta(x)\Theta(s)\} = \{\Theta(x)\Theta(s)f(x,s), \Theta(x)\Theta(s)e^{itp_s}\Theta(x)\Theta(s)\}$ and has center $\{\Theta(x)\Theta(s)f(x+s)\}$. The automorphism τ_ρ can be extended over $\mathcal{A}$ to an automorphism $\hat{\tau}_\rho$ via

$$\hat{\tau}_t\rho \widehat{f(x,t,z)} = e^{2\rho} \widehat{f(e^\rho x, e^\rho t}, e^\rho z). \tag{15.18}$$

On $\mathcal{A}$ we can define a trace by

$$Tr\widehat{f(x,t,z)} = \int f(x+s,0,z)\Theta(s-z)\Theta(s)\Theta(x)\Theta(z)dsdxdz. \tag{15.19}$$

This expression can be obtained as

$$\lim_{a\to\infty} \int db < \Omega_{a,b}|\widehat{f(x,t,z)}|\Omega_{a,b} >, \Omega_{a,b}(x,s) = e^{isb}\Theta(a-x)\Theta(a-s). \tag{15.20}$$

With respect to the automorphism group $\hat{\tau}_\rho$ it satisfies

$$Tr\hat{\tau}_\rho \widehat{f(x,t,z)} = e^{-\rho} Tr\widehat{f(x,t,z)}. \tag{15.21}$$

The algebra is a type I_∞ algebra with an unbounded trace and an automorphism that is not inner on the center. This automorphism is unitarily implemented whereas the horocyclic action as endomorphism is not.

It remains to see whether the clustering properties remain the same as in the classical situation. As for $L^\infty(R^+, dx)$ we cannot find a subalgebra but only an ideal, namely the algebra imbedded between projectors that converge strongly to 0. Possibilities are $\widehat{P}(t)$ with $\widehat{P}(t)\Omega(x,s) = \Theta(t-s)\Theta(s)\Theta(x)\Omega(x,s)$ or $\widehat{\Theta(a-x)}$ with $\widehat{\Theta(a-x)}\Omega(x,s) = \Theta(a-x+s)\Omega(x,s)$. Triviality of the tail follows because st-lim $\hat{\tau}_\rho\widehat{\Theta(a-x)} = \widehat{\Theta(a-e^\rho x)} = 0$.

Our main interest concerns clustering behavior. Now we study $trA\hat{\tau}_\rho B$. We notice that we can translate the proof in Theorem 1 to the present situation, since the trace is invariant under $\sigma_s, s > 0$. However we have to control in addition that $trA_1(f) < \infty$, which holds if we choose $A_1 \in \mathcal{A}_a$ such that we can estimate

$$\frac{|trA_1(f)\hat{\tau}_\rho A_2(f)|}{tr|A_1(f)|} \leq e^{-\lambda\rho}\frac{|tr|\delta A_1(f)|}{tr|A_1(f)|}||F||_1 \quad \forall A_2, trA_2(f) = 0. \tag{15.22}$$

Nevertheless we have one essential difference: In the general classical framework we started with two horocyclic actions. These two actions are related to time reflection. Consider on the abelian $L^\infty(R^+, dx)$ the map $\gamma f(x) = f(1/x)$ such that $\gamma\tau_t\gamma f(x) = f(e^{-t}x) = \tau_{-t}f(x)$. Acting on the horocyclic action we have $\gamma\sigma_s^+\gamma = \sigma_s^-$ with $\sigma_s^- f(x) = \gamma f(\frac{1}{x-s}) = f(\frac{x}{1-xs})$. Therefore the second horocyclic action appears naturally. This changes on the noncommutative level. Passing to the crossed product neither time reflection nor σ_s^- can be extended. Though $\hat{\tau}_\rho$ is a group there is a natural asymmetry with respect to time direction.

15.5 The Type *III* Case

The type *III* case offers an example in which Kolmogorov structure and Anosov structure imply one another, however again with the modification that we permit endomorphisms. In addition this example appears in the physical context of quantum field theory. First however we observe a no go theorem:

Theorem 2. *Let $\mathcal{A}$ be a type III algebra and τ_t its modular automorphism group with respect to a state ω. If $\tau_t \sigma_s \tau_{-t} = \sigma_{e^{\lambda t} s}$ for some automorphism group σ_s, then $\lambda = 0$.*

Proof. Both τ_t and σ_s can be unitarily implemented. This implementation is unique if we demand $JU_sJ = U_s, JV_tJ = V_t$ with J the modular conjugation. Therefore $U_tV_sU_{-t} = V_{e^{\lambda t}s}$, so that also the unitaries satisfy the Anosov relations. It follows that also ω is invariant under σ_s. A general theorem tells us that τ and σ have to commute, which reduces the Lyapunov exponent to 1. □

Already in the abelian situation we also considered the case when σ_s is an endomorphism. Let therefore $\sigma\mathcal{M} = \mathcal{N} \subset \mathcal{M}$ for algebras $\mathcal{M}, \mathcal{N}$. This permits us to pass from Kolmogorov systems to Anosov systems and from Anosov systems to Kolmogorov systems. These are results starting in [5], continued in [20] and just recently extended in [1]. (Compare also [6].)

Theorem 3. *Let $\mathcal{M}, \mathcal{N}$ be two von Neumann algebras with common cyclic and separating vector Ω. Denote the corresponding modular operators and conjugations by $\Delta_\mathcal{M}, \Delta_\mathcal{N}, J_\mathcal{M}, J_\mathcal{N}$. Let the endomorphism σ mapping $\mathcal{M}$ into $\mathcal{N}$ be implemented by the unitary V. Let $V|\Omega>=|\Omega>$.*

Then $V(t) = \Delta_\mathcal{M}^{-it} V \Delta_\mathcal{N}^{it}$ is strongly continuous in $t \in R$ and it can analytically be extended to the strip $S(0,1/2) = \{t \in C, 0 < Im\, t < 1/2\}$. In this strip $||V(\tau)|| \leq 1$. On the boundary $V(t+i/2) = J_\mathcal{M} V(t) J_\mathcal{N}$.

The theorem has a converse:

Theorem 4. *Let $\mathcal{M}, \mathcal{N}$ be two von Neumann algebras with common cyclic and separating vector Ω. Let $W(s)$ be a family of operators that satisfy the following requirements: $W(s)|\Omega>=|\Omega>$, $W(s)$ is analytic and bounded with $||W(\sigma)|| \leq 1$ in the strip $S(0,1/2)$, unitary at the boundary of the strip. Further $W(s)\mathcal{N}W(s)^* \subset \mathcal{M}$, $\quad W(s+i/2)\mathcal{N}'W(s+i/2)^* \subset \mathcal{M}'$. It follows that*

$$\Delta_\mathcal{M}^{it} W(s) \Delta_\mathcal{N}^{-it} = W(s-t), \quad J_\mathcal{M} W(s) J_\mathcal{N} = W(s+i/2). \tag{15.23}$$

As a special example for V in Theorem 3 we can take $V = J_\mathcal{M} J_\mathcal{N}$, the operator that implements the Longo shift [12]. It acts as $V^*\mathcal{M}V = \bar{\mathcal{N}} \subset \mathcal{N}$ and we can replace $\mathcal{N}$ by $\bar{\mathcal{N}}$ in Theorem 3.

The analyticity properties for $V(s)$ become more precise if we know in addition that $\mathcal{N} \subset \mathcal{M}$ form a Kolmogorov system, i.e. that $\Delta_\mathcal{M}^{it} \mathcal{N} \Delta_\mathcal{M}^{-it} \subset \mathcal{N} \forall t \in R^+$.

Theorem 5. *Let $\mathcal{M}, \mathcal{N}$ be two von Neumann algebras with common cyclic and separating vector Ω and assume $\Delta_\mathcal{M}^{it} \mathcal{N} \Delta_\mathcal{M}^{-it} \subset \mathcal{N} \forall t \in R^+$. Then*

$$\frac{1}{2\pi}(\log \Delta_\mathcal{N} - \log \Delta_\mathcal{M}) = G \tag{15.24}$$

is defined as an essentially selfadjoint operator that satisfies with $U(s) = e^{iGs}$:

a) *G is positive*
b) $\Delta_\mathcal{M}^{-it} U(s) \Delta_\mathcal{M}^{it} = \Delta_\mathcal{N}^{-it} U(s) \Delta_\mathcal{N}^{it} = U(e^{2\pi t}s)$
c) $J_\mathcal{M} U(s) J_\mathcal{M} = J_\mathcal{N} U(s) J_\mathcal{N} = U(-s) \quad s \in R$
d) $U(s)\mathcal{M}U(s)^* \subset \mathcal{M} \quad \forall s \in R^+$.

Therefore the Kolmogorov structure implies the Anosov structure.

On the other hand we can start with $U(s)$. Assume that $U(s)\mathcal{M}U(-s) \subset U(s')\mathcal{M}U(-s') \forall s > s'$. If in addition $\Delta_{\mathcal{M}}^{-it}U(s)\Delta_{\mathcal{M}}^{it} = U(e^{2\pi t}s)$, then with

$$\mathcal{N}_s = U(s)\mathcal{M}U(-s)$$

we have

$$\begin{aligned}\Delta_{\mathcal{M}}^{-it}\mathcal{N}_s\Delta_{\mathcal{M}}^{it} &= \Delta_{\mathcal{M}}^{-it}U(s)\mathcal{M}U(-s)\Delta_{\mathcal{M}}^{it} = U(e^{2\pi t}s - s)U(s)\mathcal{M}U(-s)U(-e^{2\pi t}s + s)\\ &= U(e^{2\pi t}s - s)\mathcal{N}_s U(-e^{2\pi t}s + s) \subset \mathcal{N}_s \quad \forall t \in R^+.\end{aligned}$$

So the Anosov structure guarantees that we have a Kolomorov system, and the two structures with respect to the modular evolution are equivalent for a type III_1 system.

If we compare the situation with that of the abelian algebra, we can observe an essential difference: there we constructed an inverse to the endomorphism that we needed in the crossed product construction. We could also look at the inverse as an adjoint $\sigma_s^* = \sigma_s$ that satisfies

$$\omega(A\sigma_s B) = \omega(\sigma_s^* A \circ B) \quad \forall A, B \in L^\infty(R^+, dx).$$

We can do the corresponding construction for the type III case using the natural correspondence between normal states and operators of the (antiisomorphic) commutant:

$$\omega_A(B) = <\Omega|JAJB|\Omega> = <\Omega|A^*\Delta^{1/2}B|\Omega> .$$

We define the adjoint with respect to the state ω as

$$\omega_A(\sigma_s B) = \omega_{\sigma_s^* A}(B) = <\Omega|A^*\Delta^{1/2}U(s)B|\Omega> = <\Omega|A^*U(-s)\Delta^{1/2}B|\Omega> = \omega_{\sigma_s A} \tag{15.25}$$

according to Theorem 5. Therefore under the assumptions of Theorem 5, with σ_s implemented by $U(s)$, we have $\sigma_s^* = \sigma_s$.

The above observation can again be taken as a definition for an Anosov structure for the following reason:

Theorem 6. *Let σ_s be a continuous endomorphism group such that with respect to the state ω we have $\omega \circ \sigma_s = \omega, \sigma_s = \sigma_s^*$. Then τ_t, the modular automorphism with respect to ω, and σ_s define an Anosov system.*

Proof. We implement the endomorphism by e^{iGs}. According to our assumption and (25) we have

$$e^{-iGs}\Delta^{1/2} = \Delta^{1/2}e^{iGs}.$$

This holds on the domain on which $\Delta^{1/2}$ is essentially selfadjoint, namely on $\mathcal{M}|\Omega>$. If we smear A to $A_f = \int_0^\infty ds f(s)\sigma_s(A)$ for functions that satisfy $f, f' \in L^1, f(0) = 0$, then these operators are strongly dense in $\mathcal{M}$ and the corresponding domain is still sufficiently large for $\Delta^{1/2}$. By partial integration G is defined on this domain and we have $G\Delta^{1/2} = -\Delta^{1/2}G$. We want to calculate $e^{-iGs}\Delta^{1/2}e^{iGs} = \Delta_s^{1/2}$ respectively (compare Theorem 5) $\log e^{-iGs}\Delta^{1/2}e^{iGs}$. It suffices to calculate $[G, \log \Delta^{1/2}]$ which we can obtain as

$$[G, \int_0^\infty d\alpha(\frac{1}{\Delta^{1/2} + \alpha} - \frac{1}{1 + \alpha})]$$

where we can choose the path such that the integrand apart from $\alpha = 0$ remains bounded. We argue, always on the appropriate domain with $\epsilon > 0$,

$$
\begin{aligned}
0 &= [G, \frac{1}{\Delta^{1/2}+\alpha}(\Delta^{1/2}+\alpha)] \\
&= [G, \frac{1}{\Delta^{1/2}+\alpha}](\Delta^{1/2}+\alpha) + \frac{1}{\Delta^{1/2}+\alpha}[G,(\Delta^{1/2}+\alpha)] \\
&= [G, \frac{1}{\Delta^{1/2}+\alpha}](\Delta^{1/2}+\alpha) + \frac{1}{\Delta^{1/2}+\alpha}2G\Delta^{1/2}.
\end{aligned}
$$

Therefore

$$
\begin{aligned}
[G, \frac{1}{\Delta^{1/2}+\alpha}] &= -2\frac{1}{\Delta^{1/2}+\alpha}G\frac{\Delta^{1/2}}{\Delta^{1/2}+\alpha} \\
&= -2G\frac{\Delta^{1/2}}{(\Delta^{1/2}+\alpha)^2} + 2[G, \frac{1}{\Delta^{1/2}+\alpha}]\frac{\Delta^{1/2}}{(\Delta^{1/2}+\alpha)^2}.
\end{aligned}
$$

We continue to

$$
[G, \frac{1}{\Delta^{1/2}+\alpha}](1 - \frac{2\Delta^{1/2}}{\Delta^{1/2}+\alpha}) = [G, \frac{1}{\Delta^{1/2}+\alpha}]\frac{\alpha-\Delta^{1/2}}{\Delta^{1/2}+\alpha} = -2G\frac{\Delta^{1/2}}{\Delta^{1/2}+\alpha},
$$

$$
[G, \frac{1}{\Delta^{1/2}+\alpha}] = 2G\frac{\Delta^{1/2}}{(\Delta^{1/2}+\alpha)(\Delta^{1/2}-\alpha)}.
$$

If we integrate over a path in the lower half plan (which is dictated from $s \geq 0$) we obtain

$$
\int_{-\infty-i\epsilon}^{\infty-i\epsilon} \frac{\Delta^{1/2}}{(\Delta^{1/2}+\alpha)(\Delta^{1/2}-\alpha)}d\alpha = 1/2\int_{-\infty-i\epsilon}^{\infty-i\epsilon} d\alpha(\frac{1}{\Delta^{1/2}+\alpha} - \frac{1}{\Delta^{1/2}-\alpha}) = i\pi.
$$

In fact we obtain $\log \Delta_s^{1/2} = \log \Delta^{1/2} + \pi G$ as in Theorem 5. □

Finally we remark that the estimate of Theorem 1 on the clustering properties can be applied, since we have an invariant state and the possibility of smearing.

15.6 Application to Quantum field Theory

A typical example of a quantum Anosov system is the Rindler wedge. As shown by [3, 4] the modular automorphism group of the Rindler wedge is the boost. The corresponding endomorphism is light-like translation. The translated Rindler wedge is stable under positive boosts. Therefore the Poincare group can be constructed from the imbeddings of the Rindler wedge [8]. Especially $\sigma_s^* = \sigma_s$ appears as a natural fact, corresponding to the reflection along one light ray.

Another model that is close to an Anosov system but does not satisfy all requirements is the de Sitter space. It is the hyperbolic manifold defined as the submanifold S^4 of the five-dimensional space $\int d^5x\delta(x_0^2 - x_1^2 - x_3^2 - x_4^2 + R^2)$. This imbedding into a higher dimensional space can be used to construct a quantum field theory over the de Sitter space [7]. The analog to the Rindler wedge is the reduction of the five dimensional wedge to the de Sitter space $W_j = \{x \in R^5; x_j \geq |x_0|\} \cap S^4, j = 1, 2, 3, 4$. Geometrical automorphisms of the quantum field theory over the de Sitter space are induced by Killing vector fields, whose existence however is rather limited. They are determined by a Lie algebra $L_{\mu\nu}$ where $L_{0\mu}$ are the generators of the modular evolution in a wedge, which coincides with the time evolution of a geodesic observer. The L_{ij} correspond to rotations. They satisfy the Anosov commutation relations

$$[L_{0i}, L_{0j} - L_{ij}] = i(L_{0i} - L_{ij}).$$

However $L_{0i} - L_{ij}$ does not leave W_1 invariant. Up to order ϵ $(i = 1, j = 2)$,

$$x_0 \to x_0 + \epsilon x_1 + \frac{\epsilon^2}{2}(x_0 + x_2), \ \ x_1 \to x_1 + \epsilon x_0 + \epsilon x_2, \ \ x_2 \to x_2 - \epsilon x_1 - \frac{\epsilon^2}{2}(x_0 + x_2).$$

The shift transfers again the wedge along the light ray. If therefore an operator is sufficiently localized in the interior of the wedge, still $\int ds f(s)\sigma_j(s)A$ will remain in the wedge. We have only to assume that $f(s)$ is twice differentiable and has compact support in a sufficiently small region. With this modification we again obtain from Theorem 1 exponential clustering of the correlation functions.

References

1. H. Araki and L. Zsido: *Extension of the structure theorem of Borchers and its application to half-sided modular inclusions.* arXiv:math.OA/0412061.
2. V.I. Arnold and A. Avez: *Problemes Ergodiques de la Mecanique classique.* Gauthier-Villars, Paris, 1967.
3. J. Bisognano and E.H. Wichmann: On the duality condition for a Hermitian scalar field. *J. Math. Phys.* 16:985-1007 (1975).
4. J. Bisognano and E.H. Wichmann: On the duality condition of quantum fields. *Jour. Math. Phys.* 17:303-312 (1976).
5. H.-J. Borchers: On modular inclusion and spectral condition. *Lett. Math. Phys.* 27:311-324 (1993).
6. H.-J. Borchers: *On the revolutionization of quantum field theory by Tomita's modular theory.* Preprint ESI, 1999.
7. J. Bros, H. Epstein and U. Moschella: Analyticity properties and thermal effects for general quantum field theory on de Sitter space-time. *Commun. Math. Phys.* 196:535-570 (1998).
8. D. Buchholz, O. Dreyer, M. Florig and S.J. Summers: Geometric modular action and space-time symmetry groups. *Rev. Math Phys.* 12:475-560 (2000).
9. G.G. Emch, H. Narnhofer, G.L. Sewell and W. Thirring: Anosov actions on noncommutative algebras. *J. Math. Phys.* 35/11:5582-5599 (1994).
10. L.D. Faddeev: Discrete Heisenberg-Weyl group and modular group. *Lett. Math. Phys.* 34:249-254 (1995).
11. B.O. Koopman and J. von Neumann: Dynamical systems of continuous spectra. *N.A.S. Proc.* 18:255-263 (1932).
12. R. Longo: Simple injective subfactors. *Adv. math.* 63:152-171 (1987).
13. H. Narnhofer and W. Thirring: Quantum K-Systems. *Commun. Math. Phys.* 125:565-577 (1989).
14. H. Narnhofer: Kolmogorov systems and Anosov systems in quantum theory. *IDAQP* 4/1:85-119 (2001).

15. H. Narnhofer: The Pauli principle for a quantum theory on T^2 with magnetic field. *Rep. Math. Phys.* 53:91-102 (2004).
16. J. von Neumann: Einige Sätze über messbare Abbildungen. *Ann. Math.* 33:574-586 (1932).
17. J. von Neumann: On rings of operators III. *Ann. Math.* 41:94-161 (1940).
18. M.A. Rieffel: C^*-algebras associated with irrational rotations. *Pac. J. Math.* 93:415-429 (1981).
19. M. Takesaki: Duality for crossed products and the structure of von Neumann algebras of type *III*. *Acta Math.* 131:249-310 (1973).
20. H.J. Wiesbrock: Half sided modular inclusions of von Neumann algebras. *Commun. Math. Phys.* 157:83 (1993). Erratum *Commun. Math. Phys.* 184:683 (1997).

16

DFR Perturbative Quantum Field Theory on Quantum Space Time, and Wick Reduction

Gherardo Piacitelli*

I.P. "De Sandrinelli", Via Paolo Veronese 3, 34131 Trieste, Italia
gherardo@piacitelli.org

Summary. We discuss the perturbative approach *à la Dyson* to a quantum field theory with nonlocal self–interaction $:\phi\star\cdots\star\phi:$, according to Doplicher, Fredenhagen and Roberts (DFR). In particular, we show that the Wick reduction of *nonlocally* time-ordered products of Wick monomials can be performed as usual, and we discuss a very simple Dyson diagram.

Dedicated to Jacques Bros on the occasion of his 70th birthday.

16.1 Introduction

During the XXth century, locality has been so valuable a principle in the development of high energy physics, that it is strongly encoded in our minds. Sometimes we advocate locality even when it is not strictly necessary.

I once heard Daniel Kastler telling a story about a sixty or so pages long paper on a certain topic in commutative algebra; the astonishing fact was that the only necessary change in order to generalize the results of that paper to the non-abelian case was the removal of every occurrence of the word "abelian". The conclusion of Kastler's tale was that commutativity comes so naturally to our mind, that sometimes even an expert might overlook the generality of some arguments.

Something similar happens with locality. We will see that some aspects and methods of the perturbation theory of a certain class of nonlocal theories are exactly the same as in the local case.

Indeed, we will consider the nonlocal theory which naturally arises when attempting a perturbative quantum field theory *à la* Dyson on the flat DFR quantum space-time. It could have been easily recognized that the usual diagrammatic representation of the correction terms to the trivial scattering matrix has nothing to do with locality. The interesting point, however, is that it has *not* been easy to recognize this, probably because of some psychological obstruction of the kind mentioned by Daniel Kastler.

*Research partially supported by MIUR and GNAMPA-INdAM

In the next section, we will give a short description of the machinery underlying the DFR model of a flat quantum space-time, and of a quantum field theory built on top of it. We will take the occasion to clarify the relations between the original DFR notation and that which is now current in the literature (see also the appendix).

In Section 16.3, we will describe how to derive the Dyson diagrams for a (possibly) nonlocal perturbation theory using precisely the same methods which were developed in the late 1940s. An analogous discussion can be done for the Feynman diagrams, see [18] for details.

We will draw some conclusions in Section 16.4.

16.2 DFR Quantum Space-time, and All That

16.2.1 The Underlying Philosophy

In their seminal paper [9], Doplicher, Fredenhagen and Roberts proposed to derive a simple model of space-time coordinates quantization, stemming from first principles endowed with an operational meaning.

Indeed, the idea of quantizing the coordinates was quite old [22]. The idea that non-commuting coordinates would produce a coarse-grained space-time also was far from new. However, the spirit of the DFR proposal was rather original. While Snyder's space coordinates have discrete spectra so as to induce a covariant analogue of lattice discretization, the DFR coordinates all have continuous spectra, so that no limitations arise on the precision of the localization *in one coordinate.* Limitations arise instead on the precision of simultaneous localization in two or more non-commuting coordinates.

It is common folk lore[2] that limitations on localization in the small should arise since, according to our understanding of high energy physics, the localization process requires that a certain amount of energy be transferred to the geometric background: the smaller the localization region, the higher the energy density induced in the localization region. If the localization process reaches a sufficiently small length scale (typically the Planck length scale $\lambda_P = \sqrt{G\hbar c^{-3}}$), a closed horizon might trap the region under observation, preventing any information escaping from it.

This classical argument has been invoked for example to claim that it is not possible to localize with a precision below the Planck length scale (see e.g. [8,14]). A moment's thought, however, would make it clear that such a statement is not really supported by the above argument. Indeed, one might envisage to localizing below the Planck length scale in one space dimension, at the cost of sufficiently delocalizing in the remaining space dimensions. The resulting admissible localization region would be very large (compared with λ_P) and very thin; thin enough to obtain localization much below the Planck length (in one direction), and large enough to keep the energy density sufficiently low to avoid black hole formation. It was precisely this remark that led DFR to cast limitations on the admissible localization regions in the form of uncertainty relations among the coordinates. Lacking more direct motivations,

[2]Apparently the first who made this remark (in a slightly different form, namely revisiting Heisenberg's microscope *gedankenexperiment*) was C. Alden Mead as early as 1959, but his paper underwent referee troubles and was not published until 1964 [14]. See the interesting letter of Mead to Physics Today [15].

they decided to reproduce the path which, long ago, led to canonical quantization; namely to find commutation relations inducing precisely the required uncertainty relations.

16.2.2 The DFR Basic Model

A heuristic analysis led DFR to postulate a very simple toy model of a flat, fully covariant quantum space-time, described by four quantum coordinates, i.e. four selfadjoint operators q^μ on the infinite dimensional, separable Hilbert space $\mathfrak{H}$. Setting $\lambda_P = 1$ (in suitable units) and

$$Q^{\mu\nu} = -\mathrm{i}[q^\mu, q^\nu],$$

the commutation relations are $[q^\mu, Q^{\varrho\sigma}] = 0$, or equivalently[3]

$$\exp(\mathrm{i}k_\mu q^\mu)\exp(\mathrm{i}h_\mu q^\mu) = \exp\Big(-\frac{\mathrm{i}}{2}k_\mu Q^{\mu\nu}h_\nu\Big)\exp(\mathrm{i}(h+k)_\mu q^\mu), \tag{16.1}$$

to be complemented with the statement that the joint spectral values $\sigma^{\mu\nu}$ of the pairwise commuting operators $Q^{\mu\nu}$ define precisely the set Σ of the antisymmetric matrices σ fulfilling[4]

$$\sigma_{\mu\nu}\sigma^{\mu\nu} = 0, \quad (\sigma^{\mu\nu}(*\sigma)_{\mu\nu})^2 = 16.$$

The requirement that $[q^\mu, Q^{\varrho\sigma}] = 0$ is a simplifying, otherwise arbitrary ansatz; once this ansatz is accepted, the limitations on the set Σ stem out of the DFR stability condition of space-time under localization, together with the quest for covariance. We will not give the details; the interested reader is referred to the original paper.

The coordinates are covariant: there is a unitary representation U of the Poincaré group $\mathscr{P}$, such that

$$U(\Lambda, a)q^\mu U(\Lambda, a)^{-1} = \Lambda^\mu{}_\nu q^\nu + a^\mu, \quad (\Lambda, a) \in \mathscr{P}.$$

In strict analogy with Weyl quantization [23], one may consider the quantization of an ordinary function $f = f(x)$ of $\mathbb{R}^4$ defined by

$$f(q) = \int_{\mathbb{R}^4} dk\, \check{f}(k)e^{ikq}, \tag{16.2}$$

where $kq = k_\mu q^\mu$, and

$$\check{f}(k) = \frac{1}{(2\pi)^4}\int_{\mathbb{R}^4} dx\, f(x)e^{-ikx}.$$

At this point, one might wish to follow the suggestion of Weyl (for example as von Neumann did [16]) and describe the operator product $f_1(q)f_2(q)$ in terms of a suitable product of the ordinary functions f_1, f_2. Unfortunately, the set of all the

[3] More precisely, equation (16.1) gives the formal relations $[q^\mu, Q^{\varrho\sigma}] = 0$ the precise mathematical status of regular, strong commutation relations.

[4] We recall that the Hodge dual $*\sigma$ of an antisymmetric 2-tensor σ is given by $(*\sigma)_{\mu\nu} = (1/2)\varepsilon_{\mu\nu\lambda\varrho}\sigma^{\lambda\varrho}$.

operators of the form $f(q)$ is not closed under operator products. It is then necessary to preliminarily extend the DFR quantization to a larger class of functions, namely the functions $F = F(\sigma, x)$ of $\Sigma \times \mathbb{R}^4$. The full DFR quantization of a function of the form

$$F(\sigma, x) = \sum_i g_i(\sigma) f_i(x)$$

(a DFR symbol) is given by

$$F(Q, q) = \sum_i g_i(Q) f_i(q);$$

above, $f_i(q)$ is understood as in (16.2), while $g_i(Q)$ is the joint functional calculus of the pairwise commuting operators $Q^{\mu\nu}$.

We may now define a covariant product $\star$ of two DFR symbols (the DFR twisted product) by requiring that

$$F_1(Q, q) F_2(Q, q) = (F_1 \star F_2)(Q, q).$$

By standard computations, one easily finds

$$(F_1 \star F_2)(\sigma, x) = \frac{1}{\pi^4} \int_{(\mathbb{R}^4)^2} da db\, F_1(\sigma, x + a) F_2(\sigma, x + b) \exp\big(- 2ib_\mu (\sigma^{-1})^{\mu\nu} a_\nu \big). \tag{16.3}$$

An asymptotic expansion of the (reduced) DFR product is widely known as the Moyal product. See the appendix for more details.

Following [9], we may define two maps, which, for self-evident reasons, we will denote by $\int_{\{q^0=t\}} d^3q$ and $\int d^4q$, respectively:

$$\int_{\{q^0=t\}} d^3q\, F(Q, q) = \int_{\mathbb{R}^3} d^3\boldsymbol{x}\, F(Q, (t, \boldsymbol{x})), \qquad \int d^4q\, F(Q, q) = \int_{\mathbb{R}^4} d^4x\, F(Q, x).$$

These maps are positive: for all F's and t's,

$$\int_{\{q^0=t\}} d^3q\, F(Q, q) F(Q, q)^* \geqslant 0, \qquad \int d^4q\, F(Q, q) F(Q, q)^* \geqslant 0.$$

In particular, the positivity of the map $\int_{\{q^0=t\}} d^3q$ is compatible with the uncertainty relations, since the latter allow for exact localization in q^0, at the cost of total delocalization in the remaining coordinates. Note also that, for any $x \in \mathbb{R}^4$ fixed, the map $F(Q, q) \mapsto F(Q, x)$ is *not* positive.

16.2.3 Field Theory

A (Wightman, say) quantum field $\phi(x)$ on ordinary Minkowski space-time is a (generalized) function taking values (morally) in the field algebra $\mathfrak{F}$, namely it is (morally) in $\mathcal{C}(\mathbb{R}^4, \mathfrak{F}) \simeq \mathcal{C}(\mathbb{R}^4) \otimes \mathfrak{F}$. Here $\mathcal{C}(\mathbb{R}^4)$ is the localization algebra.

Hence, it is natural to replace the classical localization algebra $\mathcal{C}(\mathbb{R}^4)$ with its quantized counterpart, the algebra $\mathcal{E}$ generated by the quantum coordinates q^μ. In this "semiclassical" model, it is natural to seek for quantum fields on quantum space-time as elements of (morally) the algebra $\mathcal{E} \otimes \mathfrak{F}$. By analogy with the quantization of ordinary functions, DFR proposed the following quantization of the free Klein–Gordon field:

$$\phi(q) := \int_{\mathbb{R}^4} dk\, e^{ikq} \otimes \check{\phi}(k).$$

Then they made the following, fundamental remark. Let $\mathscr{H}_0(\phi(x), \partial_\mu\phi(x))$ be the free Hamiltonian density. It is well known that the full free Hamiltonian

$$H_0 = \int d^3\boldsymbol{x}\; \mathscr{H}_0(\phi(t,\boldsymbol{x}), \partial_\mu\phi(t,\boldsymbol{x}))$$

does not depend on the time t. Then, it was shown in [9] that

$$\int_{\{q^0=t\}} d^3q\; \mathscr{H}_0(\phi(q), \partial_\mu\phi(q)) = H_0$$

(as a constant function of σ). To put it in a more explicit way, with

$$\mathscr{H}_0(\phi(x), \partial_\mu\phi(x)) = \frac{1}{2} {:}\left(\sum_\mu (\partial_\mu\phi)^2(x) + m^2\phi^2(x)\right){:},$$

we have

$$\int_{\{q^0=t\}} d^3q\; \mathscr{H}(\phi(q), \partial_\mu\phi(q))$$
$$= \frac{1}{2}\int_{\mathbb{R}^3} d\boldsymbol{x}\; {:}\left(\sum_\mu ((\partial_\mu\phi)\star(\partial_\mu\phi))(t,\boldsymbol{x}) + m^2(\phi\star\phi)(t,\boldsymbol{x})\right){:} = H_0$$

(as a constant function of σ).

This remark[5] is the starting point for defining an effective perturbation theory on the ordinary Minkowski quantum space-time, in the so-called interaction representation; of course, this ansatz should be taken with a grain of salt. Since the Hamiltonian is unaffected by the replacement of ordinary pointwise products with twisted products, we may consider perturbations of the usual free fields. The only remaining difficulty is the dependence on σ; to get rid of it in view of an effective theory on ordinary space-time, we need to integrate it out by means of some measure

[5]This result is independent from the well-known *general* fact that, for any two admissible functions f, g, not necessarily solutions of the Klein–Gordon equation, $\int d^4x\, (f \star g)(x) \equiv \int d^4x\, (fg)(x)$. The latter result is commonly taken as the starting point for a perturbative approach to nonlocal theories in the Euclidean setting, since it implies that the free action is unchanged by replacing the ordinary pointwise product with some twisted product $*$. Note however that the Wick rotation of a twisted product is ill defined (namely no well defined $*$ may be reached by Wick rotating the DFR product $\star$), so that at present (and to the best of the author's knowledge) the only known relation between Euclidean and Minkowskian "twisted theories" is a weak, indirect formal analogy.

on Σ. Unfortunately, there is no Lorentz invariant measure on Σ; the best we can find is a rotation invariant measure $d\sigma$. Quite unfortunately, this will destroy the covariance of this class of models under Lorentz boosts.

By analogy with the case of the local ϕ^n interaction, we may consider the interaction Hamiltonian

$$H_I(t) = \int_{q^0=t} :\phi(q)^n: = \int_{\mathbb{R}^3} d\boldsymbol{a} \; :(\phi\star\cdots\star\phi)(Q,(t,\boldsymbol{a})):,$$

where $:AB\cdots:$ denotes the normal (i.e. Wick) ordering of operator products. To obtain an effective interaction Hamiltonian on the classical Minkowski space-time, we integrate the σ dependence out, getting

$$H_I^{\text{eff}}(t) = = \int_\Sigma d\sigma \int_{\mathbb{R}^3} d\boldsymbol{a} \; :(\phi\star\cdots\star\phi)(\sigma,(t,\boldsymbol{a})): = \int_{\mathbb{R}^3} d\boldsymbol{a} \int_{\mathbb{R}^{4n}} dx_1\cdots dx_n \, W_{(t,\boldsymbol{a})}(x_1,\ldots,x_n) :\phi(x_1)\cdots\phi(x_n):$$

for a suitable kernel[6] W_x. Following Dyson [10], the (formal) S-matrix is given by

$$S = I + \sum_{N=1}^{\infty} \frac{(-\mathrm{i}g)^N}{N!} \int_{\mathbb{R}^N} dt_1\cdots dt_N \, T[H_I^{\text{eff}}(t_1),\ldots,H_I^{\text{eff}}(t_N)],$$

where

$$T[H_I^{\text{eff}}(t_1),\ldots,H_I^{\text{eff}}(t_N)] = \sum_\pi H_I^{\text{eff}}(t_{\pi(1)})\cdots H_I^{\text{eff}}(t_{\pi(N)}) \prod_{k=1}^{N-1} \theta(t_{\pi(k+1)} - t_{\pi(k)})$$

is the product of the $H_I^{\text{eff}}(t_j)$'s taken in the order of decreasing times (the sum over π running over all permutations of $(1,\ldots,N)$). Note that, contrary to the usual conventions, we wrote $T[A,B,\ldots]$ instead of the traditional $T[AB\cdots]$.

As pointed out in [9], the time ordering acts on the overall times $t_1,\ldots,t_n$ of the interaction Hamiltonians, not on the time arguments of the fields which appear

---

[6] If C_σ is the kernel such that

$$(f_1\star\cdots\star f_n)(\sigma,x) = \int_{(\mathbb{R}^4)^n} dx_1\cdots dx_n \, C_\sigma(x-x_1,\ldots,x-x_n) f_1(x_1)\cdots f_n(x_n),$$

then

$$W_x(x_1,\ldots,x_n) = \int_\Sigma d\sigma \, C_\sigma(x-x_1,\ldots,(x-x_n).$$

in the definition of the interaction Hamiltonian. Since the perturbation theory described above is built on top of a Hamiltonian model, the S-matrix of the DFR $\phi^{\star n}$ interaction is (formally) unitary by construction. More recent concerns about possible unitarity violation were the consequence of a too naive way of performing the time ordering prescription (see [3], and references therein; see also [1, 20, 21]). The ultraviolet regularity of a $\phi^{\star 3}$ DFR model has recently been proved by Bahns [2], under a weaker prescription for averaging over σ.

At this point it might be natural to convince ourselves that the trick of absorbing the time ordering into some analogue of the Stueckelberg–Feynman propagator is not possible any more. Indeed, in the local case the Stueckelberg–Feynman propagators allow us to consider one diagram, describing at once all possible arrangements of the time of the vertices [12, 19]; this feature might seem apparently lost in the present nonlocal case. We will see in the next section how, on the contrary, things are bound to go exactly the same way as in the local case.

Before closing this section, let us recall that different perturbative approaches, which are equivalent in the local case, may well fail to be such in the nonlocal case. As an example of this situation we mention the noncommutative analogue of the Yang–Feldman equation proposed in [3], and developed in [1, 5]. Finally, see [4, 17] for a different generalization of the Wick product, based on optimal localization; the resulting model is free of ultraviolet divergences.

16.3 Nonlocal Dyson Diagrams

In [7], Denk and Schweda showed that the above mentioned concerns were wrong; indeed, they were able to show that it was possible to absorb the time ordering in the definition of a simple generalization of the Stueckelberg–Feynman propagator, namely

$$\mathscr{D}(x;\tau) = \frac{1}{\mathrm{i}}\left(\Delta_+(x)\theta(\tau) + \Delta_+(-x)\theta(-\tau)\right).$$

When the theory is local, then the above general propagator (the "contractor", according to Denk and Schweda) is always evaluated at $\tau = x^0$, in which case it reproduces the usual Stueckelberg–Feynman propagator:

$$\mathscr{D}(x;x^0) = \Delta_{SF}(x).$$

The original argument (which is cast in the case of non-Wick ordered interactions, and worked out in the framework of the Gell–Mann & Low formula) is rather involved: it relies on a clever, though very tricky manipulation consisting of Wick reducing the *ordinary* pointwise products of fields, and recombining the products of two–point functions and θ functions by hand.

There is, however, a profound reason why such a clever rearrangement of terms leads to the desired result: indeed, one can instead reproduce exactly the same line of reasoning that can be found in any standard textbook on local quantum field theory, since the Wick reduction of *time ordered* products of Wick monomials can be performed in the nonlocal setting considered here, too [18]. In other words, the second Wick theorem is not local; this is so deeply true that even the original proof of Wick does not rely on locality [24].

Let us look at this in the case of the Dyson diagrams. Consider the Nth order contributions to the S-matrix:

$$S^{(N)} = \frac{(-\mathrm{i}g)^N}{N!} \int_{\mathbb{R}^N} dt_1 \cdots dt_N \, T[H_I^{\text{eff}}(t_1), \ldots, H_I^{\text{eff}}(t_N)].$$

We introduce the following shorthand:

$$\underline{x}_j = (x_{j1}, \ldots, x_{jn}) \in \mathbb{R}^{4n}, \quad d\underline{x}_j = \prod_{k=1}^{n} dx_{jk}, \quad \phi^{(n)}(\underline{x}_j) = \phi(x_{j1}) \cdots \phi(x_{jn}),$$

so that

$$H_I^{\text{eff}}(t) = \int_{\mathbb{R}^3} d\boldsymbol{a} \int_{\mathbb{R}^{4n}} d\underline{x} \, W_{(t,\boldsymbol{a})} :\phi^{(n)}(\underline{x}):,$$

and a short, standard computation gives[7]

$$\begin{aligned} S^{(N)} =& \frac{(-\mathrm{i}g)^N}{N!} \int_{(\mathbb{R}^4)^N} da_1 \cdots da_N \int_{(\mathbb{R}^{4n})^N} d\underline{x}_1 \cdots d\underline{x}_N \, W_{a_1}(\underline{x}_1) \cdots W_{a_N}(\underline{x}_N) \\ & \times T^{{a_1}^0, \ldots, {a_N}^0}[:\phi^{(n)}(\underline{x}_1):, \ldots, :\phi^{(n)}(\underline{x}_N):], \end{aligned} \tag{16.4}$$

where we introduced the following, natural notation:

$$T^{\tau_1, \tau_2, \ldots, \tau_k}[A_1, A_2, \ldots, A_k] = \sum_{\pi} A_{\pi(1)}) \cdots A_{\pi(k)} \prod_{j=1}^{k-1} \theta\big(\tau_{\pi(j+1)} - \tau_{\pi(j)}\big),$$

namely the product of the A_j's is taken in the order of decreasing τ_j's. Note that this definition may be given in general; there is no need for any a priori relation between the factors A_j and the parameters τ_j which we may wish to attach to those factors; we will call the above a *general* time ordered product, to highlight this fact.

The key remark here is that the mechanism for the Wick reduction of a general time ordered product works as usual. The only difference with respect to the local case is that, here, we have to keep in mind that to each field there corresponds a parameter driving its position in the time ordered product; this was implicit in the local case, where the time parameter corresponding to each field $\phi(x^0, \boldsymbol{x})$ was precisely x^0.

In view of this remark, we need a notation which indicates this correspondence explicitly, e.g.

$$:\overbrace{\phi(x_{11}) \cdots \phi(x_{1n})}^{{a_1}^0} \overbrace{\phi(x_{21}) \cdots \phi(x_{2n})}^{{a_2}^0} \cdots \overbrace{\phi(x_{N1}) \cdots \phi(x_{Nn})}^{{a_N}^0}: \tag{16.5}$$

Wick contractions[8], then, may be defined in the obvious way with respect to the

[7] We rename t_j as ${a_j}^0$, so that $\int_{\mathbb{R}} dt_j \int_{\mathbb{R}^3} d\boldsymbol{a}_j = \int_{\mathbb{R}^4} da$.

[8] Some reader might be surprised (I was surprised) to discover that the standard graphic notation for contractions is not due to Wick; it was first introduced by Houriet and Kind in *Helv. Phys. Acta* 22:319 (1949). In his paper, Wick politely complains that he was forced to abandon that very convenient notation for typographical reasons. Hence, he writes e.g. : $U^{\cdot}VW^{\cdot\cdot}X^{\cdot\cdot}Y^{\cdot}Z$: instead of :$UVWXYZ$:.

above correspondence:

$$
\begin{aligned}
&:\phi(x_{11})\cdots\underbrace{\phi(x_{iu})\cdots\phi(x_{j_v})}\cdots\phi(x_{Nn}):\\
=\quad &:\phi(x_{11})\cdots\widehat{\phi(x_{iu})}\cdots\widehat{\phi(x_{j_v})}\cdots\phi(x_{Nn}):\mathscr{D}(x_{iu}-x_{j_v};{a_i}^0-{a_j}^0),
\end{aligned}
$$

where a caret indicates omission. The reader is invited to read the beautiful original paper of Wick [24], to convince herself that the proof does not rely on the requirement that the time associated to the field is the same as the time argument of the field (see also [18] for a more detailed, "nonlocal" discussion). Hence, we still may state the second general Wick theorem, according to which

The general time ordered product

$$
T^{{a_1}^0,\ldots,{a_N}^0}[:\phi^{(n)}(\underline{x}_1):,\ldots,:\phi^{(n)}(\underline{x}_N):]
$$

equals the sum of the terms obtained by applying all possible choices of any number (including none) of allowed general contractions to (16.5), where no contraction is allowed, which involves two fields associated with the same time parameter ${a_j}^0$.

By applying the Wick reduction to (16.4), we obtain a certain sum of integrals, which we may label by means of Dyson diagrams, namely diagrams consisting of N (= the order in perturbation theory) vertices, each of which is the origin of no more than n (possibly none) lines for a $\phi^{\star n}$ interaction; each line connects two distinct vertices (no loops, no external lines). This is possible because a Wick monomial $:\phi(x_1)\cdots\phi(x_n):$ is totally symmetric in the arguments $x_1,\ldots,x_n$, so that we may safely replace each kernel $W_a(x_1,\ldots,x_n)$ by its totally symmetric part in (16.4). To see how to proceed, let us consider the following second-order contribution in the case of a $\phi^{\star 3}$ interaction:

$$
\begin{aligned}
&\frac{-g^2}{2}\int_{(R^4)^2} dadb \int_{(R^{4n})^2} d\underline{x}d\underline{y}\; W_a(\underline{x})W_b(\underline{y}) :\phi(x_1)\phi(x_2)\phi(x_3)\phi(y_1)\phi(y_2)\phi(y_3):\\
&\qquad=\frac{-g^2}{2}\int_{(R^4)^2} dadb \int_{(R^{4n})^2} d\underline{x}d\underline{y}\; W_a(\underline{x})W_b(\underline{y}) :\phi(x_3)\phi(y_1):\\
&\qquad\times\mathscr{D}(x_1-y_2;a^0-b^0)\mathscr{D}(x_2-y_3;a^0-b^0).
\end{aligned}
$$

It is clear that, up to renaming the integration variables and using the total symmetry of W_a, W_b and $:\phi(x_3)\phi(y_1):$, the above integral is exactly the same that we would obtain from the contribution

$$
\frac{-g^2}{2}\int_{(R^4)^2} dadb \int_{(R^{4n})^2} d\underline{x}d\underline{y}\; W_a(\underline{x})W_b(\underline{y}) :\phi(x_1)\phi(x_2)\phi(x_3)\phi(y_1)\phi(y_2)\phi(y_3): .
$$

Hence, the only information which is needed to write it down are that there are two Wick monomials and two contractions between those two monomials. To give this information, it is sufficient to draw a diagram with two vertices (the two Wick monomials) and two lines connecting them (the contractions). Of course, one also has to count the multiplicity of a diagram, namely the number of different contraction schemes that would lead to the same integral up to dummy integration variables.

16.4 Conclusions

We have seen that the usual diagrammatic expansion of the Dyson series is not special to local interactions. Indeed, diagrams are a consequence of two nonlocal tools, namely the Dyson perturbation series and the normal (Wick) ordering of products of creation and annihilation parts. Both these tools are completely unrelated to locality, hence this result should not have come out as a surprise. Feynman diagrams require an additional tool, the Gell–Mann & Low formula [13], which also has nothing to do with locality; indeed, it was shown in [18] that the dear old Feynman diagrams also arise naturally in the reduction of the nonlocal Green functions of the DFR perturbative model.

Actually, we did not introduce any really new argument, everything was already virtually contained in the papers of Dyson, Wick, and Gell–Mann & Low. The case of the usual local ϕ^n interaction can be reobtained as a special case, by setting $W_a(x_1, \ldots, x_n) = \prod_j \delta^{(4)}(x_j - a)$ in (16.4).

The unified treatment of both local and nonlocal ϕ^n interactions may be of some practical interest, since it allows us to study the convergence of the large scale limit diagramwise. Moreover, it may allow for developing a renormalization scheme for nonlocal theories with a strict correspondence between local and nonlocal subtractions. In the large scale limit, it might be natural to expect that nonlocal, possibly finite, subtractions converge to the infinite subtractions of the local renormalized theory. However, it should be kept in mind that a different point of view may well be taken. For example, in [5], a different prescription for the admissible subtractions was investigated, in order to only select those subtractions which are divergent already at the Planck length scale.

We close with the following remark. Many years ago, Caianiello raised some concerns about what he regarded as a too naive interpretation of Feynman diagrams as pictorial representations of actual scattering processes (see [6]). He remarked that, even if we were inclined to accept such a view prior to renormalization, some paradox might arise after implementing some subtraction prescription (e.g. in the fermionic case, strong interference among free field modes belonging to different diagrams might produce violations of the exclusion principle). Here, we found that, even prior to renormalization, diagrams seem to be nothing more than a graphic representation of the CCR algebra combined with ordinary quantum mechanical perturbation theory; it might be misleading to try to see more than that.

Appendix. Twisted Products

Actually, for both historical and technical reasons, the DFR twisted product was laid down in Fourier space[9]. To avoid confusion, in this appendix we will reserve

[9] Twisted products first arose in the framework of canonical quantization, in the late 1920s. Their use was first advocated by Weyl [23], who however did not publish explicit equations; following Weyl's suggestion, von Neumann [16] laid down the twisted product in Fourier space (twisted convolution). The twisted product in position space first appeared (in the form of an asymptotic expansion) in a paper by Grönewold; the integral form was first used by Baker and explicitly written down

the symbol $\times$ to indicate the ordinary convolution product, and $\tilde{\times}$ to indicate the twisted convolution product (twisted product in Fourier space). Setting

$$(\varphi_1 \tilde{\times} \varphi_2)(\sigma, k) = \int_{\mathbb{R}^4} dh\, \varphi_1(\sigma, h)\varphi_2(\sigma, k-h) \exp\left(\frac{\mathrm{i}}{2} k_\mu \sigma^{\mu\nu} h_\nu\right),$$

$$\varphi_j(\sigma, \cdot) \in L^1(\mathbb{R}^4),$$

one easily finds

$$F_1 \star F_2 = \widehat{\check{F}_1 \tilde{\times} \check{F}_2}, \qquad F_j(\sigma, \cdot) \in L^1(\mathbb{R}^4) \cap \widehat{L^1(\mathbb{R}^4)}.$$

A (formal) asymptotic expansion of the product $\star$ can be easily obtained by standard Fourier theory[10]. Indeed, by replacing the exponential factor $\exp[(\mathrm{i}/2)k\sigma h] = \exp[-(\mathrm{i}/2)h\sigma(k-h)]$ with the corresponding exponential series and (formally) exchanging the sum and the integration, one obtains

$$\begin{aligned}
&(F_1 \star F_2)(\sigma, x) \\
&= \int_{\mathbb{R}^4} dk\, e^{ikx} \int_{\mathbb{R}^4} dh\, \check{F}_1(\sigma, h)\check{F}_2(\sigma, k-h) \exp\left\{\frac{\mathrm{i}}{2}\sigma^{\mu\nu} k_\mu (h-k)_\nu\right\} \\
&\text{“=”}\ F_1(\sigma, x)F_2(\sigma, x) \\
&\quad + \sum_{n=1}^{\infty} \frac{(-i/2)^n}{n!} \sigma^{\mu_1\nu_1} \cdots \sigma^{\mu_n\nu_n} \big((\partial_{\mu_1} \cdots \partial_{\mu_n} F_1)\partial_{\nu_1} \cdots \partial_{\nu_n} F_2\big)(\sigma, x).
\end{aligned}$$

The above formal series gives a precise meaning to the more compact definition

$$\mathscr{A}[F_1 \star F_2](\sigma, x) = \left(\exp\left\{-\frac{\mathrm{i}}{2}\sigma^{\mu\nu}\partial_\mu \otimes \partial_\nu\right\} F_1 \otimes F_2\right)(\sigma, x) \tag{16.6}$$

of the asymptotic expansion $\mathscr{A}[F_1 \star F_2]$ of $F_1 \star F_2$; here, $\otimes$ is a tensor product of functions defined fibrewise over σ by

$$(F_1 \otimes F_2)(\sigma, x, y) = F_1(\sigma, x)F_2(\sigma, y),$$

and m is the fibrewise multiplication map

$$m(F_1 \otimes F_2)(\sigma, x) = F_1(\sigma, x)F_2(\sigma, x).$$

---

by Pool. The first rigorous results on asymptotic expansions of twisted products are probably due to Antonet, and a comprehensive investigation can be found in [11], to which we also refer for the bibliographical coordinates missing in this footnote. The seminal work of Weyl and von Neumann inspired Wigner to define the so-called Wigner transform; Wigner's work in turn led Moyal to define the so-called Moyal bracket or sine–bracket $\{f, g\}_\star = f \star g - g \star f$; the Moyal bracket then played a fundamental role in a seminal paper by Bayen et al about geometric quantization of phase manifolds. The covariant version of the twisted product was first introduced by DFR in order to quantize the space-time. For some strange reason, in the current literature about QFT on noncommutative space-time, the DFR variant of the Weyl–von Neumann twisted product is widely known as the Moyal product.

[10] With our conventions, we have $(\widehat{\partial_\mu f})(k) = -\mathrm{i}k_\mu \hat{f}(k)$ and $\hat{f} \times \hat{g} = (2\pi)^4 \widehat{fg}$.

Some authors take (16.6) as their definition of twisted product.

This asymptotic expansion, however, may be rather misleading. Assume that, for a fixed value of σ, the functions $F_j(\sigma, x)$ of x are $\mathcal{C}^\infty$ and have compact support, $j = 1, 2$; moreover, assume that the supports are disjoint. Since derivatives cannot enlarge the supports, we have $\mathscr{A}[F_1 \star F_2](\sigma, \cdot) \equiv 0$. In this precise sense, the asymptotic expansion of the twisted product defines (in the sense of formal power series) a *local*, non-commutative product. This is rather unsatisfactory from the point of view of space-time quantization, since on general grounds we expect the quantum geometry to be nonlocal.

Note also that, in the special case envisaged right above, there is no reason why we should have $F_1 \star F_2 \equiv 0$, and there are counterexamples, indeed. In other words, to naively rely on the above asymptotic expansion amounts to work with a completely different algebra than that of "true" twisted products. This also shows that the series may fail to converge (if it converges at all) to the actual twisted product $F_1 \star F_2$.

While for large classes of functions the asymptotic expansion truncated at order n agrees with the twisted product up to terms of order $\lambda_P^{2(n+1)}$, the issue of convergence is rather delicate, and there are very few general results (see [11], and references therein). Certainly, the asymptotic expansion converges to (the Antonet extension of) $F_1 \star F_2$ if $\check{F}_1, \check{F}_2$ have compact supports; in this case, F_1, F_2 are real–analytic, i.e. they are nonlocal in the sense that they cannot be deformed locally while preserving analiticity. In a sense, we may equivalently describe the nonlocality of quantized space-time using either (16.3) as a nonlocal product of local objects (L^1 functions), or (16.6) as a local product of nonlocal objects (analytic functions). As a (formal) product of smooth functions, (16.6) is irremediably local, and should be dismissed.

References

1. D. Bahns: *Perturbative Methods on the Noncommutative Minkowski Space.* Doctoral thesis, DESY-THESIS-04-004, Hamburg, 2003.
2. D. Bahns: *Ultraviolet Finiteness of the Averaged Hamiltonian on the Noncommutative Minkowski Space.* To be published. [arXiv:hep-th/0405224].
3. D. Bahns, S. Doplicher, K. Fredenhagen and G. Piacitelli: On the Unitarity Problem in space/time noncommutative theories. *Phys. Lett. B* 533:178 (2002). [arXiv:hep-th/0201222].
4. D. Bahns, S. Doplicher, K. Fredenhagen and G. Piacitelli: Ultraviolet finite quantum field theory on quantum space-time. *Commun. Math. Phys.* 237:221–241 (2003). [arXiv:hep-th/0301100].
5. D. Bahns, S. Doplicher, K. Fredenhagen and G. Piacitelli: Field theory on noncommutative space-times: Quasiplanar Wick products. *Phys. Rev. D* 71:25022 (2005). [arXiv:hep-th/0408204].
6. E.R. Caianiello: *Combinatorics and Renormalization in quantum Field Theory.* W.A. Benjamin, Reading, Mass., 1973.
7. S. Denk and M. Schweda: Time ordered perturbation theory for non-local interactions: Applications to NCQFT. *JHEP* 309:32 (2003). [arXiv:hep-th/0306101].
8. B. DeWitt: Quantizing Geometry. In: *Gravitation: An Introduction to Current Research* (L. Witten, ed.). Wiley, New York, 1962.

9. S. Doplicher, K. Fredenhagen and J.E. Roberts: The quantum structure of space-time at the Planck scale and quantum fields. *Commun. Math. Phys.* 172:187–220 (1995). [arXiv:hep-th/0303037].
10. F.J. Dyson: The Radiation Theories Of Tomonaga, Schwinger, and Feynman. *Phys. Rev.* 75:486–502 (1949).
11. R. Estrada, J.M. Gracia-Bondia and J. C. Varilly: On Asymptotic Expansions Of Twisted Products. *J. Math. Phys.* 30:2789–2796 (1989).
12. R.P. Feynman: Space-Time Approach To Quantum Electrodynamics. *Phys. Rev.* 76:769–789 (1949).
13. M. Gell-Mann and F. Low: Bound States in Quantum Field Theory. *Phys. Rev.* 84:350–354 (1951).
14. C.A. Mead: Possible Connection Between Gravitation and Fundamental Length. *Phys. Rev. B* 135:849–862 (1964).
15. C.A. Mead: Walking the Planck Length through History. Letter to the editor, *Phys. Today* 54 N. 11 (2001).
16. J. von Neumann: Die Eindeutigkeit der Schrödingerschen Operatoren. *Math. Ann.* 104:570 (1931).
17. G. Piacitelli: *Normal ordering of Operator Products on Noncommutative Space Time and Quantum Field Theory.* Doctoral thesis, 2001.
18. G. Piacitelli: Non local theories: New rules for old diagrams. JHEP 0408 031 (2004). [arXiv:hep-th/0403055].
19. D. Rivier and E.C.G. Stueckelberg: A Convergent Expression for the Magnetic Moment of the Neutron. Letter to the Editor of *Phys. Rev.* 74:218 (1948), and references 2,3,4 therein.
20. Y. Liao and K. Sibold: Time-ordered perturbation theory on noncommutative space-time: Basic rules. *Eur. Phys. J. C* 25:469 (2002). [arXiv:hep-th/0205269].
21. Y. Liao and K. Sibold: Time-ordered perturbation theory on noncommutative space-time. II. Unitarity. *Eur. Phys. J. C* 25:479 (2002). [arXiv:hep-th/0206011].
22. H.S. Snyder: Quantized Space-Time. *Phys. Rev.* 71:38 (1947).
23. H. Weyl: *Gruppentheorie und Quantenmechanik.* Hirzel, Leipzig 1928.
24. G.C. Wick: The Evaluation of the Collision Matrix. *Phys. Rev.* 80:268–272 (1950).

17

On Local Boundary CFT and Non-Local CFT on the Boundary

Karl-Henning Rehren

Institut für Theoretische Physik, Universität Göttingen,
37077 Göttingen, Germany
rehren@theorie.physik.uni-goe.de

Summary. The holographic relation between local boundary conformal quantum field theories (BCFT) and their non-local boundary restrictions is reviewed, and non-vacuum BCFT's, whose existence was conjectured previously, are constructed. (Based on joint work [18] with R. Longo.)

17.1 Introduction

This contribution highlights some aspects of a recent analysis of relativistic conformal QFT in the presence of a boundary [18]. The main result in [18] is that the local observables of a conformal field theory on the half-space $x > 0$ of two-dimensional Minkowski space-time ("local boundary conformal QFT" or BCFT for short) can be reconstructed from their restrictions to the boundary, which define a non-local chiral conformal QFT ("non-chiral local fields arise from non-local chiral fields"), and vice versa. This fact may be regarded as a "holographic" relation between quantum field theories in different space-time dimensions.

A more detailed statement of the above is the following: A local conformal QFT on the half-space $M_+ = \{(t,x) : x > 0\}$ contains a subalgebra of chiral fields, which may be naturally identified with its restriction to the boundary, i.e., a local CFT on the real line. Restricting the full BCFT to the boundary, one obtains a chiral CFT which is non-local, but relatively local with respect to its chiral subtheory. Conversely, every chiral CFT which is relatively local with respect to a given local chiral subtheory, induces a local BCFT on M_+. Under some (rather natural) technical assumptions, if one restricts a BCFT to the boundary and induces another BCFT from the restriction, then the latter is the maximal local extension of the original theory on the same Hilbert space. In particular, every BCFT is a subtheory of one arising by induction from some non-local chiral CFT.

This fact gives fresh motivation to study non-local chiral quantum field theories. E.g., to regard the non-local chiral theories on the boundary as the primary objects, opens a new route to classification of boundary CFT. Structure results on non-local chiral QFT have immediate consequences for the induced BCFT's (and their sub-theories).

While the statement of the main result amounts to the very simple identity (2.8) below, its proof is rather involved. It turns out advantageous to swap between the two-dimensional (BCFT) and the one-dimensional (chiral CFT) point of view. This "holographic" attitude is particularly powerful in combination with Modular Theory ([5,22], see below). In order to emphasize this situation, we reorganize in these notes the line of argument and focus the attention on the holographic and modular aspects in the interplay between chiral CFT and boundary CFT, rather than doubling the approach in [18].

The Modular Theory of von Neumann algebras [22] (briefly reviewed in the appendix) is most naturally tied to the algebraic approach to QFT [9]. Many results of great generality about local QFT have been obtained, when this theory is applied to the algebras of bounded local observables in suitable space-time regions in conjunction with the vacuum vector [4]. The most prominent is the Bisognano-Wichmann (BW) property [1] in local QFT, which states that the modular group Δ^{it} associated with the von Neumann algebra of observables in a wedge region and the vacuum vector, coincides with the unitary group of Lorentz boosts which preserve the wedge, hence the boosts are of modular origin. In fact the entire Poincaré group (including positivity of the energy spectrum [23]) can be constructed from modular groups of local algebras [13]. These methods are particularly powerful in local chiral QFT where the conformal group was found to be of modular origin [8].

For non-local chiral theories, some nontrivial results concerning the modular group of interval algebras have been previously obtained [5] for $\mathbb{Z}_2$-graded local (i.e., fermionic) theories, including the Bisognano-Wichmann property. We show that these results generalize to non-local chiral CFT's, whenever these are relatively local with respect to a subtheory (which is automatically local) which is contained with finite index; in particular, this is true whenever the subtheory is the fixed point subalgebra under the action of a finite internal symmetry group.

In Sect. 17.4 of this contribution, we construct positive-energy representations of chiral extensions associated with "nimreps" (non-negative integer matrix representations) of the fusion rules of the underlying chiral observables. By the holographic relation, these theories give rise to local "non-vacuum" boundary CFT's which were conjectured to exist in [18], while special cases were constructed explicitly in terms of chiral exchange fields [20]. Their general construction strengthens the parallelism between local BCFT and Euclidean BCFT in Statistical Mechanics, where analogous theories are widely considered for the remarkable properties of their partition functions Tr $\exp -\beta L_0$ and their relation to matrix elements of a two-dimensional heat kernel $\exp -(2\pi^2/\beta)(L_0^+ + L_0^-)$ at inverse temperature, between pairs of states from a finite family of distinguished improper "boundary states" [24].

17.2 Algebraic boundary conformal QFT

We denote by M_+ the half-space $M_+ = \{(t,x) : x > 0\}$, and by $I \times J$ the set $\{(t,x) : t+x \in I,\ t-x \in J\}$. If I and J are open intervals in $\mathbb{R}$, then $O = I \times J$ is a double-cone in Minkowski space-time, and $O \subset M_+$ iff $I > J$ elementwise.

A BCFT is a local quantum field theory on M_+, covariant under the subgroup of Möb $\times$ Möb (acting on $t+x$ and $t-x$ separately) which preserves the boundary. This is the diagonal subgroup, and will be identified with the Möbius group itself.

A BCFT contains certain chiral observables, e.g., the stress-energy tensor or conserved currents. Due to a boundary condition at $x = 0$, the left and right chiral fields of the BCFT coalesce and are identified with a local chiral field defined on the boundary $\mathbb{R}$. Let $A(I)$ denote the von Neumann algebras generated by the latter chiral observables smeared over an interval $I \subset \mathbb{R}$. The inclusion-preserving assignment $I \mapsto A(I)$ is called a *chiral net.* The chiral net A is local and Möbius covariant (and in fact extends to a net on the circle, into which $\mathbb{R}$ is embedded via a Cayley transformation). We shall henceforth assume that the chiral net A is *completely rational* [15], i.e., it is split ($A(I) \vee A(J) \simeq A(I) \otimes A(J)$ if I and J are disjoint without common boundary points) and the four-interval subfactor has finite index, implying strong additivity [19].

The von Neumann algebras generated by the chiral observables of the BCFT smeared in a double-cone $O = I \times J \subset M_+$ are

$$A_+(O) = A(I) \vee A(J). \tag{17.1}$$

This formula expresses the fact that smearing the BCFT chiral fields over $O = I \times J$ is the same as smearing the chiral fields on $\mathbb{R}$ over $I \cup J$.

In general, a BCFT will contain local observables beyond the chiral ones: A BCFT is a Möbius covariant local net $O \mapsto B_+(O)$ where $O \subset M_+$ and $B_+(O)$ contain $A_+(O)$. These von Neumann algebras act on a Hilbert space which is in general reducible as a representation of A_+. We call this representation π, thus

$$\pi(A_+(O)) \subset B_+(O). \tag{17.2}$$

The representation π of A_+ is at the same time a representation of the chiral net $I \mapsto A(I)$, and we require it to be a covariant positive-energy representation. We also assume that the algebra generated by $\pi(A)$ together with any single local algebra $B_+(O)$ acts irreducibly on the Hilbert space of B_+ ("joint irreducibility"), expressing the physical property that the stress-energy tensor (contained in A) locally generates the translations of the BCFT, and hence together with a single local algebra $B_+(O)$ generates the entire net.

Our aim is to understand the structure of the extension (17.2). It is the most remarkable conclusion of our work that this is possible in terms of (local or non-local) *chiral* nets which contain the chiral observables.

A *chiral extension* of A is a Möbius covariant net of inclusions

$$\pi(A(I)) \subset B(I) \tag{17.3}$$

in the vacuum representation of B, such that B and A are relatively local, i.e., $B(I_1)$ commute with $\pi(A(I_2))$ whenever I_i are disjoint. B contains A irreducibly if the inclusions (17.3) are irreducible. In this case, $B(I)$ are factors and (17.3) are subfactors.

The passage between chiral and boundary CFT makes use of two basic operations: "restriction" and "induction". Restriction associates a chiral net (over the intervals $I \subset \mathbb{R}$) with a given BCFT net (over the double-cones $O \subset M_+$), and induction associates a BCFT net with a given chiral net.

Every interval $I \subset \mathbb{R}$ defines a left wedge $W_L(I) = \{(t,x) \in M_+ : t+x \in I, t-x \in I\}$, and a right wedge $W_R(I) := W_L(I)'$ (the interior of the causal complement within M_+).

Let a BCFT B_+ be given. For either wedges, let $B_+(W) := \bigvee_{O\subset W} B_+(O)$. The *boundary net* $I \mapsto \partial B_+(I)$ is defined by *restriction* of B_+ to the left wedges,

$$\partial B_+(I) := B_+(W_L(I)). \tag{17.4}$$

For the time being, we also introduce the net $I \mapsto \partial' B_+(I)$ defined by

$$\partial' B_+(I) := B_+(W_R(I))' \tag{17.5}$$

but we shall later see that this net coincides with ∂B_+. Both chiral nets $B = \partial B_+$ or $= \partial' B_+$ are chiral extensions of A; joint irreducibility of B_+ implies irreducibility of the chiral extensions. But although B_+ is local, ∂B_+ and $\partial' B_+$ are in general non-local nets.

Conversely, let a chiral extension $\pi(A) \subset B$ be given. For $O = I \times J \subset M_+$ and $J = (a,b) < I = (c,d)$, let K and L denote the intervals $K = (b,c)$ and $L = (a,d)$. The *induced net* $O \mapsto B_+^{\rm ind}(O)$ is defined by

$$B_+^{\rm ind}(O) := B(K)' \cap B(L). \tag{17.6}$$

Even if B may be non-local, $B_+^{\rm ind}$ is local. If B contains A irreducibly, then $B_+^{\rm ind}$ satisfies joint irreducibility. It follows that $B_+^{\rm ind}$ is a BCFT net.

One has the following relations between the two constructions.

Theorem 1. *Let B be an irreducible local extension of A, and let B_+ be a BCFT net. Then:*

(T1) $\partial' B_+ = \partial B_+$.

(T2) $\partial(B_+^{\rm ind}) = B$.

(T3) $(\partial B_+)_+^{\rm ind} = (B_+)^{\rm dual}$.

The dual net in (T3) is defined standardly: since O' is the union of a left wedge W_L and a right wedge W_R, let $B_+(O') := B_+(W_L) \vee B_+(W_R)$. Then

$$(B_+)^{\rm dual}(O) := B_+(O')'. \tag{17.7}$$

The dual of a local net extends the local net but need not be local itself; if it is local, then it equals its own dual, i.e., it is self-dual.

Let us first discuss the far-reaching implications of this theorem.

Corollary 1.

(C1) *Every BCFT satisfies* wedge duality, *i.e., $B_+(W') = B_+(W)'$ for any wedge W.*

(C2) *Every irreducible chiral extension arises by restriction of some BCFT.*

(C3) *Every self-dual BCFT net arises by induction from some irreducible chiral extension.*

(C4) *The dual net of a BCFT is local, hence self-dual.*

(C5) *Every induced BCFT net is self-dual.*

(C6) *The maps $B \mapsto B_+^{\rm ind}$ (induction) and $B_+ \mapsto \partial B_+$ (restriction) give a bijection between self-dual BCFT nets and irreducible chiral extensions.*

Proof. (C1) is equivalent to (T1) by the definitions. (C2) is obvious from (T2), and (C3) from (T3). (C4) follows from (T3) because every induced net is local. (T2) and (T3) together imply (C5) because $(B_+^{\rm ind})^{\rm dual} = (\partial B_+^{\rm ind})_+^{\rm ind} = B_+^{\rm ind}$, as well as (C6) because $\partial(B_+^{\rm ind}) = B$ and $(\partial B_+)_+^{\rm ind} = (B_+)^{\rm dual} = B_+$. □

In particular, for self-dual BCFT nets, one has

$$B_+(O) = B(K)' \cap B(L) \tag{17.8}$$

($B = \partial B_+$ the corresponding boundary net, and K and L as in (17.6) before). One may regard the identity (17.8) as a "holographic relation" between a two-dimensional CFT and a one-dimensional (chiral) CFT: The inclusion (17.2) is completely determined by the chiral extension (17.3), see Sect. 17.3 below.

By (C6), classification of self-dual BCFT's is equivalent to classification of irreducible chiral extensions. This is a finite problem, because completely rational nets possess only finitely many irreducible chiral extensions. E.g., for Virasoro nets with $c < 1$ and for $SU(2)$ chiral current algebras, complete classifications of local chiral extensions have been obtained in [14], and non-local ones can be classified along the same lines.

A general BCFT B_+ is intermediate between $\pi(A_+)$ and $(B_+)^{\rm dual}$. Because one can show that $\pi(A_+(O)) \subset (B_+)^{\rm dual}(O)$ given by (17.8) has finite index, the number of intermediate nets is also finite.

We disentangle the proof of the theorem by a series of lemmas, which are of more or less interest of their own.

Lemma 1. (All nets and extensions here are assumed to be Möbius covariant. As before, chiral extensions are relatively local. Otherwise, locality of the nets is stated explicitly if assumed, and so is complete rationality.)

(L1) *If a local chiral net A is completely rational, then every irreducible chiral extension of A has finite index.*

(L2) *If B extends A with finite index and A is split, then B is split. (The split property for non-local nets is $B(K) \vee B(L)' \simeq B(K) \otimes B(L)'$ if K and L are disjoint intervals without common boundary points.)*

(L3) *If a chiral net is split, then $\bigcap_{K\subset L}[B(K) \vee B(L)'] = [\bigcap_{K\subset L} B(K)] \vee B(L)' = B(L)'$.*

(L4) *Every chiral net B has a maximal relatively local net $C \subset B$. C is unique, hence covariant and invariant under the gauge group of B. Clearly, C is local, and if $A \subset B$ is relatively local then $A \subset C \subset B$.*

(L5) *If $C \subset B$ is invariant under the gauge group of B, then there are local vacuum-preserving conditional expectations $B(I) \to C(I)$.*

(L6) *If $A \subset C$ and C is local, then there are local vacuum-preserving conditional expectations $C(I) \to A(I)$.*

(L7) *The local conditional expectations $B(I) \to A(I)$ of a chiral extension (existing according to* (L4)–(L6)*) are consistent, i.e., the expectations for two intervals coincide on the intersection of the corresponding local algebras, and are implemented by the projection on the subspace $\overline{A\Omega}$.*

(L8) *If $A \subset B$ has finite index and there are local vacuum-preserving conditional expectations $B(I) \to A(I)$, then B has the BW property if A does.*

(L9) *If a chiral extension B of A has finite index, then B has the BW property.*

Proof. (L1) [11] and (L2) [18] employ subfactor theory techniques (the generation of the larger algebra by the smaller and certain isometries). Proving (L3) and (L4) is elementary [18]. (L5–L9) invoke Modular Theory.

We focus on the steps involving Modular Theory (referring to the appendix for its results used here). (L5) and (L6) follow with Takesaki's Theorem about conditional expectations and modular stability of subalgebras as well as the characterization of the departure from the BW property in non-local theories, based on Borchers' modular commutation relations: in (L5), $C(I)$ is invariant under the modular group of $(B(I), \Omega)$ by (A.5), and in (L6), C, being local, satisfies the BW property, hence $A(I)$ is invariant under the modular group of $(C(I), \Omega)$. So in both cases (i) $\Rightarrow$ (ii) in Takesaki's Theorem (see the appendix) proves the claim. (L7) uses the restriction and implementation properties in Takesaki's Theorem: since A is local, the cocycle for A is trivial, hence the cocycle $z(t)$ for B is trivial on the cyclic subspace of A, hence the modular group of $B(I)$ acts trivially on $A(I)$. (L7) follows because the projection on $\overline{A(I)\Omega}$ is independent of I (Reeh-Schlieder theorem). By the same argument, it follows in (L8) that the fixed point subalgebra $B(I)^z$ of $B(I)$ under the one-parameter group $\mathrm{Ad}_{z(t)}$ of automorphisms contains $A(I)$, hence the index of the fixed point subalgebra is finite. Because $\mathbb{R}$ has no nontrivial finite quotients, we must have $B(I)^z = B(I)$, hence $B(I)$ commutes with $z(t)$. Ω being cyclic for $B(I)$; this implies that $z(t)$ is trivial, hence (L8). Combining (L4)–(L8), gives (L9) because A, being local, has the BW property. □

Now the theorem is proven easily. (L1)–(L3) mean that $\bigvee_{K\subset L} B_+^{\mathrm{ind}}(O) = B(L)$, which is the statement of (T2). By (L1) and (L9) we conclude that $\partial B_+(I)$ and $\partial' B_+(I)$ have the same modular group (namely the dilations of I). Because $\partial B_+(I) \subset \partial' B_+(I)$ and Ω is cyclic for both algebras, the implementation property in Takesaki's Theorem implies equality. This is (T1). (T3) is then obvious by writing (17.7) as $\partial B_+(K) \cap \partial' B_+(L)$. □

Along the way, we proved the following proposition (assembling results on chiral extensions) and its corollary (assembling the implications for BCFT):

Proposition 1.

(P1) *Every chiral extension has a consistent family of vacuum-preserving conditional expectations* $\mathcal{E}^I : B(I) \to A(I)$.

(P2) *If a chiral net B is a chiral extension with finite index of a local net A, then B satisfies the BW property. In particular, this is the case for all irreducible chiral extensions of completely rational nets.*

(P3) *The split property is upward hereditary for chiral extensions of finite index.*

Proof. (P1) is the combination of (L4)–(L7). (P2) is (L8), using the fact that a local chiral net has the BW property. (P3) is (L2). □

The existence of a consistent family of vacuum-preserving conditional expectations has been a crucial assumption, expressing a generalized symmetry principle, in the general structure theory for chiral extensions [17]. We see from (P1–P3) that it is automatic for boundary nets of a BCFT which are irreducible extensions of a completely rational chiral net. Thus, subfactor methods as developed in [17] may be applied.

Corollary 2. (C7) *Every BCFT satisfies the split property for wedges.*

(C8) *The index of* $\pi(A_+(O)) \subset B_+(O)$ *is universal for self-dual BCFT nets* B_+ *(i.e., it depends only on the chiral observables* A*).*

(C9) *A self-dual BCFT is strongly additive, i.e., if* O_i *are two spacelike separated double-cones which touch each other in one point and* $O = I \times J$ *is the double-cone spanned by* O_1 *and* O_2*, then* $B_+(O_1) \vee B_+(O_2) = B_+(O)$*.*

(C10) *A self-dual BCFT satisfies Haag duality for finite unions of spacelike separated double-cones.*

(C11) *A self-dual BCFT net has no DHR sectors.*

Proof. (C7) follows for induced BCFT nets by the definition because B is split by (P3) because A is completely rational. Then (C7) is true for any BCFT because B_+ is contained in $(B_+)^{\mathrm{dual}}$, which is an induced net. For (C8), let $B = \partial B_+$ denote the boundary net, and consider the chain of inclusions

$$\pi(A(K)) \vee \pi(A(L')) \subset \pi(A(K)) \vee B(L)' \subset B(K) \vee B(L)' \subset \pi(A(I))' \cap \pi(A(J))'.$$

By the split property (P3) for B, the first two inclusions both have index $[B : A]$, and by the general theory in [15], the total index equals $d(\pi)^2 \mu_A$ where $d(\pi) = [B : A]$ and μ_A is the "dimension" of the DHR superselection category of A, i.e., the sum of the squares of the dimensions of all irreducible DHR sectors of A. Hence, the index of the last inclusions is μ_A, and by passing to the commutants, this is the index of the subfactor in (C8).

(C8) implies (C9) by standard subfactor methods [16], relying only on the finiteness of the index in (C8). By similar methods as for (C8), the index of $B_+(E) \subset B_+(E')'$ when E is a finite union of spacelike separated double-cones, is computed [18] to be a power of $[(B_+)^{\mathrm{dual}} : B_+]$, hence it is 1 if B_+ is self-dual, implying (C10). (C10) implies (C11) by standard arguments. □

17.3 The charge structure of BCFT fields

The holographic formula (17.8) allows for an explicit computation of the local algebras $B_+(O)$ of a self-dual boundary CFT in terms of the chiral extension $A \subset B$. It is found [18] that $B_+(O)$ is generated by $A_+(O)$ along with a finite system of charged BCFT operators, carrying a product of chiral charges which is bi-localized in I and in J if $O = I \times J$.

The notion of chiral charge here refers to the irreducible superselection sectors (generalized charges) of the theory A. The bi-localized charge structure expresses itself in the commutation relations with the local fields on the boundary: A charged operator $\psi \in B_+(O)$ is an intertwiner for $\sigma \circ \bar{\tau}$ where σ and $\bar{\tau}$ are DHR endomorphisms of A localized in I and in J, respectively. Hence ψ carries a charge $[\sigma]$ localized in I and a charge $[\bar{\tau}]$ localized in J.

The precise combinatorial structure (including the pairing of chiral charges) of these local charged intertwiners in terms of non-local "chiral vertex operators" can be determined by solving a certain eigenvalue problem, depending on the chiral extension $A \subset B$, within the DHR superselection category of A.

The admissible pairing of chiral charges is described by a modular invariant matrix $Z_{\sigma,\tau}$. It is also proven in [18] that the induced BCFT $B_+^{\rm ind}$ is locally isomorphic to a local two-dimensional CFT B_2 on the entire Minkowski space-time (whose Hilbert space is given by the same modular invariant matrix Z), previously obtained by a "generalized quantum double construction" [21]. There is thus a common (sub)net, defined sufficiently far away from the boundary, of which B_2 and $B_+^{\rm ind}$ may be regarded as different quotients (representations) distinguished by the absence or presence of the boundary.

17.4 Nimreps and non-vacuum BCFT

This section closes a gap which was left open in [18]. We canonically associate with a given non-local chiral extension $\pi(A) \subset B$ a family of chiral extensions B_b along with a family of positive-energy representations π^{ab} of B_b where the labels a and b ("boundary conditions") run over the same finite set. The holographic construction yields a corresponding family of BCFT's $B_{+,ab}$.

The multiplicities n^ρ_{ab} of the irreducible DHR sectors $[\rho]$ of the chiral observables A within π^{ab} form a normalized "nimrep" (non-negative integer matrix representation) of the fusion rules of the DHR sectors of A,

$$n^\sigma \cdot n^\rho = \textstyle\sum_\tau N^\tau_{\sigma\rho}\, n^\tau \qquad \text{such that} \qquad n^0_{ab} = \delta_{ab}. \tag{17.9}$$

In particular, unless $a = b$, these representations do not contain the vacuum sector. The property (17.9) implies that the partition functions $\mathrm{Tr}\,\exp -\beta L_0$ of $B_{+,ab}$ coincide with matrix elements of the heat kernel $\exp -(2\pi^2/\beta)(L_0^+ + L_0^-)$ of the associated Minkowski theory B_2 between pairs of states $|a\rangle$ and $|b\rangle$ from a finite family of distinguished improper "boundary states" [24].

Our construction starts by "acting with the DHR sectors of A" on $\pi(A) \subset B$ as in [2, 18]: Let ι denote the inclusion homomorphism $\iota : A(I) \to B(I)$ for a fixed interval I. Then consider homomorphisms $\iota \circ \sigma : A(I) \to B(I)$ where σ runs over the DHR endomorphisms of A localized in I, and let X be the (finite) set of all equivalence classes $[a]$ of irreducible subhomomorphisms[1] $a \prec \iota \circ \sigma$ as σ varies. For every pair a, b in X and $\bar{a} : B(I) \to A(I)$ a conjugate[2] homomorphism of a, the product $\bar{a} \circ b : A(I) \to A(I)$ is (the local restriction of) a DHR endomorphism localized in I. It follows that the numbers[3]

$$n^\rho_{ab} = \mathrm{dimHom}(\rho, \bar{a} \circ b) \equiv \mathrm{dimHom}(a \circ \rho, b) \tag{17.10}$$

form a nimrep as above, including the normalization.

[1] A homomorphism $\beta : N \to M$ is a subhomomorphism ($\beta \prec \alpha$) of $\alpha : N \to M$, if the intertwiner space $\mathrm{Hom}(\beta, \alpha) := \{t \in M : t\beta(n) = \alpha(n)t \;\forall\, n \in N\}$ contains an isometry, $t^*t = 1$; hence $\beta(n) = t^*\alpha(n)t$.

[2] A homomorphism $\bar{\alpha} : M \to N$ is conjugate to $\alpha : N \to M$, if $\mathrm{id}_N \prec \bar{\alpha} \circ \alpha$ and $\mathrm{id}_M \prec \alpha \circ \bar{\alpha}$. Since A is assumed to be completely rational, all homomorphisms in the sequel decompose into finitely many irreducibles, and possess conjugates.

[3] Strong additivity of A guarantees equivalence between local and global intertwiners, hence the notions of equivalence and subendomorphisms are the same for DHR endomorphisms and for their local restrictions.

The construction of the family of chiral CFT's and hence BCFT's associated with this nimrep now relies on the following equivalence.

Theorem 2. [17] *There is a 1 : 1 correspondence (up to unitary resp. algebraic equivalences) between*

(i) *irreducible covariant chiral extensions $\pi(A) \subset B$, defined on the vacuum Hilbert space of B.*
(ii) *irreducible subfactors $A_1 \subset A(I_0)$ (for any fixed interval I_0) whose canonical endomorphism*[4] *is (the local restriction of) a DHR endomorphism θ of A localized in I_0.*

More precisely [17, Cor. 3.3 and Prop. 3.4]*: a chiral extension gives an inclusion homomorphism $\iota : A \to B$ such that $\pi = \pi^0 \circ \iota$, where π^0 is the defining vacuum representation of B. For any interval I_0, one can construct a conjugate homomorphism $\bar{\iota} : B \to A$ such that $\bar{\iota}\restriction_{B(I)}: B(I) \to A(I)$ is conjugate to $\iota\restriction_{A(I)}: A(I) \to B(I)$ whenever $I \supset I_0$, and $\theta = \bar{\iota} \circ \iota$ is a DHR endomorphism of A localized in I_0. Then π^0 is unitarily equivalent to $\pi_0 \circ \bar{\iota}$, where π_0 is the defining vacuum representation of A, and consequently π is unitarily equivalent to $\pi_0 \circ \theta$. This unitary equivalence turns the covariance $\mathfrak{U}$ of B into the covariance*[5] *$\mathfrak{U}_\theta$ of A in the DHR representation $\pi_0 \circ \theta$.*

Proof. The 1:1 correspondence in the theorem is given by the subfactor

$$A_1 := \bar{\iota}(B(I_0)) \subset A(I_0). \tag{17.11}$$

From the subfactor $A_1 \subset A(I_0)$, one can reconstruct $A(I_0) \subset B(I_0)$ by the Jones basic construction [12], along with a conditional expectation which extends the vacuum state on $A(I_0)$ to a state on $B(I_0)$. The construction of the entire net B is then achieved with the help of charge transporters or using the Möbius covariance [17, Thm. 4.9]. The vacuum state extends to the net, and its GNS representation is the vacuum representation of B.

Now, let a chiral extension $\pi(A) \subset B$ be given. Fix an interval I_0 and let $\iota : A(I_0) \to B(I_0)$ denote the local inclusion homomorphism. Then $\theta = \bar{\iota} \circ \iota$ is (the local restriction of) a DHR endomorphism localized in I. Let σ be a DHR endomorphism localized in I_0 and $a \prec \iota \circ \sigma$ an irreducible sub-homomorphism as before. Then the subfactor $A_{1,a} := \bar{a}(B(I_0)) \subset A(I_0)$ has canonical endomorphism $\theta_a = \bar{a} \circ a$ which is (the local restriction of) a DHR endomorphism because $\theta = \bar{\iota} \circ \iota$ is (the local restriction of) a DHR endomorphism and $\bar{a} \circ a \prec \bar{\sigma} \circ \theta \circ \sigma$. By (iii)$\Rightarrow$(i), the subfactor $A_{1,a} \subset A(I)$ defines a chiral extension $\pi_a(A) \subset B_a$.

This construction results in a finite family of chiral extensions in their vacuum representations (called the "DHR orbit" in [18]; note that inequivalent $a \in X$ may result in equivalent extensions, e.g., if $a_1 = a_2 \circ \tau$ where τ is a DHR automorphism).

Each element of this family generates the whole family (and the same nimrep) by the same construction [2, 3]: namely, starting from $A(I_0) \subset B(I_0)$, let $c \in X$, and construct $A(I_0) \subset B_c(I_0)$ as before with inclusion homomorphism

[4]If $N \subset M$ with inclusion homomorphism ι and conjugate $\bar{\iota}$, then $\bar{\iota}\circ\iota \in \mathrm{End}(N)$ is the canonical endomorphism, and $\iota \circ \bar{\iota} \in \mathrm{End}(M)$ the dual canonical endomorphism.

[5]The covariances here are the unitary representations of the Möbius group such that $\mathrm{Ad}_{\mathfrak{U}(g)} B(I) = B(gI)$ and $\mathrm{Ad}_{\mathfrak{U}_\theta(g)} \circ \pi_0 \circ \theta(A(I)) = \pi_0 \circ \theta(A(gI))$, respectively.

$\iota_c : A(I_0) \to B_c(I_0)$ and dual canonical endomorphism $\theta_c = \bar\iota_c \circ \iota_c = \bar c \circ c$. Then because $\bar c(B(I_0)) = A_{1,c} = \bar\iota_c(B_c(I_0))$, the mapping $\varphi_c := \bar c^{-1} \circ \bar\iota_c : B_c(I_0) \to B(I_0)$ is well defined, and is an algebraic isomorphism between $B_c(I_0)$ and $B(I_0)$. Using $A(I_0) \subset B_c(I_0)$ as the starting point instead, denote by X_c the set of equivalence classes of irreducible subhomomorphisms $a_c : A(I_0) \to B_c(I_0)$ of $\iota_c \circ \sigma$ for DHR endomorphism σ localized in I_0. Then $\varphi_c \circ a_c : A(I_0) \to B(I_0)$ is contained in $\bar c^{-1} \circ \bar\iota_c \circ \bar\iota_c \circ \sigma = \bar c^{-1} \circ \bar c \circ c \circ \sigma = c \circ \sigma \prec \iota \circ \sigma' \circ \sigma$ because c is contained in $\iota \circ \sigma'$ for some σ'. Hence $\varphi_c \circ a_c$ belongs to X. Conversely, if $a \prec \iota \circ \sigma$ belongs to X, then $a \prec c \circ \bar\sigma' \circ \sigma$ for some σ', hence $\varphi_c^{-1} \circ a \prec \bar\iota_c^{-1} \circ \bar c \circ c \circ \bar\sigma' \circ \sigma = \iota_c \circ \bar\sigma' \circ \sigma$ belongs to X_c.

This means that φ_c provides by left composition a bijection between the irreducible subhomomorphisms of $\iota \circ \sigma : A(I_0) \to B(I_0)$ as σ varies, and those of $\iota_c \circ \sigma$: $A(I_0) \to B_c(I_0)$ as σ varies. Moreover, $\bar a_c(B_c(I_0)) = \bar a_c \circ \varphi_c^{-1}(B(I_0)) = \bar a(B(I_0))$ give the same subfactors $A_{1,a} \subset A(I_0)$, and hence the same family of chiral extensions. Obviously, the bijection preserves the nimrep $n^\rho_{ab} = \dim\mathrm{Hom}(\rho, \bar a \circ b)$.

We now turn to positive-energy representations of chiral extensions, using α-induction [17]. α-induction covariantly extends a DHR endomorphism σ of A to an endomorphism of its extension B, and hence to a covariant induced representation of B. The latter is unitarily equivalent to $\pi_0 \circ \sigma \circ \bar\iota$ [17]. This representation of B restricts to the DHR representation $\pi_0 \circ \sigma \circ \bar\iota \circ \iota = \pi_0 \circ \sigma \circ \theta$ of A. Since the covariance of B in the representation $\pi_0 \circ \bar\iota$ is $\mathfrak{U}_\theta$, the covariance in induced representation $\pi_0 \circ \sigma \circ \bar\iota$ is $\mathfrak{U}_\sigma \sigma(\mathfrak{U}_0^* \mathfrak{U}_\theta) = \mathfrak{U}_{\sigma\circ\theta}$ [6]. The induced representation is in general reducible.

Let $a, b \in X$. Then $\bar a \circ b$ is the local restriction of a DHR endomorphism θ_{ab} with multiplicities n^ρ_{ab}. Moreover, there is a DHR endomorphism σ of A localized in I_0 and an isometric local intertwiner $t \in \mathrm{Hom}(\bar a, \sigma \circ \bar b) \subset A(I_0)$. We want to show that t reduces the induced representation $\pi_0 \circ \sigma \circ \bar\iota_b$ of B_b. Namely, t is an intertwiner between the local restrictions of the DHR endomorphisms $\theta_{ab} = \bar a \circ b$ and $\sigma \circ \theta_b = \sigma \circ \bar b \circ b$, and hence it is also a global intertwiner. The projection tt^* thus commutes with $\sigma \circ \bar b(B(I_0)) = \sigma \circ \bar\iota_b(B_b(I_0))$ (i.e., locally) and with $\sigma \circ \bar\iota_b \circ \iota_b(A)$ (globally). Since $B_b(I_0)$ and A generate the entire net B_b, tt^* commutes globally with $\sigma \circ \bar\iota_b(B_b)$.

Thus tt^* defines a subrepresentation $\hat\pi^{ab} = \pi_0 \circ \mathrm{Ad}_{t^*} \circ \sigma \circ \bar\iota_b$ of the induced representation of B_b, which restricts to the DHR representation $\pi_0 \circ \mathrm{Ad}^*_t \circ \sigma \circ \theta_b = \pi_0 \circ \theta_{ab}$ of A. Because $t \in \mathrm{Hom}(\theta_{ab}, \sigma \circ \theta_b)$ also intertwines the covariances of the DHR endomorphisms [6], $\hat\pi^{ab}$ is covariant with covariance $\mathfrak{U}_{\theta_{ab}}$.

By choosing a unitary operator U to transport $\pi_0 \circ \theta_{ab}$ to a representation π_{ab} of A on $\mathcal{H}_{ab} = \bigoplus_\rho n^\rho_{ab} \mathcal{H}_\rho$, we obtain the desired covariant extensions

$$\pi_{ab}(A(I)) \subset \pi^{ab}(B_b(I)) \tag{17.12}$$

with

$$\pi^{ab} := \mathrm{Ad}_U \circ \pi_0 \circ \mathrm{Ad}_{t^*} \circ \sigma \circ \bar\iota_b. \tag{17.13}$$

These theories are positive-energy representations (subrepresentations of α-induced representations) of the chiral extensions B_b, which were previously defined in their vacuum representations. □

Note that in (17.13) we cannot write $\mathrm{Ad}_{t^*} \circ \sigma \circ \bar\iota_b = \mathrm{Ad}_{t^*} \circ \sigma \circ \bar b \circ \varphi_b = \bar a \circ \varphi_b$ because φ_b is only defined on $B_b(I_0)$ while $\bar\iota_b$ extends to the net B_b. Nevertheless, π^{ab} depends only on a and b, and not on the choice of σ and $t \in \mathrm{Hom}(\bar a, \sigma \circ \bar b)$. Namely, this is true for $\pi^{ab}(B_b(I_0)) = \mathrm{Ad}_U \circ \pi_0 \circ \bar{a} \circ \varphi_b(B_b(I_0)) = \mathrm{Ad}_U \circ \pi_0 \circ \bar{a}(B(I_0))$, and $\pi^{ab}(B_b(I))$ is obtained by acting with the covariance of $\mathrm{Ad}_U \circ \pi_0 \circ \theta_{ab}$ on $\pi^{ab}(B_b(I_0))$.

In the case $a = b$, we may choose σ trivial and $t = \mathbf{1}$ in (17.13). Thus $\pi^{aa}(B_a)$ coincides with the chiral extension B_a in its vacuum representation.

Corollary 3. *The same "holographic construction" as in* (17.6)*:*

$$B_{+,ab}(O) := \pi^{ab}(B_b(K))' \cap \pi^{ab}(B_b(L)) \tag{17.14}$$

for all $a, b \in X$ defines a family of covariant BCFT's containing $\pi_{ab}(A_+)$ on the Hilbert spaces $\mathcal{H}_{ab}$ with multiplicities given by the nimrep (17.10)*.*

Proof. The corollary is obvious by the results of [18] mentioned in Sect. 17.2. □

17.5 Appendix: Modular Theory in QFT and in BCFT

We assemble those aspects of Modular Theory, which are particularly relevant for BCFT (and for QFT in general). We recommend also [4] and [22].

The fundamental result of Modular Theory is the following.

Theorem 3 (Tomita). [22, Chap. VI, Thm. 1.19] *Let M be a von Neumann algebra on a Hilbert space $\mathcal{H}$ with a cyclic and separating vector Ω. Then the anti-linear operator $S : m\Omega \mapsto m^*\Omega$ is closable, and its closure has the polar decomposition $\overline{S} = J\Delta^{\frac{1}{2}}$ where $J = J_{(M,\Omega)}$ is an anti-unitary involution and $\Delta = \Delta_{(M,\Omega)} \geq 0$ is an invertible positive self-adjoint operator such that Δ^{it} is a unitary one-parameter group. These "modular data" have the properties*

(i) $\Delta J = J\Delta^{-1}$, $\Delta^{it} J = J\Delta^{it}$.

(ii) Δ^{it} *implements a one-parameter group of automorphisms σ_t of M, i.e.,*

$$\sigma_t(M) = M \qquad \text{where} \qquad \sigma_t(m) := \Delta^{it} m \Delta^{-it} \quad (m \in M). \tag{17.15}$$

(iii) *The conjugation J maps M onto its commutant, i.e.,*

$$j(M) = M' \qquad \text{where} \qquad j(m) := JmJ \quad (m \in M). \tag{17.16}$$

(iv) *The state $\omega = (\Omega, \cdot\, \Omega)$ is a KMS state of inverse temperature $\beta = 1$ on M with respect to the "dynamics" given by the inverse modular automorphism group σ_{-t}. The modular automorphism group is determined by this property.*

The modular data contain highly nontrivial information about the "position" of a von Neumann algebra in its Hilbert space relative to the distinguished state. In local QFT, this information is of dynamical nature. The Bisognano-Wichmann theorem [1] states that, if Ω is the vacuum vector and M the von Neumann algebra of observables localized in a wedge region, then the modular conjugation J is a CPT operator (which in asymptotically complete QFT is related to the scattering matrix), and the modular group Δ^{it} coincides with the unitary group of Lorentz boosts which preserve the wedge:

$$\Delta^{it}_{(A(W),\Omega)} = U(\Lambda_W(-2\pi t)). \tag{17.17}$$

By the statement (iv), the BW property (17.17) explains the Unruh effect, that an accelerated observer whose "dynamics" is given by the boosts in the Rindler wedge senses the vacuum state like a thermal state. Also the Hawking effect can be explained in this way, because an observer of a black hole "at rest" in constant distance from the surface, is in fact an accelerated observer.

Since, as the wedges vary, the corresponding Lorentz boosts generate the Poincaré group, the BW property implies that the unitary representation of the Poincaré group is of modular origin. In a more ambitious program [13], one attempts to derive the Poincaré group from the modular data of a finite set of wedge algebras, which are required to be in a suitable "relative modular position" such as to ensure the correct relations among their respective modular groups in order to generate the Poincaré group.

In chiral theories, wedges are replaced by $\mathbb{R}_+$ and the Lorentz boosts by the dilations. By conformal covariance, this situation is transported to any interval and the subgroup of Möb which preserves the interval. Thus, the BW property is the statement that the modular group of an interval algebra in the vacuum coincides with the unitary representation of this subgroup. Since, as the interval varies, these subgroups generate the Möbius group, the BW property implies that the conformal group is of modular origin.

Of relevance for BCFT are mainly the following two profound theorems.

Theorem 4 (Takesaki). [22, Chap. IX, Thm. 4.2] *Let $N \subset M$ be a pair of von Neumann algebras on a Hilbert space $\mathcal{H}$ and $\Omega \in \mathcal{H}$ a vector which is cyclic and separating for M. Then the following are equivalent:*

(i) *N is globally invariant under the modular automorphism group of (M, Ω), i.e., $\sigma_t^{(M,\Omega)}(N) = N$.*
(ii) *There exists a conditional expectation $\mathcal{E} : M \to N$ which preserves the state $\omega = (\Omega, \cdot\, \Omega)$, i.e., $\omega\restriction_N (\mathcal{E}(m)) = \omega(m)$.*

In this case, $\mathcal{E}$ is implemented by the projection E onto the cyclic subspace $\overline{N\Omega}$, i.e., $EmE = \mathcal{E}(m)E$. Moreover, the modular data associated with (M, Ω) restrict on the cyclic subspace to those associated with (N, Ω), i.e., $E\Delta_{(M,\Omega)} = \Delta_{(M,\Omega)}E = \Delta_{(N,\Omega)}$ and $EJ_{(M,\Omega)} = J_{(M,\Omega)}E = J_{(N,\Omega)}$. In particular, if Ω is cyclic also for N, then $E = \mathbf{1}$ and $N = M$.

Theorem 5 (Borchers). [4, Thm. II.5.2] *Let M be a von Neumann algebra on a Hilbert space $\mathcal{H}$ and $\Omega \in \mathcal{H}$ a vector which is cyclic and separating for M. Let $U(s)$ be a unitary group preserving Ω such that $U(s)MU(-s) \subset M$ for $s > 0$. If the generator of U is positive, then one has*

$$\Delta^{it}U(s)\Delta^{-it} = U(e^{-2\pi t}s). \tag{17.18}$$

Choosing $M = B(\mathbb{R}_+)$ in a Möbius covariant chiral QFT, Ω the vacuum, and $U(s)$ the translation subgroup of the Möbius group, the spectrum condition implies by Borchers' theorem that the modular group has the same commutation relations with the translations as the dilations (scaled by a factor of -2π). This implies [5,7] that the modular group coincides with the unitary representers of the scaled dilation subgroup up to a unitary cocycle $z(t)$ taking values in the center of the gauge group of B,

$$\Delta^{it}_{(B(I),\Omega)} = U(\Lambda_I(-2\pi t)) \cdot z(t). \tag{17.19}$$

The BW property is the statement that this cocycle is trivial.

If B is local, it has been shown previously [7] that the BW property holds, by extending the theory to a net on the circle on which the Möbius group acts globally, and using the triviality of $U(R(2\pi))$ where R is the subgroup of rotations of the circle. This argument has been generalized [5] to $\mathbb{Z}_2$-graded (fermionic) theories, but it fails for more general non-local nets for which $U(R(4\pi)) \neq 1$. Our result (P2) provides an alternative sufficient condition for the BW property of non-local chiral nets which does not depend on the spectrum of the rotations.

References

1. J.J. Bisognano and E. H. Wichmann: On the duality condition for quantum fields. *J. Math. Phys.* 17:303–321 (1976).
2. J. Böckenhauer and D.E. Evans: Modular invariants and subfactors. *Fields Inst. Commun.* 30:11–37 (2001). [arXiv:math.OA/0008056]
3. D.E. Evans: Fusion rules of modular invariants. *Rev. Math. Phys.* 14:709–732 (2002).
4. H.-J. Borchers: On revolutionizing QFT with modular theory. *J. Math. Phys.* 41:3604–3673 (2000).
5. C. D'Antoni, R. Longo and F. Radulescu: Conformal nets, maximal temperatur and models from free probability. *J. Oper. Theory* 45:195–208 (2001). [arXiv:math.OA/9810003].
6. K. Fredenhagen, K.-H. Rehren and B. Schroer: Superselection sectors with braid group statistics II. *Rev. Math. Phys.* SI1 (Special issue):113–157 (1992).
7. D. Guido and R. Longo: The conformal spin and statistics theorem. *Commun. Math. Phys.* 181:11–35 (1996).
8. D. Guido, R. Longo and H.-W. Wiesbrock: Extensions of conformal nets and superselection structures. *Commun. Math. Phys.* 192:217–244 (1998).
9. R. Haag: *Local Quantum Physics.* Springer Verlag, Berlin, Heidelberg, New York, 1996.
10. M. Izumi and H. Kosaki: On a subfactor analogue of the second cohomology. *Rev. Math. Phys.* 14:733–757 (2002).
11. M. Izumi, R. Longo and S. Popa: A Galois correspondence for compact groups of automorphisms of von Neumann algebras with a generalization to Kac algebras. *J. Funct. Anal.* 155:25–63 (1998).
12. V. Jones: Index for subfactors. *Invent. Math.* 72:1–25 (1983).
13. R. Kähler and H.W. Wiesbrock: Modular theory and the reconstruction of four-dimensional quantum field theories. *J. Math. Phys.* 42:74–86 (2001).
14. Y. Kawahigashi and R. Longo: Classification of local conformal nets: case $c < 1$. To appear in *Ann. Math.* [arXiv:math-ph/0201015].
15. Y. Kawahigashi, R. Longo and M. Müger: Multi-interval subfactors and modularity of representations in conformal field theory. *Commun. Math. Phys.* 219:631–669 (2001).
16. R. Longo: Conformal subnets and intermediate subfactors. *Commun. Math. Phys.* 237:7–30 (2003).
17. R. Longo and K.-H. Rehren: Nets of subfactors. *Rev. Math. Phys.* 7:567–597 (1995).

18. R. Longo and K.-H. Rehren: Local fields in boundary conformal QFT. *Rev. Math. Phys.* 16:909–960 (2004).
19. R. Longo and F. Xu: Topological sectors and a dichotomy in conformal field theory. *Commun. Math. Phys.* 251:321–364 (2004).
20. U. Pennig: Diploma thesis, Göttingen, work in progress.
21. K.-H. Rehren: Canonical tensor product subfactors. *Commun. Math. Phys.* 211:395–406 (2000).
22. M. Takesaki: *Theory of Operator Algebras II*. Springer Encyclopedia of Mathematical Sciences, Vol. 125, 2003.
23. H.-W. Wiesbrock: A comment on a recent work of Borchers. *Lett. Math. Phys.* 25:157–159 (1992).
24. J.-B. Zuber: CFT, BCFT, ADE and all that. *Contemp. Math.* 294:230–266 (2002). [arXiv:hep-th/0006151].

18

Algebraic Holography in Asymptotically Simple, Asymptotically AdS Space-times

Pedro Lauridsen Ribeiro *

Departamento de Física Matemática, Instituto de Física,
Universidade de São Paulo, CP 66.318, 05315-970 São Paulo, SP, Brazil
pribeiro@fma.if.usp.br

This work is dedicated to Professor Jacques Bros, on the occasion of his 70th birthday.

Summary. We describe a general geometric setup allowing a generalization of Rehren duality to asymptotically anti-de Sitter space-times whose classical matter distribution is sufficiently well-behaved to prevent the occurence of singularities in the sense of null geodesic incompleteness. We also comment on the issues involved in the reconstruction of an additive and locally covariant bulk net of observables from a corresponding boundary net in this more general situation.

18.1 Introduction

The inception of quantum field theory in curved space-time, about forty years ago, brought into evidence a host of new conceptual problems hitherto absent or left unnoticed due to the peculiarities of Minkowski space-time, such as the very definitions of the notions of vacuum, particle, S-matrix, etc.. One hopes that the clarification of such issues may bring some new insights into the deeper problem of the quantization of gravity. An example of a possible interface between QFT in curved space-time — based on the *principle of locality* — and a would-be quantum theory of gravity – where such a principle is likely to be only macroscopically valid — is black hole thermodynamics. The idea that a stationary black hole is a "black" object in the *quantum* sense of the word — i.e., it produces a thermal bath with a certain universal temperature – suggests, together with the peculiar geometrical behaviour of its event horizon, some remarkable consequences, such as: 1) the notion of (thermodynamical) entropy is no longer extensive as in usual thermodynamics, but leads to a quantity depending linearly on the *area* of the event horizon (Bekenstein-Hawking entropy) 2) A black hole can "evaporate", i.e., lose all its mass by thermally

*Talk given during the Symposium "Rigorous Quantum Field Theory" in the honour of Prof. J. Bros, SPhT-Saclay, Paris, July 19-21, 2004. This project is supported by FAPESP under grant no. 01/14360-1.

radiating it to infinity (Hawking radiation), in finite time, which leads to a complete decoherence of an initially pure global state through entanglement with the partial state inside the horizon of the vanishing black hole. To explain these phenomena without violating basic postulates of quantum mechanics, 't Hooft and Susskind have put forward the *holographic principle* — namely, that the horizon has already *all* physical degrees of freedom, in the sense that one can completely reconstruct the physical data contained in a (*bulk*) volume from the physical system living on the *boundary* of this volume, in pretty much the same way a tridimensional picture is rebuilt from a two-dimensional hologram.

A concrete implementation of this principle in string theory was conjectured by Maldacena [22] and Witten [31] — the famous *AdS-CFT correspondence*, which triggered an impressive amount of theoretical development afterwards. Surprisingly, although the holographic principle was thought to be inconsistent with the principle of locality [6], it is possible to *rigorously* prove that the essentials of the AdS-CFT correspondence — more precisely, the peculiar geometry of the space-times involved — allow the reformulation of this correspondence in a manner *consistent with the principle of locality*, that is, within the context of QFT in curved space-time. Such a result is proven in Rehren's paper [25], which is formulated within the framework of Local Quantum Physics (Algebraic Quantum Field Theory). It states that "theories of local observables in Anti-de Sitter (*AdS*) space-time that are covariant under global (rigid) isometries can be put in a one-to-one correspondence to theories of local observables in *AdS*'s boundary — that is conformal to Minkowski space-time of one dimension less — that are covariant under global (rigid) conformal transformations". This result did not get the attention it deserved outside the realm of Local Quantum Physics, being misinterpreted as a "fake proof" of Maldacena's AdS-CFT correspondence, and therefore deemed useless by string theorists; this comes as a consequence of the fact that only *rigid* isometries are implemented, i.e., the quantum observables are completely decoupled from the gravitational degrees of freedom — there is no clue to how the bulk quantum system transforms under arbitrary diffeomorphisms of space-time, let alone how it reacts to arbitrary, but compactly localized, changes of the metric, and how these changes manifest themselves in the holographic dual theory.

Here we understand Rehren's theorem — called heretofore *algebraic holography* or *Rehren duality* — as an *independent* result, that, at the same time, poses questions with a counterpart in the "stringy" AdS-CFT correspondence, and issues deeply related to the foundations of relativistic quantum theory itself. It is from this perspective that the author's work starts.

In Section 18.2, after recalling some basic definitions and results in Lorentzian geometry that will be needed in the sequel, we will extend Rehren's geometrical setup to asymptotically simple, asymptotically AdS space-times of any dimension greater than 2, based in the simple, but crucial remark: wedges in AdS are simply diamonds with both tips belonging to the conformal infinity. This not only renders Rehren duality quite natural, but also shows that it depends *essentially* on the global causal structure of AdS's conformal infinity, therefore begging for a generalization to space-times that share these properties. We will see, however, in Section 18.3, that there are some subtle, but important aspects in this more general setting. Namely, one needs some global constraints on the classical matter distribution (which can be put into strictly geometrical terms) in order to utilize algebraic holography to preserve causality when going from the bulk to the boundary. We will see that these conditions also open up the possibility of encoding bulk gravitational effects in a non-geometrical way at the boundary — namely, in the form of spontaneous symmetry breaking (breakdown of Haag duality for diamonds at the boundary), if the bulk theory is causal and Haag dual. This has remarkable consequences, due to previous results by Brunetti, Guido and Longo about modular covariance in conformal QFT [9]. These same conditions raise, on the other hand, great difficulties when it comes to reconstruct the (compact) localization of the bulk observables using only boundary CFT data and the (bulk wedge ⇔ boundary diamond) correspondence. It can be shown, nevertheless, that for sufficiently small bulk diamonds this reconstruction can indeed be done. This is just enough for *additive* bulk theories, which can thus be "holographically rebuilt". Section 18.4 closes with some remarks on open problems and further work to be done by the author.

The developments to be presented in what follows are, first and foremost, geometric. We will focus on two essential aspects: causality and localization (in the sense of manifold topology — see Section 18.3). We will refrain from studying covariance aspects of our construction in detail, as they demand a separate paper of their own for a proper discussion and limit ourselves to some remarks at the end.

18.2 Doing away with coordinates in Rehren duality

18.2.1 Some tools in Lorentzian geometry

Let us recapitulate some definitions. For details, see the monographs of Wald [30], Hawking and Ellis [19], O'Neill [23] and Beem, Ehrlich and Easley [4]. By a *space-time* it will be understood a pair $(\widehat{\mathscr{M}}, \widehat{g})$[2], where $\widehat{\mathscr{M}}$ is a paracompact, connected and orientable $\mathscr{C}^\infty$ manifold, and $\widehat{g}$ is a time-orientable, Lorentzian $\mathscr{C}^\infty$ metric, with Levi-Civita connection ∇_a.

Let $\mathscr{U} \subset \widehat{\mathscr{M}}$ be an open set, and $p \in \mathscr{U}$. The *chronological* (resp. *causal*) *future* of p with respect to $\mathscr{U}$, denoted by $I^+(p,\mathscr{U})$ (resp. $I^+(p,\mathscr{U})$) is given by the following sets:

$$\begin{aligned} I^+(p,\mathscr{U}) \doteq \{x \in \mathscr{U} : \exists\gamma : [0,a] \xrightarrow{\mathscr{C}^\infty} \mathscr{U} \text{ timelike} \\ \text{and future such that } \gamma(0) = p,\ \gamma(a) = x\}; \quad (18.1) \\ J^+(p,\mathscr{U}) \doteq \{x \in \mathscr{U} : x = p \text{ ou } \exists\gamma : [0,a] \xrightarrow{\mathscr{C}^\infty} \mathscr{U} \text{ causal} \\ \text{and future such that } \gamma(0) = p,\ \gamma(a) = x\}. \quad (18.2) \end{aligned}$$

Exchanging future with past, one can define in a dual fashion the *chronological* (resp. *causal*) *past* $I^-(p,\mathscr{U})$ (resp. $J^-(p,\mathscr{U})$) of p with respect to $\mathscr{U}$. It follows from these definitions that $I^\pm(p,\mathscr{U})$ is open and $int(J^\pm(p,\mathscr{U})) = I^\pm(p,\mathscr{U})$. Using such sets we can define chronology and causality relations between two points. Let $p,q \in \mathscr{U} \subset \mathscr{V}$. We say that p *chronologically* (resp. *causally*) *precedes* q with respect

[2]The use of hats follows the convention for the interior of the conformal completion (see later).

to $\mathscr{U}$ if $p \in I^-(q, \mathscr{U})$ (resp. $p \in J^-(q, \mathscr{U})$). We denote this relation by $p \ll_{\mathscr{U}} q$ (resp. $p \leq_{\mathscr{U}} q$). If $p \leq_{\mathscr{U}} q$ and $p \neq q$, we write $p <_{\mathscr{U}} q$. If $p \not\ll_{\mathscr{U}} q$ and $q \not\ll_{\mathscr{U}} q$ (resp. $p \not<_{\mathscr{U}} q$ e $p \not>_{\mathscr{U}} q$), we say that p and q are *chronologically* (resp. *causally*) *disjoint* — in this case, we write $p \curlywedge_{\mathscr{U}} q$ (resp. $p \perp_{\mathscr{U}} q$). All chronology and causality relations defined above, as well as the notions of chronological/causal future/past, are defined for arbitrary nonvoid sets in an obvious way. If $p, q \in \mathscr{U} \subset \mathscr{V}$ implies that $p \curlywedge_{\mathscr{U}} q$ (resp. $p \perp_{\mathscr{U}} q$), we say that $\mathscr{U}$ is *achronal* (resp. *acausal*) with respect to $\mathscr{U}$. For instance, given $\mathscr{O} \subset \mathscr{U}$, the set $\partial I^{+/-}(\mathscr{O}, \mathscr{U}) = \partial J^{+/-}(\mathscr{O}, \mathscr{U})$ is said to be the *future/past achronal boundary* of $\mathscr{O}$ with respect to $\mathscr{U}$, and constitutes an achronal, topological submanifold of $\mathscr{V}$ such that every $p \in \partial I^{+/-}(\mathscr{O}, \mathscr{U})$ belongs to a (necessarily unique) null geodesic segment, achronal with respect to $\mathscr{U}$ and contained in $\partial I^{+/-}(\mathscr{O}, \mathscr{U})$, which is either past/future inextendible or possesses a past/future endpoint in $\mathscr{O}$. Such geodesics are the *null generators* of $\partial I^{+/-}(\mathscr{O}, \mathscr{U})$.

Let now $\mathscr{S} \subset \mathscr{V}$ be closed and achronal. The *future* (resp. *past*) *domain of dependence* of $\mathscr{S}$, denoted by $D^+(\mathscr{S})$ (resp. $D^-(\mathscr{S})$) is given by:

$$\begin{aligned} D^{+/-}(\mathscr{S}) \doteq \{p \in \mathscr{V} : \forall \gamma : [0, a) \longrightarrow \mathscr{V} \text{ past/future inextendible,} \\ \text{causal such that } \gamma(0) = p, \exists b < a \text{ such that } \gamma(b) \in \mathscr{S}\}. \end{aligned} \tag{18.3}$$

$D(\mathscr{S}) \doteq D^+(\mathscr{S}) \cup D^-(\mathscr{S})$ is said to be the *domain of dependence* or *Cauchy development* of $\mathscr{S}$. The *edge* of $\mathscr{S}$ (notation: $\dot{\mathscr{S}}$) is given by the points $p \in \mathscr{S}$ such that *any* open neighborhood of p possesses points $q \in I^-(p)$, $r \in I^+(p)$ and a timelike curve γ linking q to r, and with empty intersection with $\mathscr{S}$. If S is not only achronal but also acausal, then the set $int(D(\mathscr{S}))$ is globally hyperbolic (in such a case, we say that S is a *Cauchy surface* for $int(D(\mathscr{S}))$). The closed, achronal set $H^+(\mathscr{S}) \doteq \overline{D^+(\mathscr{S})} \setminus I^-(D^+(\mathscr{S}))$, denoted *future Cauchy horizon* of $\mathscr{S}$, possesses the following property: any $p \in H^+(\mathscr{S})$ is contained in an achronal, null geodesic segment contained in $H^+(\mathscr{S})$, which is either past inextendible or has a past endpoint in $\dot{\mathscr{S}}$. An analogous property holds for $H^-(\mathscr{S})$, the *past Cauchy horizon* of $\mathscr{S}$; the *(full) Cauchy horizon* $H(\mathscr{S}) \doteq H^+(\mathscr{S}) \cup H^-(\mathscr{S})$ equals $\partial D(\mathscr{S})$.

Now, for the notion of *conformal infinity*:

Definition 1. The *conformal infinity* or *conformal boundary* of an n-dimensional space-time $(\widehat{\mathscr{M}}, \widehat{g})$ is an $n-1$-dimensional semi-Riemannian manifold $(\mathscr{I}, b)$ such that there exists an n-dimensional, Lorentzian manifold-with-boundary $(\mathscr{M}, g)$ (the *conformal closure* or *conformal completion* of $(\widehat{\mathscr{M}}, \widehat{g})$) satisfying:

- $\mathscr{I} \equiv \partial\mathscr{M}$; there is a diffeomorphism Φ of $\widehat{\mathscr{M}}$ onto $\Phi(\widehat{\mathscr{M}}) = \mathscr{M} \setminus \partial\mathscr{M}$;
- b is the (possibly degenerate) semi-Riemannian metric induced by g in $\mathscr{I}$;
- There exists a *conformal (Weyl) factor* (that is, a real-valued, positive $\mathscr{C}^\infty$ function Ω in $\widehat{\mathscr{M}}$, that admits a $\mathscr{C}^\infty$ extension to $\mathscr{M}$ such that $\Omega \restriction_{\mathscr{I}} \equiv 0$ and $\Delta\Omega \restriction_{\mathscr{I}} \neq 0$ em $\mathscr{I}$) satisfying $g_{ab} = \Omega^2 \widehat{g}_{ab}$.

Note that, if $n_a := \nabla_a \Omega \neq 0$ in $\mathscr{I}$ (n_a is the *normal (co)vector* to $\mathscr{I}$; this condition can be made to hold if the Einstein equations are satisfied in a neighborhood of $(\mathscr{I}, b)$ and the classical matter fields satisfy certain decay conditions in this neighborhood), then Ω can be chosen in such a way that the extrinsic curvature (second fundamental form) $K_{ab} := \nabla_a n_b$ vanishes in $\mathscr{I}$. It is enough to multiply Ω by a real-valued, positive, nowhere vanishing $\mathscr{C}^\infty$ function ω in $\mathscr{M}$ — the new factor Ω still satisfies all the conditions in Definition 1. Nonetheless, this choice by no means constrains the values ω can take in $\mathscr{I}$ [3]. Therefore, $(\mathscr{I}, b)$ *can be taken totally geodesic, regardless of the choice of representative of the conformal structure of* b.

Definition 2. Let $(\widehat{\mathscr{M}}, \widehat{g})$ be an n-dimensional space-time with conformal infinity $(\mathscr{I}, b)$. We say that $(\widehat{\mathscr{M}}, \widehat{g})$ is *asymptotically simple* if any inextendible null geodesic $(\widehat{\mathscr{M}}, \widehat{g})$ has a unique extension to $(\mathscr{M}, g)$ such that $\mathscr{I}$ contains precisely both of its endpoints.

Obviously, this is only possible if $(\widehat{\mathscr{M}}, \widehat{g})$ is *null geodesically complete.* Actually, when $(\mathscr{I}, b)$ is timelike (and therefore a space-time in its own right), one can say more, justifying the name "asymptotically simple":

Theorem 1. *If $(\widehat{\mathscr{M}}, \widehat{g})$ is asymptotically simple and has a timelike conformal infinity, then it is causally simple, that is, $J^{\pm}(p, \widehat{\mathscr{M}})$ is closed in $\widehat{\mathscr{M}}$ (and therefore equal to $\overline{I^{\pm}(p, \widehat{\mathscr{M}})}$) for all $p \in \widehat{\mathscr{M}}$.*

Proof. First, notice that if $p, q \in \widehat{\mathscr{M}}$ are such that $p \not\ll_{\widehat{\mathscr{M}}} q$, then $p \not\ll_{\mathscr{M}} q$ and likewise exchanging future with past, for a timelike curve in $\mathscr{M}$ linking p to q can always be slightly deformed so as to give a timelike curve *contained in* $\widehat{\mathscr{M}}$ and linking p to q. Now, suppose that $p \in \partial I^-(q, \widehat{\mathscr{M}})$ and $p \notin J^-(q, \widehat{\mathscr{M}})$. By the reasoning above, we have $p \in \partial I^-(q, \mathscr{M})$. Moreover, by hypothesis, a null generator γ of $\partial I^-(q, \widehat{\mathscr{M}})$ must reach its future endpoint at infinity without crossing q before this. Let r be such an endpoint. Then, $r \in \partial I^-(q, \mathscr{M})$ since this set is closed. Since the infinity is totally geodesic, γ must hit it transversally and thus any future causal extension of γ must be broken. Therefore, if one extends γ slightly to the future by a null generator segment γ' of $\partial I^-(q, \mathscr{M})$ crossing r (say, by setting the affine parameter t of γ' equal to zero in r and extending up to $t = \epsilon > 0$), then there is a timelike curve in $\mathscr{M}$ linking p to $\gamma'(\epsilon)$ [19], which violates the achronality of $\partial I^-(q, \mathscr{M})$. Repeat the argument exchanging future with past. □

An asymptotically simple space-time, however, need not be globally hyperbolic — a prime example is AdS space-time, which will be studied in the next subsection.

18.2.2 Asymptotically Anti-de Sitter (AAdS) space-times and Rehren duality

We will recapitulate some definitions given in [27]. Recall that n-dimensional AdS space-time (notation: AdS_n, $n \geq 3$) is given by the hyperquadric in $\mathbb{R}^{n+1}$ ($\mathbf{X} = (X^1, X^2, \ldots, X^{n-2})$)

$$-X^0X^0 + \mathbf{X} \cdot \mathbf{X} + X^{n-1}X^{n-1} - X^nX^n = A^2,\ A > 0, \tag{18.4}$$

where the $X^0 - X^n$ plane determines the time orientation. AdS_n is a homogeneous space for the isometry group $SO(2, n-1)$. Consider now the region $AdS_n^+ \doteq \{X \in AdS_n : X^{n-1} + X^n > 0\}$. One can build a chart for this region (denoted *horocyclic* or *Poincaré parametrization*) with the parameters (x, z), where $x \in \mathbb{R}^{1,n-2}$ and $z \in \mathbb{R}_+$:

$$\begin{cases} X^\mu &= \frac{A}{z} x^\mu \ (\mu = 0, \dots, n-2), \\ \frac{1}{A} X^{n-1} &= \frac{1-z^2}{2z} + \frac{1}{2z} x_\mu x^\mu, \\ \frac{1}{A} X^n &= \frac{1+z^2}{2z} - \frac{1}{2z} x_\mu x^\mu. \end{cases} \tag{18.5}$$

One can see from the formulae (18.5) that each timelike hypersurface given by $z = const.$ is conformal to $\mathbb{R}^{1,n-2}$ by a factor

$$(X^{n-1} + X^n)^2 = \frac{A^2}{z^2}. \tag{18.6}$$

In this chart, the AdS_n metric is written as

$$\Delta s^2 = \frac{A^2}{z^2}(\Delta x_\mu \Delta x^\mu + \Delta z^2), \tag{18.7}$$

that is, AdS_n^+ is a semi-Riemannian *"warped product"* of $\mathbb{R}^{1,n-2}$ with $\mathbb{R}_+^*$. The universal covering of AdS_n (notation: $\widetilde{AdS_n}$) is asymptotically simple, and possesses the Einstein static universe (ESU) $\mathscr{I} = \mathbb{R} \times S^{n-2}$ as conformal infinity. Specializing to the Poincaré chart, we see that in AdS_n^+ the conformal factor is given by (18.6), that is, $z = 0$ corresponds to the conformal embedding of Minkowski space-time into the Einstein static universe. $\widetilde{AdS_n}$ satisfies the empty space Einstein equations with (negative) cosmological constant $\Lambda = -\frac{(n-1)(n-2)}{2A^2}$.

Now, let us define, for $p, q \in \mathscr{I}, p \ll_{\mathscr{I}} q$:

$$\mathscr{W}_{p,q} \doteq (I^-(p, \overline{AdS_n}) \cap I^+(q, \overline{AdS_n})) \cap \widetilde{AdS_n} \quad \text{((bulk) } \textit{wedge}\text{);} \tag{18.8}$$

$$\begin{aligned} \mathscr{D}_{p,q} &\doteq (I^-(p, \overline{AdS_n}) \cap I^+(q, \overline{AdS_n})) \cap \mathscr{I} \\ &= I^-(p, \mathscr{I}) \cap I^+(q, \mathscr{I}) \ \text{((boundary) } \textit{diamond}\text{).} \end{aligned} \tag{18.9}$$

Let $p \in \mathscr{I}$. All future null geodesics emanating from p will focus at a single point of $\mathscr{I}$, which is the future endpoint of all null generators of $\partial I^+(p, \mathscr{I})$. This point is denoted *antipodal* of p (notation: $\bar{p}$). The antipodal of p has the following properties:

$$\partial I^+(p, \mathscr{I}) = \partial I^-(\bar{p}, \mathscr{I}); \tag{18.10}$$

$$\partial I^+(p, \overline{AdS_n}) = \partial I^-(\bar{p}, \overline{AdS_n}). \tag{18.11}$$

Let $\bar{\bar{p}} \doteq \overline{(\bar{p})}$, and define $\mathscr{M}in(p) \doteq \mathscr{D}_{p,\bar{\bar{p}}}$, the *Minkowski domain* to the future of $p \in \mathscr{I}$. This region corresponds to the conformal embedding of $\mathbb{R}^{1,n-2}$ into $\mathscr{I}$ such that p corresponds to the past timelike infinity of $\mathbb{R}^{1,n-2}$. $\mathscr{P}oi(p) \doteq \mathscr{W}_{p,\bar{\bar{p}}}$ corresponds to the domain of a Poincaré chart in $\widetilde{AdS_n}$, therefore being denominated the *Poincaré domain* to the future of p. Given the objects defined above, we can define the geometrical setup for Rehren duality as follows:

1. The isometry group of $\widetilde{AdS_n}$ *acts transitively* on the collections $\mathfrak{W} \doteq \{\mathscr{W}_{p,q} : p, q \in \mathscr{I}\}$ of bulk wedges and $\mathfrak{D} \doteq \{\mathscr{D}_{p,q} : p, q \in \mathscr{I}\}$ of boundary diamonds.
2. $\mathscr{W}_{p,q}$ and $\mathscr{D}_{p,q}$ share the *same isotropy subgroup*.

3. From (18.10) and (18.11), it follows respectively that $\mathfrak{W}$ and $\mathfrak{D}$ are closed under causal complements. More precisely (by $\mathscr{O}'_{\mathscr{U}}$ we mean the causal complement of $\mathscr{O}$ with respect to $\mathscr{U} \supset \mathscr{O}$), we have, for all $p, q \in \mathscr{M}in(r)$, $r \in \mathscr{I}$, $p \ll_{\mathscr{I}} \bar{q}$,

$$\overline{\mathscr{W}_{q,\bar{p}}} \cap \widetilde{AdS_n} = (\mathscr{W}_{p,\bar{q}})'_{\widetilde{AdS_n}} \tag{18.12}$$

and

$$\overline{\mathscr{D}_{q,\bar{p}}} = (\mathscr{D}_{p,\bar{q}})'_{\mathscr{I}}. \tag{18.13}$$

4. As a consequence of the above statements, the *Rehren bijection*

$$\begin{aligned} \rho : \mathfrak{W} &\longrightarrow \mathfrak{D} \\ \mathscr{W}_{p,q} &\mapsto \alpha(\mathscr{W}_{p,q}) \doteq \mathscr{D}_{p,q} \end{aligned} \tag{18.14}$$

is one-to-one and onto, preserves inclusions and causal complements, and intertwines the action of $\widetilde{AdS_n}$'s isometry group, which is also the conformal group of $(\mathscr{I}, b)$ and the universal covering of the conformal group of Minkowski space-time. *Rehren duality* = *algebraic holography* is simply the *transplantation* (change of index set) of theories of local observables under the map ρ[3].

Now, what happens if we "perturb" AdS space-time in such a way that *we still have the ESU* $(\mathscr{I}, b)$ *as conformal infinity?* This corresponds to the class of *asymptotically AdS* space-times. More precisely, by employing a definition similar to the ones given in [3] and [2], one can write:

Definition 3. An n-dimensional space-time $(n \geq 3)$ $(\widehat{\mathscr{M}}, \widehat{g})$ with conformal infinity $(\mathscr{I}, b)$ is said to be *asymptotically anti-de Sitter* (notation: AAdS) if:

1. It satisfies Einstein's equations $\widehat{R}_{ab} - \frac{1}{2}\widehat{R}\widehat{g}_{ab} - \Lambda\widehat{g}_{ab} = 8\pi G_{(n)}\widehat{T}_{ab}$, where $G_{(n)}$ is the n-dimensional Newton's constant, and the cosmological constant Λ is < 0 (one can attribute an "AdS radius" to such space-times, by setting $A = \sqrt{-\frac{(n-1)(n-2)}{\Lambda}}$);
2. $(\mathscr{I}, b)$ is globally conformally diffeomorphic to the $(n-1)$-dimensional Einstein static universe;
3. The (classical) energy-momentum tensor $\widehat{T}_{ab}$ of $(\widehat{\mathscr{M}}, \widehat{g})$ decays fast enough close to $\mathscr{I}$ for $\Omega^{2-n}\widehat{T}^a_b$ to possess a $\mathscr{C}^\infty$ extension to the conformal closure $(\mathscr{M}, g)$.

The condition on the decay of $\widehat{T}_{ab}$ is motivated by considering the behaviour of classical fields emanating from compactly localized sources in AdS_n, especially massless fields (electromagnetic, Yang-Mills). The global condition on the conformal infinity makes sense in general because solutions of Einstein's equations with negative cosmological constant possess a timelike conformal infinity.

In what follows, we shall make two additional demands on the class of AAdS space-times we will deal with:

[3]This coordinate-free form of the Rehren bijection, which solely makes use of causal relationships in the conformal closure, is based in a suggestion from K.-H. Rehren [26], and was employed in this form by Bousso and Randall [7] for studying qualitative aspects of the AdS-CFT correspondence.

Asymptotic simplicity. This is indispensable for rebuilding bulk localization from wedges. The existence of a large class of asymptotically simple AAdS space-times was proven by Friedrich [11, 13].

Global focusing of null geodesics. More precisely, it is demanded that all inextendible null geodesics shall possess a pair of conjugate points (recall that a pair of points p, q in a null geodesic γ are said to be *conjugate* if there is a Jacobi field on γ — i.e., a vector field that satisfies the geodesic deviation equation on each point of γ — nowhere vanishing on the open segment of γ linking p to q but vanishing at both p and q. It is well known [4,19,30] that, in this case, any point of γ to the future (resp. past) of q (resp. p) can be linked to p (resp.q) by a timelike curve). It is precisely this condition that guarantees the Rehren bijection will preserve causality, and it also ends up playing an important role in the reconstruction of bulk localizatiion. Even if one does not require asymptotic simplicity, one can still show that any chronological space-time which satisfies this focusing condition is strongly causal [4], and therefore its topology is generated by diamonds. AdS does not satisfy this condition, but it follows from energy conditions on the energy-momentum tensor as weak as the NEC, ANEC and the Borde energy condition [5][4], and does look natural from the viewpoint of certain *rigidity theorems* for asymptotically simple space-times: for asymptotically flat and de Sitter space-times which satisfy, say, NEC, it was proven by Galloway [14, 15], by employing the stability results of Friedrich [11, 12], that if such space-times possess a so-called *null line* (a complete, achronal null geodesic), then they are *globally isometric* to Minkowski space-time (resp. de Sitter space-time, with radius determined by the value of the cosmological constant appearing in the Einstein equations). From the viewpoint of *stability of the (conformal) [11] mixed Cauchy/boundary problem*, one can see that the occurrence of null lines is an *unstable* feature of such space-times, i.e., any arbitrarily small perturbation of Cauchy data that preserves boundary conditions at the conformal infinity *destroys all null lines*, i.e., all complete null geodesics acquire a pair of conjugate points. There is still no similar result for asymptotically simple AAdS space-times in the sense of Definition 3, yet the the global structure of conformal infinity suggests that this may still be true. If so, our analysis complements and extends Rehren's.

We shall now study how this framework behaves in the more general situation of AAdS space-times complying with the conditions above.

18.3 Properties of the Rehren bijection in AAdS space-times

18.3.1 Causality (bulk-to-boundary)

In principle, gravitational effects deep inside the bulk may produce *causal shortcuts* through the bulk linking causally disjoint point at the boundary, i.e., it may

[4]If one wants to extend our considerations to *semiclassical* AAdS space-times, i.e., with quantum backreaction, we remark that quantum energy inequalities seem to be capable of guaranteeing that a "Planck-scale coarse-grained (i.e., transversally smeared)" ANEC holds for the renormalized quantum energy-momentum tensor [10], but whether this implies, say, the Borde energy condition, and thus gives rise to the needed focusing theorems [5], or not, is still an open question.

happen that $I^+(p, \mathscr{I}) \cap I^-(q, \mathscr{I}) \subsetneqq (I^+(p, \mathscr{M}) \cap I^-(q, \mathscr{M})) \cap \mathscr{I}$, rendering the second identity in (18.9) *false*. Such a thing would be ruinous to the Rehren bijection to keep preserving causality in AAdS space-times. We shall show now that, luckily, (18.9) still holds under our set of hypotheses. Here, we make use of the notion of *gravitational time delay* [16,24,32] of complete null geodesics in AAdS space-times. The Einstein static universe $(\mathscr{I}, b)$ is globally hyperbolic; let it be given a foliation in Cauchy surfaces such that the orbits of the global time function t (supposed to be oriented in the same way as the time orientation of $(\mathscr{I}, b)$) generating the foliation are complete timelike geodesics, and the values of the global time function correspond to a common affine parametrization of this family of geodesics. Now, let γ be a complete null geodesic in $\mathscr{M}$ traversing $\widehat{\mathscr{M}}$, with past endpoint p and future endpoint q belonging to the orbits T_p (resp. T_q), and γ' a null geodesic segment *in* $\mathscr{I}$ starting at p and ending at, say, $q' \in T_q$. The gravitational time delay of γ with respect to $\mathscr{I}$ is given by the difference $\Delta t = t(q) - t(q')$ (notice that, due to the properties of null geodesics in ESU, it follows that any other null geodesic segment in $\mathscr{I}$ starting at p that crosses T_q afterwards will necessarily do it at q'.). Although this value depends on the choice of foliation, the *sign* of Δt (< 0, $= 0$, > 0) does not. Under our set of hypotheses, the gravitational time delay in AAdS space-times is always *posititive*:

Theorem 2. [5] *Let $(\widehat{\mathscr{M}}, \widehat{g})$ be an asymptotically simple AAdS space-time, such that every inextendible null geodesic has a pair of conjugate points, and $p \in \mathscr{I}$. Then, every null geodesic segment emanating from p which does not belong to $\mathscr{I}$ has its future endpoint in $I^+(p, \mathscr{I})$.*

Proof. Let γ be a null geodesic segment emanating from p and traversing the bulk, and let p' be the future endpoint of γ in $\mathscr{I}$. Since we have assumed that $(\widehat{\mathscr{M}}, \widehat{g})$ is causal, one can see that $p' \notin \overline{I^-(p, \mathscr{M})}$. Now, we prove two lemmata:

Lemma 1 (Absence of causal shortcuts). *Let $p, p' \in \mathscr{I}$. If $p \perp_{\mathscr{I}} p'$, then there is no causal curve in $(\mathscr{M}, g)$ linking p to p'.*

Proof. Suppose that $p' >_{\mathscr{M}} p$ (the opposite case is treated analogously). We will prove that the gravitational time delay implied by the presence of a pair of conjugate points contradicts the causal disjointness of p and p' with respect to $\mathscr{I}$. Denote by $T(p')$ the timelike generator of $(\mathscr{I}, b)$ containing p'.

Note that $\partial I^+(p, \mathscr{I}) \doteq \Sigma$ is a closed, achronal surface that cuts $(\mathscr{I}, b)$, in two disjoint subsets $I^+(p, \mathscr{I}) \doteq A$ and $\mathscr{I} \setminus \overline{I^+(p, \mathscr{I})} \doteq B$ and intersects each timelike generator of $\mathscr{I}$ in precisely one point, as every timelike generator has points in $I^+(p, \mathscr{I})$ and $I^-(\bar{p}, \mathscr{I})$. By hypothesis, $p' \in B$. Moreover, $T(p')$ must cross Σ at some instant of time. Therefore, there exists $p'' \in T(p')$ such that $p'' \mathbf{g}_{\mathscr{I}} p'$ and $p'' \in \Sigma$. Let γ be a null generator of Σ that contains p''. As the segment of γ that links p to p'' is null and achronal, γ is necessarily the fastest curve in $(\mathscr{I}, b)$ linking p to $T(p')$.

[5] The method of proof was communicated to me by Sumati Surya [29]. It is analogous to the proof of a positive mass theorem for asymptotically flat space-times due to Penrose, Sorkin and Woolgar [24] and for AAdS space-times due to Woolgar [32]. Another proof of this, using a somewhat different strategy, was provided by Gao and Wald [16].

Now, consider the achronal boundary $\partial I^+(p,\mathscr{M}) = \partial J^+(p,\mathscr{M}) \doteq \overline{\Sigma}$. $\overline{\Sigma} \cap \mathscr{I}$ is closed, achronal and intersects each timelike generator of $\mathscr{I}$ in precisely one point, as every timelike generator has points in $I^+(p,\mathscr{M})$ and $I^-(p,\mathscr{M})$, and $\overline{\Sigma}$ separates $\mathscr{M}$ in two disjoint open sets (from the viewpoint of a manifold-with-boundary, of course) $I^+(p,\mathscr{M}) \doteq \dot{A}$ and $\mathscr{M} \setminus \overline{I^+(p,\mathscr{M})} \doteq \dot{B}$. Thus, $T(p')$ must cross $\overline{\Sigma}$ in, say, p'''. Since $p <_{\mathscr{M}} p'$, we must have $p''' \ll_{\mathscr{I}} p'$ or $p''' = p'$. In both cases, we have $p''' \ll_{\mathscr{I}} p''$, which implies that any null generator $\overline{\gamma}$ of $\overline{\Sigma}$ containing p''' is strictly faster than γ. As γ was the fastest curve in $(\mathscr{I}, qb)$ linking p to $T(p')$, $\overline{\gamma}$ necessarily traverses $\widehat{\mathscr{M}}$. Hence, we have built a complete and achronal null geodesic in $(\widehat{\mathscr{M}},\widehat{g})$. However, such a geodesic cannot exist since it must have a pair of conjugate points and therefore cannot be achronal. □

Lemma 2. *Let $p,p' \in \mathscr{I}$. If $p' \in \partial I^+(p,\mathscr{I})$ and $p' \neq \overline{p}$, then there is no null geodesic segment in $(\mathscr{M},g)$ that does not belong to $\mathscr{I}$ and links p to p'.*

Proof. Let γ be the (necessarily unique) null generator of $\partial I^+(p,\mathscr{I})$ linking p to p'. Suppose that there is another null geodesic, traversing $\widehat{\mathscr{M}}$ and linking p to p'. Then, if one picks any point p'' in γ after p', there is a *broken* null geodesic segment linking p to p'', which in turn implies that there is a timelike curve in $(\mathscr{M},g)$ linking p to p'' [19].

Let $T(p'')$ be the timelike generator of $(\mathscr{I},b)$ containing p''. Now, consider $\overline{\Sigma}$ as in the preceding lemma. Again, $T(p'')$ must cross $\overline{\Sigma}$, say, in p'''. Thus, necessarily $p''' \ll_{\mathscr{I}} p''$. But this implies that, since $(\mathscr{M},g)$ is causal, $p \perp_{\mathscr{I}} p'$. That contradicts the preceding lemma. □

Both lemmata above imply that, if p' is not dragged inside $I^+(p,\mathscr{I})$ by gravitational time delay, then p' coincides with $\overline{p}$, just as in AdS. But, even in such a case, the presence of conjugate points in γ implies that there is a timelike curve traversing the bulk and linking p to $\bar{p}$. Repeating the argument in Lemma 2, the result follows. □

All arguments above can be repeated exchanging future with past, and p with its antipodal $\bar{p}$, yielding a similar result in the opposite time orientation. It is now immediate to show that, as a direct consequence of Theorem 2, (18.9) holds. Thus, we shall keep the same notation for the conformal infinity, for bulk wedges, boundary diamonds and the Rehren bijection, as well as for Minkowski and Poincaré domains, when dealing with AAdS space-times.

Remark 1. Our set of hypotheses, however, entails that, although the Rehren bijection preserves causality, it no longer does so in a *maximal* way — the collection of wedges is no longer closed under causal complements. More precisely:

Proposition 1. *Let $(\widehat{\mathscr{M}},\widehat{g})$ be an AAdS space-time satisfying the hypotheses of Theorem 2. Define $\Xi^+_p \doteq \partial I^+(p,\mathscr{M}) \setminus \{p,\overline{p}\}$ and $\Xi^-_{\overline{p}} \doteq \partial I^-(\overline{p},\mathscr{M}) \setminus \{p,\overline{p}\}$. Then:*

(i) $\Xi^+_p \cap I^-(\overline{p},\mathscr{M}) = \Xi^-_{\overline{p}} \cap I^+(p,\mathscr{M}) = \varnothing$.

(ii) $\Xi^+_p \cap \Xi^-_{\overline{p}} \cap \widehat{\mathscr{M}} = \varnothing$.

Proof. (i) We know that $\Xi^+_p \cap \mathscr{I} = \Xi^-_{\overline{p}} \cap \mathscr{I}$, so let us concentrate only at the bulk. Namely, suppose that there is $q \in \widehat{\mathscr{M}}$ such that $q \in \Xi^-_{\overline{p}}$ and $q \notin \Xi^+_p$. If $q g p$, that contradicts the fact that there is no timelike curve linking p to $\overline{p}$. Repeat the argument exchanging past with future, and the roles of p and $\overline{p}$.

(ii) If Ξ_p^+ and $\Xi_{\overline{p}}^-$ coincide in some point $q \in \widehat{M}$, then there is an at least broken null geodesic segment linking p to $\bar{p}$, which implies that there is a timelike curve through the bulk doing the same. The result easily follows by repeating the reasoning at the end of the proof of Theorem 2. □

Note that the argument above is symmetric with respect to time orientation. An important consequence of Proposition 1 is that our global focusing condition entails a nontrivial "shrinking" of the AAdS wedges towards the boundary, as a side effect of the gravitational time delay. As a consequence of the latter, it follows that, although by Theorem 2, $\bar{p}$ still satisfies (18.10), it will certainly violate (18.11). Moreover, $\overline{\mathscr{W}_{q,\bar{p}}} \cap \widehat{\mathscr{M}} \subsetneqq (\mathscr{W}_{p,\bar{q}})'_{\widehat{\mathscr{M}}}$ and $\mathscr{P}oi(r) \subsetneqq \{r\}'_{\mathscr{M}} \cap \widehat{M}$. Hence, property 3 for wedges no longer remains valid, and therefore the collection of bulk wedges cannot be closed by causal complements in the sense of (18.12). The strict inclusions above suggest that Haag duality for boundary diamonds cannot be satisfied *without violating bulk causality* for local observables, as the local (von Neumann, for concreteness) algebra $\mathfrak{A}(\mathscr{W}(p,\bar{q})')$ may be *strictly larger* than $\mathfrak{A}(\mathscr{W}(q,\bar{p})) = \mathfrak{A}(\mathscr{D}(q,\bar{p}))$ (here the Borchers timelike tube theorem does not remove the strictness of the inclusion, as here one would need to extend the localization beyond the set of points causally between p and $\bar{q}$). Hence, in such a case, if $\mathfrak{A}(\mathscr{D}(p,\bar{q}))' = \mathfrak{A}(\mathscr{D}(q,\bar{p})) = \mathfrak{A}(\mathscr{D}(p,\bar{q})')$, then the algebra $\mathfrak{A}(\mathscr{W}(p,\bar{q})')$ necessarily has elements that do not commute with $\mathfrak{A}(\mathscr{W}(p,\bar{q})) = \mathfrak{A}(\mathscr{D}(p,\bar{q}))$, therefore violating local causality. That means that the local algebras also "shrink" from the viewpoint of the boundary — in this way, the boundary net can "feel" bulk gravitational effects as a spontaneous symmetry breaking that necessarily follows from the breakdown of Haag duality [18]. We will say more on this at the end of this contribution.

18.3.2 Reconstruction (boundary-to-bulk)

Knowing the wedge localization of bulk observables may not be enough for the full reconstruction of the bulk quantum theory using only boundary data and the Rehren bijection ρ. We need to be able to specify the localization of the local procedures in arbitrarily small open regions, or, which amounts to the same thing, the localization with respect to a basis of the bulk topology. This can, in principle, be performed by taking intersections of wedges, but it is by no means clear whether these give a basis for the bulk topology or not. This must be done in a more precise way. In AdS, we are fortunate, because any *relatively compact AdS diamond* (for the purpose of generating the manifold topology, these suffice) *can be enveloped by AdS wedges*: given

$$\mathscr{O}_{p,q} \doteq I^+(p,\widetilde{AdS_n}) \cap I^-(q,\widetilde{AdS_n}),\ p \ll_{\widetilde{AdS_n}} q, \tag{18.15}$$

we can write

$$\mathscr{O}_{p,q} = \bigcap_{r\in\partial I^-(p,\overline{AdS_n})\cap\mathscr{I},\ s\in\partial I^+(q,\overline{AdS_n}),\ r,s\in\mathscr{M}in(u)} \mathscr{W}_{r,s}. \tag{18.16}$$

We will see shortly that the achronality of the inextendible null geodesics in AdS is crucial for the precise enveloping. In an AAdS space-time as in Theorem 2, the issue is much more delicate, because of the following

Proposition 2. *Let $(\widehat{\mathscr{M}}, \widehat{g})$ be an AAdS space-time satisfying the hypotheses of Theorem 2, and $p \in \widehat{\mathscr{M}}$. Then, $\partial I^+(p, \mathscr{M})$ intersect each timelike generator if $(\mathscr{I}, b)$* precisely once.

Proof. By the achronality of $\partial I^+(p, \mathscr{M})$, it intersects every timelike generator of $(\mathscr{I}, b)$ *at most once.* Suppose that the thesis is false. Then, given a timelike generator T, we have the following possibilities:

(i) $\boxed{T \subset I^+(p, \mathscr{M})}$ – Consider a complete null geodesic γ crossing p, and let q be the *past* endpoint of γ. Then, there exists a value t of the affine parameter of T such that $T(t) \ll_{\mathscr{I}} q$. Therefore, $T(t) \ll_{\mathscr{M}} p$, which is absurd since $(\mathscr{M}, g)$ is chronological.

(ii) $\boxed{T \cap J^+(p, \mathscr{M}) = \varnothing}$ – Consider a complete null geodesic γ crossing p, and let r be the *future* endpoint of γ. Then, there exists a value t of the affine parameter of T such that $T(t) \mathbf{g}_{\mathscr{I}} r$. Therefore, $T(t) \mathbf{g}_{\mathscr{M}} p$, contradicting the hypothesis. □

Proposition 3. *Let $(\widehat{\mathscr{M}}, \widehat{g})$ be an AAdS space-time which satisfies the hypotheses of Theorem 2, $q, r \in \widehat{\mathscr{M}}$ such that $r \in \partial I^-(q, \widehat{\mathscr{M}})$, and γ a null generator of $\partial I^-(q, \widehat{\mathscr{M}})$ to which r belongs. Let $s_1(r), s_2(r), s_3(r) \in \mathscr{I}$ be defined as:*

- *$s_1(r)$ is the future endpoint of γ;*
- *$s_2(r)$ is the point where $\partial I^+(q, \mathscr{M})$ intersects the timelike generator $T(s_1(r))$ to which $s_1(r)$ belongs;*
- *$s_3(r)$ is the point where $\partial I^+(r, \mathscr{M})$ intersects $T(s_1(r))$.*

Then:

(i) *$s_3(r) \leq_{\mathscr{I}} s_2(r) \leq_{\mathscr{I}} s_1(r)$.*
(ii) *$s_3(r) = s_2(r) = s_1(r)$ if and only if the segment of γ linking r to $s_1(r)$ is achronal.*

Proof. (i) Immediate, as is (ii) $\Rightarrow$. It remains to prove (ii) $\Leftarrow$. Namely, suppose that $s_3(r)$ equals $s_2(r)$. Then, the null geodesic segment linking q to $s_2(r)$ must belong to γ, for otherwise there would be a broken geodesic segment linking r and $s_3(r)$, contradicting the definition of the latter (this, in particular, proves that $s_1(r) = s_3(r)$ even if we just assume $s_2(r) = s_3(r)$). If γ is not achronal, once more we have a contradiction with the definition of $s_3(r)$. □

Similar results are valid if we exchange future with past. Now, in an AAdS space-time satisfying the hypotheses of Theorem 2, let $\mathscr{O}_{p,q}$ be a *relatively compact diamond with a contractible Cauchy surface* — any sufficiently small diamond satisfies both conditions. Let us now consider the region

$$\mathscr{Q}_{p,q} = \bigcap_{r \in \partial I^-(p,\mathscr{M}) \cap \mathscr{I}, s \in \partial I^+(q,\mathscr{M}) \cap T(r)} \mathscr{W}_{r,s}. \tag{18.17}$$

It follows naturally from the definition that $\mathscr{Q}_{p,q} \supset \mathscr{O}_{p,q}$, it is *causally complete*, as it is the intersection of causally complete regions, and

$$\mathscr{Q}_{p,q} \cap J^+(q, \mathscr{M}) = \mathscr{Q}_{p,q} \cap J^-(p, \mathscr{M}) = \varnothing. \tag{18.18}$$

In AdS, $\mathscr{Q}_{p,q} = \mathscr{O}_{p,q}$ [25]. For AAdS space-times as in Theorem 2, however, it may happen that $\mathscr{Q}_{p,q} \supsetneqq \mathscr{O}_{p,q}$. Likewise, defining $\mathscr{E}_{p,q} \doteq \partial I^+(p, \widehat{\mathscr{M}}) \cap \partial I^-(q, \widehat{\mathscr{M}})$, let us start from

$$\widetilde{\mathscr{Q}}_{p,q} = \bigcap_{r \in \mathscr{E}_{p,q}} \mathscr{W}_{s_3'(r), s_3(r)}, \tag{18.19}$$

where $s_3'(r)$ corresponds to $s_3(r)$ if we exchange future with past in the statement of Proposition 3. Here, $\widetilde{\mathscr{Q}}_{p,q} \subset \mathscr{Q}_{p,q}$ is again causally complete, *if nonvoid.* However, if the space-time metric deep inside the bulk is sufficiently "distorted", and causing a sufficient number of null generators of, say, $\partial I^-(q, \widehat{\mathscr{M}})$ to acquire a pair of conjugate points between $\mathscr{E}_{p,q}$ and $\mathscr{I}$, for all we know (Proposition 3) $\widetilde{\mathscr{Q}}_{p,q}$ could very well be empty (the intersection of the corresponding algebras may even be nonvoid, but then we will not be able to attribute any localization whatsoever to this algebra). This is suggested by the following remarks:

1. In a causally simple space-time, any relatively compact diamond $\mathscr{O}_{p,q}$ is a globally hyperbolic region, for which any Cauchy surface has a boundary equal to $\mathscr{E}_{p,q}$;
2. Any causally complete region $\mathscr{U}$ has the following property: if $\mathscr{S} \subset \mathscr{U}$ is a closed, achronal set with respect to $\mathscr{U}$, then $D(\mathscr{S}) \subset \mathscr{U}$.

Both remarks together show that, if $r \in \mathscr{E}_{p,q}$ is such that a null generator of, say, $\partial I^-(q, \widehat{\mathscr{M}})$ crossing r acquires a pair of conjugate points between r and $s_1(r)$, then, by causal simplicity, it follows that there is a neighbourhood of q that is causally disjoint from $s_3(r)$, and, therefore, $I^-(s_3(r), \mathscr{M})$ cannot contain a Cauchy surface for $\mathscr{O}_{p,q}$. Since, on the other hand, this does not exclude the possibility that $\widetilde{\mathscr{Q}}_{p,q}$ may contain points outside $\mathscr{O}_{p,q}$ either, it is by no means clear whether the collections of $\mathscr{Q}_{p,q}$ and $\widetilde{\mathscr{Q}}_{p,q}$ give bases for the topology of $\widehat{\mathscr{M}}$ or not.

One way to circumvent these problems could be to restrict ourselves to sufficiently small diamonds, such that none of the null generators of $\partial I^+(q, \mathscr{M})$ can travel far enough beyond q in order to develop a pair of conjugate points. But there is a situation such that, no matter how small the extension, it will always cease to be achronal: it is when q *itself* is conjugate to $s_1(r)$. In this limiting case, $s_1(r) = s_2(r)$ but $s_2(r) \neq s_3(r)$.

We show now that the key to these problems is to try to build a region similar to $\widetilde{\mathscr{Q}}_{p,q}$, but employing, instead of the points $s_3(r)$, $s_3'(r)$ for $r \in \mathscr{E}_{p,q}$, the points for which the problem, entailed by Proposition 3 and mentioned above, is, in a certain sense, "minimized". To perform this task, we start from a different viewpoint, which will also eventually show that the critical situation in the preceding paragraph is ruled out by null geodesic completeness. First, notice that, using an argument similar to the one used in [19] and [30] to prove the existence of a (Lipschitz) topological manifold structure for achronal boundaries, one can show that $\mathscr{E}_{p,q}$ is locally the graph of an $\mathbb{R}$-valued, locally Lipschitz function of $n-2$ real arguments, and therefore it is a compact, acausal, embedded (Lipschitz) topological submanifold of $\widehat{\mathscr{M}}$, with codimension 2. Notice as well that one can smoothly parametrize the family of timelike generators of $(\mathscr{I}, b)$ by a latter's Cauchy surface $\mathscr{S}$ which is homeomorphic to S^{n-2} and thus also compact. Let t be the common affine parametrization of the timelike generators of $(\mathscr{I}, b)$ mentioned in the previous subsection. Define the function

$$\tau : \mathscr{E}_{p,q} \times \mathscr{S} \ni (r,\theta) \mapsto \tau(r,\theta) \in \mathbb{R}, \tag{18.20}$$

where

$$\partial I^+(R,\mathscr{M}) \cap T(\theta) = \{T(\theta)(\tau(r,\theta))\}. \tag{18.21}$$

Proposition 2 shows that the definition of τ is not empty. Moreover:

Proposition 4. *τ is upper semicontinuous in r for fixed θ.*

Proof. Let $\epsilon > 0$. r lies in the chronological past of the point $T(\theta)(\tau(r,\theta)+\epsilon)$, and thus there is an open neighborhood U of r in $\mathscr{E}_{p,q}$ which lies in the chronological past of $T(\theta)(\tau(r,\theta)+\epsilon)$. Therefore, for all $r' \in U$, we must have $\tau(r',\theta) < \tau(r,\theta)+\epsilon$. □

One can actually prove that τ is Lipschitz continuous in θ for fixed r, but this will not be used in the sequel. The function $\tau(.,\theta)$ will be called *future Fermat potential* with respect to θ. The name is a remnant of the Huygens-Fermat principle of geometrical optics (see, for instance, pages 249-250 of [1]). Now, extend the definition of $\tau(.,\theta)$ to the closure $\overline{\mathscr{F}_{p,q}}$ of some Cauchy surface $\mathscr{F}_{p,q}$ for $\mathscr{O}_{p,q}$, denoting it by the same symbol, since no confusion arises here. By the same argument employed in Proposition 4, $\tau(.,\theta)$ is upper semicontinuous in $\overline{\mathscr{F}_{p,q}}$. Since both $\mathscr{E}_{p,q} = \partial\mathscr{F}_{p,q}$ and $\overline{\mathscr{F}_{p,q}}$ are closed subsets of the compact set $\overline{\mathscr{O}_{p,q}}$, they are compact themselves. By a standard result of analysis (see, for instance, pages 110–111 of [21]), $\tau(.,\theta)$ has a maximum value both in $\overline{\mathscr{F}_{p,q}}$ and $\mathscr{E}_{p,q}$. The next theorem shows that $\tau(.,\theta)$ has indeed a distinguishing property of potentials:

Theorem 3 (Maximum principle for the future Fermat potential). *The maximum value of $\tau(.,\theta)$ in $\overline{\mathscr{F}_{p,q}}$ is achieved at $\mathscr{E}_{p,q}$.*

Proof. Let r be a point of $\mathscr{E}_{p,q}$ where $\tau(.,\theta)$ achieves its maximum in $\mathscr{E}_{p,q}$, and let r' be a point of $\mathscr{F}_{p,q}$ such that $\tau(r',\theta) \geq \tau(r',\theta)$. In such a case, it is obvious that $\mathscr{E}_{p,q}$ lies in the causal past of $T(\theta)(\tau(r',\theta))$. Pick a curve segment in $\overline{\mathscr{F}_{p,q}}$ starting at r', initially pointing outside $J^-(T(\theta)(\tau(r',\theta)),\mathscr{M})$ and ending in some point of $\mathscr{E}_{p,q}$. Then, any such a curve segment must cross $\partial I^-(T(\theta)(\tau(r',\theta)),\mathscr{M})$ at least once more after r', and before or at $\mathscr{E}_{p,q}$. This shows that $\partial I^-(T(\theta)(\tau(r',\theta)),\mathscr{M}) \cap \overline{\mathscr{F}_{p,q}}$ encloses an open subset X of $\mathscr{F}_{p,q}$ outside the causal past of $T(\theta)(\tau(r',\theta)$.

The remainder of the proof is analogous to the proof of Penrose's singularity theorem [19, 30]: namely, we will show that the properties of ∂X imply that there must exist an incomplete null geodesic in $(\widetilde{\mathscr{M}},\widetilde{g}_{ab})$. First, we will show that the closed, acausal set $\partial X = \partial I^-(T(\theta)(\tau(r',\theta)),\mathscr{M}) \cap \overline{\mathscr{F}_{p,q}}$ is *past trapped*, i.e., $\partial I^-(X,\widehat{\mathscr{M}})$ is *compact*. The past "ingoing" null geodesics of ∂X constitute the past Cauchy horizon of X, which is thus contained in $\overline{\mathscr{O}_{p,q}}$ and therefore compact, as it is closed [19,30]. The past "outgoing" null geodesics are precisely the null generators of $\partial I^-(T(\theta)(\tau(r',\theta)),\mathscr{M})$ that cross ∂X. Given a common affine parametrization to the null generators of $\partial I^-(T(\theta)(\tau(r',\theta)),\mathscr{M})$ such that the zero value of the affine parameter corresponds to ∂X. Then, let t_0 the largest value of affine parameter for which a past endpoint of $\partial I^-(T(\theta)(\tau(r',\theta)),\mathscr{M})$ is achieved. It must be finite, for each inextendible null geodesic must acquire a pair of conjugate points before reaching infinity, although the value of the affine parameter at a past endpoint of the null generator segment starting at, say, $r'' \in \partial X$ can be zero if r'' happens to be itself a past endpoint. Anyway, the portion of $\partial I^-(T(\theta)(\tau(r',\theta)),\mathscr{M})$ in the causal past of ∂X, being closed, has a closed inverse image in the compact set $[0,t_0]\times\partial X$ under the chosen parametrization of the null generators, and is therefore compact. Hence, the set $\partial I^-(\partial X,\widehat{\mathscr{M}})=H^-(X)\cup\partial X\cup(\partial I^-(T(\theta)(\tau(r',\theta)),\mathscr{M})\cap J^-(\partial X,\widehat{\mathscr{M}}))$ is a compact, achronal subset of $\widehat{\mathscr{M}}$, as asserted.

However, any causally simple space-time is stably causal [4]. That is, one can smoothly foliate $\widehat{\mathscr{M}}$ by "constant-time", spacelike surfaces of codimension 1. By the structure of the conformal infinity, such surfaces (*leaves*) cannot be compact. Moreover, each timelike *orbit* of the foliation crosses an achronal set at most once. By following these orbits, one can continuously map $\partial I^-(\partial X,\widehat{\mathscr{M}})$ into a spacelike leaf of this foliation. As the image of this map is compact, it must have a nonvoid boundary. But it is known that a set of the form $\partial I^-(Y,\widehat{\mathscr{M}}), Y\subset\widehat{\mathscr{M}}$ is a topological submanifold without boundary of $\widehat{\mathscr{M}}$, and, as such, it cannot have a boundary. This shows that some null generator of $\partial I^-(T(\theta)(\tau(r',\theta)),\mathscr{M})$ must terminate at a singularity before reaching its past endpoint. But this conflicts with the null geodesic completeness of $\widehat{\mathscr{M}}$, entailed by asymptotic simplicity. Hence, no point in $\mathscr{F}_{p,q}$ can achieve a maximum for $\tau(.,\theta)$ in $\overline{\mathscr{F}_{p,q}}$ — this maximum always takes place at $\mathscr{E}_{p,q}$. □

Proposition 4 and Theorem 3 together show that, for each θ, there will always be an $r\in\mathscr{E}_{p,q}$ such that, given *any* Cauchy surface $\mathscr{F}_{p,q}$ for $\mathscr{O}_{p,q}$, the set $\overline{\mathscr{F}_{p,q}}$ will always lie in the causal past of $T(\theta)(\tau(r,\theta))$. By Proposition 3 and the remarks above, this can only happen if the achronal null geodesic segment $\gamma(r,\theta)$ linking r to $T(\theta)(\tau(r,\theta))$ crosses q. Thus, this maximum point is *unique:* suppose otherwise. Then, there would be another $r'\in\mathscr{E}_{p,q}$ such that there is an achronal null geodesic segment $\gamma(r',\theta)$ linking r' to $T(\theta)(\tau(r',\theta)) = T(\theta)(\tau(r,\theta))$ and crossing q. Now consider the curve segment $\gamma'(r,\theta)$ which coincides with $\gamma(r,\theta)$ from r to q, and coincides with $\gamma(r',\theta)$ from q to $T(\theta)(\tau(r,\theta))$. This segment is necessarily broken, which conflicts with the achronality of $\gamma(r,\theta)$. Exchanging the roles of r and r', one sees that this argument also conflicts with the achronality of $\gamma(r',\theta)$. Notice, however, that an arbitrary $r\in\mathscr{E}_{p,q}$ *need not* maximize $\tau(.,\theta)$ for some θ. Two instances where this cannot occur are:

1. r is conjugate to q along a null generator of $\partial I^-(q,\widehat{\mathscr{M}})$ — any future extension of this generator beyond q will not be achronal;
2. q is conjugate to $s_2(r)$ along a null generator of $\partial I^+(q,\mathscr{M})$, by the remarks made above.

The second instance, however, is excluded by our line of reasoning, because it renders impossible, by Proposition 3 and Theorem 3, to $\tau(.,\theta)$ to achieve a maximum value in $\mathscr{E}_{p,q}$. This cannot happen, since for every θ a maximum must exist by Proposition 4. The first instance can be circumvented by picking $\mathscr{O}_{p,q}$ contained, say, in a convex normal neighbourhood, which can always be done, as here $(\widehat{\mathscr{M}},\widehat{g})$ is strongly causal. One can go further and take $\mathscr{O}_{p,q}$ sufficiently small (yet nonvoid) so that every $r\in\mathscr{E}_{p,q}$ is a maximum point of $\tau(.,\theta)$ for some θ, as the only obstacle to this would be the second instance above, which is excluded by the above argument. All results above have a past counterpart, by exchanging q with p and reversing the time orientation.

Summing up, we have showed that *sufficiently small $\mathscr{O}_{p,q}$ can always be precisely enveloped by wedges,* by means of the prescription (18.16). In such a case, any point not belonging to $\overline{\mathscr{O}_{p,q}}$ lies either in the chronological future of $\partial I^-(q,\widehat{\mathscr{M}})$ or in the chronological past of $\partial I^+(p,\widehat{\mathscr{M}})$, and, as such, will fail to belong to some wedge enveloping $\mathscr{O}_{p,q}$. Since the points at $\partial\mathscr{O}_{p,q}$ are already excluded from the intersection by construction, one concludes that $\mathscr{O}_{p,q} = \mathscr{Q}_{p,q}$ for sufficiently small $\mathscr{O}_{p,q}$. Moreover, in such a situation, each wedge in the definition (18.17) of $\mathscr{Q}_{p,q}$ is guaranteed to be contained in some Poincaré domain. Therefore, one can even restrict to a Poincaré domain and perform the bulk reconstruction there starting from a boundary CFT in Minkowski space-time.

18.4 Perspectives and open problems

For *additive* local quantum theories, it suffices to specify the localization of the procedures for a basis of the manifold topology. Therefore, the results in the previous section indicate that one can completely recover the *bulk* quantum theory by just employing localization data from the *boundary* quantum theory and the Rehren bijection, and this theory is guaranteed to be causal if its holographic dual is. In situations where the *boundary theory* is additive, then all compactly localized bulk observables are necessarily multiples of the identity [25]. In such a case, it suffices to have just wedge localization in the bulk.

The covariance issue is obviously more complicated than in the AdS case. For a proper implementation of conformal covariance in the boundary theory, two diffeomorphisms which are "asymptotic isometries" [3] which differ only by a diffeomorphism which is an "asymptotic identity" (i.e., acting as the identity on the boundary) should differ, from the viewpoint of the boundary theory, only by an internal (non-geometric) symmetry. The lack of bulk isometry groups cries out for a locally covariant formalism for local quantum physics, such as the one developed in [8]. Algebraic holography then maps *the realization of a locally covariant* quantum theory in the bulk to a *globally conformally covariant* quantum theory at the boundary, where the latter has, in principle, an enormous amount of internal symmetry. For the conformal group to be unitarily implementable in some GNS representation, these internal symmetries should not be generating a non-trivial cohomological obstruction. If the state asssociated to the GNS representation satisfies the Reeh-Schlieder property, it follows from Proposition 1, the discussion following it, and the work of Brunetti, Guido and Longo [9] that, due to the breakdown of Haag duality, the Tomita-Takesaki modular groups associated with the diamond von Neumann algebras cannot unitarily implement the isotropy groups of the respective diamonds. From this, it follows that either (or both) 1) The unitary representation of the conformal group cannot be of positive energy, or 2) The conformal group is spontaneously broken. Both scenarios are of the greatest interest for further study, as well as the possibility that such a spontaneous breaking has a cohomological structure stemming from the nontrivial asymptotic identities, and possible connections with the phenomenon of holographic Weyl anomalies [20]. This may even reveal a holographic encoding of bulk gravitational degrees of freedom into the modular structure of the boundary theory.

All the reasoning in Subsection 18.3.2 applies equally well if one wants to rebuild the bulk localization using (sufficiently small) bulk *regular diamonds* [17,28] instead of ordinary ones. This makes it also a good starting point for studying how the superselection sector structure is holographically mapped between both theories. This problem will be attacked in forthcoming work.

Acknowledgements

I am greatly indebted to Karl-Henning Rehren for his comments and suggestions built over a critical reading of my former paper [27], and to Jens Mund for sharing with me his insights on the intrinsic characterization of wedges in curved space-times. I would also like to thank Sumati Surya and A.P. Balachandran regarding their help with Subsection 18.3.1, and Fernando Auil for numerous discussions.

References

1. V.I. Arnold: *Mathematical Methods of Classical Mechanics.* Springer Verlag, 2nd ed., 1989.
2. A. Ashtekar and S. Da Asymptotically Anti-de Sitter Space-times: Conserved Quantities. *Class. Quant. Grav.* 17:L17-L30 (2000).
3. A. Ashtekar and A. Magnon: Asymptotically Anti-de Sitter Space-times. *Class. Quant. Grav.* 1:L39-L44 (1984).
4. J.K. Beem, P.E. Ehrlich and K.L. Easley: *Global Lorentzian Geometry.* M. Dekker, 2nd ed., 1996.
5. A. Borde: Geodesic Focusing, Energy Conditions and Singularities. *Class. Quant. Grav.* 4:343-356 (1987).
6. R. Bousso: The Holographic Principle. *Rev. Mod. Phys.* 74:825-874 (2002).
7. R. Bousso and L. Randall: Holographic Domains of Anti-de Sitter Space. *JHEP* 0204:057 (2002).
8. R. Brunetti, K. Fredenhagen and R. Verch: The Generally Covariant Locality Principle — A New Paradigm for Local Quantum Physics. *Commun. Math. Phys.* 237:31-68 (2003).
9. R. Brunetti, D. Guido and R. Longo: Modular Structure and Duality in Conformal Quantum Field Theory. *Commun. Math. Phys.* 156:201-219 (1993).
10. C.J. Fewster and T.A. Roman: Null Energy Conditions in Quantum Field Theory. *Phys. Rev. D* 67:044003 (2003).
11. H. Friedrich: Cauchy Problems for the Conformal Vacuum Field Equations in General Relativity. *Commun. Math. Phys.* 91:445-472 (1983).
12. H. Friedric Existence and Structure of Past Asymptotically Simple Solutions of Einstein's Field Equations with Positive Cosmological Constant. *J. Geom. Phys.* 3:101-117 (1986).
13. H. Friedrich: Einstein Equations and Conformal Structure: Existence of Anti-de Sitter-type Space-Times. *J. Geom. Phys.* 17:125-184 (1995).
14. G.J. Galloway: Maximum Principles for Null Hypersurfaces and Null Splitting Theorems. *Annales Henri Poincaré* 1:543-567 (2000).

15. G.J. Galloway: Some global results for asymptotically simple space-times. In: *Proceedings of the Workshop "The conformal structure of space-times: geometry, analysis, numerics"*, Tübingen, April 2001 (J. Frauendiener and H. Friedrich, eds.). Lect. Notes Phys. 604:51-60 (2002).
16. S. Gao and R.M. Wald: Theorems on gravitational time delay and related issues. *Class. Quant. Grav.* 17:4999-5008 (2000).
17. D. Guido, R. Longo, J.E. Roberts and R. Verch: Charged sectors, spin and statistics in quantum field theory on curved space-times. *Rev. Math. Phys.* 13:125-198 (2001).
18. R. Haag: *Local Quantum Physics.* Springer Verlag, 2nd ed., 1996.
19. S.W. Hawking and G.F.R. Ellis: *The Large Scale Structure of Space-Time.* Cambridge, 1973.
20. M. Henningson and K. Skenderis: Holography and the Weyl Anomaly. *Fortsch. Phys.* 48:125-128 (2000).
21. A.N. Kolmogorov and S.V. Fomin: *Introductory Real Analysis* (Translation by R.A. Silverman). Dover, 1975.
22. J. Maldacena: The Large N Limit of Superconformal Field Theories and Supergravity. *Adv. Theor. Math. Phys.* 2:231-252 (1998).
23. B. O'Neill: *Semi-Riemannian Geometry.* Academic Press, 1983.
24. R. Penrose, R.D. Sorkin and E. Woolgar: A Positive Mass Theorem Based on the Focusing and Retardation of Null Geodesics. arXiv:gr-qc/9301015.
25. K.-H. Rehren: Algebraic Holography. *Annales Henri Poincaré* 1:607-623 (2000).
26. K.-H. Rehren: Private communication.
27. P.L. Ribeiro: Renormalization Group Flow in Algebraic Holography. In: *Proceedings of the International Conference on Renormalization Group and Anomalies in Gravity and Cosmology* (M. Asorey and I.L. Shapiro, eds.). Nucl. Phys. B – Proc. Suppl. 127:193-195 (2004). [arXiv:hep-th/0306024].
28. G. Ruzzi: Punctured Haag duality in locally covariant quantum field theories. arXiv:math-ph/0402027.
29. S. Surya: Private communication.
30. R.M. Wald: *General Relativity.* Chicago University Press, 1984.
31. E. Witten: *Anti de Sitter Space and Holography. Adv. Theor. Math. Phys.* 2:253-291 (1998).
32. E. Woolgar: The Positivity of Energy for Asymptotically Anti-de Sitter Spacetimes. *Class. Quant. Grav.* 11:1881-1900 (1994).

19

Non-Commutative Renormalization

Vincent Rivasseau and Fabien Vignes-Tourneret

Laboratoire de Physique Théorique, Université Paris-Sud XI, Orsay Cedex, France
Vincent.Rivasseau@th.u-psud.fr

Summary. We review the recent approach of Grosse and Wulkenhaar to the perturbative renormalization of non-commutative field theory and suggest a related constructive program. This paper is dedicated to J. Bros on his 70th birthday.

19.1 Introduction

Non-commutative field theory has attracted interest in recent years from several different point of views [1]. It is a concrete example of non-commutative geometry which, according to mathematicians like A. Connes, should be in a broad sense the correct framework to quantize gravity. It can be generated as an effective limit of string theory in certain cases. And it is related to condensed matter problems such as two-dimensional Fermi systems in the presence of strong magnetic transverse fields, i.e. the quantum Hall effect.

Quantizing gravity is considered the "Holy Grail" of theoretical physics, and should lead to new insights on the ultimate nature of space, time, and the universe. In this direction a more and more insistent thread of the recent years is the possibility of some duality or mixing between short and long distance physics. A priori this idea runs against the well established tradition of determinism, the modern embodiment of which is the standard philosophy of the renormalization group (RG). In this philosophy, iterated integration over short range degrees of freedom of a bare "fundamental" microscopic action leads to an effective action for macroscopic long range variables. Infrared/ultraviolet duality or mixing clash with this philosophy but would fit well within the context of increased links between particle physics and astrophysics or cosmology. These links are certainly a major trend of physics in the last decade, which saw supernovas and cosmic background radiation studies boost the concept of the universe and the big bang as the ultimate laboratory and experiment.

A paradigmatic example for the possibility of infra-red/ultraviolet mixing is the $R \to 1/R$ duality for strings compactified on circles of radius R. If something of this kind occurs in the real universe, we might have to change the usual renormalization group picture in which microscopic observations solely determine fundamental laws. We could view ourselves as observing the universe from some kind of a "middle scale", which could be called the "ultrared-infraviolet". Our observations of microscopic ultraviolet scales through e.g. accelerators and of macroscopic infrared scales through various kinds of "telescopes" might more and more reveal entangled aspects of physics. Both should be considered inputs of some ultraviolet-infrared bare laws of physics and then combined into a modified RG analysis that would lead to effective laws of physics (and perhaps of chemistry and biology as well?) for the ultrared-infraviolet middle scale. This may smell like the "anthropic" point of view. But what are the scales involved in this grand mixing? The quantization scale for gravity is the Planck scale of about 10^{-33} meters. The observable present radius of the universe is about ten giga light-years or 10^{26} meters. Although there may be nothing fundamental with this value, if we follow the fashionable guess that new physics just lies "a factor of 10 around the corner of present observations", we obtain 10^{27} meters as a possible fundamental macroscopic length. A geometric ratio betwen these microscopic and macroscopic scales leads to a fundamental "middle scale" of about a millimeter. Therefore we should perhaps call this point of view the "antropic" rather than anthropic principle...

More seriously, from the mathematical physicist point of view, it seems a long way before a completely rigorous string or M-like version of quantized gravity can be developed. In the meantime, it would be good to have simpler mathematical toy models of ultraviolet-infrared mixing in which we could begin to grasp the phenomenon. On this road simple Lagragian non-commutative field theories stand as the first natural station.

Indeed the simplest way to generalize the algebra of ordinary commuting space time coordinates is to add a nonzero constant commutator between them. This means the relation

$$[x^\mu, x^\nu] = i\theta^{\mu\nu} \tag{19.1}$$

should hold, with θ a constant antisymmetric tensor. As in symplectic geometry we could by a linear transformation put θ under standard symplectic form. In this way space-time coordinates would occur as "symplectic pairs" and we would have the simple notion of non-commutative $\mathbb{R}^2$ or $\mathbb{R}^4$ for which $\theta^{12} = \theta^{34} = -1$, $\theta^{21} = \theta^{43} = +1$ and all other $\theta^{\mu\nu}$'s are zero[1]. Obviously this new tensor breaks Lorentz invariance.

It also breaks locality. Indeed the ordinary product of functions is replaced by a non-commutative product called star product or Moyal product. By the Campbell-Hausdorff formula we can compute the star product of two plane-waves:

$$e^{ik.x} \star e^{ik'.x} = e^{-\frac{i}{2} k_\mu \theta^{\mu\nu} k'_\nu} e^{i(k+k').x} \tag{19.2}$$

from which one can deduce by linearity and Fourier analysis a kernel which explicitly displays the non-locality of the star multiplication:

$$f \star g(z) = \int dx dy K(z; y, x) f(x) g(y) \;, \tag{19.3}$$

[1]Of course we can also consider e.g. a non-commutative $\mathbb{R}^3$ in which there is just one symplectic pair and a remaining commutative coordinate but this is not very appealing.

$$K(z;y,x) = \delta(z-x)\star\delta(z-y) = \frac{1}{(2\pi)^d}\int dkdk' e^{ik(z-x)} \star e^{ik'(z-y)} \qquad (19.4)$$
$$= \frac{1}{(2\pi)^d \det\theta} e^{i(z-x)\theta^{-1}(z-y)}.$$

19.2 The UV-IR Problem for ϕ^4_4

The ordinary action for the simplest commutative renormalizable bosonic field theory, the ϕ^4_4 theory is:

$$S = \int d^4x \frac{1}{2}(\partial^\mu\phi\partial_\mu\phi + m^2\phi^2) + \frac{\lambda}{(4!)}\phi^4 , \qquad (19.5)$$

which is easily generalized to a naive non-commutative ϕ^4_4 theory by replacing ordinary products by the star product and integration by the non-commutative integration, noted $\int_{nc}$, which is a kind of combination of ordinary integration plus a trace [1]:

$$S = \int_{nc} \frac{1}{2}(\partial^\mu\phi\star\partial_\mu\phi + m^2\phi\star\phi) + \frac{\lambda}{(4!)}\phi\star\phi\star\phi\star\phi . \qquad (19.6)$$

If we rewrite this action in momentum space and use the rule (19.2) for the star product of plane waves, we get

$$S = \frac{1}{2}\int \frac{d^4p}{(2\pi)^4}(p^2+m^2)\phi(p)\phi(-p)$$
$$+\frac{\lambda}{(4!)}\int\prod_{i=1}^{4}\frac{d^4p_i}{(2\pi)^4}(2\pi)^4\delta(\sum_{i=1}^{4}p_i)\exp(-\frac{i}{2}\sum_{i<j}p_i^\mu\theta_{\mu\nu}p_j^\nu)\prod_{i=1}^{4}\phi(p_i) . \qquad (19.7)$$

Remark that this formula is identical to the commutative case except for the oscillating vertex factor

$$\exp(-\frac{i}{2}\sum_{i<j}p_i^\mu\theta_{\mu\nu}p_j^\nu) . \qquad (19.8)$$

The form of this factor shows that we cannot order fields arbitrarily at a vertex but that we have to respect cyclic ordering, and that the Feynman graphs of this theory are ribbon graphs, as in matrix models. In fact this *is* a matrix model if we introduce the correct functional basis. Let us recall that experts concluded that the model is not renormalizable because of infrared/ultraviolet entanglement [2–4].

Indeed planar graphs have amplitudes identical to ordinary ϕ^4 graphs, since the oscillating factors all cancel out in the planar case. Still this means the theory has an infinite number of divergent graphs which require renormalization. But on the other hand, for non-planar graphs the oscillating factors (19.8) make the amplitudes finite but divergent when external momenta become small. This leads in turn to infrared divergences of arbitrary strength when these non-planar subgraphs are inserted into loops. No simple idea solved this "entanglement".

For instance for the planar tadpole graphs of Figure 19.1 A, the amplitude is the standard diverging one:

$$\frac{\lambda}{6}\int\frac{d^4k}{(2\pi)^4}\frac{1}{k^2+m^2} , \qquad (19.9)$$

while for the non-planar tadpole of Figure 19.1 B, it is

$$\frac{\lambda}{12}\int \frac{d^4k}{(2\pi)^4)}\frac{e^{ip^\mu k^\nu \theta_{\mu\nu}}}{k^2+m^2} = \frac{\lambda}{48\pi^2}\sqrt{\frac{m^2}{\tilde{p}^2}}K_1(\sqrt{m^2\tilde{p}^2}) \; , \tag{19.10}$$

where K_1 is a standard special function and $\tilde{p}_\mu = \theta_{\mu\nu}p^\nu$. This amplitude behaves as $\tilde{p}^{-2}$ for small p. It does not seem possible to improve such a finite amplitude by an ultraviolet subtraction. Furthermore if we insert n of these non-planar tadpoles in a closed loop as in Figure 19.2, we get divergence of the integral at small p of arbitrarily high power when n gets large.

A B

Fig. 19.1. Planar and twisted tadpoles.

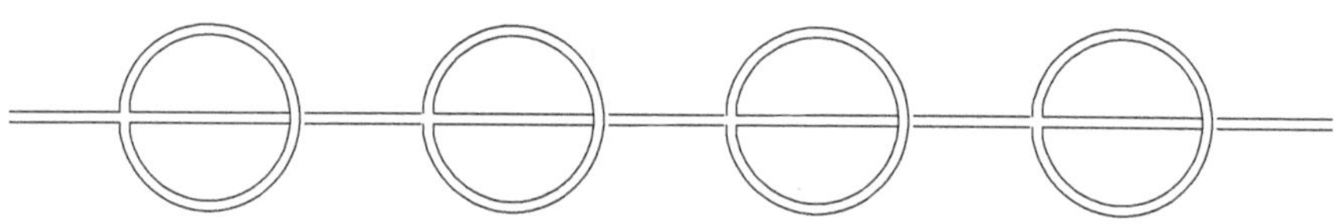

Fig. 19.2. Chain of twisted tadpoles.

19.3 The Covariant Theory

The solution to this problem was recently obtained by Grosse and Wulkenhaar. In a beautiful series of papers [5–7, 20], they claim that action (19.6) is indeed not renormalizable but simply in the sense of an anomalous theory, because it is not covariant under the Langmann-Szabo duality [8]. They found the correct Langmann-Szabo covariant version of the theory by modifying the quadratic part of the action, and proved the BPHZ theorem, namely renormalizability of the theory at all orders in perturbation. Their proof still assumes some reasonable bounds on the propagator that they checked numerically. We have now proved these bounds using a different type of cutoffs [9], so that the proof is complete even for "purists".

To understand why action (19.6) is not correct but anomalous, we should introduce the Langmann-Szabo duality:

$$p_\mu \to \tilde{x}_\mu = 2(\theta^{-1})_{\mu\nu}x^\nu \; , \tag{19.11}$$

and define the Fourier transform of the field with a cyclic sign:

$$\hat{\phi}(p) = \int d^4x e^{(-1)^a i p_{a,\mu} x_a^\mu} \phi(x_a) \ , \tag{19.12}$$

where the index $a = 1, 2, 3, 4$ follows a cyclic ordering at the vertex.

Then if we write this interaction in terms of ordinary products and kernels:

$$\begin{aligned} S_{int} = \int_{nc} \phi \star \phi \star \phi \star \phi(x) &= \int \prod_{a=1}^{4} d^4x_a \phi(x_1)\phi(x_2)\phi(x_3)\phi(x_4) V(x_1, x_2, x_3, x_4) \\ &= \int \prod_{a=1}^{4} \frac{d^4p_a}{(2\pi)^4} \hat{\phi}(p_1)\hat{\phi}(p_2)\hat{\phi}(p_3)\hat{\phi}(p_4)\hat{V}(p_1, p_2, p_3, p_4) \end{aligned} \tag{19.13}$$

we find

$$V(x_1, x_2, x_3, x_4) = \frac{1}{\pi^4 \det\theta} \delta^4(x_1 - x_2 + x_3 - x_4) \cos[2\theta^{-1}_{\mu\nu}(x_{1\mu}x_{2\nu} + x_{3\mu}x_{4\nu}], \tag{19.14}$$

$$\hat{V}(p_1, p_2, p_3, p_4) = (2\pi)^4 \delta(p_1 - p_2 - p_3 + p_4) \cos \frac{\theta^{\mu\nu}}{2} [p_{1\mu}p_{2\nu} + p_{3\mu}p_{4\nu}] \tag{19.15}$$

and this interaction has Langmann-Szabo duality. But the free part of the action is not covariant under this duality; the local mass term, like the interaction, is covariant but the p^2 Laplacian is not because it has no $\tilde{x}^2$ counterpart. So Grosse and Wulkenhaar proposed as correct Langmann-Szabo covariant action for non-commutative ϕ^4 the generalized action with a harmonic potential dual to the kinetic term:

$$S_{free} = \int_{nc} \frac{1}{2} [\partial_\mu \phi \star \partial^\mu \phi + \Omega^2 (\tilde{x}_\mu \phi \star \tilde{x}^\mu \phi + m^2 \phi \star \phi] \ . \tag{19.16}$$

Then

$$S_{free} + S_{int} = \int_{nc} \frac{1}{2} [\partial_\mu \phi \star \partial^\mu \phi + \Omega^2 (\tilde{x}_\mu \phi \star \tilde{x}^\mu \phi + m^2 \phi \star \phi] + \frac{\lambda}{(4!)} \phi \star \phi \star \phi \star \phi \tag{19.17}$$

is covariant under duality $p_\mu \to \tilde{x}_\mu$:

$$S(\phi, m, \lambda, \Omega) \to \Omega^2 S(\phi, \frac{m}{\Omega}, \frac{\lambda}{\Omega^2}, \frac{1}{\Omega}) \ . \tag{19.18}$$

The theory has one additional parameter, namely Ω. It is completely invariant at $\Omega = 1$, where there is complete equivalence between space coordinates and momenta. It is this theory that is renormalizable at all orders of perturbation theory [7].

The best rigorous understanding of renormalization is neither in pure space nor in pure momentum space, but in phase-space, the only frame in which it can be understood even in a constructive or non-perturbative sense. But what is the equivalent of phase space here? An important observation is that in a non-commutative space one cannot simultaneously localize all x coordinates with arbitrary precision. There should be analogs of the Heisenberg uncertainty relations but now between coordinates. This may be grasped using Gaussian wave packets. The star multiplication of such Gaussians is again Gaussian, as in the commutative case. But in contrast with that commutative case, if we try to focus more and more a Gaussian by star multiplying with itself, there is a limiting scale at which this focusing process must stop. More precisely we have for a two-dimensional symplectic pair with commutator $[x^1, x^2] = \theta$,

$$e^{-\frac{x^2}{a^2}} \star e^{-\frac{x^2}{b^2}} = ce^{-\frac{x^2}{d^2}} \tag{19.19}$$

with $c = [1 + \theta^2/(ab)^2]$, $d = \sqrt{\frac{a^2b^2+\theta^2}{a^2+b^2}}$, so that $d < a$ if $a > \sqrt{\theta}$, $d > a$ if $a < \sqrt{\theta}$ and there is a fixed point at $a = \sqrt{\theta}$.

This means there is a canonical best focused Gaussian which is

$$f_0(x_1, x_2) = 2e^{-\frac{x_1^2+x_2^2}{\theta}} . \tag{19.20}$$

This suggests introducing creation and annihilation operators

$$a = \frac{x_1 + ix_2}{\sqrt{2}}; \ \bar{a} = \frac{x_1 - ix_2}{\sqrt{2}} \tag{19.21}$$

and the base functions

$$f_{mn}(x) = \frac{1}{\sqrt{n!m!\theta^{m+n}}} \bar{a}^{\star m} \star f_0 \star a^{\star n} \tag{19.22}$$

in which the star product becomes an ordinary matrix product:

$$f_{mn} \star f_{kl} = \delta_{nk} f_{ml} \ . \tag{19.23}$$

This is called the matrix base. Any function of x or its Fourier transform in p which belongs to Schwartz space, hence is smooth with rapid decrease, can be analyzed as a series in the f_{mn} modes:

$$a(x) = \sum_{m,n=0}^{\infty} a_{mn} f_{mn}(x) \ , \tag{19.24}$$

$$a \star b(x) = \sum_{m,n=0}^{\infty} (ab)_{mn} f_{mn}(x) \ ; \ (ab)_{mn} = \sum_{k=0}^{\infty} a_{mk} b_{kn} \ . \tag{19.25}$$

The condition of smoothness and rapid decay is equivalent to rapid decay of the a_{mn} coefficients:

$$\forall k \ \sum_{m,n} \left[(2m+1)^{2k} (2n+1)^{2k} |a_{mn}|^2 \right]^{1/2} < \infty \ , \tag{19.26}$$

and the non-commutative integral becomes an ordinary trace:

$$\int_{nc} f_{mn}(x) = 2\pi\theta\delta_{mn} \ . \tag{19.27}$$

In this theory, to cut the large values of m and n, for instance through a condition such as $m, n < \Lambda$, corresponds to imposing a cutoff in both position and momentum space, so it is an infrared-ultraviolet cutoff and the remaining modes $m, n < \Lambda$ could be called the ultrared-infraviolet modes below Λ. As we said above the usual RG philosophy is now to integrate slice by slice the high m, n modes to get the effective theory for the remaining ultrared-infraviolet modes. The renormalization condition instead of being formulated at zero or low momentum are now formulated at the fundamental mode $m = n = 0$, which means the fundamental f_0 Gaussian wave packet.

It is well known that in practice there are different technical ways to implement this RG philosophy, and in particular to create the different "momentum slices" for the "fluctuation fields" that are to be integrated out. The best way to do this when we have a positive quadratic form as free action is to invert it and cut the slices directly on the propagator. This leads to factorizing the Gaussian measure into independent measures so that the slice fields are orthogonal with respect to this decomposition.

This is what Grosse and Wulkenhaar do in [6, 7]. For that they have first to compute the free action (19.16) in the matrix base, which is not too difficult. One can perform first the computation for a single pair of coordinates, so in $\mathbb{R}^2$ since the two pairs are in fact independent. The result is

$$S_{free} = 2\pi\theta \sum_{m,n,k,l \in \mathbb{N}} \frac{1}{2} G_{m,n;k,l} \phi_{mn} \phi_{kl}, \tag{19.28}$$

$$\begin{aligned} G_{m,n;k,l} &= m^2 + \frac{2}{\theta}(1+\Omega^2)(n+m+1)\delta_{nk}\delta_{ml} \\ &-\frac{2}{\theta}(1-\Omega^2)(\sqrt{(n+1)(m+1)}\delta_{n+1,k}\delta_{m+1,l} + \sqrt{nm}\delta_{n-1,k}\delta_{m-1,l}) \, . \end{aligned} \tag{19.29}$$

Then one has to invert this quadratic form to get the propagator of the theory in the matrix base. This is a beautiful non-trivial computation of Grosse and Wulkenhaar, which uses Meixner polynomials [7, 10], which are some kind of hypergeometric functions. But we are really interested in the scaling properties of this propagator. It is not necessary to know anything about Meixner polynomials or special functions of any kind to understand or use the result, which is based an integral representation of the propagator analog to Feynman-Schwinger parametric representation. The propagator $\Delta_{m,n;k,l}$ is zero unless $m+l = n+k$, so we can express it in terms of three multi-integers $\Delta_{m,m+\alpha;l+\alpha,l}$, and the result is

$$\begin{aligned} \Delta(m, m+\alpha; l+\alpha, l) = \frac{(1+\Omega)^2}{4\Omega} \int_0^1 dz \, z^{\frac{\mu_0^2 \theta}{8\Omega}} \frac{\theta}{8\Omega(1+B)^2} \sum_{u=0}^{\min(m,l)} \mathcal{A} \left(\frac{\sqrt{z}}{1+B}\right)^{2u+\alpha} \\ \times \left(\frac{B(1+\Omega)}{(1+B)(1-\Omega)}\right)^{m+l-2u} \end{aligned} \tag{19.30}$$

where $\mathcal{A}(m,l,\alpha,u) = \sqrt{\binom{m}{m-u}\binom{m+\alpha}{m-u}\binom{l}{l-u}\binom{l+\alpha}{l-u}}$ and $B(z,\Omega) = \frac{(1-\Omega)^2(1-z)}{4\Omega}$. Indices such as m, l, α and u have two components, one for each symplectic pair of $\mathbb{R}^4$, hence run over $\mathbb{N}^2$. Sums or products apply to these two components.

Remark that every contribution in these sums and integrals is positive, so we should think of this representation as the non-commutative analog of the heat kernel or path representation of the propagator in the commutative case.

Grosse and Wulkenhaar then go on, slicing the propagator with sharp conditions on the maximum of the four indices $m, n; k, l$ appearing in the propagator. For instance the i-th slice of the RG in their framework would be

$$\Delta_i(m,n;k,l) = \Delta(m,n;k,l)\chi(M^i \le \max(m,n,k,l) < M^{i+1}) \tag{19.31}$$

where M is some fixed number, e.g. $M = 2$, and $\chi(X)$ is the characteristic function of the event X considered.

Then they check that the theory with full action is renormalizable, using the popular Polchinski's scheme which feeds inductive bounds directly into the RG equations [11]. This is a good way of proving perturbative renormalizability and even large order bounds.

Let us however recall that as of now there is no constructive version of this scheme. Indeed it seems very difficult to iterate bounds in the RG expansions of constructive field theory because many potential Wick contractions remain hidden in functional integrals. For example in the simplest cases of Fermionic models, constructive field theory simply amounts to expanding a tree of Wick contractions connecting external sources and to keep half of the contractions (those of the loops) inside a determinant [12]. This determinant is bounded in the end through some Gram's inequality. But intermediate bounds à la Polchinski on earlier kernels at an intermediate stage of the expansion seem to inevitably destroy the cancellations in the determinant that are responsible for constructive i.e. non-perturbative convergence of the scheme.

19.4 Smooth slices

We would like to introduce an improvement on [6, 7] original scheme, by using sharp cutoffs in Feynman parametric space rather than directly on the indices m and n. As is known e.g. in constructive theory, sharp cutoffs in direct or momentum space for ordinary field theory have bad decay properties in the dual Fourier variable, and this leads to difficulties for phase space analysis, whence cutoffs implemented through the parametric representation are usually much more convenient. So we suggest picking a number $M > 1$ and to defining the i-th slice of the RG analysis through the sliced propagator

$$\begin{aligned}&\Delta_i(m, m+\alpha; l+\alpha, l)\\&= \int_{1-M^{-i+1}}^{1-M^{-i}} dz z^{\frac{\mu_0^2\theta}{8\Omega}} \frac{\theta}{8\Omega(1+B)^2} \{\frac{\sqrt{z}}{1+B}\}^{m+l+\alpha} \sum_{u=0}^{\min(m,l)} \mathcal{A}\{\frac{B(1+\Omega)}{\sqrt{z}(1-\Omega)}\}^{m+l-2u}.\end{aligned} \tag{19.32}$$

We can then use bounds such as

$$\mathcal{A}(m,l,\alpha,u) \le \frac{\sqrt{m(\alpha+m)}^{m-u}\sqrt{l(\alpha+l)}^{l-u}}{(m-u)!(l-u)!} \le \frac{(m+\alpha/2)^{m-u}(l+\alpha/2)^{l-u}}{(m-u)!(l-u)!} \tag{19.33}$$

in the ultraviolet region, which corresponds to z near 1.

Let us consider a slice $M^{-i} \le 1-z \le M^{-i+1}$. For i large enough we have

$$\begin{aligned}&\Delta_i(m, m+\alpha; l+\alpha, l)\\&= \int_{1-M^{-i+1}}^{1-M^{-i}} dz z^{\frac{\mu_0^2\theta}{8\Omega}} \frac{\theta}{8\Omega(1+B)^2} \{\frac{\sqrt{z}}{1+B}\}^{m+l+\alpha} \sum_{u=0}^{\min(m,l)} \mathcal{A}\{\frac{B(1+\Omega)}{\sqrt{z}(1-\Omega)}\}^{m+l-2u}\\&\le \int_{1-M^{-i+1}}^{1-M^{-i}} dz z^{\frac{\mu_0^2\theta}{8\Omega}} \frac{\theta}{8\Omega(1+B)^2} \{\frac{\sqrt{z}}{(1+B)}\}^{m+l+\alpha} \sum_{u=0}^{\min(m,l)} \frac{X^{m-u}Y^{l-u}}{(m-u)!(l-u)!}\end{aligned} \tag{19.34}$$

where $X = \frac{B(1+\Omega)(m+\alpha/2)}{\sqrt{z}(1-\Omega)}$ and $Y = \frac{B(1+\Omega)(l+\alpha/2)}{\sqrt{z}(1-\Omega)}$.

Since obviously $\sum_{u=0}^{\min(m,l)} \frac{X^{m-u}Y^{l-u}}{(m-u)!(l-u)!} \leq \exp(X+Y)$:

$$\Delta_i(m, m+\alpha; l+\alpha, l) \leq \int_{1-M^{-i+1}}^{1-M^{-i}} dz z^{\frac{\mu_0^2 \theta}{8\Omega}} \frac{\theta}{8\Omega(1+B)^2} \{\frac{\sqrt{z}}{(1+B)}\}^{m+l+\alpha}$$
$$\exp\{\frac{B(1+\Omega)(m+l+\alpha)}{\sqrt{z}(1-\Omega)}\} \tag{19.35}$$

Expanding to first order in $1 - z = \epsilon$, since $\sqrt{z} \simeq 1 - \epsilon/2$ and $B = C\epsilon$ with $C(\Omega) = (1-\Omega)^2/4\Omega$, one finds, since $C + (1/2) - C\frac{1+\Omega}{1-\Omega} = \Omega/2$.

$$\Delta_i(m, m+\alpha; l+\alpha, l) \leq \frac{K\theta}{\Omega} \int_{1-M^{-i+1}}^{1-M^{-i}} dz \exp\{(-C - 1/2 + C\frac{1+\Omega}{1-\Omega})\epsilon(m+l+\alpha)\}$$
$$\leq \frac{K}{\Omega} M^{-i} \prod_{j=1,2} e^{-M^{-i}\Omega(\alpha+m+l)/2} \tag{19.36}$$

for some constants c and K.

In the same vein, we shall prove in [9] bounds such as

$$\sum_l \max_\alpha \Delta_i(m, m+\alpha; l+\alpha, l) \leq \frac{K}{\Omega} M^{-i} \prod_{j=1,2} e^{-cM^{-i}m} \tag{19.37}$$

for a certain range of parameters Ω including a neighborhood of $\Omega = 1$. These bounds establish the regular non-local matrix model behavior in the sense of [5]. This should complete for purists the corresponding renormalization theorem at all orders in perturbation theory of [7].

19.5 Towards Non-commutative Constructive Field Theory

We would like to extend these results of perturbative field theory to the non-perturbative or constructive level. For this we have roughly speaking to perform the sum over all Feynman diagrams. This is usually possible for Bosonic models which are both stable and asymptotically safe, and for Fermionic models which are asymptotically safe. As remarked already, one can no longer directly use the Polchinski inductive scheme, but there are similar constructive renormalization group schemes in phase space (see [13,14] for reviews on general methods and results on constructive field theory).

19.5.1 Gross-Neveu

One of the easiest models of constructive field theory is the Gross-Neveu model of Fermions with N colors and a quartic vector-like interaction. In two dimensions this model has the same counting power as ϕ_4^4. It shows asymptotic freedom in the ultraviolet and mass generation in the infrared, all these features having been rigorously established at the constructive level ([12]– [16] and references therein).

There is an analogous model on commutative $\mathbb{R}^2$ which is studied at large N in [15]. However if we rewrite the model in the matrix base, the regime of large matrix indices is not the one studied in [15]. The non-commutative Gross Neveu model on $\mathbb{R}^2$ ought to be the easiest candidate for rigorous non-perturbative construction of a non-commutative field theory. But we expect at least one difficulty which is the following. In Fermionic vector models the easiest constructive analysis of [12] can be summarized as follows. Instead of developing the full set of Wick contractions that leads to the series of all Feynman graphs, one develops only "half of them", namely those Wick contractions of a spanning tree in each diagram, and the other contractions are kept in a Grassmann unexpanded integral, i.e. a determinant. Gram's bound is then applied on this determinant, which does not generate the dangerous factorial leading usually to divergence of perturbation theory. This scenario works well in vector theories, such as for instance the interacting Fermi liquid in two dimensions [17,18], because Gram's bound leads to a single sum over the vector index at each vertex.

Here however we have matrix indices for each field or anti-field. After picking an explicit tree a naive Gram's bound on the remaining determinant may lead to two index sums at each vertex, which is one more than what power counting can afford. In short, in an N vector model Pauli's principle stops perturbation theory roughly at order N but in an $N \times N$ matrix model, Pauli's principle stops perturbation theory only at order $\simeq N^2$. This is somewhat reminiscent of the situation in three-dimensional condensed matter models. Hence the constructive analysis of non-commutative models may require some heavier techniques, in fact the "Bosonic techniques" of multi-scale cluster expansions, and Hadamard's bound instead of Gram's bound, as in [19].

19.5.2 ϕ^4_4 at $\Omega = 1$

The ordinary ϕ^4_4 model, as well known, is disappointing from the constructive point of view because for the repulsive sign of the coupling constant it lacks ultraviolet asymptotic freedom and for the attractive sign it lacks stability!

However in the non-commutative case there should be ultraviolet-infrared mixing so we expect the asymptotic freedom in the infrared regime to come partly to the rescue of the ultraviolet regime. At $\Omega = 1$ there should be perfect symmetry between the infrared and ultraviolet sides. Therefore we expect the β function of ϕ^4_4 should vanish at $\Omega = 1$. This has been checked at least at one loop in [20].

This means that at $\Omega = 1$ the "running coupling constant" does not run at all, and the bare coupling therefore equals the renormalized one. The model is not asymptotically free but asymptotically safe. As a result there is no Landau ghost problem, and no reason for which this theory should not exist in a non-perturbative sense, probably being the Borel sum of its perturbative series. The same difficulties as for the Gross-Neveu model should be also expected here, namely the constructive analysis should overcome the difficulties linked to a matrix model. However building non-perturbatively a ϕ^4_4 theory, even of a non-commutative type, is a tantalizing perspective. After so many years it would partly realize one of the dreams of the founding fathers of constructive theory, although this non-commutative model should obviously not be expected to fulfill Wightman's axioms, in particular Lorentz invariance.

References

1. M. R, Douglas and N. A. Nekrasov: Non commutative Field Theory. arXiv:hep-th0106048.
2. S. Minwalla, M. Van Raamsdonk and N. Seiberg: Non commutative perturbative dynamics. *JHEP* 0002:020 (2000).
3. I. Chepelev and R. Roiban: Renormalization of quantum field theories on non-commutative $\mathbb{R}^d$. I: Scalars. *JHEP* 0005:037 (2000).
4. I. Chepelev and R. Roiban: Convergence Theorems for non-commutative Feynman graphs and renormalization. *JHEP* 0103:001 (2001).
5. H. Grosse and R. Wulkenhaar: Power Counting Theorems for non -local matrix models and renormalization. arXiv:hep-th/0305066.
6. H. Grosse and R. Wulkenhaar: Renormalization of ϕ^4 theory on noncommutative $\mathbb{R}^2$ in the matrix base. arXiv:hep-th/0307017.
7. H. Grosse and R. Wulkenhaar. Renormalization of ϕ^4 theory on noncommutative $\mathbb{R}^4$ in the matrix base. arXiv:hep-th/0401128.
8. E. Langmann and R.J. Szabo: Duality in scalar field theory on noncommutative phase space. *Phys Lett. B* 553:168 (2002).
9. V. Rivasseau, F. Vignes-Tourneret and R. Wulkenhaar: To appear.
10. R. Koekoek and R. F. Swarttouw: The Askey-scheme of hypergeometric orthogonal polynomials and its q-analogue. arXiv:math.CA/9602214.
11. J. Polchinski: Renormalization and Effective Lagrangians. *Nucl. Phys B* 231:269 (1984).
12. M. Disertori and V. Rivasseau: Continuous Constructive Fermionic Renormalization. *Ann. Henri Poincaré* 1:1 (2000).
13. V. Rivasseau: *From perturbative to constructive renormalization.* Princeton University Press, 1991.
14. V. Rivasseau (ed.): *Constructive Physics.* Proceedings of the International Workshop at Ecole Polytechnique, Palaiseau, July 1994. Lecture Notes in Physics 446, Springer Verlag, 1995.
15. E.T. Akhmedov, P. de Boer and G.W. Semenoff: Non Commutative Gross-Neveu Model at large N. arXiv:hep-th/0103199.
16. C. Kopper, J. Magnen and V. Rivasseau: Mass Generation in the Large N Gross-Neveu Model. *Commun. Math. Phys.* 169:121 (1995).
17. M. Disertori and V. Rivasseau: Interacting Fermi liquid in two dimensions at finite temperature, Part I and II. *Commun. Math. Phys.* 215:251 and 291 (2000).
18. Joel Feldman, H. Knörrer and E. Trubowitz: A two dimensional Fermi Liquid. *Commun. Math. Phys.* 247:1-319 (2004); and *Reviews in Math. Physics* 15(9):949-1169 (2003).
19. M. Disertori, J. Magnen and V. Rivasseau: Interacting Fermi liquid in three dimensions at finite temperature, part I: Convergent Contributions. *Ann. Henri Poincaré* 2:733-806 (2001).
20. H. Grosse and R. Wulkenhaar: The β-function in duality-covariant noncommutative ϕ^4-theory. arXiv:hep-th/ 0402093.

20

New Constructions in Local Quantum Physics

Bert Schroer

CBPF, Rua Dr. Xavier Sigaud 150, 22290-180 Rio de Janeiro, Brazil
and
Institut für Theoretische Physik, FU-Berlin, Germany
schroer@cbpf.br

Summary. Among several ideas which arose as consequences of modular localization there are two proposals which promise to be important for the classification and construction of QFTs. One is based on the observation that wedge-localized algebras may have particle-like generators with simple properties and the second one uses the structural simplification of wedge algebras in the holographic lightfront projection. Factorizable $d = 1+1$ models permit us to analyse the interplay between particle-like aspects and chiral field properties of lightfront holography.

20.1 How modular theory entered particle physics

The following introductory remarks about the history and the physical content of modular theory are intended to be helpful to understand the recent role of modular localization in the classification and construction of models of QFT without the use of Lagrangian quantization.

20.1.1 Remarks about history of modular localization

The beginnings of modular theory date back to the second half of the 1960s when two independent ideas, one from mathematics and one from particle physics, merged together [5]. On the mathematical side the Japanese mathematician Tomita generalized a concept, which before was only studied in the special context of the Haar measure ("unimodular") in group algebra theory, to the general setting of von Neumann algebras. At the same time three physicists, Haag, Hugenholtz and Winnink [16], found a conceptual framework for the direct field theoretic description of the thermodynamic limit ("open systems") in terms of operator algebras and their commutants [15]. Their important contribution, which became immediately incorporated by Takesaki into Tomita's modular theory, was the realization that the KMS condition (introduced by Kubo, Martin and Schwinger as a computational tool) acquired a fundamental conceptual significance in their new thermal setting.

It took another decade in order to appreciate the geometric significance of this modular formalism for the problem of localization of algebras and states in QFT [4].

This was preceded by an important mathematical application in the classification of type III von Neumann algebras by A. Connes [12] and followed up by a theorem of Driessler [13] stating that wedge-localized algebras are factors of type III_1. As a consequence double cone algebras in conformal invariant theories inherit this property[1]. Later refinements supported the idea that compactly localized subalgebras in QFT are isomorphic to the unique hyperfinite type III_1 factor. For more detailed reviews of modular theory from the mathematical physics viewpoint we refer to [5] [37] [40]

Although hyperfinite type III_1 algebras appear at first sight (as a result of this uniqueness) in a certain sense as void of structure as points in geometry, they are in other aspects much richer since they contain subalgebras of all types and one can form nontrivial intersections from copies placed into different positions within a common Hilbert space H. In fact we know from later developments of algebraic QFT that the full richness of a model of QFT is encoded in the notion of a net of space-time-indexed von Neumann algebras as subalgebras of $B(H)$ [15].

The net result of this thread of ideas, which culminated in the mathematical identification of simple building blocks of QFT, is interesting from many viewpoints. From a philosophical standpoint it tells us that the algebraic aspects of QFT comply perfectly with Leibniz's dictum that reality emerges from the relation between indecomposable entities ("monades") and not from their individual position with respect to an absolute outside reference.

This is not the first time that philosophical ideas of Leibniz became relevant in physics. In Einstein's "hole argument" [30] it played a significant role in the birth of general relativity; in particular it helped Einstein (and independently Hilbert) to overcome a misconception about how the local covariance of the Einstein-Hilbert field equations and the Newtonian limit fit together. By upholding the local covariance principle, i.e. the idea that local isomorphism classes of isometric diffeomorphisms replace the global notion of an absolute Minkowski space-time inertial frame of special relativity, Einstein realized that his difficulties in obtaining agreement with the Newtonian limit came from a computational misconception.

In fact it was shown recently that the Leibniz viewpoint of physical reality emerging from relations between entities rather than from positions in a pre-assigned absolute "inertia ether" can actually be extended in order to combine the quantum algebraic modular aspects with the classical covariance principle into a "local (quantum) covariance principle" [8]. This places QFT in CST much closer to a still elusive background-independent quantum gravity than ever before.

In the following we will argue that the "monades" of QFT are the wedge-localized algebras which (thanks to Driessler's work [13]) are known to be isomorphic copies of hyperfinite type III_1 factor. In order to avoid lengthy terminology we will refer to the basic hyperfinite type III_1-von Neumann algebra as the monade algebra (MA). The most convincing affirmation of this way of viewing QFT as arising from wedge localized MA is the fact that models of quantum field theory can be completely specified by positioning a finite number of operator algebra copies of the MA into suitably chosen relative positions within a common Hilbert space[2] [21].

---

[1]In conformal theories double cone algebras are conformally equivalent to wedge algebras and therefore inherit the hyperfinite typ III_1 property.

[2]We will use the terminology MA also for the positioned operator algebra copies of the basic MA.

This way of looking at QFT permits a particularly natural intrinsic formulation in low dimensional QFT. In the case of $d = 1+1$ the "modular inclusion" of two MAs specifies all data needed to characterize a specific QFT in terms of the structure of its Poincaré covariant nets and for $d = 1 + 2$ one achieves a complete characterization in terms of "modular intersection" of three MAs. Modular positions which are associated with a characterization of higher dimensional QFT models are also known [21], but in their present formulation they appear less natural i.e. more concocted in order to generate the desired Poincaré symmetry structure of Minkowski space-time.

Accepting the philosophical, conceptual and mathematical implications, one may ask the question whether this approach guided by Leibniz's philosophy is just an esoteric new way of looking at particle physics or if it also has constructive clout, i.e. whether one can actually classify models of QFT and elaborate a realistic scenario of their construction along those lines. Admittedly the apparent simplicity of generating QFT from the positioning of a finite number of MAs is somewhat deceiving; the problem of an intrinsically formulated positioning of MAs is actually quite hard since the appropriate concepts and mathematical tools are to a large extent still missing. Already the characterization of one MA in Hilbert space i.e. in the setting of local quantum physics of massive particles, the description $A(W) \subset_P B(H)$ is a difficult problem; here $A(W)$ denotes a wedge-localized MA, $B(H)$ is the algebra of bounded operators on Fock space of massive particles obtained by scattering theory (assuming asymptotic completeness) and the subscript P indicates that the inclusion is meant in the extended sense that also the action of the Poincaré group on it (which creates a family of wedge-localized MAs) is known.

There are two situations in which this positioning of the MA is reasonably simple and the construction of the net (and its generating pointlike field coordinatizations) can actually be carried out. These are the interactionless theories whose one-particle components are described in terms of Wigner representations of elementary systems and $d = 1+1$ factorizing models. For general interacting theories the idea of lightfront holography is very helpful because it suggests that we classify and construct wedge algebras in terms of their lightfront holographies. These problems will be addressed in this paper.

In the remainder of this introduction the modular approach to the interaction-free QFTs will be given; as a result of its simplicity it also serves well as a pedagogical introduction into the setting of modular localization.

20.1.2 Modular construction of interaction-free QFT

This construction via modular localization proceeds in three steps as follows [9, 14, 26, 29]

1. Fix a reference wedge region, e.g. $W_R = \{x \in \mathbb{R}^4; x^1 > |x^0|\}$ and use the Wigner positive energy representation of the W_R-affiliated boost group $\Lambda_{W_R}(\chi)$ and the $x^0 - x^1$–reflection j_W[3] along the edge of the wedge j_{W_R} in order to define

[3] The reflection on the edge of the wedge is related to the total TCP reflection by a π-rotation around the x_1-axis. Hence in certain cases the irreducible representation has to be doubled. This is always the case for zero mass finite helicity representations and more generally if particles are not selfconjugate.

the following antilinear unbounded closable operator (with $closS = clos\Delta^{\frac{1}{2}}$). Retaining the same notation for the closed operators, one defines

$$\begin{aligned} S_{W_R} &:= J_{W_R}\Delta^{\frac{1}{2}}, \\ J_{W_R} &:= U(j_{W_R}),\ \Delta^{it} := U(\Lambda_{W_R}(2\pi t)). \end{aligned} \tag{20.1}$$

The commutativity of J_{W_R} with Δ^{it} together with the antiunitarity of J_{W_R} yield the property which characterizes a Tomita operator[4] $S_{W_R}^2 \subset 1$ whose domain is identical to its range. Such operators are completely characterized in terms of their $+1$ real eigenspaces which in the present context amounts to real standard subspace $K(W_R)$ of the Wigner representation space H,

$$\begin{aligned} K(W_R) &:= \{\psi \in H,\ S_{W_R}\psi = \psi\}, \\ \overline{K(W_R) + iK(W_R)} &= H,\ K(W_R) \cap iK(W_R) = 0, \\ J_R K(W_R) &= K(W_R)^{\perp} =: K(W_R)'. \end{aligned} \tag{20.2}$$

$K(W_R)$ is closed in H whereas the complex subspace spanned together with the -1 eigenspace $iK(W_R)$ is the dense domain of the Tomita operator S_{W_R} and forms a Hilbert space in the graph norm of S_{W_R}. The denseness in H of this span $K(W_R) + iK(W_R)$ and the absence of nontrivial vectors in the intersection $K(W_R) \cap iK(W_R)$ is called "standardness". The right-hand side in the third line refers to the symplectic complement i.e. a kind of "orthogonality" in the sense of the symplectic form $Im(\cdot,\cdot)$. The application of Poincaré transformations to the reference situation generates a consistent family of wedge spaces $K(W) = U(\Lambda, a)K(W_R)$ if $W = (\Lambda, a)W_R$. These subspaces carry a surprising amount of information about local quantum physics; their structure even preempts the spin-statistics connection by producing a mismatch between the symplectic and the geometric complement (W' denotes the causal complement in terms of Minkowski space geometry) which is related to the spin-statistics factor [14, 26]

$$\begin{aligned} K(W)' &= ZK(W'), \\ Z^2 &= e^{2\pi is}. \end{aligned} \tag{20.3}$$

Another surprising fact is that the modular setting prepares the ground for the field theoretic on-shell crossing property, since the equation characterizing the real modular localization subspaces in more detail reads

$$\left(J\Delta^{\frac{1}{2}}\psi\right)(p) = V\overline{\psi_c(-p)} = \psi(p), \tag{20.4}$$

i.e. the complex conjugate of the analytically continued wave function (but now referring to the charge-conjugate situation) is up to a constant matrix V which acts on the spin indices equal to the original wave function.

[4]Operators with this property are the corner stones of the Tomita-Takesaki modular theory [38] of operator algebras. Here they arise in the spatial Rieffel-van Daele spatial setting [31] of modular theory from a realization of the geometric Bisognano-Wichmann situation within the Wigner representation theory.

2. The sharpening of localization is obtained by intersecting wedges in order to obtain real subspaces as causally closed subwedge regions:
$$K(\mathcal{O}) := \cap_{W\supset\mathcal{O}} K(W). \tag{20.5}$$
The crucial question is whether they are "standard". According to an important theorem of Brunetti, Guido and Longo [9] standardness universally holds for spacelike cones $\mathcal{O} = \mathcal{C}$ in all positive energy representations. In case of finite spin/helicity representations the standardness also holds for intersections leading to (arbitrary small) double cones $\mathcal{D}$. In those cases where the double cone localized spaces with pointlike "cores" are trivial (massless infinite spin [28], massive d=1+2 anyons [27]), the smallest localization regions are spacelike cones with semi-infinite strings as cores. Without loss of generality one may restrict localization regions to causally complete regions.
3. In the absence of interactions the transition from free particles to localized operator algebras is most appropriately done in a functorial way by applying the Weyl (CCR) (or in case of half-integer spin the CAR functor) to the localization K-spaces[5]:
$$\begin{aligned} \mathcal{A}(\mathcal{O}) &:= alg\left\{Weyl(\psi)|\ \psi \in K(\mathcal{O})\right\}, \\ Weyl(f) &:= expi\left\{a^*(\psi) + h.a.\right\}. \end{aligned} \tag{20.6}$$
where $a^{\#}(\psi)$ are the creation/annihilation operators of particles in the Wigner wave function ψ. The functorial relation between real subspaces and von Neumann algebras preserves the causal localization structure [25] and commutes with the process of improvement of localization through the formation of intersections.

For later purposes we introduce the following definition [32].

Definition 1. *A vacuum-polarization-free generator (PFG) for a region O is an operator affiliated with the algebra $\mathcal{A}(\mathcal{O})$ which created a vacuum-polarization-free one-particle vector*
$$\begin{aligned} &G\,\eta\,\mathcal{A}(\mathcal{O}), \\ &G\Omega = 1 - particle. \end{aligned} \tag{20.7}$$

Since these wedge algebra-affiliated operators G are generally unbounded, one has to comment on their domain properties. We will adopt the natural assumptions that they admit dense domains which, similar to smeared Wightman fields, are stable under translations. This definition permits us to characterize the presence of interactions by the interaction induced vacuum polarization as a result of the following statement.

Proposition 1. *The existence of subwedge-localized PFGs characterizes interaction-free theories.*

[5] To maintain simplicity we limit our presentation to the bosonic situation and refer to [14, 26] for the general treatment.

The proof uses the fact that PFGs are on-shell (weak solutions of the Klein-Gordon equation [6]) and subwedge-localized; the analytic argument is completely analogous to that of the theorem about the equality of a two-point function with that of a free field implying the equality of the associated covariant field with a free field [35, 36] (the restriction to covariance and pointlike localization is easily seen to be not necessary). The existence of wedge-localized PFGs $G\eta\mathcal{A}(W)$ is a consequence of modular theory, but their domain $dom(G)$ is generally not stable under all translations (but only under those translations which transform the wedge into itself). Such PFGs do not admit a Fourier transform i.e. they are not tempered [6]. Hence in the presence of interactions the particle localization through the application of localized operators to the vacuum is weakened; according to the previous proposition the QFT cannot localize particles in subwedge regions. Accordingly the functorial relation between particle and field localization breaks down and one has to look for a substitute.

In the next section we will show that the requirement that wedge-localized G fulfill the domain properties of the definition (i.e. are "tempered") leads to an explicit characterization of the associated wedge-localized algebras in terms of a simple algebraic structure of their generators. This amounts to the complete knowledge of the QFT in the sense of its algebraic net. Namely it can be shown that the knowledge of generators of wedge algebras together with the knowledge of how Poincaré transformations act on this reference wedge algebra and generate the family of all wedge algebras in different space-time position is sufficient to build up the complete net of algebras through the formation of intersections of wedge algebras (in analogy to (20.5)). Examples of tempered wedge-localized PFGs are obtained by Fourier transforming generating operators of Zamolodchikov-Faddeev algebras (20.8) and there are reasons to believe that the $d+1+1$ factorizing models exhaust the possibilities for tempered PFGs. Knowing the PFG generators explicitly as one does in these models, one can construct the net and its local field generators which are of course much more involved than the non-local wedge generators.

In the third section the idea of lightfront holographic projection will be used in order to classify wedge algebras in terms of extended chiral algebras. The unsolved problems of inverse lightfront holography i.e. the problem of reconstructing ambient algebras from their holographic projections, is the main obstacle in the general classification and construction. Here again the restriction to $d = 1+1$ factorizing models is very helpful.

20.2 Modular localization and the bootstrap-formfactor program

The various past attempts at S-matrix theories which aimed at direct constructions of scattering data without the intermediate use of local fields and local observables provide illustrations of what is meant by an "on-shell" approach to particle physics. The motivation behind such proposals was first spelled out by Heisenberg [18]. It consisted in the hope that by limiting oneself to particles and their mass-shells, one avoids (integration over) fluctuations on a scale of arbitrarily small spacelike distances causing ultraviolet divergencies which at the pre-renormalization days of Heisenberg's S-matrix proposal appeared to be an incurable disease of QFT. The main purpose of staying close to particles and using scattering concepts ("on-shell") is the avoidance of inherently singular objects as pointlike fields in calculational steps. This is certainly a reasonable aim independent of whether one believes or not that a formulation of interactions in terms of singular pointlike fields exists for $d = 1 + 3$ QFT in the mathematical physics sense.

Since the early 1950s, in the aftermath of renormalization theory, the relation between particles and fields received significant elucidation through the derivation of time-dependent scattering theory. In the course of this it also became clear that Heisenberg's S-matrix proposal had to be amended by the addition of the crossing property i.e. a prescription of how to analytically continue particle momenta on the complex mass shell in order to relate matrix elements of local operators between incoming ket and outgoing bra states with a fixed total sum of incoming and outgoing particles as different boundary values of one analytic "masterfunction". In physical terms crossing allows us to relate matrix elements describing real particle creation with particles in both the incoming ket- and outgoing bra-states to the vacuum polarization matrix elements where the ket-state (or the bra state) is the vacuum vector.

Whereas Heisenberg's requirements of Poincaré invariance, unitarity and cluster factorization on a relativistic S-matrix can also be implemented in a "direct particle interaction" scheme [11, 33], the implementation of crossing is conceptually related to the presence of vacuum polarization for which QFT with its micro-causality is the natural arena.

The LSZ time-dependent scattering theory and the associated reduction formalism relates such a matrix element (referred to as a generalized formfactor) in a natural way to one in which an incoming particle becomes "crossed" into an outgoing anti-particle on the backward real mass shell; it is at this point where analytic continuation from a positive energy physical process enters. In this setting the S-matrix is the formfactor of the identity operator.

The important remark here is that the use of particle states requires the restriction of the analytic continuation to the complex mass shell ("on-shell"). It was Bros[6] in collaboration with Epstein and Glaser [7] who gave the first rigorous proofs of crossing in special configurations. In the special case of the elastic scattering amplitude, the crossing of only one particle from the incoming state has to be accompanied by a reverse crossing of one of the outgoing particles in order to arrive at a physical process allowed by energy-momentum conservation[7].

A derivation of crossing in the setting of QFT for general multi-particle scattering configurations and for formfactors (as one needs it for the derivation of a bootstrap-formfactor program, see later) from the general principles of local quantum physics does not yet exist. It is not clear to me whether the present state of art in algebraic QFT would permit one to go significantly beyond the old but still impressive results quoted before.

The crossing property became the cornerstone of the so-called bootstrap S-matrix program and several ad hoc representations of analytic scattering amplitudes

---

[6]Since the issue of crossing constitutes an important property of the present paper, it is particularly appropriate to dedicate this work to Jacques Bros on the occasion of his 70th birthday.

[7]This crossing of a pair of particles from the in/out elastic configuration is actually the origin of the terminology "crossing" and was the main object of rigorous analytic investigations.

were proposed (Mandelstam representation, Regge poles...) in order to incorporate crossing in a more manageable form.

The algebraic basis of the bootstrap-formfactor program for the special family of $d = 1+1$ factorizable theories is the validity of a momentum space Zamolodchikov-Faddeev algebra [41]. The operators of this algebra are close to free fields in the sense that their Fourier transform is on-shell (see (20.8) in the next section), but unlike the latter they are not local in the pointlike sense. A closer look reveals that they are localizable in the weaker sense[8] of space-time wedge regions [23, 32]. In fact the existence of such Fourier transformable ("tempered") wedge-localized PFGs, which implies the absence of real particle creation through scattering processes [6], turns out to be the prerequisite for the success of the bootstrap-formfactor program for factorizable models in which one uses only formfactors and avoids (short-distance singular) correlation functions.

According to an old structural theorem which is based on certain analytic properties of a field theoretic S-matrix [1, 6], interaction-induced vacuum polarization without real particle creation is only possible in $d = 1+1$ theories. This in principle leaves the possibility of direct 3- or higher- particle elastic processes beyond two particle scattering. But an argument by Karowski based on formfactor crossing[9] shows that the nonvanishing of higher connected elastic contributions would be inconsistent with the absence of real particle creation. In this sense the Z-F algebra structure, which is at the heart of factorizing models, turns out to be a consequence of special properties of modular wedge-localized PFGs, a fact which places the position of the factorizing models within general QFT into sharper focus. The crossing property of the two-particle scattering amplitude is a consistency prerequisite for the formfactor crossing. Providing a special illustration of the previous general unicity argument of inverse scattering based on crossing, the bootstrap formfactor approach associates precisely one QFT in the sense of one local equivalence class of fields (or one net of localized operator algebras) to a prescribed factorizing S-matrix.

In agreement with the philosophy underlying AQFT, which views pointlike fields as coordinatizations for generators of localized algebras, the bootstrap-formfactor construction for $d = 1+1$ factorizing models primarily aims to determine coordinatization-independent double-cone algebras by computing intersections of wedge algebras. The nontriviality of a theory is then tantamount to the nontriviality ($\neq C1$) of such intersections. The computation of a basis of pointlike field generators of these algebras is analogous to (but more involved than) the construction of a basis of composites of free fields in the form of Wick polynomials. As we saw before for noninteracting theories the functorial description of the algebras (20.6) based on modular localization is conceptually simpler than the use of free fields and their local equivalence class of Wick-ordered composites e.g. one is not obliged to introduce a non-intrinsic Wick basis.

The crossing property is crucial for linking scattering data with off-shell operator spaces. As explained in the previous section, it relates the multi-particle component of vectors obtained by one-time application of a localized (at least wedge-localized)

[8] An operator which is localizable in a certain causally closed space-time region is automatically localized in any larger region but not necessarily in a smaller region. The unspecific terminology "non-local" in the literature is used for any non-pointlike localized field.

[9] I am indebted to M. Karowski for this argument.

operator to the vacuum with the connected formfactors of that operator. It is important to note that in factorizing models, crossing is not an assumption but rather follows from the properties of tempered PFGs for wedge algebras similar to crossing of formfactors for composite operators of free fields [34].

In the following some of the details of wedge-localized PFGs and their connections with the Zamolodchikov-Faddeev algebra structure are presented. In the simplest case of a scalar chargeless particle without bound states[10] the wedge generators are of the form [32]

$$\begin{aligned} \phi(x) &= \frac{1}{\sqrt{2\pi}} \int (e^{ip(\theta)x(\chi)} Z(\theta) + h.c.) d\theta, \\ Z(\theta)Z^*(\theta') &= S^{(2)}(\theta - \theta')Z^*(\theta')Z(\theta) + \delta(\theta - \theta'), \\ Z(\theta)Z(\theta') &= S^{(2)}(\theta' - \theta)Z(\theta')Z(\theta). \end{aligned} \tag{20.8}$$

Here $p(\theta) = m(ch\theta, sh\theta)$ is the rapidity parametrization of the $d = 1 + 1$ mass-shell and $x = r(sh\chi, ch\chi)$ parametrizes the right-hand wedge in Minkowski space-time. $S^{(2)}(\theta)$ is a structure function of the Z-F algebra which is a nonlocal *-algebra generalization of canonical creation/annihilation operators. The notation preempts the fact that $S^{(2)}(\theta)$ is the analytic continuation of the physical two-particle S-matrix $S^{(2)}(|\theta|)$ which via the factorization formula determines the general scattering operator S_{scat} (20.11). The unitarity and crossing of S_{scat} follows from the corresponding two-particle properties which in terms of the analytic continuation are $S^2(z)^* = S^{(2)}(-z)$ (unitarity) and $S^{(2)}(z) = S^{(2)}(i\pi - z)$ (crossing) [22]. The $Z^*(\theta)$ operators applied to the vacuum in the natural order $\theta_1 > \theta_2 > \cdots > \theta_n$ are by definition equal to the outgoing canonical Fock space creation operators, whereas the re-ordering from any other ordering has to be calculated according to the Z-F commutation relations e.g.

$$Z^*(\theta)a^*(\theta_1)\cdots a^*(\theta_n)\Omega = \prod_{i=1}^{k} S^{(2)}(\theta - \theta_i)a^*(\theta_1)\cdots a^*(\theta_k)a^*(\theta)a^*(\theta_{k+1})\cdots a^*(\theta_n)\Omega \tag{20.9}$$

where $\theta < \theta_i$ $i = 1, \ldots, k$, $\theta > \theta_i$ $i = k + 1, \ldots, n$. The general Zamolodchikov-Faddeev algebra is a matrix generalization of this structure.

It is important not to identify the Fourier transform in (20.8) of the momentum with a localization variable. Although the x in $\phi(x)$ behaves covariantly under Poincaré transformations, it is not marking a causal localization point; in fact it is a non-local variable in the sense of the standard use of this terminology[11]. It is however wedge-localized in the sense that the generating family of operators for the right-hand wedge W Wightman-like (polynomial) algebra $alg\{\phi(f), suppf \subset W\}$ commutes with the TCP transformed algebra $alg\{J\phi(g)J, suppg \subset W\}$ which is

[10] A situation which in case of factorizing models with variable coupling (e.g. the massive Thirring model) can always be obtained by choosing a sufficiently small coupling. Bound state poles in the physical θ-strip require nontrivial changes of the algebraic formalism.

[11] The word local is reserved for "commuting for spacelike distances". In this work we are dealing with non-local fields which are nevertheless localized in causally complete subregions (wedges, double cones) of Minkowski space-time.

the left wedge algebra [23]

$$[\phi(f), J\phi(g)J] = 0, \qquad (20.10)$$
$$J = J_0 S_{scat}.$$

Here J_0 is the TCP symmetry of the free field theory associated with $a^{\#}(\theta)$ and S_{scat} is the factorizing S-matrix which on (outgoing) n-particle states has the form

$$S_{scat} a^*(\theta_1) a^*(\theta_2) ... a^*(\theta_n)\Omega = \prod_{i<j} S^{(2)}(\theta_i - \theta_j) a^*(\theta_2) ... a^*(\theta_n)\Omega \qquad (20.11)$$

if we identify the $a^{\#}(\theta)$ with the incoming creation/annihilation operators. It is then possible to give a rigorous proof [23] that the Weyl-like algebra generated by exponential unitaries is really wedge-localized and fulfills the Bisognano-Wichmann property

$$\mathcal{A}(W) = alg\left\{e^{i\phi(f)} \mid suppf \subset W\right\}, \qquad (20.12)$$
$$\mathcal{A}(W)' = J\mathcal{A}(W)J = \mathcal{A}(W'),$$

where the dash on operator algebras is the standard notation for their von Neumann commutant and the dash on space-time regions stands for the causal complement. Within the modular setting the relative position of the causally disjoint $A(W')$ depends via S_{scat} on the dynamics. The TCP operator J is the (antiunitary) "angular" part of the polar decomposition of Tomita's algebraically defined unbounded antilinear S-operator with the characterization

$$SA\Omega = A\Omega, \; A \in \mathcal{A}(W), \qquad (20.13)$$
$$S = J\Delta^{\frac{1}{2}}, \; \Delta^{it} = U(\Lambda(-2\pi t)),$$

with $\Lambda(\chi)$ being the Lorentz boost at the rapidity χ.

At this point the setup looks like a relativistic quantum mechanics since the $\phi(f)$ (similar to genuine free fields if applied to the vacuum) do not generate vacuum polarization clouds. The advantage of the algebraic modular localization setting is that interaction-caused vacuum polarization is generated by algebraic intersections, which is in agreement with the intrinsic definition of the notion of interaction presented in terms of PFGs in the previous section

$$\mathcal{A}(D) \equiv \mathcal{A}(W) \cap \mathcal{A}(W_a') = \mathcal{A}(W) \cap \mathcal{A}(W_a)', \qquad (20.14)$$
$$D = W \cap W_a'.$$

This is the operator algebra associated with a double cone D (which is chosen symmetric around the origin by intersecting suitably translated wedges and their causal complements). Note the difference from the quantization approach, where pointlike localized fields are used from the outset and the sharpening of localization of smeared products of fields is simply achieved by the classical step of restricting the space-time support of the test functions. The problem of computing intersected von Neumann algebras is in general not only difficult (since there are no known general computational techniques) but also very unusual as compared to methods of standard quantization. There is a well-founded hope that one can solve the existence problem of factorizing models by showing the nontriviality of the intersections $A(D)$ [10].

The problem becomes more amenable if one considers instead of operators their formfactors i.e. their matrix elements between incoming ket and outgoing bra state vectors. In the spirit of the LSZ formalism one can then make an Ansatz in form of a power series in $Z(\theta)$ and $Z^*(\theta) \equiv Z(\theta - i\pi)$ (corresponding to the power series in the incoming free field in LSZ theory). In a shorthand notation which combines both frequency parts we may write

$$A = \sum \frac{1}{n!} \int_C \cdots \int_C a_n(\theta_1, \ldots, \theta_n) : Z(\theta_1) \cdots Z(\theta_n) : d\theta_1 \cdots d\theta_n \tag{20.15}$$

where each integration path C extends over the upper and lower part of the rim of the $(0, -i\pi)$ strip in the complex θ-plane. The strip-analyticity of the coefficient functions a_n expresses the wedge-localization of A.[12] It is easy to see that these coefficients on the upper part of C (the annihilation part) are identical to the vacuum polarization form factors of A

$$\langle \Omega | A | p_n, \ldots, p_1 \rangle^{in} = a_n(\theta_1, \ldots, \theta_n), \quad \theta_n > \theta_{n-1} > \cdots > \theta_1 \tag{20.16}$$

whereas the crossing of some of the particles into the left-hand bra state (see the previous section) leads to the connected part of the formfactors

$$^{out}\langle p_1, \ldots, p_k | A | p_n, \ldots, p_{k+1} \rangle^{in}_{conn} = a_n(\theta_1 + i\pi, \ldots \theta_k + i\pi, \theta_{k+1}, \ldots, \theta_n) \tag{20.17}$$

Hence the crossing property of formfactors is encoded into the notation of the operator formalism (20.15) in that there is only one analytic function a_n which describes the different possibilities of placing θ on the upper or lower rim of C.

The presence of bound states (poles in the physical θ-strip) leads to a weakening of the wedge localization in the sense that the wedge commutativity (20.10) only holds between states from the subspace generated from the "elementary" states linearly related to (20.8). This requires considerable modifications of the algebraic formalism which goes beyond the modest aims of this paper.

The essential advantage of this algebraic formalism over the calculation of formfactors of individual fields is expected to appear if one tries to secure the existence of the theory. Whereas the conventional way via controlling Wightman functions and checking their properties appears hopelessly complicated (the mathematical control of the convergence of the formfactor series (20.15) has not even been achieved in simple models), the "modular nuclearity property" of wedge algebras in d=1+1 which secures the nontriviality of the intersected algebras $A(D)$ [10] seems to be well in reach [24].

20.3 Constructive aspects of lightfront holography

In the previous sections it was shown how modular theory together with on-shell concepts can be used to analyze special wedge algebras in the presence of

---

[12] Compact localization leads to coefficient functions which are meromorphic outside the open strip [2].

interactions. The constructive use was limited to the presence of so-called tempered PFGs which in turn restricted computable models to $d = 1+1$ factorizing theories. In this section I will present a recent proposal which also uses modular localization ideas but tries to analyze wedge algebras in terms of (extended) chiral theories by means of "algebraic lightfront holography" (ALH). Again we limit ourselves to some intuitively accessible remarks mainly emphasizing analogies as well as differences with the standard formalism of QFT; for a more detailed mathematical description we again refer to the literature [34].

The following comparison with the canonical formalism turns out to be helpful. The ETCR formalism tries to classify and construct QFTs by assuming the validity of canonical equal time-commutation relations (ETCR). The shortcomings of that approach are well-known. Even if one ignores the fact that the ETC structure is inconsistent with the presence of strictly renormalizable interactions[13], the usefulness of the ETCR is still limited by its insensitivity with respect to interactions. One would prefer to start with a structure which senses the presence of interactions and is capable of utilizing the enormous amount of knowledge and structural richness which has been obtained in studying chiral theories by providing a concept of rich universality classes for higher dimensional QFT (instead of just one ETCR class).

Lightfront holography tries to address this imbalance by replacing the ETCR by the richer structure of (extended) chiral theories on the lightfront. Its main aim is to shift the cut between kinematics and dynamics in such a way that what has been learned by studying low dimensional theories can be used as a kinematical input for higher dimensional models.

The holographic projection turns out to map many different interacting ambient theories to the same holographic image; in this respect there is a certain similarity to the better known scale invariant short distance universality classes which are the key to the understanding of critical phenomena. But in contradistinction to scaling universality classes which change the theory to an associated massless theory, holographic projections live in the same Hilbert space as the ambient theory; in fact they just organize the space-time aspects of a shared algebraic structure in a radically different way.

Let us briefly recall some salient points of ALH[14].

ALH may be viewed as a kind of conceptually and mathematically refined "lightcone quantization" (or "$p \to \infty$ frame" description). Whereas the latter never faced up to the question of how the lightfront quantized fields are related to the original local fields i.e. in which sense the new description addresses the original problems posed by the ambient theory, the ALH is conceptually precise and mathematically rigorous on this point.

It turns out that the idea of restricting fields to the lightfront is limited to free fields and certain superrenormalizable interacting models with finite wave function renormalization. Theories with interaction-caused vacuum polarization which leads to Källén-Lehmann spectral functions with diverging wave function renormalization do not permit lightfront restrictions for the same reason as they do not have equal

[13]Only superrenormalizable interactions (finite wave function renormalization) as the polynomial scalar models in $d = 1+1$ have fields which restrict to equal times

[14]We add this prefix "algebraic" in order to distinguish the present notion of holography from the gravitational holography of t'Hooft [19]. More on similarities and differences between the two can be found in the concluding remarks,

time restrictions; e.g. for scalar fields one has[15]

$$\langle A(x)A(y)\rangle = \int_0^\infty \rho(\kappa^2) i\Delta^{(+)}(x-y,\kappa^2)d\kappa^2, \tag{20.18}$$
$$\langle A(x)A(y)\rangle\,|_{LF} \sim \int_0^\infty \rho(\kappa^2)d\kappa^2\delta(x_\perp - y_\perp)\int_0^\infty \frac{dk}{k}e^{-ik(x_+-y_+)},$$

where in passing to the second line we used the correct rule for lightfront restriction; this is obviously not the naive one obtained by simply restricting the coordinates in the Kallén-Lehmann representation. To obtain the second line, which replaces the free field Δ^+ function by the transverse $\delta(x_\perp - y_\perp)$ delta function times the longitudinal chiral function in the x_+ lightray variable, one starts from the free field representation in terms of momentum space creation/annihilation operators. In the z-t wedge region this field may be parametrized in terms of rapidities χ, θ as follows:

$$A(x) = \frac{1}{(2\pi)^{\frac{3}{2}}}\int\int (e^{im_{eff}rch(\chi-\theta)+\vec{p}_\perp\vec{x}_\perp}a^*(p) + h.c.)\frac{d\theta}{2}dp_\perp, \tag{20.19}$$

$$\left[a(p), a^*(p')\right] = 2\delta(\theta-\theta')\delta(p_\perp - p'_\perp),\; m_{eff} = \sqrt{\vec{p}_\perp^2 + m^2}, \tag{20.20}$$
$$x = (r\sinh\chi, \vec{x}_\perp, \cosh\chi),\; p = (\cosh\chi, \vec{p}_\perp, m_{eff}\sinh\theta).$$

The limit $r \to 0$ together with a compensating limit $\chi = \hat{\chi} - \ln m_{eff}r$ provides a finite lightfront limit in terms of the same creation/annihilation operators and hence takes place in the same Hilbert space (unlike the scaling limit used for critical phenomena) and leads to the desired result

$$A(x)|_{LF} = \frac{1}{(2\pi)^{\frac{3}{2}}}\int_0^\infty\int (e^{ip_-x_+ + ip_\perp x_\perp}a^*(p) + h.c.)\frac{dp_-}{2p_-}dp_\perp, \tag{20.21}$$
$$p_- \simeq e^{-\theta},$$

which yields the above formula for the two-point function. The infrared-divergence in the longitudinal factor is spurious if one views the lightfront localization in the setting of modular wedge localization[16]. On the other hand the obstruction resulting from the large κ divergence of the K-L spectral function (short distance regime of interaction-caused vacuum polarization) is shared with that which limits the range of validity of the ETCR formalism. But whereas equal time restricted interacting fields in $d \geqslant 1+2$ simply do not exist, there is no such limitation on the short distance properties of generalized chiral conformal fields which turn out to generate the ALH. What breaks down is the idea that these lightfront generating fields can be gotten simply by restricting the fields of the ambient theory, as was the case in the example of free fields.

[15] It is important to realize that LF restriction is not a pointwise local procedure. This becomes clearer within the setting of modular localization.

[16] By re-expressing the rapidity testfunction space in terms of the p_+ integration variable, one obtains the vanishing of the testfunctions at $p_+ = 0$. The same argument also shows that an additive modification of $\hat{\chi}$ (a multiplicative change of p_-) does not change the result in the appropriate test function setting.

It turns out that modular theory provides a useful tool to analyze the connection between the ambient theory and its holographic projection. Although the ambient theory may well be given in terms of pointlike fields and the ALH may also allow a pointlike description (see 20.25), there is no known direct relation between these fields. This was of course precisely the problem of lightcone quantization which remained unresolved. Even in the above interaction-free case when the restriction method in the sense of (20.21) works, the ALH net of algebras turns out to be non-local relative to the ambient algebra and hence the recovery of the ambient from the ALH involves non-local steps which the standard formalism cannot handle. Whereas lightcone quantization was not able to address those subtle problems, ALH solves them.

The intuitive physical basis of this algebraic approach is a limiting form of the causal closure property. Let O be a space-time region and O'' its causal closure (the causal disjoint subsequently taken twice) then the causal closure property is the equality

$$\mathcal{A}(\mathcal{O}) = \mathcal{A}(\mathcal{O}''). \tag{20.22}$$

In the case of free fields this abstract algebraic property[17] is inherited via quantization from the Cauchy propagation in the classical setting of hyperbolic differential equations. The lightfront is a limiting case (characteristic surface) of a Cauchy surface. Each lightray which passes through O'' either must have passed or will pass through O. For the case of a $x^0 - x^3$ wedge W and its $x^0 - x^3 = 0$ (upper) causal lightfront boundary $LFB(W)$ (which covers half of a lightfront) the relation

$$\mathcal{A}(LFB(W)) = \mathcal{A}(W) \subset B(H) \tag{20.23}$$

is a limiting situation of the causal shadow property; a lightlike signal which goes through this boundary must have passed through the wedge (or in the terminology of causality, the wedge is the backward causal completion of its lightfront boundary). Classical data on the lightfront define a characteristic initial value problem and the smallest region which generates data localized in an open ambient region is half the lightfront as in (20.23); for any transversely not two-sided infinite extended subregion O_{LF} on the lightfront, as well as for any region on the lightfront which is bounded in the lightray direction, the causal completion is trivial i.e. $O_{LF} = O''_{LF}$. This unusual behavior of the lightfront is related to the fact that as a manifold with its metric structure inherited from the ambient Minkowski space-time it is not even locally hyperbolic.

Several symmetries which the lightfront inherits from the ambient Poincaré group are obvious; it is clear that the lightlike translation together with the two-transverse translation and the transverse rotation are leaving the lightfront invariant and that the longitudinal Lorentz boost, which leaves the wedge invariant, acts as a dilatation on the lightray in the lightfront. There are however two additional invariance transformations of the lightfront which are less obvious. Their significance in the ambient space is that of the two "translations" in the 3-parametric Wigner little group $E(2)$ of the light ray in the lightfront (a Euclidean subgroup of the 6-parametric Lorentz group). Projected into the lightfront these "translations" look like transverse Galilei transformations in the various $(x_\perp)_i - x_+$ planes.

Modular concepts (in particular modular inclusions and intersections) provide a firm operator algebraic basis for the interplay between the ambient causality and

[17]This equality is the local version of the "time slice property" [17].

the localization structure as well as the 7-parametric symmetry of the lightfront holography[18]. Among the many structural consequences we only collect those which are important for the constructive use of holography:

- The Poincaré group $P(4)$ and hence also the 7-parametric subgroup $G_{LF} \subset P(4)$ which leaves the lightfront invariant are of modular origin. The full lightfront symmetry is much larger and includes the Moebius group extension of the 2-parametric longitudinal translation-dilation group which is also of modular origin.
- The lightfront algebra has no vacuum fluctuations in transverse direction i.e. the operator algebra of a longitudinal infinitely extended cylindrical region $\Xi = \{x_\perp \in Q, -\infty < x_+ < \infty\}$ with finite transverse extension Q is a tensor factor of the full lightfront algebra which is identical to the full ambient algebra $B(H)$,

$$\mathcal{A}(LF) = B(H) = \mathcal{A}(\Xi) \otimes \mathcal{A}(\Xi)',\ \mathcal{A}(\Xi)' = \mathcal{A}(\Xi'). \tag{20.24}$$

  In longitudinal direction the cyclicity of the vacuum (the Reeh-Schlieder property) prevents such a factorization i.e. the lightfront holography "squeezes" the field theoretic vacuum fluctuation into the lightray direction so that the transverse structure becomes purely quantum mechanical. As a result of the Moebius covariance along the lightray and the quantum mechanical factorization in transverse direction, the lightfront holography has the structure of an (quantum mechanically) extended chiral QFT.

For the derivation of the local net structure of the lightfront theory in the longitudinal and transverse directions we refer to [3,34]; this is the part which requires the use of modular localization concepts (modular inclusions and modular intersections of wedge algebras, relative commutants) which differ significantly from concepts of the standard approach to QFT. The following remarks are only intended to facilitate understanding and highlight some consequences.

Although there is presently no rigorous proof, the structural analogy of the lightfront holographic projection with chiral theory leads one to expect that similar group theoretical arguments as in [20] provide the existence of covariant pointlike generators. In cases where they exist, their commutation relations are severely restricted; the transverse quantum mechanical nature only permits a delta function without derivative and the balance in the scaling dimensions restricts the longitudinal singularity structure to that of Lie fields known from chiral current or W-algebras.

$$\begin{aligned}&[\psi_i(x_\perp, x_+), \psi_j(x'_\perp, x'_+)] \\ &= \delta(x_\perp - x'_\perp)\left\{\delta^{(n_{ij})}(x_+ - x'_+) + \sum_k \delta^{(n_{ijk})}(x_+ - x'_+)\psi_k(x_\perp, x_+)\right\}.\end{aligned} \tag{20.25}$$

As in the pure chiral case one may hope for rational situations in which there exists a finite set of generating fields.

The difficult part of a constructive proposal of the lightfront holography is the "inverse holography" i.e. the reconstruction of an ambient theory from its holographic

[18] For the inverse holography the information from the fundamental lightfront inclusion $A(LF(W)) \subset_{G_{LF}} A(LF) = B(H)$ has to be complemented by the action of the x_-translation (similar to the Hamiltonian input in the ETCR approach).

projection. Apart from the interaction-free case which is characterized by a c-number commutator, the kinematical holographic information is insufficient. The analogy with the canonical formalism suggests expecting that the action of the x_- lightray translation on the lightfront net or on its generating fields (20.25) should select a particular ambient model from the holographic equivalence class.

The remainder of this section contains some comments on the inverse holography of factorizing models, where as a result of the two-dimensionality the transverse structure is absent and the holographic projection is a bona fide chiral theory. The on-shell aspect of covariant PFG generators for wedge algebras (20.8) trivializes the passing between the ambient theory and its holographic projection; within the setting of factorizing models the holographic inversion is unique and amounts to representing the action of the x_- translation (similar to the case of free fields) by the multiplication with $e^{ip_+x_-}$, $p_+ = \frac{m^2}{p_-}$ in the formfactor representation (20.15). The reason for this uniqueness is that the covariance property of the particle-like Z-F creation/annihilation operators implicitly fix the transformation properties of the full Poincaré group i.e. including the LF-changing transformations beyond G_{LF}.

The holographic restriction of factorizing models also highlights a new interesting aspect of chiral theories. At least those chiral models which originate in this way permit a formal representation in terms of PFGs (20.15) inherited from ambient theory. Although this basis of $Z^{\#}(\theta)$ operators loses its particle interpretation in the chiral holographic projection, it still continues to provide an unexpected simple "on-shell representation" simplicity for these chiral algebras. In this representation the Moebius rotations applied to states $Z^*(\theta)\Omega$ "dress" the latter with a vacuum polarization cloud. Again we refer for more details to [3, 34]. Analogous to the free field case the inverse holography in this particular representation just amounts to multiplication of the formfactors a_n with the appropriate $e^{ip_+x_-}$ translation factors.

This inverse holography also raises an interesting question about the possible dynamical role of modular generated non-local ambient symmetries beyond the local vacuum preserving Poincaré transformations. This is part of a quest for a more profound future understanding of the relation between particle aspects of the ambient theory and chiral field aspects of its holographic projection.

We conclude with some remarks about the difference in the underlying philosophy as compared to the standard approach to QFT which is based on quantizing classical field theories i.e. on the idea that important models of particle physics can be constructed by subjecting the classical Lagrangian formalism to quantization rules. This setting leads to a finite number of possibilities of renormalizable local coupling between higher dimensional ($d \geq 1+2$) covariant fields which contains the important standard gauge theory model of electro-weak interactions.

The modular based approach advocated here disposes of the parallelism to classical field theories; instead of quantizing concrete classical field models it aims at a classifying of models according to their intrinsic algebraic structure. The prototype situation is that of chiral models on the lightray which are classified by their Lie-type commutation structure or alternatively by analyzing the possible modular position of three MAs. The underlying philosophy is that of universality classes as it is successfully used in the condensed matter physics of critical phenomena. Instead of trying to find a unique model of particle physics by quantizing a selected classical Lagrangian, one classifies holographic equivalence classes and refines the search for the best mathematical description of particle physics in terms of local quantum theory by adding additional dynamical information.

References

1. S. Aks: *J. Math. Phys.* 6:516 (1965).
2. H.M. Babujian, A. Fring, M. Karowski and A. Zappletal: *Nucl. Phys. B* 538:535 (1999).
3. J. Barata, J. Mund and B. Schroer: In preparation.
4. J.J. Bisognano and E.H. Wichmann: *J. Math. Phys.* 16:985 (1975).
5. H.-J. Borchers: *J. Math. Phys.* 41:3604 (2000).
6. H.-J. Borchers, D. Buchholz and B. Schroer: *Commun. Math. Phys.* 219:125 (2001). arXiv:hep-th/0003243.
7. J. Bros, H. Epstein and V. Glaser: *Commun. Math. Phys.* 1:240 (1965).
8. R. Brunetti, K. Fredenhagen and R. Verch: *Commun. Math. Phys.* 237:31 (2003).
9. R. Brunetti, D. Guido and R. Longo: *Rev. Math. Phys.* 14:759 (2002).
10. D. Buchholz and G. Lechner: Modular Nuclearity and Localization. arXiv:math-ph/0402072.
11. F. Coester: *Helv. Physica Acta* 38:7 (1965).
12. A. Connes: *Ann. Inst. Fourier* 24:121 (1974).
13. W. Driessler: *Commun. Math. Phys.* 53:295 (1977).
14. L. Fassarella and B. Schroer: *J. Phys. A: Math. Gen.* 35:9123 (2002).
15. R. Haag: *Local Quantum Physics.* Springer, 1996.
16. R. Haag, N. Hugenholtz and M. Winnink: *Commun. Math. Phys.* 5:215 (1967).
17. R. Haag and B. Schroer: *J. Math. Phys.* 5:248 (1962).
18. W. Heisenberg: *Zeitschr. für Naturforschung* 1:608 (1946).
19. G. 't Hooft: In: *Salam-Festschrift* (A. Ali et al., eds.). World Scientific, 1993, p. 284.
20. M. Jörss: *Lett. Math. Phys.* 38:257 (1996).
21. R. Kaehler and H.-W. Wiesbrock: *J. Math. Phys.* 42:74 (2001).
22. M. Karowski, H.J. Thun, T.T. Truong and P. Weisz: *Phys. Lett. B* 67:321 (1977).
23. G. Lechner: *Lett. Math. Phys.* 64:137 (2003).
24. G. Lechner: On the Existence of Local Observables in Theories with a Factorizing S-Matrix. arXiv:math-ph/0405062.
25. P. Leyland, J. Roberts and D. Testard: *Duality for quantum free fields.* Unpublished notes, 1978.
26. J. Mund: *Ann. H. Poinc.* 2:907 (2001).
27. J. Mund, B. Schroer and J. Yngvason: *J. Math. Phys.* 44:2037 (2003).
28. J. Mund, B. Schroer and J. Yngvason: *Phys. Lett. B* 596:156 (2004).
29. J. Mund, B. Schroer and J. Yngvason: *String-localized Quantum Fields and Modular Localization.* In preparation.
30. J. Norton: *The Hole Argument.* The Stanford Encyclopedia of Philosophy, Spring 2004 edition (Edward N. Zalta, ed.). `http://plato.stanford.edu/archives/spr2004/entries/space-time-holearg/`
31. M. Rieffel and A. van Daele: *Pacific J. of Math.* 69:187 (1977).
32. B. Schroer: *Annals of Physics* 295:190 (1999).

33. B. Schroer: An anthology of non-local QFT and QFT on noncommutative space-time. arXiv:hep-th/040510.
34. B. Schroer: Constructive proposals for QFT based on the crossing property and on lightfront holography. arXiv:hep-th/0406016.
35. O. Steinmann: *Commun. Math. Phys.* 87:259 (1982).
36. R.F. Streater and A.S. Wightman: *PCT Spin&Statistics and All That.* Benjamin, 1964.
37. S.J. Summers: Tomita-Takesaki Modular Theory. To appear in the Encyclopedia of Mathematical Physics, Elsevier Publ.
38. M. Takesaki: *Tomita's Theory of Modular Hilbert Algebras and its Applications.* Lecture Notes in Mathematics Vol. 128, Springer Verlag, Berlin, Heidelberg, New York, 1970.
39. H.-W. Wiesbrock: *Comm. Math. Phys.* 158:537 (1993).
40. J. Yngvason: The role of type III factors in quantum field theory. arXiv:math-ph/0411058.
41. A.B. Zamolodchikov: *Int. J. of Mod. Phys. A* 1:1235 (1989).

21

Physical Fields in QED

Othmar Steinmann

Fakultät für Physik, Universität Bielefeld, D-33501 Bielefeld, Germany
stema@Physik.Uni-Bielefeld.de

Dedicated to Jacques Bros, an esteemed colleague and friend

Summary. The connection between the Gupta-Bleuler formulation and the Coulomb gauge formulation of QED is discussed. It is argued that the two formulations are not connected by a gauge transformation. Nor can the state space of the Coulomb gauge be identified with a subspace of the Gupta-Bleuler space. Instead a more indirect connection between the two formulations via a detour through the Wightman reconstruction theorem is proposed.

21.1 Introduction

This article is concerned with a major unsolved problem of QED, that of an exact formulation of the notion of gauge invariance, more especially the problem of an exact characterization of gauge transformations and their uses. Why are such seemingly disparate formulations as the Gupta-Bleuler (GB) formalism and the Coulomb gauge (C gauge) description physically equivalent, and how are they connected? Needless to say, this problem will not be solved here or even fully described. I will merely put forward a few possibly useful remarks and suggestions. Attention will be restricted to the two most widely used 'gauges' already mentioned, the GB and the C gauges. And since there still does not exist a rigorous formulation of QED in any gauge, these results will be based on the experience gained in perturbation theory (PT). Any result which is valid in every order of PT has a good chance of describing a feature present in a possibly existing exact theory. That this statement must be taken with a grain of salt will become apparent later on.

For the structural studies we have in mind, the Wightman functions are more convenient tools than the Green's functions of the traditional formulations. The PT of the Wightman functions of QED has been developed in [8]. Our results are based on the rules derived there. But the reader's acquaintance with these rules will not be assumed. The claims made will therefore in general be substantiated by somewhat heuristic arguments rather than full proofs. The emphasis is on statements of facts and the discussion of their significance, rather than on proofs. However, all the results claimed *can* be rigorously proved to all orders of PT. And they are usually plausible enough as they stand.

21.2 Formal Considerations

The basic fields of QED with charged particles of spin 1/2 are the electromagnetic potentials $A_\mu(x)$ and the Dirac spinors $\psi(x)$ and $\bar{\psi}(x) = \psi^*(x)\gamma^0$.

The *GB formalism* has the advantage of working with local, covariant, fields A_μ, ψ, as we like them. But it also has two grave drawbacks. First, its state space $\mathcal{V}_{GB}$ is equipped with an indefinite scalar product, hence it is not a Hilbert space. This contradicts the basic rules of quantum mechanics, and it is mathematically inconvenient. Second, the Maxwell equations

$$\partial_\nu F^{\nu\mu}(x) = j^\mu(x) , \tag{21.1}$$

with

$$F^{\nu\mu}(x) = \partial^\nu A^\mu(x) - \partial^\mu A^\nu(x) , \qquad j^\mu(x) = e\,\bar{\psi}(x)\gamma^\mu\psi(x) , \tag{21.2}$$

are not satisfied as operator equations on $\mathcal{V}_{GB}$, even after renormalization. This means that $\mathcal{V}_{GB}$ contains unphysical states, that is vectors which do not describe states ever encountered in a laboratory. This raises the question of how to characterize the physical states, and the second question whether $\mathcal{V}_{GB}$ can be defined such that it contains sufficiently many physical states to give a full description of reality. The standard textbook answer to the first question is this: the divergence

$$B(x) = \partial_\mu A^\mu(x) \tag{21.3}$$

solves the free field equation $\partial^\mu\partial_\mu B(x) = 0$, hence can be split into a creation part $B^-(x)$ and an annihilation part $B^+(x)$. And the physical subspace $\mathcal{V}_{ph} \subset \mathcal{V}_{GB}$ is defined to be the kernel of B^+:

$$B^+(x)\,\mathcal{V}_{ph} = 0 . \tag{21.4}$$

This condition ensures the validity of the Maxwell equations on $\mathcal{V}_{ph}$. The second question is, as a rule, simply ignored. A clean definition of $\mathcal{V}_{GB}$ and thus of $\mathcal{V}_{ph}$ is hardly ever given.[1]

The *C gauge* has complementary advantages and drawbacks. Its state space $\mathcal{V}_C$ is a Hilbert space and contains only physical states. The Maxwell equations are satisfied. But the basic fields A_μ, ψ, $\bar{\psi}$, are neither local (i.e. they do not commute at spacelike distances) nor covariant. This is very inconvenient for a detailed *ab ovo* elaboration of the theory, especially for a convincing formulation of renormalization. And the lack of manifest covariance also raises the question of how observational Lorentz invariance emerges from the theory.

The complementary aspects of the two methods suggests a joining of their forces. This may consist in first working out the theory of the GB-fields A_μ, ψ, as fully as possible, and then identifying in this framework "physical fields" $\mathcal{A}_\mu$, Ψ, which

[1] There exist more thoughtful treatments of the problem outside of textbooks, for example [3, 7, 12]. These approaches make essential use of the notion of asymptotic fields, which is not unproblematic in QED and will be avoided in the present work.

generate $\mathcal{V}_{ph}$ out of the vacuum Ω, yielding the C gauge formulation. Formally, such "C-fields" are obtained from the GB-fields by the Dirac ansatz

$$\begin{aligned} \mathcal{A}_\mu(x) &= A_\mu(x) - \frac{\partial}{\partial x^\mu} \int dy\, r^\rho(x-y)\, A_\rho(y)\,, \\ \Psi(x) &= \exp\left\{ie \int dy\, r^\rho(x-y)\, A_\rho(y)\right\} \psi(x)\,, \\ \bar{\Psi}(x) &= \Psi^*(x)\,\gamma^0\,, \end{aligned} \tag{21.5}$$

with

$$r^0 = 0\,, \qquad r_j(x) = -r^j(x) = \delta(x^0)\,\partial_j \frac{1}{|\vec{x}|}\,, \tag{21.6}$$

for $j = 1, 2, 3$. The auxiliary 'functions' r^μ satisfy

$$\partial_\mu r^\mu(x) = \delta^4(x)\,, \tag{21.7}$$

or in momentum space

$$p_\mu \tilde{r}^\mu(p) = i \tag{21.8}$$

with $\tilde{r}_\mu(p) = \int dx \exp\{ipx\}\, r^\mu(x)$. Notice that *formally* the definitions (21.5) have the form of a gauge transformation with the operator-valued and field-dependent gauge function

$$G(x) = -\int dy\, r^\rho(x-y)\, A_\rho(y)\,. \tag{21.9}$$

That the fields (21.5) really are the desired C-fields follows from the fact that they satisfy

$$\big[B(x), \Psi(y)\big] = \big[B(x), \bar{\Psi}(y)\big] = \big[B(x), \mathcal{A}_\mu(y)\big] = 0\,, \tag{21.10}$$

which equations then also hold for the annihilation part B^+. And this together with $B^+\Omega = 0$ shows that states of the form

$$\Phi = \mathcal{P}(\Psi, \bar{\Psi}, \mathcal{A}_\mu)\,\Omega\,, \tag{21.11}$$

with $\mathcal{P}$ a polynomial — or a more general function, if definable — of fields averaged over test functions, are physical in the sense of (21.4). The property (21.10) has been called "strict gauge invariance" by Symanzik [11] and, following him, by Strocchi and Wightman [10]. A proof of (21.10) is found in the first of these references.

21.3 Problems, and a Solution

The contents of Sect. 21.2 were purely formal. In order to give the definition (21.5) a rigorous meaning we must start from a rigorous description of the GB formalism, especially of $\mathcal{V}_{GB}$. The use of a suitably adapted version of the Wightman formalism [1, 6, 9] suggests itself. Of course, the naturally defined scalar product being indefinite, $\mathcal{V}_{GB}$ cannot be a Hilbert space. But we assume it to be equipped with a non-degenerate scalar product. And we wish to retain all the other Wightman axioms. In particular, the fields A_μ, ψ, $\bar{\psi}$, are to be operator-valued tempered distributions. And we assume the subspace $\mathcal{V}_0$ spanned by tempered field polynomials applied to the vacuum Ω to be dense in $\mathcal{V}_{GB}$ in the weak topology induced by the scalar product. $\mathcal{V}_0$ is called the space of "local states". The density of $\mathcal{V}_0$

is important because it allows us to construct $\mathcal{V}_{GB}$ from the Wightman functions by the reconstruction theorem. The Wightman functions are easily calculated in PT and other schemes of approximation. And their properties, as properties of tempered distributions, are easier to investigate than those of unbounded operators in a space with indefinite metric. On $\mathcal{V}_0$ the Wightman functions determine the scalar product.

Right at the start we are confronted with a slightly disturbing fact. Let the charge operator Q be defined by

$$[Q, \psi(x)] = -e\,\psi(x)\,, \quad [Q, \bar{\psi}(x)] = e\,\bar{\psi}(x)\,, \quad [Q, \mathcal{A}_\mu(x)] = 0\,, \quad Q\,\Omega = 0\,, \tag{21.12}$$

with e the charge of the positron. Define the state Φ to be *physical* if the Gauß law

$$Q\,\Phi = \int_{x^0=t} d^3x\,\nabla\vec{E}(x)\,\Phi \tag{21.13}$$

holds on it[2], with $\vec{E}$ the electric field strength. Then it is known [2] that $\mathcal{V}_0$ contains no charged physical states. This means that the construction of charged physical states in $\mathcal{V}_{GB}$ as limits of local states is a non-trivial task.[3]

Let us now turn to the problem of giving a rigorous meaning to the equations (21.5) defining the C-fields in terms of the GB-fields, supposing a rigorous theory of the latter to be at hand as just explained. In this attempt we encounter two problems, an ultraviolet (UV) one and an infrared (IR) one.

The UV problem is this: The factor $\psi(x)$ in $\Psi(x)$ is a distribution, not a function, and so is the exponential factor. Notice that the auxiliary 'functions' r^j are not functions in the strict sense of the word, let alone test functions, so that the exponent does not exist as a function. This problem can be solved by standard renormalization procedures, most easily by subtraction at $p = 0$ in momentum space ('intermediate renormalization'). Unfortunately such a subtraction destroys the product form of Ψ, because in p-space the product $a(x)b(x)$ becomes the convolution $\int dk\,\tilde{a}(p-k)\,\tilde{b}(k)$ (the tilde denotes the Fourier transform). This reads in its subtracted form

$$\int dk\,\left[\tilde{a}(p-k)\,\tilde{b}(k) - \tilde{a}(-k)\,\tilde{b}(k)\right],$$

which is no longer a convolution. Hence the mapping $(\psi, A) \to (\Psi, \mathcal{A})$ has no longer the form of a gauge transformation. It is very unlikely that this problem can be solved by a more sophisticated method of renormalization. *It is my opinion that we should not bemoan this fact, but accustom ourselves to the idea that the notion of gauge transformations is not as useful in QED as it is in classical electrodynamics, but is only of a heuristic value.*[4] The true problem is to find fields satisfying the condition (21.10) of strict gauge invariance, no matter whether or not they can be derived from the GB-fields by a gauge transformation. That the renormalized Dirac

[2] As is well known, in this crude version of the condition the right-hand side does not make sense. Suitable regularizations in space and time are necessary. But this problem is immaterial to our present purposes.

[3] That it is not necessarily an unsolvable task has recently been re-emphasized by Morchio and Strocchi [13].

[4] This scepticism does not extend to gauge transformations of the first kind, that is global transformations with an x-independent real gauge function.

ansatz yields a formulation of QED which satisfies all the necessary requirements, fully justifies its use.

The IR problem is this: Are Ψ, $\mathcal{A}_\mu$, after renormalization, defined as fields on $\mathcal{V}_{GB}$? In other words, is the space $\mathcal{V}_C$ generated from the vacuum Ω by the C-fields a subspace of $\mathcal{V}_{GB}$? At first, the answer gleaned from PT is a plain 'no'! The scalar product (Φ_0, Φ) of the physical state

$$\Phi = \int dx\, f(x)\, \Psi(x)\, \Omega \quad \in \mathcal{V}_C \tag{21.14}$$

and the local state

$$\Phi_0 = \int dy\, \overline{g(y)}\, \bar{\psi}^*\, \Omega \quad \in \mathcal{V}_0\,, \tag{21.15}$$

with f and g tempered test functions, can be shown to diverge already in second order of PT (see end of Appendix). This divergence is caused by the divergence at large y of the exponent $\int dy\, r^\rho(x-y)\, A_\rho(y)$ in (21.5): it is an IR problem.

But this result is misleading. As is well known, the generic S-matrix element of QED calculated with the LSZ reduction formula is in general IR divergent in finite orders of PT. But these divergences can be isolated and summed over all orders, yielding a *vanishing* result. And this result is expected to be closer to the truth than the finite-order divergences, because it agrees with information obtained from other sources, especially the Bloch-Nordsieck model. Something similar might happen for the mixed 2-point function

$$F(x,y) = (\Omega,\, \bar{\psi}(y)\, \Psi(x)\, \Omega) \tag{21.16}$$

occurring in (Φ_0, Φ). And, indeed, it does happen! The IR divergences in F can be isolated in all orders of PT and summed to yield

$$F(x,y) \equiv 0\,. \tag{21.17}$$

For a sketch of the proof we refer to the Appendix. Again, we expect the result (21.17), rather than the finite-order divergences, to correspond to the true situation. But, as in the case of the S-matrix, this does not solve our problem. In the same way as (21.17) it can be generally shown that Φ is orthogonal to all local states. The state Φ of charge $-e$ is orthogonal to $\mathcal{V}_0$, hence it cannot be the weak limit of a sequence of local states, hence $\mathcal{V}_C$ cannot be a subspace of $\mathcal{V}_{GB}$, in which we assumed $\mathcal{V}_0$ to be dense.

As a result we obtain:

The state space $\mathcal{V}_C$ of the Coulomb gauge cannot be obtained as a subspace of an extension $\mathcal{V}_{GB}$ of $\mathcal{V}_0$, if the scalar product defined on $\mathcal{V}_0$ can be extended to $\mathcal{V}_{GB}$ in such a way that $\mathcal{V}_0$ is weakly dense in $\mathcal{V}_{GB}$.

The problems discussed in this section lead to the following conclusions. The C-fields Ψ, $\mathcal{A}_\mu$, cannot be defined as fields acting on $\mathcal{V}_{GB}$. Nor are they related to the GB-fields ψ, A_μ, by a gauge transformation. The traditional explanations of the connection between the two formulations do not work. A working method of relating the two formalisms has been derived in Chap. 12 of [8]. It will be briefly described without giving proofs. The method makes essential use of the Wightman reconstruction theorem. At first it is demonstrated (in PT) that the Wightman functions of the C-fields can be obtained from those of the GB-fields by a limiting procedure, like this: Replace the auxiliary functions $r^j(x)$ in (21.5) and (21.6) by the regularized version

$$r^j_\xi(x) = \chi(\xi x)\, r^j(x)\,, \qquad 0 < \xi < \infty\,, \tag{21.18}$$

with $\chi(u)$ a test function with compact support, satisfying $\chi(0) = 1$. The resulting fields Ψ_ξ, A^μ_ξ *are* definable as acting on a slight extension of $\mathcal{V}_0$. Their Wightman functions can be computed. And the limits $\xi \to 0$ of the latter exist and define a field theory via the reconstruction theorem, which has all the desired properties of QED in the C gauge. It *is* QED in the C gauge! But the states and the fields of this theory have no discernible direct connection with the states and fields of the GB formalism.

Let us end this section with a few remarks on how observational Lorentz invariance emerges from the not manifestly covariant C gauge formalism. The problem has not yet been discussed in depth. I can therefore only state a program rather than results. This program is based on the spirit of the local observables approach [4] to quantum field theory, according to which only observables localized in bounded regions of space-time are true observables. The first problem facing us is finding an exact characterization of the operators representing observables. Since the C-fields should describe the theory fully, an observable in the bounded domain $\mathcal{R}$ must be a function of the fields with arguments in $\mathcal{R}$. Observables should be represented by hermitian operators (self-adjointness is hard to handle in PT), and they should satisfy the axioms of local quantum physics. In particular, two observables localized in relatively spacelike domains should commute. And the algebras of observables in the domain $\mathcal{R}$ and its image $\mathcal{R}_\Lambda$ under the Lorentz transformation Λ should be related by an automorphism α_Λ, which we can, of course, not expect to be unitarily implemented. The traditional requirement that observables should be gauge invariant is difficult or impossible to formulate in C gauge, on account of the doubtful status of gauge transformations. We propose to solve this conundrum as we did for fields, by starting from the GB formalism. There, observables must satisfy the same requirements as stated for the C gauge. But the additional requirement of gauge invariance can now be interpreted to denote strict gauge invariance. This means that observables must commute with $B(x)$.[5] The C gauge equivalent of such a GB observable A can then be defined as a bilinear form by

$$(\Omega, \bar{\Psi}(x)\, A\, \Psi(y)\, \Omega) = \lim_{\xi \to 0} (\Omega, \bar{\Psi}_\xi(x)\, A\, \Psi_\xi(y)\, \Omega) \tag{21.19}$$

and obvious generalizations. Under exactly what conditions on A this limit exists has not yet been investigated. It is clear that the observable fields $F_{\mu\nu}$ and j_μ belong to this class.

Given the local observables of the theory, the second problem is that of describing observational Lorentz invariance without using a non-existent unitary representation of the Lorentz group. The formulation proposed here is based on the important insights put forward by Haag and Kastler in [5] concerning the relevance of Fell's theorem (a purely mathematical statement) to quantum mechanics. This relevance rests on the observation that in any given experiment we can only measure a *finite number* of *local* observables with a *finite accuracy*. In view of this we can formulate observational invariance as follows.

[5] On $\mathcal{V}_C$ B vanishes identically.

Let $A_1, \ldots, A_n$, be a finite set of local observables with measuring accuracies ϵ_i, $A_i^\Lambda = \alpha_\Lambda(A_i)$ their images under the Lorentz transformation Λ, and let Φ be a (physically preparable) state in $\mathcal{V}_C$. Then there exists a state $\Phi_\Lambda \in \mathcal{V}_C$ such that

$$|(\Phi_\Lambda,\, A_i^\Lambda \,\Phi_\Lambda) - (\Phi,\, A_i\,\Phi)| < \epsilon_i\,, \tag{21.20}$$

for $i = 1, \ldots, n$.

If Λ is a boost, then Φ_Λ represents the original state as seen by a moving observer, as far as the experiment in question is concerned. That this form of Lorentz invariance is satisfied in $\mathcal{V}_C$ is at the moment still a conjecture. But it has a good chance of being true. Partial results in this direction can be found at the end of Chap. 12 in [8].

Appendix: Proving Equation (21.17)

Using the methods of [8] a rigorous proof of the claimed result (21.17) can be given by summing the IR relevant parts of F in all orders of PT. In this appendix we will not give the full proof, but only sketch its essential ideas.

We denote the vacuum expectation value $(\Omega, \ldots, \Omega)$ by $\langle \cdots \rangle$. The expression of interest is

$$\begin{aligned} A &= \langle \bar{\psi}(y)\, \Psi(x) \rangle \\ &= \langle \bar{\psi}(y)\, \exp\{ie \int du\, r^j(x-u)\, A_j(u)\}\, \psi(x) \rangle \end{aligned} \tag{A.1}$$

with

$$r^j(v) = \delta(v^0)\, \partial^j \frac{1}{|\vec{v}|}\,. \tag{A.2}$$

We are only interested in the IR problems connected with this expression, ignoring the UV problems which can, however, easily be taken into account. The reader is free to get rid of them by a suitable UV regularization. This means that our problem is the existence of the u-integral in (A.1) *at large* u. Assume x, y, to be restricted to bounded regions by integrating them over test functions with compact support. Again, this restriction is not essential. Then, because of the δ-factor in (A.2), u can tend to infinity only in spacelike directions. For large u we can then neglect $\vec{x}$ with respect to $\vec{u}$ in $r^j(x-u)$. Using the cluster property we find

$$A \sim \langle \psi(x)\, \bar{\psi}(y) \rangle \langle \exp\{ie \int du\, r^j(-u)\, A_j(u)\} \rangle\,, \tag{A.3}$$

where the symbol $\sim$ means that only the potentially IR dangerous contributions to A are considered. The first factor depending on x and y is IR harmless. We need therefore only discuss the second factor, which we call X. Going over to momentum space we find

$$X = \Big\langle \exp\Big\{e \int dk^0\, d^3k\, \frac{k^j}{|\vec{k}|^2}\, \tilde{A}_j(k)\Big\}\Big\rangle \tag{A.4}$$

with $\tilde{A}_j$ the Fourier transform of A_j (the tilde will be omitted in the sequel). The existence of the k^0-integral is part of the UV problem which we ignore. The same goes for the existence of the $\vec{k}$-integral at $|\vec{k}| \to \infty$. Our only concern is the possible divergence of the $\vec{k}$-integral at $\vec{k} = 0$. In PT the exponential in (A.4) is defined as a power series. (Notice the factor e in the exponent.) Only the even terms in this expansion survive because the Wightman functions of an odd number of A's and no ψ or $\bar{\psi}$ vanish. Hence we find

$$X = \sum_{\sigma=0}^{\infty} \frac{e^{2\sigma}}{(2\sigma)!} \left\langle \left(\int dk \, \frac{k^j}{|\vec{k}|^2} A_j(k) \right)^{2\sigma} \right\rangle . \tag{A.5}$$

This expression contains terms of the form

$$\left\langle \prod_{\omega=1}^{2\sigma} A_{j_\omega}(k_\omega) \right\rangle$$

which must be evaluated in PT, and its IR divergences isolated. The graph rules for these functions are similar to, but somewhat more complicated than, the familiar Feynman rules for Green's functions. As in this case it is found that connected components with ≥ 4 external lines of a graph give convergent contributions to X, because they vanish sufficiently strongly at $k_\omega = 0$ to overcome the $1/|\vec{k_\omega}|$ singularities in (A.5). This is a consequence of the Ward identities, which for these fermion-less connected graphs take the form $k^\mu A_\mu(k) = 0$ in all orders except the 0th. These contributions can be factored out. The IR problem is concentrated in graphs consisting exclusively of 2-variable components. But such a 2-point function is assumed to satisfy the renormalization condition

$$\langle A_\mu(k_1)\, A_\nu(k_2) \rangle = \delta^4(k_1 + k_2) \left(g_{\mu\nu} \delta_+(k_1) + \text{IR harmless terms} \right) \tag{A.6}$$

with $\delta_+(k) = \theta(k_0)\, \delta(k^2)$. The IR divergences are entirely due to the zero-order term $g_{\mu\nu}\delta_+$. Inserting this into the expansion (A.5) and doing the combinatorics right we find

$$X \sim (\text{finite factor}) \cdot \exp\left\{ -e^2 \int \frac{d^3k}{2|\vec{k}|^3} \right\} . \tag{A.7}$$

The k-integral diverges positively at $\vec{k} = 0$. The UV divergence at $|\vec{k}| \to \infty$ is in a more detailed treatment removed by renormalization and does not concern us here. The result is, then:

$$X \sim 0 \tag{A.8}$$

which proves (21.17). Notice that the e^2 term in the power series expansion of (A.7) diverges, as was claimed in Sect. 3.

References

1. N.N. Bogolubov, A.A. Logunov, A.I. Oksak and I.T. Todorov: *General Principles of Quantum Field Theory*. Kluwer, Dordrecht, 1990.
2. R. Ferrari, L. Picasso and F. Strocchi: *Commun. Math. Phys.* 35:25 (1974).
3. J. Fröhlich, G. Morchio, and F. Strocchi: *Ann. of Phys.* 119:2441 (1979).
4. R. Haag: *Local Quantum Physics*. Springer, Berlin, 1993.
5. R. Haag and D. Kastler: *J. Math. Phys.* 5:848 (1964).

6. R. Jost: *The General Theory of Quantized Fields.* Am. Math. Soc., Providence, RI, 1965.
7. G. Morchio and F. Strocchi: *Nucl. Phys. B* 211:471 (1983).
8. O. Steinmann: *Perturbative QED and Axiomatic Field Theory.* Springer, Berlin, 2000.
9. R. F. Streater and A. S. Wightman: *PCT, Spin & Statistics, and All That.* Benjamin/Cummings, Reading, MA, 2nd ed., 1978.
10. F. Strocchi and A. S. Wightman: *J. Math. Phys.* 15:21 (1974).
11. K. Symanzik: *Lectures on Lagrangian Quantum Field Theory.* Internal report DESY T71-1, 1971.
12. D. Zwanziger: *Phys. Rev. D* 14:2570 (1976).
13. G. Morchio and F. Strocchi: *J. Math. Phys.* 44:5569 (2003).

Complex Angular Momentum Analysis and Diagonalization of the Bethe–Salpeter Structure in Axiomatic Quantum Field Theory

Giovanni Alberto Viano

Dipartimanto di Fisica, Università di Genova
Istituto Nazionale di Fisica Nucleare, sez. di Genova, Via Dodecaneso, 33
16146 Genova, Italy
viano@ge.infn.it

Summary. Recent experimental measurements at HERA of the *Structure Functions* in lepton-hadron inelastic scattering started to bridge the gap between the *hard* and the *soft* regimes of strong interactions. In this scenario the Complex Angular Momentum (CAM) theory, which plays a relevant role in describing soft processes, which are intrinsically non-perturbative, is widely adopted as a phenomenological model. It is therefore of some interest to explore up to what extent one can give a theoretical foundation to CAM methods, starting from the basic axiomatic principles of Quantum Field Theory (QFT). In this review we shall try to expose the main results obtained recently on this topic.

22.1 Introduction: Phenomenological Motivation

Hadronic processes are traditionally classified in two distinct classes: soft processes and hard processes.

The soft processes are characterized by an energy scale of the order of the hadron size R ($\sim$ 1 fm). The squared momentum transfer is generally small: $|t| \sim 1/R^2 \sim$ (few hundred MeV)^2. The t-dependence of the cross-sections is exponential $d\sigma/dt \sim \exp(-R^2|t|)$. Classical examples of soft processes are elastic hadron-hadron scattering and diffractive dissociation. From a theoretical point of view, perturbative quantum chromodynamics (QCD) is inadequate to describe these processes. The presence of a large length scale R makes them intrinsically non-perturbative. The approach which has been adopted since the 1960s to describe soft processes is the complex angular momentum (CAM) theory [1].

The hard processes are characterized by two (or more) energy scales: one is still the hadron size, the other is a *hard* energy scale. The momentum transfer is of the order of this scale, and therefore large ($\gtrsim$ 1 GeV^2). The high value of the momentum transfer allows one to use perturbative QCD. Part of the process, however, is still of non-perturbative origin. The main novelty of the last years is the discovery of diffractive processes which have soft and hard components at the same time. A typical process of this type is the diffractive deep inelastic scattering in which two different energy scales coexist: a soft and a hard one [1].

For a more precise understanding of this point it is convenient to look at the behavior of the so-called *structure functions*, which can be measured in the deep inelastic lepton-hadron scattering, and parametrize the structure of the target as *seen* by the virtual photon. They are usually denoted by $F_i(x, Q^2)$ ($i = 1, 2$), and the kinematic variables are: $Q^2 = -q^2$, $q^\mu = k^\mu - k'^\mu$ is the momentum transfer (k^μ and k'^μ being the incoming and the outgoing lepton four momenta, respectively), and the Bjorken variable $x = Q^2/2\nu$, $\nu = p \cdot q$ ($p^2 = M^2$, M being the nucleon mass). Bjorken (1969) argued that in the limit $\nu, Q^2 \to \infty$, $x = Q^2/2M\nu$ fixed, F_1 and F_2 should (approximately) scale, i.e. depend on x only [1]. Furthermore, assuming the Callan–Gross relation, F_1 and F_2 are proportional to each other and are given by $F_2(x) = 2xF_1(x)$ [1]. Therefore we shall refer only to the behavior of F_2 as a function of x. The results can be summarized as follows. The CAM methods are believed to apply to the low-Q^2 region: i.e., $Q^2 \lesssim 1$ GeV^2. In this region the observed behavior of $F_2(x)$ is close to the one dictated by the Donnachie–Landshoff formula [12], that is $F_2(x) \sim x^{-0.08}$. On the other hand, as soon as one enters the perturbative regime (i.e. for $Q^2 > 1$ GeV^2) the CAM fit fails; QCD predicts that if F_2 is parametrized as $F_2 \propto x^{-\lambda}$, the exponent of x should change with Q^2: one finds $\lambda \simeq 0.25 \div 0.35$ at large Q^2 ($\sim (10 \div 100)$ GeV^2). Unfortunately, from the theoretical point of view the transition from the CAM picture to the QCD picture is not well understood: i.e., the transition from soft processes to hard processes is still unclear.

The CAM methods, which were widely used in the 1960s, lost ground in the 1970s. The main reason was due to the explosion of interest for the gauge field theory and the birth of QCD. But, then, in the 1980s and 1990s the interest in diffractive scattering was reignited mainly for the experimental results at HERA and at the Tevatron. In the same years some relevant theoretical results were obtained: on one hand A. Donnachie and P.V. Landshoff [12] proposed an empirically simple CAM model able to reproduce a variety of high-energy processes; on the other hand, a perturbative QCD picture of the hard diffraction was developed by various authors (notably by Lipatov and co-workers [16]). We can thus say that a basic problem of the present day's debate is the study of the interplay of soft and hard interactions.

In this scenario another relevant problem emerges: the CAM model (i.e., the Regge theory) has been rigorously proved in the non-relativistic potential scattering for a well-defined class of potentials (the Yukawian class), but a rigorous proof of its extension to the relativistic dynamics was missing. More precisely, a tentative formulation of this extension was given in the approach of the so-called *S-matrix* theory based on the general, but rather loose concept, of *maximal analyticity* [11]. We can thus say that the relativistic CAM theory was, up to a certain extent, a phenomenological theory whose validity was essentially tested on the fits which reproduce the experimental data. One is then, quite naturally, led to pose the following questions:

Problem: i) Is it possible to prove that the existence of analytic structures in the CAM variables is a property of QFT satisfying the basic properties of *local commutativity, Poincaré invariance, spectral condition and temperate ultraviolet behavior*?

ii) Is it possible, assuming these properties of QFT, to perform a CAM diagonalization of the Bethe–Salpeter equation (diagonalization with respect to the angular variables), which generates Regge-type singularities in the CAM plane?

In [9, 10] an answer to these questions has been given by the use of peculiar mathematical methods (i.e., harmonic analysis on complex hyperboloids [8]), which will be rapidly sketched in the next section. The physical results will be reported and briefly commented in Section 22.3.

22.2 Harmonic Analysis on Complex Hyperboloids

22.2.1 The Complex Quadric $\mathbf{X}^{(c)}_{d-1}$

Consider the space $\mathbb{C}^{(d)}$ of the complex variables $z = (z^{(0)}, \vec{z}, z^{(d-1)})$ with $\vec{z} = (z^{(1)}, \dots, z^{(d-2)})$, and the complex quadric $X^{(c)}_{d-1}$ with equation:

$$s(z) = z^{(0)\,2} - \vec{z}^{\,2} - z^{(d-1)\,2} + 1 = 0, \tag{22.1}$$

where

$$\vec{z}^{\,2} = z^{(1)\,2} + \cdots + z^{(d-2)\,2}. \tag{22.2}$$

Two submanifolds of $X^{(c)}_{d-1}$ are relevant:

a) the one-sheeted hyperboloid $X_{d-1} = \mathbb{R}^{(d)} \cap X^{(c)}_{d-1}$;
b) the *Euclidean sphere* $S_{d-1} = \left(\mathrm{i}\mathbb{R} \times \mathbb{R}^{(d-1)}\right) \cap X^{(c)}_{d-1}$, ($\mathrm{i}\mathbb{R}$ refers to the coordinate $z^{(0)}$ of z).

22.2.2 Froissart–Gribov Equalities

A) Case $d = 2$

On the complex hyperbola $X^{(c)}_1 = (z^{(0)} = -\mathrm{i}\sin\theta, z^{(1)} = \cos\theta, \theta = u + \mathrm{i}v)$ one considers the domain $D = X^{(c)}_1 \setminus (\Sigma^{(c)}_+ \cup \Sigma^{(c)}_-)$, whose representation in the θ-plane is the periodic cut-plane $\Pi = \mathbb{C} \setminus (\sigma_+ \cup \sigma_-)$ (the cuts σ_+, σ_- are embedded respectively in $[+1, +\infty)$ and $(-\infty, -1])$, and the corresponding subsets $\Sigma^{(c)}_+$, $\Sigma^{(c)}_-$ of $X^{(c)}_1$ are given by:

$$\Sigma^{(c)}_\pm = \left\{ z = (z^{(0)}, z^{(1)}) \in X^{(c)}_1 ; \pm z^{(1)}_1 \in [\cosh v_\pm, +\infty) \right\}.$$

Note that the circle $S_1 = \{z = (\mathrm{i}y_0, x_1) : y_0^2 + x_1^2 = 1; y_0, x_1 \text{ real}\}$ of $X^{(c)}_1$, represented by the periodic u-axis in Π, is contained in the domain D. Next, one defines the following Laplace-type transform:

$$\tilde{F}(\lambda) = \int_\gamma f(\theta)\, e^{\mathrm{i}\lambda\theta}\, d\theta, \tag{22.3}$$

where the functions $f(\theta)$ are the even periodic functions, holomorphic in the cut-plane Π, bounded by $|f(\theta)| \leqslant C\, e^{m|v|}$, and the contour γ encloses the components of σ_+, σ_- inside the half-strip $v > 0$, $-\alpha < u < 2\pi - \alpha$, with $0 < \alpha < \pi$. Then, replacing the contour γ by a pair of contours (γ_+, γ_-) enclosing respectively the cuts

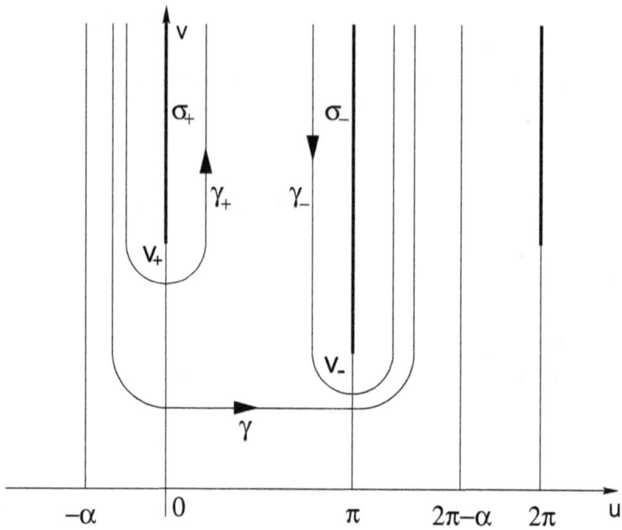

Fig. 22.1. Integration paths in the cut-plane Π.

σ_+, σ_-, and then from flattening them (in a folded way, see Fig. 22.1) onto the cuts, from (22.3) one obtains:

$$\tilde{F}(\lambda) = \tilde{F}_+(\lambda) + e^{\mathrm{i}\pi\lambda}\,\tilde{F}_-(\lambda), \tag{22.4}$$

$$\tilde{F}_\pm(\lambda) = \int_0^{+\infty} e^{-\lambda v}\,\Delta f_\pm(v)\,dv, \tag{22.5}$$

$\Delta f_\pm$ being the corresponding jumps of $\mathrm{i}f$ across the cuts, i.e.

$$\Delta f_+ = \mathrm{i}\lim_{\epsilon\to 0}\left[f(\epsilon + \mathrm{i}v) - f(-\epsilon + \mathrm{i}v)\right], \tag{22.6a}$$

$$\Delta f_- = \mathrm{i}\lim_{\epsilon\to 0}\left[f(\pi + \epsilon + \mathrm{i}v) - f(\pi - \epsilon + \mathrm{i}v)\right]. \tag{22.6b}$$

Note that we have assumed that the bound given by $e^{m|v|}$ is uniform in Π, and therefore applies to the discontinuity functions $\Delta f_\pm$ and, accordingly, $\tilde{F}_\pm(\lambda)$ are holomorphic in $C_+^{(m)} = \{\lambda \in \mathbb{C}; \operatorname{Re}\lambda > m\}$.

Then, from choosing $\gamma = \gamma_\alpha$ with support $(-\alpha + \mathrm{i}\infty, -\alpha] \cup [-\alpha, 2\pi - \alpha] \cup [2\pi - \alpha, 2\pi - \alpha + \mathrm{i}\infty)$ and taking into account the 2π-periodicity of $f(\theta)$, one obtains for $\lambda = \ell$ integer:

$$\tilde{F}(\ell) = f_\ell \text{ for all integers such that } \ell > m, \tag{22.7}$$

where

$$f_\ell = \int_{-\alpha}^{2\pi-\alpha} e^{\mathrm{i}\ell u}\,f(u)\,du. \tag{22.8}$$

Finally, from (22.4) and (22.8) it follows that

$$f_\ell = \tilde{F}_+(\ell) + \tilde{F}_-(\ell) \quad (\ell \text{ even} > m), \tag{22.9a}$$

$$f_\ell = \tilde{F}_+(\ell) - \tilde{F}_-(\ell) \quad (\ell \text{ odd} > m). \tag{22.9b}$$

Next, combining the symmetric and antisymmetric parts, i.e. $\tilde{F}^{(s)} = \tilde{F}_+ + \tilde{F}_-$, and $\tilde{F}^{(a)} = \tilde{F}_+ - \tilde{F}_-$, we get:

$$\tilde{F}^{(s)}(2\ell) = f_{2\ell}, \tag{22.10a}$$

$$\tilde{F}^{(a)}(2\ell+1) = f_{2\ell+1}. \tag{22.10b}$$

Equalities (22.10) (called Froissart–Gribov equalities [11]) give a bijection between classes of functions $f(\theta)$, holomorphic in Π, bounded by $Ce^{m|v|}$, and corresponding classes of functions $\tilde{F}(\lambda) = \tilde{F}_+(\lambda) + e^{i\pi\lambda}\tilde{F}_-(\lambda)$ holomorphic in the half-plane $C_+^{(m)} = \{\lambda \in \mathbb{C}, \operatorname{Re}\lambda > m\}$; moreover, the functions $\tilde{F}^{(s),(a)}(\lambda)$ are the unique Carlsonian interpolations in the half-plane $C_+^{(m)}$ of the respective sets of even and odd Fourier coefficients of $f|_{\mathbb{R}}$.

Let us note that analogous results hold when f admits boundary values in the sense of distributions on $i\mathbb{R}$ and $\pi + i\mathbb{R}$, and therefore the discontinuities are defined as distributions with respective supports σ_+, σ_- [9].

B) Case $d > 2$

We now consider the complexified unit hyperboloid $X_{d-1}^{(c)}$: the analyticity domain D is the preimage of the cut-plane $\mathbb{C} \setminus (\sigma_+ \cup \sigma_-)$ in $X_{d-1}^{(c)}$ (through the mapping $z \to z^{(d-1)} = \cos\theta \to \pm\theta$). The sphere is embedded in D and projects onto the interval $[-1, +1]$ in the $z^{(d-1)}$-plane; the cuts σ_+ and σ_- are the images of the subsets Σ_+ and Σ_- of X_{d-1}, defined respectively by the conditions $z^{(0)} > 0, z^{(d-1)} \geqslant \cosh v_+$ (i.e., $\theta = iv, v \geqslant v_+$) and $z^{(0)} < 0, z^{(d-1)} \leqslant -\cosh v_-$ (i.e., $\theta = iv + \pi, v \geqslant v_-$). The functions $\mathcal{F}(z)$ considered, are holomorphic in the cut-domain D of $X_{d-1}^{(c)}$, invariant under the stabilizer G_{z_0} (isomorphic to $SO_0^{(c)}(1, d-2)$) of the base point $z_0 = (0, \ldots, 0, 1)$; therefore they only depend on $z^{(d-1)} = \cos\theta$, so that one can again put $\mathcal{F}(z) = f(\theta)$, with f even, 2π-periodic and holomorphic in the cut-plane $\Pi = \mathbb{C} \setminus (\sigma_+ \cup \sigma_-)$. Furthermore, a moderate growth condition of the form $|\mathcal{F}(z)| \leqslant \text{const.}\,|z_{d-1}|^m$ is assumed. We now write a Fourier–Laplace-type transform similar to formula (22.3)

$$\tilde{F}(\lambda) = \omega_{d-2} \int_\gamma e^{i(\lambda + \frac{d-2}{2})\theta} \left(\mathcal{A}_d^{(c)} f\right)(\theta)\, d\theta, \tag{22.11}$$

where $\omega_{d-2} = 2\pi^{(d-2)/2}/\Gamma(\frac{d-2}{2})$ = area of S_{d-3} $(d \geqslant 3)$, γ is the same contour as for the case $d = 2$, and the definition of $\mathcal{A}_d^{(c)}$ requires the following procedure. We introduce a decomposition of $f(\theta)$ of the form $f(\theta) = f_+(\theta) + f_-(\theta)$, where f_+ and f_- have the same analyticity and symmetry properties as f, but enjoy the following additional property: f_+ (resp. f_-) admits a single cut, namely σ_+ (σ_-) across which its discontinuity coincides with the corresponding one of f, denoted by $\Delta_+ f(v)$ (resp. $\Delta_- f(v)$). We then define:

$$\left(\mathcal{A}_d^{(c)} f\right)(\theta) = \left(\mathcal{A}_{d+}^{(c)} f_+\right)(\theta) + \left(\mathcal{A}_{d-}^{(c)} f_-\right)(\theta), \tag{22.12}$$

where $\mathcal{A}_{d+}^{(c)} f_+$ and $\mathcal{A}_{d-}^{(c)} f_-$ are respectively defined as holomorphic functions in the periodic cut-planes $\mathbb{C} \setminus \sigma_+$ and $\mathbb{C} \setminus \sigma_-$ by the following integrals:

$$\left(\mathcal{A}^{(c)}_{d+} f_+\right)(\theta) = -\int_{\gamma(\pi,\theta)} f_+(\tau)\left[2\left(\cos\theta - \cos\tau\right)\right]^{(d-4)/2} \sin\tau \, d\tau, \quad (22.13a)$$

$$\left(\mathcal{A}^{(c)}_{d-} f_-\right)(\theta) = -\int_{\gamma(0,\theta)} f_-(\tau)\left[2\left(\cos\theta - \cos\tau\right)\right]^{(d-4)/2} \sin\tau \, d\tau. \quad (22.13b)$$

The path $\gamma(\pi,\theta)$ (resp. $\gamma(0,\theta)$) with end-points π and θ (resp. 0 and θ) has to belong to the domain $\mathbb{C}\setminus\sigma_+$ ($\mathbb{C}\setminus\sigma_-$), and the function $[2(\cos\theta-\cos\tau)]^{(d-4)/2}$ is determined by the condition that it is positive for $\theta = \mathrm{i}v$, $\tau = \mathrm{i}w$, $0 < w < v$. By using the same contour distorsion argument as for the case $d = 2$, one then obtains:

$$\tilde{F}(\lambda) = \tilde{F}_+(\lambda) + e^{\mathrm{i}\pi\lambda}\tilde{F}_-(\lambda), \quad (22.14)$$

where

$$\tilde{F}_+(\lambda) = \omega_{d-2}\int_{v_+}^{+\infty} e^{-\lambda v} e^{-(\frac{d-2}{2})v} \mathcal{A}_d f_+(v)\, dv, \quad (22.15a)$$

$$\mathcal{A}_d f_+(v) = \int_{v_+}^{v} f_+(w)\left[2\left(\cosh v - \cosh w\right)\right]^{(d-4)/2} \sinh w \, dw, \quad (22.15b)$$

and

$$\tilde{F}_-(\lambda) = \omega_{d-2}\int_{v_-}^{+\infty} e^{-\lambda v} e^{-(\frac{d-2}{2})v} \mathcal{A}_d f_-(v)\, dv, \quad (22.16a)$$

$$\mathcal{A}_d f_-(v) = \int_{v_-}^{v} f_-(w)\left[2\left(\cosh v - \cosh w\right)\right]^{(d-4)/2} \sinh w \, dw. \quad (22.16b)$$

The functions $\tilde{F}_\pm(\lambda)$ can be rewritten in a form closer to the standard expression of the Fourier–Laplace transform by the use of two systems of local coordinates on X_{d-1}, equally valid in a neighborhood of the set $\Sigma_+ = \{z \in X_{d-1}; z^{(d-1)} \geqslant \cosh v_+, z^{(0)} > 0\}$, namely:

i) The polar coordinates:

$$z^{(0)} = \sinh w \cosh\varphi, \; z^{(d-1)} = \cosh w,$$
$$[\vec{z}] = (z^{(1)},\ldots,z^{(d-2)}) = \sinh w \sinh\varphi\, [\vec{\alpha}],$$
$$[\vec{\alpha}] \in S_{d-3}.$$

ii) The horocyclic coordinates:

$$z^{(0)} = \sinh v + \frac{1}{2}\|\vec{x}\|^2 e^v; \; z^{(d-1)} = \cosh v - \frac{1}{2}\|\vec{x}\|^2 e^v,$$
$$[\vec{z}] = \vec{x}\, e^v, \; \vec{x} \in \mathbb{R}^{d-2}.$$

The sections $v =$ const. are paraboloids in the hyperplanes $z^{(0)} + z^{(d-1)} = e^v$, called horocycles. For classes of functions $F_+(z)$ with support in Σ_+, invariant under the stabilizer of z_0, namely $F_+(z) \equiv F_+[z^{(d-1)}] = f_+(w)$ ($\operatorname{supp} f_+ \subset [v_+, +\infty)$), and moreover satisfying a bound of the form $|F_+(z)| \leqslant \text{const.}\,|z^{(d-1)}|^m$, the Laplace transform of F_+ is defined as follows, closer to the standard form:

$$\tilde{F}_+(\lambda) = \int_{\Sigma_+} e^{-\lambda v}\, F(\cosh w)\, d\vec{x}\, dv. \quad (22.17)$$

By the use of the relationship

$$\vec{x} = e^{-v/2}\left[2\left(\cosh v - \cosh w\right)\right]^{1/2}[\vec{\alpha}] \qquad [\vec{\alpha}] \in S_{d-3}, \tag{22.18}$$

one can rewrite formula (22.17) in the form (22.15a). Analogous results hold for $\tilde{F}_{-}(\lambda)$.

For $\lambda = \ell$ integer (with $\ell > m$), one can rewrite (22.11) with the choice of the contour $\gamma = \gamma_\alpha$ (as in the case $d = 2$), and by taking the periodicity of $(\mathcal{A}_d^{(c)} f)(\theta)\exp[\mathrm{i}(\frac{d-2}{2})\theta]$ into account, one obtains from (22.14):

$$f_\ell = \tilde{F}_{+}(\ell) + (-1)^\ell \tilde{F}_{-}(\ell) \qquad (\ell > m), \tag{22.19}$$

where the terms f_ℓ, given by the equation:

$$f_\ell = \omega_{d-1}\int_0^\pi P_\ell^{(d)}(\cos\theta) f(\theta)(\sin\theta)^{(d-2)}\, d\theta \qquad (\ell \geqslant 0), \tag{22.20}$$

are the Legendre coefficients of $\mathfrak{F}|_{S_{d-1}}$, $P_\ell^{(d)}(\cos\theta)$ being the *ultraspherical Legendre polynomials* [9]. As in the case $d = 2$, we can introduce symmetric $\tilde{F}^{(s)}(\lambda) = \tilde{F}_{+}(\lambda) + \tilde{F}_{-}(\lambda)$ and antisymmetric $\tilde{F}^{(a)}(\lambda) = \tilde{F}_{+}(\lambda) - \tilde{F}_{-}(\lambda)$ combinations, and finally obtain the Froissart–Gribov equalities in the form:

$$\tilde{F}^{(s)}(2\ell) = f_{2\ell} \qquad (2\ell > m), \tag{22.21a}$$

$$\tilde{F}^{(a)}(2\ell + 1) = f_{2\ell+1} \qquad (2\ell + 1 > m). \tag{22.21b}$$

$\tilde{F}^{(s)}(\lambda)$ (resp. $\tilde{F}^{(a)}(\lambda)$) represents the unique Carlsonian interpolation of the even (odd) Legendre coefficients of $\mathfrak{F}|_{S_{d-1}}$ [9].

22.2.3 Causal Symmetric Spaces and Volterra Algebra

Since the 1960, it has been felt by various authors that the introduction of the CAM analysis should find a conceptual justification by starting from a group representation viewpoint [18]. In particular, the methods of harmonic analysis on the two-sheeted hyperboloid $SO(1,2)/SO(2)$ have been widely used for obtaining a Watson-type-resummation of the partial wave expansion, which is, indeed, one of the main ingredients of the CAM methods. But, in this framework, the corresponding Fourier-type transformation cannot yield analytic functions of λ in the half-planes $C_+^{(m)} = \{\lambda \in \mathbb{C} \ :\ \mathrm{Re}\,\lambda > m\}$ as those encountered in the Froissart–Gribov equalities. It turns out that it is the harmonic analysis on the one-sheeted hyperboloid (i.e., $X_{d-1} = SO_0(1, d-1)/SO_0(1, d-2)$) which must be considered. But then a question emerges: up to what extent is it possible to adapt to these pseudo-Riemannian spaces the methods of harmonic analysis introduced (notably by Helgason [14]) in the case of symmetric Riemannian spaces, like $SO(1,2)/SO(2)$? Even if a complete answer to this question is still a matter of research (see [15]), nevertheless some significant results have been obtained.

A relevant ingredient, which can be introduced, is the ordering on X_{d-1} induced by the ambient Minkowskian space, i.e. $\Omega_+ = \{z \in \mathbb{R}^d; z^2 \geqslant 0, z^{(0)} \geqslant 0\}$ (i.e., *the closed forward light-cone*) defines on $\mathbb{R}^d$ the following partial ordering relation: $z_1 \geqslant z_2$ (or $z_2 \leqslant z_1$) if $z_1 - z_2 \in \Omega_+$ (X_{d-1} equipped with this ordering relation is called *causal symmetric space* [15]). This order structure allows us to introduce an algebra of kernels on X_{d-1}, called *Volterra algebra* [13]. A kernel $K(z,z')$ on X_{d-1} is a Volterra kernel if its support is contained in the region $\Gamma^+_{d-1} = \{(z,z') \in X_{d-1} \times X_{d-1};\ z \geqslant z'\}$. We can thus introduce the *convolution product* $\lozenge$ of Volterra kernels:

$$(K_1 \lozenge K_2)(z,z') = \int_{\lozenge(z,z')\cap X_{d-1}} K_1(z,z'')K_2(z'',z')\,dz'', \tag{22.22}$$

where the *double-cone* region $\lozenge(z,z') = \{z'' \in X_{d-1} | z \geqslant z'' \geqslant z'\}$ is compact. Since $K_1 \lozenge K_2$ is again a Volterra kernel the space of Volterra kernels is an algebra: the *Volterra algebra*.

Next, introducing the following integral representation of the second kind Legendre function [8, III]:

$$Q^{(d)}_\lambda(\cosh w) = \frac{\omega_{d-1}}{\omega_{d-2}(\sinh w)^{d-3}} \int_w^{+\infty} e^{-(\lambda+\frac{d-2}{2})v}\,[2(\cosh v - \cosh w)]^{(d-4)/2}\,dv, \tag{22.23}$$

one can rewrite formula (22.15a) (taking into account (22.17) and (22.18)) as follows:

$$\tilde{F}_+(\lambda) = \omega_{d-1} \int_0^{+\infty} f_+(w) Q^{(d)}_\lambda(\cosh w)(\sinh w)^{d-2}\,dw. \tag{22.24}$$

One can then prove [13] that the Laplace-type transform (22.24), applied to the convolution product (22.22), transforms the convolution product into ordinary product:

$$(\widetilde{K_1 \lozenge K_2})(\lambda) = \tilde{K}_1(\lambda)\tilde{K}_2(\lambda). \tag{22.25}$$

22.2.4 Holomorphic Perikernels [7]

Besides the algebra of Volterra kernels on X_{d-1}, we may also consider the algebra of kernels on the sphere S_{d-1}. We are thus led to introduce triplets $(\mathcal{K}, \underline{K}, K)$ on the complexified hyperboloid $X^{(c)}_{d-1}$ such that:

i) $\mathcal{K}(z,z')$ is an analytic function called *perikernel*, whose domain is: $\Delta_\mu = X^{(c)}_{d-1} \times X^{(c)}_{d-1} \setminus \Sigma^{(c)}_\mu$, where $\Sigma^{(c)}_\mu$ is the *cut*:

$$\Sigma^{(c)}_\mu = \Big\{(z,z') \in X^{(c)}_{d-1} \times X^{(c)}_{d-1};\ (z-z')^2 = \rho;\ \rho \geqslant 2\,(\cosh\mu - 1)\Big\}. \tag{22.26}$$

ii) $\underline{K}$ is a kernel on S_{d-1} defined by taking the restriction of $\mathcal{K}$: $\underline{K} = \mathcal{K}|_{S_{d-1}\times S_{d-1}}$. The convolution product on the sphere is given by:

$$(\underline{K}_1 * \underline{K}_2)(z,z') = \int_{S_{d-1}} \underline{K}_1(z,z'')\underline{K}_2(z'',z')\,d\sigma(z''). \tag{22.27}$$

iii) K is a Volterra kernel on X_{d-1} defined via the jump $\Delta\mathcal{K}$ of $\mathcal{K}$ across $\Sigma^{(c)}_\mu$.

We can now introduce a convolution product on perikernels: $*^{(c)}$, i.e.

$$\mathcal{K}(z,z') = (\mathcal{K}_1 *^{(c)} \mathcal{K}_2)(z,z') = \int_{\Gamma(z,z')} \mathcal{K}_1(z,z'')\mathcal{K}_2(z'',z')\,dz''. \tag{22.28}$$

By distortion of the integration cycles, from the special cycle $\Gamma_0(z,z') \equiv S_{d-1}$, one can then prove the following proposition [7]:

Proposition 1. *Being given two perikernels $\mathcal{K}_i(z,z')$ $(i=1,2)$ on $X^{(c)}_{d-1}$, analytic in the domains Δ_{μ_i} $(\Delta_\mu = X^{(c)}_{d-1} \times X^{(c)}_{d-1} \setminus \Sigma^{(c)}_\mu)$, whose respective discontinuities on the cuts $\Sigma^{(c)}_{\mu_i}$ are the Volterra kernels $\Delta\mathcal{K}_1$, $\Delta\mathcal{K}_2$, there exists a perikernel $\mathcal{K} = \mathcal{K}_1 *^{(c)} \mathcal{K}_2$ such that:*

i) *The restriction of $\mathcal{K}$, $\mathcal{K}_1$, $\mathcal{K}_2$ to the Euclidean sphere S_{d-1} are such that:*

$$\mathcal{K}|_{S_{d-1}} = \mathcal{K}_1|_{S_{d-1}} * \mathcal{K}_2|_{S_{d-1}}. \tag{22.29}$$

ii) *The discontinuity $\Delta\mathcal{K}$ is given by the following $\Diamond$-convolution product:*

$$\Delta_+\mathcal{K} = \Delta_+\mathcal{K}_1 \Diamond \Delta_+\mathcal{K}_2, \tag{22.30}$$

(the subscript $(+)$ is because we are considering only one cut $\Sigma^{(+)}_\mu \in X_{d-1}$; the result can be extended and allows two cuts $\Sigma^{(\pm)}_\mu \in X_{d-1}$).

iii) *For every (z,z') in the analyticity domain of $\mathcal{K}$, there exists a class of cycles $\Gamma(z,z')$ such that:*

$$\mathcal{K}(z,z') = \int_{\Gamma(z,z')} \mathcal{K}_1(z,z'')\mathcal{K}_2(z'',z')\,dz'', \tag{22.31}$$

$\Gamma(z,z')$ being obtained by continuous distortion inside the analyticity domain of the integrand from the special cycle $\Gamma_0(z,z') \equiv S_{d-1}$ relevant for the Euclidean configurations $(z,z') \in S_{d-1} \times S_{d-1}$.

iv) *If $\mathcal{K}_1$, $\mathcal{K}_2$ are invariant perikernels, $\mathcal{K}$ is also an invariant perikernel.*

Next, we introduce a complexified Radon transformation $\mathcal{R}^{(c)}_d$ [8, II] acting on perikernels $\mathcal{K}$ in $X^{(c)}_{d-1}$. The result of the action of $\mathcal{R}^{(c)}_d$ will be a function $\hat{f}(\theta)$ of a single complex variable enjoying the properties of 2π-periodicity, analyticity in the cut-plane, and appropriate symmetries already seen in the case $d=2$. Then, one can perform the composition

$$\mathcal{L}_d = \mathcal{L} \circ \mathcal{R}^{(c)}_d, \tag{22.32}$$

($\mathcal{L}$ is the Laplace transform already considered in the case $d=2$). Applying the operator $\mathcal{L}_d$ to perikernels we get a mapping which transforms the $*^{(c)}$ convolution product of invariant perikernels into ordinary product of holomorphic functions: it therefore relates the properties of the Volterra convolution on X_{d-1} to those of the convolution on the sphere S_{d-1} [8, III]. The transform (22.32) is indeed an extension of the Laplace-type transform (22.24), which is a composition of the ordinary Laplace transform with an Abel transform.

22.3 Complex Angular Momentum in General Quantum Field Theory

22.3.1 Four-Point Function

Let $F(k_1,k_2;k'_1,k'_2)$ be the four-point function of a scalar Q.F. in $(d+1)$-dimensional space-time $(d \geqslant 2)$ considered through the kinematics of a two-particle channel with: incoming complex energy-momenta (k_1,k_2); outgoing complex energy-momenta (k'_1,k'_2); total energy-momentum vector $K = k_1 + k_2 = k'_1 + k'_2$; $K^2 = t$.

We assume K fixed, real, space-like: i.e., $t \leqslant 0$. We introduce also the vector variables $Z = \frac{k_1 - k_2}{2}$, $Z' = \frac{k'_1 - k'_2}{2}$. We shall consider $K \neq 0$, and choose K along the d-axis of coordinates: $K = \sqrt{-t}\, e_d$, where e_d denotes the corresponding unit vector ($e_d^2 = -1$). We also introduce the (off-shell) *scattering angle* Θ_t of the t-channel as being the angle between the two planes Π and Π' spanned respectively by the pairs of vectors (k_1, k_2) (or (Z, K)) and (k'_1, k'_2) (or (Z', K)). It is convenient to introduce (real or complex) unit vectors z, z' (uniquely determined up to a sign) orthogonal to K, belonging respectively to Π and Π', such that the following decompositions hold:

$$Z = \rho z + wK, \quad Z' = \rho' z' + w'K, \tag{22.33a}$$

$$z \cdot K = z' \cdot K = 0; \quad z^2 = z'^2 = -1. \tag{22.33b}$$

The parameters ρ, w (resp. ρ', w') can be computed in terms of the Lorentz invariant $\zeta_i = k_i^2$ (resp. $\zeta'_i = k_i'^2$), $(i = 1, 2)$ and t. One has: $w = \frac{\zeta_1 - \zeta_2}{2t}$, $w' = \frac{\zeta'_1 - \zeta'_2}{2t}$, $\rho^2 = -Z^2 + w^2 t$, $\rho'^2 = -Z'^2 + w'^2 t$.

We can now write F as $F(k_1, k_2; k'_1, k'_2)$ or as $F(\zeta, \zeta', t; cos\Theta_t)$, $\zeta = (\zeta_1, \zeta_2)$, $\zeta' = (\zeta'_1, \zeta'_2)$. From the basic axioms of *locality*, *spectrum*, and *Lorentz invariance*, we can derive the *analytic* structure of F. From the *temperateness* axiom a *power-type majorization* of F follows. Therefore one can prove the following theorem [9].

Theorem 1. i) *$F(\zeta, \zeta'; t, cos\Theta_t)$ is analytic in a cut-plane of the variable $\cos\Theta_t$ (with cuts contained in the half-lines $(-\infty, -1]$, and $[+1, +\infty)$) for all (ζ, ζ', t) belonging to a certain* off-shell *region*

$$\Delta^{(e)} = \left\{(\zeta, \zeta', t); t \leqslant 0, \zeta \in \Delta_t, \zeta' \in \Delta_t\right\}, \tag{22.34a}$$

$$\Delta_t = (\zeta_1 - \zeta_2)^2 - 2(\zeta_1 + \zeta_2)t + t^2 \leqslant 0 \qquad (\zeta = (\zeta_1, \zeta_2) \in \mathbb{R}^2). \tag{22.34b}$$

(For $K = 0$ the corresponding set Δ_0 is $\zeta_1 = \zeta_2 \leqslant 0$).

ii) *$|F(\zeta, \zeta'; t, cos\Theta_t)| \leqslant C_{\zeta,\zeta',t} |\cos\Theta_t|^N$ in the cut-plane considered.*

iii) *$\exists$ the Laplace-type transform $\tilde{F}(\zeta, \zeta'; t, \lambda)$ analytic in the corresponding half-plane*

$$C_+^{(N)} = \{\lambda \in \mathbb{C}, \operatorname{Re}\lambda > N\}, \ \forall (\zeta, \zeta', t) \in \Delta^{(e)}. \tag{22.35}$$

iv) *$\tilde{F} = \tilde{F}_+ + e^{i\pi\lambda}\tilde{F}_-$, with*

$$\tilde{F}_\pm(\zeta, \zeta'; t, \lambda) = \pm\omega_{d-1} \int_{\sigma_\pm} Q_\lambda^{(d)}(\cos\Theta_t) F_\pm(\zeta, \zeta'; t, \cos\Theta_t)(\sin\Theta_t)^{(d-3)}\, d(\cos\Theta_t), \tag{22.36}$$

where $F_\pm$ are the jumps of the function $\mathrm{i}F$ across the respective cuts σ_+ and σ_- (s-cut and u-cut, respectively) contained in the half-lines $[+1, +\infty)$ and $(-\infty, -1]$ respectively of the $\cos\Theta_t$-plane.

v) *$\tilde{F}_+$ and $\tilde{F}_-$ are analytic for $\lambda \in C_+^{(N)}$ and uniformly bounded (up to temperate factors) by $e^{-\mu_\pm \operatorname{Re}\lambda}$, where $\mu_\pm$ are positive numbers determined by the thresholds s_0 and u_0 of the s and u cuts.*

vi) *The symmetric and antisymmetric combinations $\tilde{F}^{(s)} = \tilde{F}_+ + \tilde{F}_-$ and $\tilde{F}^{(a)} = \tilde{F}_+ - \tilde{F}_-$ interpolate uniquely the even and the odd partial waves, respectively (i.e., the Froissart–Gribov equalities hold):*

$$\tilde{F}^{(s)}(\zeta, \zeta'; t, \lambda)|_{\lambda=2\ell} = f_{2\ell}(\zeta, \zeta'; t) \qquad (2\ell > N), \tag{22.37a}$$

$$\tilde{F}^{(a)}(\zeta, \zeta'; t, \lambda)|_{\lambda=2\ell+1} = f_{2\ell+1}(\zeta, \zeta'; t) \qquad (2\ell + 1 > N), \tag{22.37b}$$

with

$$f_\ell(\zeta,\zeta';t) = \omega_{d-1} \int_{-1}^{+1} P_\ell^{(d)}(\cos\Theta_t) F(\zeta,\zeta';t,\cos\Theta_t)(\sin\Theta_t)^{(d-3)}\, d(\cos\Theta_t), \tag{22.38}$$

where $P_\ell^{(d)}(\cos\Theta_t)$ *are the* ultraspherical Legendre polynomials.

22.3.2 Complex Angular Momentum Diagonalization of the Bethe–Salpeter (BS) Structure in QFT

We now denote the four-point function by $H(K;Z,Z')$: it describes the interaction of two local (and mutually local) fields Φ_1 and Φ_2. In [2,3] it has been shown that for massive QFT's satisfying the postulate of *asymptotic completeness of two-particle states* and an additional regularity assumption in the energy variable t (for $t \geqslant t_0$), the two-particle t-channel analytic structure of H is entirely encoded in any Bethe–Salpeter-type integral equation of the following form:

$$H(K;Z,Z') = B(K;Z,Z') + \int_{\Gamma(K)} B(K;Z,Z'')H(K;Z'',Z')G(K;Z'')\, dZ'', \tag{22.39}$$

where $B(K;Z,Z')$ is a four-point function satisfying the same axiomatic analyticity properties and bounds as H and, in addition, the property of *two-particle irreducibility in the t-channel* [2,3]. $G(K;Z'')$ represents a *regularized double-propagator* of the following form:

$$G(K;Z'') = \mathrm{i}\,\Pi_1^{(\mathrm{reg})}\left(\frac{K}{2}+Z''\right)\Pi_2^{(\mathrm{reg})}\left(\frac{K}{2}-Z''\right), \tag{22.40}$$

where $\Pi_i^{(\mathrm{reg})}$ $(i=1,2)$ is a regularized form of the propagator Π_i, obtained by multiplying the latter by a suitable Pauli–Villars-type factor, equal to 1 on the mass-shell. The integration cycle $\Gamma(K)$ is the Euclidean subspace $E_{d+1} = \{Z'' = (\mathrm{i}Y''^{(0)}, \vec{X}'')\}$, when K, Z, Z' are themselves taken in E_{d+1}. Equation (22.39) can be written in short:

$$H = B \circ_t H. \tag{22.41}$$

The convolution $\circ_t$ can be decomposed into an integration operation $\tilde{\circ}_t$ over the domain Δ_t, with respect to the mass variables $\zeta'' = (\zeta_1'', \zeta_2'')$, and a convolution product (in the sense of $SO(d)$) on the sphere S_{d-1}. Recalling the expression of the convolution product $*$ of $SO(d)$-invariant kernels a and b on the sphere S_{d-1}, and denoting by $\langle\omega,\omega'\rangle$ the Euclidean scalar product in $\mathbb{R}^d$, we have:

$$(a*b)(\langle\omega,\omega'\rangle) = \int_{S_{d-1}} a(\langle\omega,\omega''\rangle)b(\langle\omega'',\omega'\rangle)\, d\omega''. \tag{22.42}$$

Next, we use the *factorization property*, according to which:

$$\widetilde{(a*b)}_\ell = \tilde{a}_\ell \tilde{b}_\ell, \tag{22.43}$$

where, replacing Θ_t by θ, we have:

$$\tilde{a}_\ell = \omega_{d-1} \int_{-1}^{+1} P_\ell^{(d)}(\cos\theta)a(\cos\theta)(\sin\theta)^{(d-3)}\, d(\cos\theta). \tag{22.44}$$

We can then replace the version (22.41) of the BS equation by the following set of *BS equations for the partial waves* of H [10]:

$$\tilde{h}_\ell(\zeta,\zeta';t) = \tilde{b}_\ell(\zeta,\zeta';t) + \int_{\Delta_t} \tilde{b}_\ell(\zeta,\zeta'';t)\tilde{h}_\ell(\zeta'',\zeta';t)\underline{G}(\zeta'')\, d\mu_t(\zeta''), \tag{22.45}$$

where $\underline{G}(\zeta'') = \mathrm{i}\,\Pi_1^{(\mathrm{reg})}(\zeta_1'')\Pi_2^{(\mathrm{reg})}(\zeta_2'')$.

The BS equation (22.39) implies corresponding integral relations between the *absorptive parts* $\Delta_s H$, $\Delta_u H$ of H and $\Delta_s B$, $\Delta_u B$ of B [2,4]. The convolution $\circ_t$ is replaced by the product $\Diamond_t$. The latter can be decomposed into an integration in the mass variables and an integral in (hyperbolic) angular variables, interpretable as a certain convolution product on an appropriate orbit of the group $SO_0(1,d-1)$: namely, a one-sheeted hyperboloid. Indeed, performing suitable distortions of the integration cycle $\Gamma(K)$ (starting from the Euclidean domain E_{d+1}) and remaining inside the axiomatic domain of the integrand, one folds the cycle $\Gamma(K)$ around the supports of the absorptive parts [2,4]. One obtains:

$$\Delta_s H = \Delta_s B + \Delta_s B \Diamond_t \Delta_s H + \Delta_u B \Diamond_t \Delta_u H, \tag{22.46a}$$

$$\Delta_u H = \Delta_u B + \Delta_s B \Diamond_t \Delta_u H + \Delta_u B \Diamond_t \Delta_s H. \tag{22.46b}$$

In view of Proposition 1 (in particular, property (iii)), one can see that the BS equation can be analytically continued to all (z,z') in $X_{d-1}^{(c)} \times X_{d-1}^{(c)}$ for all $t < 0$, $(\zeta,\zeta') \in \Delta_t \times \Delta_t$ under the following form [10]:

$$\begin{aligned} &\underline{H}_{(\zeta,\zeta',t)}(z,z') \\ &= \underline{B}_{(\zeta,\zeta',t)}(z,z') - \mathrm{i}\int_{\Delta_t} \left(\underline{B}_{(\zeta,\zeta'',t)} *^{(c)} \underline{H}_{(\zeta'',\zeta',t)}\right)(z,z')\underline{G}(\zeta'')\, d\mu_t(\zeta''). \end{aligned} \tag{22.47}$$

The latter can be considered as a Fredholm resolvent equation in complex space, whose integration space is the product of Δ_t by the *floating cycle* Γ on which the complex points z, z' vary. In view of general results of [5,6] adapted to the present situation [10], the function $\underline{B}_{(\zeta,\zeta',t)}(z,z')$ is directly obtained with its full perikernel structure as a solution of the Fredholm equation (22.47). Moreover, applying the property of $*^{(c)}$-composition product of perikernels for computing side-by-side the discontinuities of (22.47) one reobtains eqs. (22.46).

Next, we apply the Laplace-type transform $\mathfrak{L}_d = \mathfrak{L} \circ \mathfrak{R}_d^{(c)}$ to (22.47). Let us recall that this transform accounts for the association of the Froissart–Gribov structure with the triplet $(\mathcal{K}, \underline{K}, K)$. Collecting all these results we can state the following theorem [10].

Theorem 2. *If* $\underline{H}_{(\zeta,\zeta',t)}(\cos\Theta_t)$ *satisfies the following bound:*

$$\left|\underline{H}_{(\zeta,\zeta',t)}(\cos\Theta_t)\right| \leqslant C_{(\zeta,\zeta',t)}^{(H)}\, \mathrm{e}^{N_H\,|\mathrm{Im}\,\Theta_t|}|\sin\mathrm{Re}\,\Theta_t|^{-n_H}, \tag{22.48}$$

where n_H *describes the maximal local order of singularity of the distribution boundary-values of* $\underline{H}_{(\zeta,\zeta',t)}$ *on the reals, and* $C_{(\zeta,\zeta',t)}^{(H)}$ *is taken equal to the uniform bound of* H *in the Euclidean subspace, namely*

$$C_{(\zeta,\zeta',t)}^{(H)} = C_E^{(H)}\left[(1+|t|^{1/2})(1+\rho)(1+|w||t|^{1/2})(1+\rho')(1+|w'||t|^{1/2})\right]^{N_H}, \tag{22.49}$$

then:

i) *It admits Laplace-type transforms* $\tilde{H}_s(\zeta,\zeta';t,\lambda_t)$ *and* $\tilde{H}_u(\zeta,\zeta';t,\lambda_t)$ *which are holomorphic with respect to the CAM variable* λ_t *in the half-plane* $C_+^{(N_H)} = \{\lambda_t \in \mathbb{C}, \operatorname{Re}\lambda_t > N_H\}$. *These Laplace-type transforms are given by:*

$$\tilde{H}_s(\zeta,\zeta';t,\lambda_t)=\omega_{d-1}\int_0^{+\infty}\Delta_s \underline{H}_{(\zeta,\zeta',t)}(\cosh v)\, Q_{\lambda_t}^{(d)}(\cosh v)(\sinh v)^{(d-2)}\, dv \quad (22.50a)$$

$$\tilde{H}_u(\zeta,\zeta';t,\lambda_t)=\omega_{d-1}\int_0^{+\infty}\Delta_u \underline{H}_{(\zeta,\zeta',t)}(-\cosh v)\, Q_{\lambda_t}^{(d)}(\cosh v)(\sinh v)^{(d-2)}\, dv \quad (22.50b)$$

ii) *The* symmetric and antisymmetric Laplace-type transforms $\tilde{H}^{(s)} = \tilde{H}_s + \tilde{H}_u$, $\tilde{H}^{(a)} = \tilde{H}_s - \tilde{H}_u$, *enjoy the following Froissart–Gribov equalities:*

$$\tilde{H}^{(s)}(\zeta,\zeta';t,2\ell) = \tilde{h}_{2\ell}(\zeta,\zeta';t) \qquad (2\ell > N_H), \tag{22.51a}$$

$$\tilde{H}^{(a)}(\zeta,\zeta';t,2\ell+1) = \tilde{h}_{2\ell+1}(\zeta,\zeta';t) \qquad (2\ell+1 > N_H). \tag{22.51b}$$

$\tilde{H}^{(s)}$ *and* $\tilde{H}^{(a)}$ *are Carlsonian (unique) interpolations of the respective sets of partial waves* $(\tilde{h}_{2\ell}, 2\ell > N_H)$ *and* $(\tilde{h}_{2\ell+1}, 2\ell+1 > N_H)$.

iii) *The corresponding interpolation in* λ_t *(*$\operatorname{Re}\lambda_t > N_H$*) of the BS equations (22.45) can be written as follows:*

$$\tilde{H}^{(s),(a)}(\zeta,\zeta';t,\lambda_t) = \tilde{B}^{(s),(a)}(\zeta,\zeta';t,\lambda_t) + \int_{\Delta_t} \tilde{B}^{(s),(a)}(\zeta,\zeta'';t,\lambda_t)\tilde{H}^{(s),(a)}(\zeta'',\zeta';t,\lambda_t)\underline{G}(\zeta'')\, d\mu_t(\zeta''), \tag{22.52}$$

which are two decoupled BS-type equations for $\tilde{B}^{(s)}$ *and* $\tilde{B}^{(a)}$ *in terms of* $\tilde{H}^{(s)}$ *and* $\tilde{H}^{(a)}$ *respectively, in which the Fredholm integration space reduces to the two-dimensional real region* Δ_t *of the plane of the squared-mass variables* $\zeta'' = (\zeta_1'', \zeta_2'')$, *while* (t,λ_t) *are parameters varying in* $\mathbb{R}^- \times C_+^{(N_H)}$.

iv) *By construction, the functions* $\tilde{B}^{(s)}$ *and* $\tilde{B}^{(a)}$ *are interpolations in the* λ_t-*plane of the corresponding Euclidean partial waves* $(\tilde{b}_{2\ell}(\zeta,\zeta';t);\ 2\ell > N_H)$ *and* $(\tilde{b}_{2\ell+1}(\zeta,\zeta';t);\ 2\ell+1 > N_H)$*; namely, there hold the following Froissart–Gribov equalities:*

$$\tilde{B}^{(s)}(\zeta,\zeta';t,2\ell) = \tilde{b}_{2\ell}(\zeta,\zeta';t) \qquad (2\ell > N_H), \tag{22.53a}$$

$$\tilde{B}^{(a)}(\zeta,\zeta';t,2\ell+1) = \tilde{b}_{2\ell+1}(\zeta,\zeta';t) \qquad (2\ell+1 > N_H). \tag{22.53b}$$

Up to now we have considered the kernel B (resp. $\tilde{B}^{(s),(a)}$) as determined by H via the Fredholm method. One may adopt a more exploratory viewpoint and assume that in the field theory under consideration the kernel B satisfies *better properties* than H. Assume that B satisfies the following bound [10]: for all u ($0 \leqslant u \leqslant \pi$ and $\pi \leqslant u \leqslant 2\pi$, the jumps at $u = 0$ and $u = \pi$ being taken into account)

$$\int_0^{+\infty} e^{-N_B v} \left|\underline{B}_{(\zeta,\zeta';t)}(\cos(u+iv))\right| dv \leqslant C^{(B)}_{(\zeta,\zeta';t)}, \tag{22.54a}$$

$$\sup_{-1\leqslant \cos\Theta_t \leqslant 1} \left|\underline{B}_{(\zeta,\zeta';t)}(\cos\Theta_t)\right| \leqslant C^{(B)}_{(\zeta,\zeta';t)}, \tag{22.54b}$$

with

$$C^{(B)}_{(\zeta,\zeta';t)} = C_E^{(B)} \left[(1+|t|^{1/2})(1+\rho)(1+|w||t|^{1/2})(1+\rho')(1+|w'||t|^{1/2})\right]^{N_B}, \tag{22.55}$$

where $\max(-1, -\frac{d-2}{2}) < N_B < N_H$, then these bounds on B imply analyticity property with respect to λ_t in $C_+^{(N_B)}$, and the following majorization for the transforms $\tilde{B}^{(s),(a)}$:

$$\left|\tilde{B}^{(s),(a)}(\zeta, \zeta'; t, \lambda_t)\right| \leqslant C^{(B)}_{(\zeta,\zeta';t)} \psi(|\mathrm{Im}\, \lambda_t|), \tag{22.56}$$

where ψ denotes a bounded positive function on $[0, \infty)$, tending to zero at infinity. Then one can consider the analytic continuation of eqs. (22.52) in the strip $N_B < \mathrm{Re}\, \lambda_t < N_H$; these equations being now regarded as Fredholm-resolvent equations defining $\tilde{H}^{(s)}$ (resp. $\tilde{H}^{(a)}$) in terms of $\tilde{B}^{(s)}$ (resp. $\tilde{B}^{(a)}$) through expression of the form [10]:

$$\tilde{H}^{(s),(a)}(\zeta, \zeta'; t, \lambda_t) = \frac{\mathcal{N}_{\tilde{B}^{(s),(a)}}(\zeta, \zeta'; t, \lambda_t)}{\mathcal{D}_{\tilde{B}^{(s),(a)}}(t, \lambda_t)}. \tag{22.57}$$

Since $\tilde{H}^{(s),(a)}$ are holomorphic in $C_+^{(N_H)}$ the resolution (22.57) provides a meromorphic continuation of $\tilde{H}^{(s),(a)}$ in the strip $N_B < \mathrm{Re}\, \lambda_t < N_H$. The poles are given respectively by the zeros of $\mathcal{D}_{\tilde{B}^{(s)}}(t, \lambda_t)$ and $\mathcal{D}_{\tilde{B}^{(a)}}(t, \lambda_t)$; the locations of the latter in the λ_t-plane are given respectively by: $\lambda_t^{(s)} = \lambda_{j^{(s)}}(t)$, $\lambda_t^{(a)} = \lambda_{j^{(a)}}(t)$.

22.3.3 Conclusions

a) The analysis, which we have reviewed, is completely worked out in the complex momentum space scenario appropriate to QFT.

b) The joint exploitation of harmonic analysis on orbital manifolds of the Lorentz group together with basic analyticity properties of QFT entails the complex angular momentum structure. The result which can be derived from these properties is that for each given two-field channel (called t-channel), with total squared energy-momentum t and (off-shell) scattering angle Θ_t, there exists an appropriate Fourier–Laplace type transform of the four-point function with certain analyticity properties in the complex angular momentum λ_t, which is the natural conjugate variable of Θ_t.

c) The analyticity domain, which was obtained in a half-plane of the form $\mathrm{Re}\, \lambda_t > N_H$, where the number N_H ($N_H > 0$) corresponds to a certain *degree of temperateness* of the four-point function at large momenta. This domain was obtained for all negative values of the squared total energy t of the channel considered.

d) The *Euclidean partial waves* admit Carlsonian (i.e., unique) analytic interpolations in a given CAM half-plane; this can be shown to be equivalent to the property of analytic continuation of the four-point function from Euclidean momentum space to Minkowskian momentum space.

e) By the use of the Fourier–Laplace transform previously introduced, the general Bethe–Salpeter structure of four-point functions of scalar fields, in a given t-channel, can be diagonalized in the corresponding CAM variable λ_t for all negative values of t, and of the squared-mass variables corresponding to Euclidean configurations in complex momentum space. More specifically, Carlsonian interpolations of the Euclidean BS equations for even and odd partial waves have been constructed in the half-plane of the form $\mathrm{Re}\, \lambda_t > N$, starting from the corresponding BS equations for the s- and u-channel absorptive parts.

f) Any information on B leading to continue analytically these interpolations in some region of the left half-plane $\mathrm{Re}\, \lambda_t \leqslant N_H$ results in the *potential generation* of Regge poles.

g) The existence of a meromorphic continuation of these interpolations in a strip of the form $N_B \leqslant \mathrm{Re}\,\lambda_t \leqslant N_H$, holds if and only if the Bethe–Salpeter kernel B satisfies bounds of the form $|\cos\Theta_t|^{N_B}$, with $N_B < N_H$.

h) More questionable is the justification of the inequality $N_B(t) < N_H(t)$. In the absence of global information on general versions of the two-particle irreducible kernel B, one can think of exploiting various contributions to B, defined in terms of generalized Feynman convolutions enjoying the graph property of *two-particle irreducibility*. The simplest ones are those associated with *particle exchange* graphs; they do satisfy the previous inequality with $N_B = -1$. However, much more complicated contributions to B are imposed by Feynman convolution structure of field theory: typical examples of those have been studied by Mandelstam [17] and others [11]. The treatment of such contributions, which these authors have given by using the methods of S-matrix theory, has led them to introduce *Regge cuts* in the λ_t-plane [11]. Although no mechanism of production of Regge cuts appears in our general field-theoretical framework, the contributions considered by these authors cannot satisfy the inequality $N_B(t) < N_H(t)$ for negative t, although they hopefully do it for positive t.

References

1. B. Barone and E. Predazzi: *High-energy particle diffraction.* Springer, Berlin, Heidelberg, New York, 2002.
2. J. Bros: Some analyticity properties implied by the two-particle structure of Green's functions in general quantum field theory. In: *Analytic methods in mathematical physics* (R.P. Gilbert and R.G. Newton, eds.). Gordon and Breach, New York, 1970.
3. J. Bros and M. Lassalle: Analiticity properties and many-particle structure in general quantum field theory. III. Two-particle irreducibility in a single channel. *Comm. Math. Phys.* 54:33–62 (1977).
4. J. Bros and M. Lassalle: Analiticity properties and many-particle structure in general quantum field theory. IV. Two-particle structural equations for the four, five and six-points functions. *Ann. Inst. H. Poincaré* 27:279–317 (1977).
5. J. Bros and D. Pesenti: Fredholm theory in complex manifolds with complex parameters: Analyticity properties and Landau singularities of the resolvent. *J. Math. Pures. Appl.* 59:375–401 (1980).
6. J. Bros and D. Pesenti: Fredholm resolvents of meromorphic kernels with complex parameters: A Landau singularity and the associated equations of type U in a non–holonomic case. *J. Math. Pures. Appl.* 62:215–252 (1983).
7. J. Bros and G.A. Viano: Connection between the algebra of kernels on the sphere and the Volterra algebra on the one-sheeted hyperboloid: holomorphic *perikernels. Bull. Soc. Math. France* 120:169–225 (1992).
8. J. Bros and G.A. Viano: Connection between the harmonic analysis on the sphere and the harmonic analysis on the one-sheeted hyperboloid: an analytic continuation viewpoint. I. *Forum Math.* 8:621–658 (1996). II. *Forum Math.* 8:659–722 (1996). III. *Forum Math.* 9:165–191 (1997).
9. J. Bros and G.A. Viano: Complex angular momentum in general quantum field theory. *Ann. Henri Poincaré* 1:101–172 (2000).

10. J. Bros and G.A. Viano: Complex angular momentum diagonalization of the Bethe–Salpeter structure in general quantum field theory. *Ann. Henri Poincaré* 4:85–126 (2003).
11. P.D.B. Collins: *An introduction to Regge theory and high energy physics.* Cambridge University Press, Cambridge, 1977.
12. A. Donnachie and P.V. Landshoff: Total cross sections. *Phys. Lett. B* 296:227–232 (1992).
13. J. Faraut and G.A. Viano: Volterra algebra and the Bethe–Salpeter equation. *J. Math. Phys.* 27:840–848 (1986).
14. S. Helgason: Lie groups and symmetric spaces. In: *Battelle Rencontres. 1967 lectures in mathematics and physics* (C.M. DeWitt and J.A. Wheeler, eds.). Benjamin, New York Amsterdam, 1967.
15. J. Hilgert and G. Olafsson: *Causal symmetric spaces, geometry and harmonic analysis.* Academic Press, San Diego, 1997.
16. L.N. Lipatov: Pomeron in quantum cromodynamics. In: *Perturbative quantum cromodynamics* (A.H. Mueller, ed.). World Scientific, Singapore, 1989.
17. S. Mandelstam: Cuts in angular momentum plane. I . *Nuovo Cimento* 30:1127–1147 (1963). II. *Nuovo Cimento* 30:1148–1162 (1963).
18. M. Toller: Group-theoretical techniques for the construction of Regge and multi-Regge models. *Riv. Nuovo Cimento* 1:403–437 (1969).

SPRINGER NATURE

GPSR Compliance

The European Union's (EU) General Product Safety Regulation (GPSR) is a set of rules that requires consumer products to be safe and our obligations to ensure this.

If you have any concerns about our products, you can contact us on ProductSafety@springernature.com

In case Publisher is established outside the EU, the EU authorized representative is:

Springer Nature Customer Service Center GmbH
Europaplatz 3
69115 Heidelberg, Germany

FSC
www.fsc.org
MIX
Papier | Fördert gute Waldnutzung
FSC® C083411

Kolibri 360 GmbH
Ferdinand-Jühlke-Straße 7
99095 Erfurt, Deutschland
produktsicherheit@kolibri360.de